MODULATION, DETECTION AND CODING

MODULATION, DETECTION AND CODING

Tommy Öberg

Signals and Systems Group, Uppsala University
Uppsala, Sweden

JOHN WILEY & SONS, LTD
Chichester · New York · Weinheim · Brisbane · Singapore · Toronto

Other Wiley Editorial Offices

John Wiley & Sons, Inc., 605 Third Avenue,
New York, NY 10158-0012, USA

WILEY-VCH Verlag GmbH
Pappelallee 3, D-69469 Weinheim, Germany

John Wiley & Sons Australia, Ltd, 33 Park Road, Milton,
Queensland 4064, Australia

John Wiley & Sons (Canada) Ltd, 22 Worcester Road
Rexdale, Ontario, M9W 1L1, Canada

John Wiley & Sons (Asia) Pte Ltd, 2 Clementi Loop #02-01,
Jin Xing Distripark, Singapore 129809

Library of Congress Cataloging-in-Publication Data

A Library of Congress catalogue record has been applied for

British Library Cataloguing in Publication Data

A catalogue record for this book is available from the British Library

ISBN 0-471-49766-5

Typeset in Times by Deerpark Publishing Services Ltd, Shannon.
Printed and bound in Great Britain by Antony Rowe Ltd, Chippenham.

This book is printed on acid-free paper responsibly manufactured from sustainable forestry in which at least two trees are planted for each one used for paper production.

CONTENTS

PREFACE

This book is intended for courses at least at Masters level to provide an introduction to the field of signal processing in telecommunications. The text therefore deals mainly with source coding, channel coding, modulation and demodulation. Adaptive channel equalisation, the signal processing aspects of adaptive antennas as well as multi-user detectors for CDMA are also included. Shorter sections on link budget, synchronising and cryptography are also included. Network aspects are not discussed and very little is given about wave propagation. The book aims to give the reader an understanding of the *fundamental signal processing functions* and of the *methods used when analysing the capacity* of a complete communication system. The field of telecommunications is developing rapidly. Therefore, a thorough understanding of analysis methods, rather than just the results of the analysis, is important knowledge which will remain up to date for a longer period and make it possible for the reader to follow the progress within this field. The presentation in the book is at the block diagram level. Hardware and software solutions are only treated sporadically, and it is therefore recommended that the student has completed basic courses in electronics and digital systems in order to be able to make the connection between real systems and the block diagrams and methods treated here. The material has a theoretical base which makes it possible for the student to continue with both deeper studies and design work with new systems not treated in this text. Completed courses in basic signal processing and probability theory are necessary for the understanding of the material presented here. The book includes: text where theory and methods are presented, completely solved examples, exercises with answers, where the reader can test his/hers understanding of the content, and short presentations of some communication systems.

ACKNOWLEDGEMENTS

The author wishes to thank all his colleagues, friends and students who inspired him, read manuscripts and in several positive ways contributed to this book.

1 TELECOMMUNICATIONS

1.1 Usage today

Today telecommunications is a technology deeply rooted in human society and has its foundation in our need to communicate with each other. It may be two people talking on the telephone, a so-called point to point connection. The end points of connection can be in the same building or on opposite sides of the earth, and may be stationary or mobile. An example of mobile connection is air communication, e.g. where the information is transferred between control tower and aircraft. Another example is coastal radio, where the information is transferred between ships at sea and stationary coastal radio stations. Mobile telecommunications is a rapidly growing application. Future telephone systems will be based on the idea of a personal telephone. A telephone number will not then lead to a particular communication terminal, such as a telephone, but to a certain person, irrespective of which terminal he/she happens to be at. Not only voices are transferred. Picture and data transfer represents a growing fraction of the information flow. Bank transfers ranging from cash points to transfers between large monetary institutions are examples of data communications which set high demands on reliability and security. Radio broadcasting is a kind of communication which, in contrast to the point to point connection, usually has one transmitter and many receivers. Apart from radio and TV entertainment, weather maps are transferred via radio broadcasting from a satellite to several receivers on Earth.

Future communication systems will continue to enlarge their capacity for transmitting information and provide greater mobility for users. Different types of information will be handled the same way in the telecommunication networks. Other aspects will make the difference in handling, such as requirements on maximum delay, maximum number of errors in the data and how much the users are willing to pay for the transmission.

1.2 History

Communication based on electrical methods (bonfires, mechanical semaphores, etc. are not included) began in 1833 when two German Professors, Gauss and Weber, established a telegraph connection using a needle telegraph. Samuel

Morse was, however, the man who further refined the methods, made them usable and developed the Morse Code of the alphabet. This is the first application of coding theory within telecommunications, where combinations of dots and dashes were used for letters; short combinations were used for the common letters. In 1837 Morse demonstrated his telegraph in Washington, and it spread quickly to become a commonly used means of communication all over the world. In February 1877, Alexander Graham Bell presented a new invention, the telephone. In a crowded lecture hall in Salem, MA, USA, the amazed audience heard the voice of Bell's assistant 30 km away. After the first Atlantic cable for telegraphy had been completed in 1864 considerable development had taken place. The first telephone cables could transfer one telephone call at a time. The introduction of coaxial cables in the 1950s made it possible to transfer 36–4000 calls per cable, using analogue carrier wave techniques. From 1965 satellites have had a considerable impact on the development of telecommunications across the world. The earliest communication satellites were Telstar and Relay which began operation in 1962. In 1963 the first geo-stationary satellite, Syncom, was launched. Its altitude must be 35,800 km to give it an orbit time of 24 h,

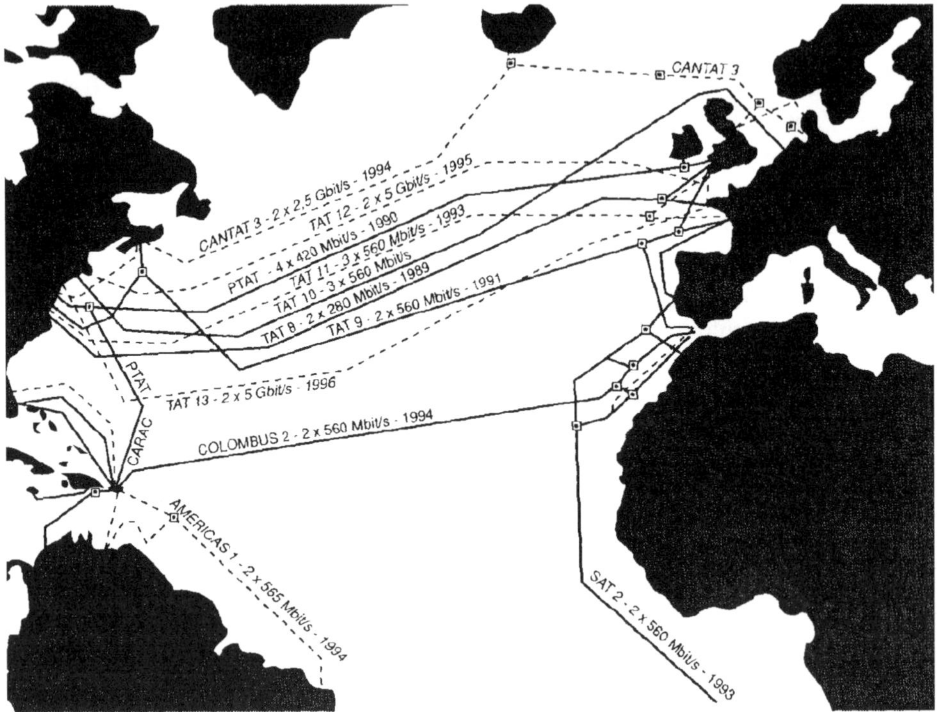

Figure 1.1   A selection of optical transmission links across the North Atlantic. © 1993 IEEE J. Thiennot; F. Pirio; J. B. Thomine, Optical Undersea Cable Systems Trends, Proceedings of the IEEE, November 1993.

making it possible to use day and night between the same continents. After 1985 optical fibres have offered the possibility of point to point connections with very high transfer capacity (cf. Figure 1.1). An optical fibre of good quality can transfer information at a rate of 400 Gbit/s, which means that one single fibre can carry more than 4,000,000 telephone calls simultaneously. With proper additional equipment an information transfer per fibre of 5–10 Tbits/s is expected in the future.

Wireless communication began in 1895. A 21 year old student, Guglielmo Marconi from Bologna, Italy, then managed to transfer Morse codes without using wires between two stations. What made this possible was a theory presented by the English mathematician Maxwell in 1867 (well known to today's physics students). He never managed to experimentally verify his theory, however. This was achieved by the German physicist Hertz in 1886. Hertz was, however, sceptical about the feasibility of electromagnetic waves for tele-graphic transfer. On 12th December 1901 Marconi managed to transfer a simple message across the Atlantic. Marconi, together with the German Ferdinand Braun, received the 1909 Nobel Prize in Physics as recognition of their work for the development of wireless telegraphy (in Russia Alexander Popov simulta-neously did successful experiments with radio transmission. However, he never obtained the industrial importance that Marconi did).

Radio offered the first real example of mass communication. The first radio broadcasting probably took place from a garage in Pittsburgh on the 2nd November 1920. The idea of radio broadcasting was launched by a Russian-American named David Sarnoff, who started as a messenger boy at Marconi's company in New York. Edwin Howard Armstrong, the great engineer, inventor and researcher, further developed radio technology at this time. He invented the regenerative receiver, the superheterodyne receiver and the FM modulation tech-nique. He invented the latter when, because of his perfectionist nature, he was dissatisfied with the sound quality that the AM modulation technique could offer. The television was not far behind. In 1926, the Scotsman John Baird succeeded in transferring a moving image from one room to another by electric means. The technology was built on the use of rotating disks with holes to successively acquire the entire image line after line. It was a Russian, Vladimir Zworykin, who fully developed electronic television. In 1931 he had designed the first electronic picture tube and the first feasible camera tube. The BBC (British Broadcasting Cooperation) started TV transmissions in London in 1932.

One of the main applications of radio communication is within the transport sector. Today's air traffic would be unthinkable without radio, radar and radio-based navigation and landing aids. Shipping was very early in utilising radio to improve safety. For the first time it was possible to call for help when in distress, no matter what the weather and sight conditions. In 1899 the American passenger ship St Paul was equipped with a Marconi transmitter of the spark type as a test. The first kind of radio communication was Morse telegraphy. Today this technique is no longer used and communication is achieved by direct speech or data, usually via the satellite-based INMARSAT (INternational MARitime SATellite organisation) system. One of the first great achievements that made this technique known was the sending of distress signals from the Titanic when she sank in 1912.

Mobile telephone systems offer the possibility of having access to the same services with mobile terminals as with stationary networks. In 1946, the American company AT&T (American Telephone and Telegraph Company) started a mobile telephone network in St Louis, USA. The system was successful and within one year mobile networks were established in another 25 cities. Each network had a base station (main station) which was designed to have as large a range as possible. The problem with such a system is that the mobile station must have a high output power to match the range of the base station, i.e. small battery-operated mobile telephones are not possible, and, furthermore, the available radio spectrum is not efficiently utilised. As early as in 1947 solutions were based on the principle of dividing the geographical area, covered by a mobile network, into a number of smaller cells arranged in a honeycomb structure. A base station having a relatively short range of operation was placed inside each cell, a so-called cellular system. A closer reuse of carrier frequencies is therefore possible and user capacity is increased. A problem which then occurred was that it was necessary to have some kind of automatic handover, i.e. when a mobile terminal passes the border between two base stations, the connection must be transferred to the other base station without interruption. As it turned out it took a long time before this problem was solved. The first cellular mobile telephone network was not in operation until the autumn of 1981. The system was called Nordic Mobile Telephone (NMT) and it had been jointly developed by the Nordic telephone authorities (Sweden, Finland, Norway, Denmark and Iceland). The first generation of mobile phone systems, such as NMT, were analogue. Today the mobile phone systems are based on digital signal processing. The third generation mobile phones are also based on Code Division Multiple Access (CDMA). A communication system based on CDMA was presented as early as 1950 by De Rosa-Rogoff. The first applications of the technique were in military systems and navigation systems. The first mobile telephony standard based on the CDMA technique was the North American IS-95 which was established in July 1993. Mobile phone systems have also taken the leap into space through systems like Iridium and Globalstar which make world wide coverage possible for hand held terminals.

The latest step in the progress of telecommunications was taken when different computer networks were connected together into an internet. The origin of what today is called the Internet is ARPANET (Advanced Research Project Agency) which was the result of a research project. The aim was to create a computer network which was able to function even if parts of the network were struck out, in for example a war. The year 1969 was a milestone when four computers in different universities in the US were connected. In another part of the world, at

the European nuclear research laboratory, CERN, they were in need of a system
for document handling. Development and standardisation of protocol and text
formats was started, which resulted in the World Wide Web (WWW), which is an
efficient way to use the Internet. In 1993 a further milestone was reached when
the computer program Mosaic was presented. The program gave a simple inter-
face to the WWW and made the resources of the network accessible for others
than computer specialists. Simultaneously the Internet was given attention from
high political level. Thereby an increase in the use of telecommunications began.

No retrospective survey can avoid mentioning the pioneering theorist Claude E.
Shannon (1916–2001), Bell Laboratories. During a time when major progress was
being made in analogue communications he laid the foundation of today's digital
communications by formulating much of the theoretical basis for secure communication. One of his best known works is the formulation of the channel transmission capacity, *A Mathematical Theory of Communication*, published in July and
October 1948. A great deal of the content in this book has its basis in his works.

For telecommunication systems to be really powerful a large operational range
is required so that many customers can be reached. It is then necessary that
different systems can communicate with each other, which requires a standard
format for communication. One of the international organisations dealing with
the standardisation of telecommunications is ITU, the International Tele Union.

1.3 Basic elements

A telecommunication system can be divided up in several different ways. One is
founded on the OSI model with its seven layers. This model formalises a hier-
archy and is not suitable for describing a telecommunication system from the
signal processing point of view, which mainly deals with layer one, the physical
layer, and layer two, the link layer. A model more suitable for this purpose is
described in Figure 1.2; this model is suitable for physical connections for speci-

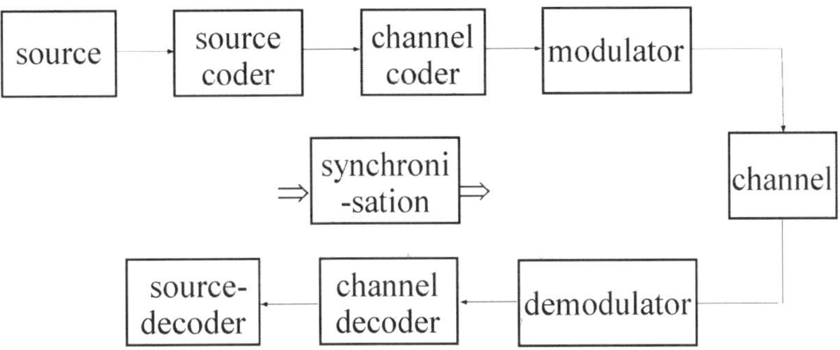

Figure 1.2   Fundamental blocks of a telecommunication system.

fic applications, so-called circuit switched connections (e.g. telephony), and for radio, TV, etc. When it comes to packet switched networks, e.g. the Internet, you cannot identify a single path between the source and the user of the information. The data find its way through where there is free capacity and different parts of the information can take different routes. In these cases, you must, from the point of view of signal processing, study single physical links in the net or use channel models which describe the connection on a word or packet level.

1.3.1 TRANSMITTER SIDE

Information is generated by some kind of source which could be the human speech organ, a data point, a picture, a computer, etc. In most cases this information has to be somehow transformed in a way suitable for the electric communication system. The transformation must be carried out in an efficient way using a minimum of expensive resources such as communication capacity. This is done in the source coder. In some cases the information is to be ciphered, which can be done after source coding. This is only briefly treated in this book, why a cipher block is not included. However, the fact that all information is in digital form makes it easy to implement communication which can prevent eavesdropping by an unauthorised user. The coded information may then require further processing to eliminate shortcomings in the channel transferring or storing the information. This is carried out by using different methods in the channel encoder. The modulator is a system component which transforms the message to a type of signal suitable for the channel.

1.3.2 THE CHANNEL

The channel is the medium over which the information is transferred or stored. Examples are: the cable between two end points in a wire bound connection, the environment between the transmitter and receiver of a wireless connection, or the magnetic medium on which the information is stored. The channel model is a mathematical model of the channel characteristics.

1.3.3 RECEIVER SIDE

After the signal has passed the channel it is distorted. In the demodulator the information is retrieved as accurately as possible. A unit which can be said to be included in the demodulator is the equaliser. To minimise the bit-error rate the equaliser should adjust the demodulation process to the changes in the signal, which can take place in the channel. Even if the demodulator is functioning as well as possible there may still be errors left in the information. Channel coding provides the possibility of correcting these errors within certain limits. The error correction or error detection is made in the channel decoding block. Since the information is coded at the source it has to be transformed in such a way that it can be understood by the receiver, for example so that a human voice is restored or a picture restored. This is done in the source decoder. For the system to function properly certain time parameters must be common for the transmitter and receiver. The receiver, for instance, must know at what rate and at what instance the transmitter changes the information symbol. These parameters are measured by the receiver in the synchronising block.

1.3.4 ANOTHER SYSTEM FOR COMMUNICATION

The same communication model can be used for the situation in which you presently are. The writer wants to transfer information to the receiver, which is you. The writer codes the information with the aid of language. Hopefully in an efficient way without unnecessary chat. Error correction is built into the system since you understand even if one or another letter should be erroneous. The book which you have in front of you is the channel. Disturbances come into the transmission, for instance in the form of different factors distracting the reader. You decode the message and obtain understanding for the subjects treated in the book.

1.3.5 THE SCOPE OF THE BOOK

This book describes the signal processing carried out in the blocks above (cf. Figure 1.3), the models that are being used and the capacity which can be obtained in the systems. The methods described give the basic ideas for the

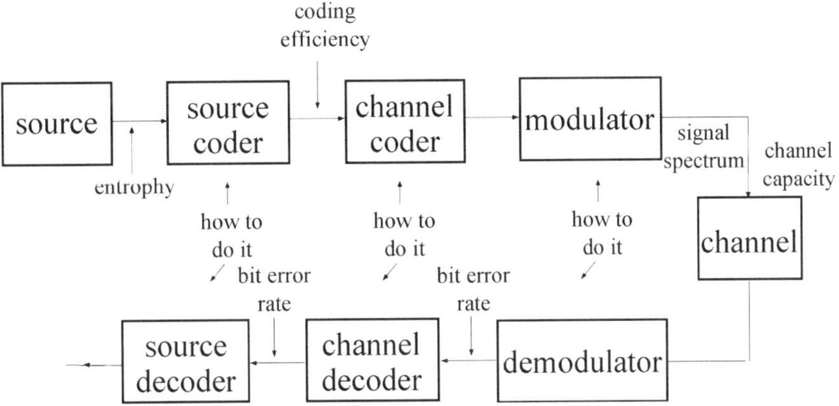

Figure 1.3 The concepts of digital communication systems illustrated above are among the items treated in this book.

respective block and provide continuity throughout the book. The limiting factor in telecommunication systems is usually the capacity of the channel, which is limited either by noise or by the available bandwidth. Methods for the calculation of channel capacity are therefore given. The channel capacity has to be put in relation to the amount of information which needs to be transferred from the source. Methods for the calculation of the amount of information or the entropy of the source are therefore treated. Perhaps demands go beyond the capacity of the channel. The book describes fundamental methods of doing the source coding and how the efficiency of the coding is calculated. Furthermore, it describes how error-correcting and error-detecting coding can be done and the capacity, in terms of bit-error rate, which can be obtained. Different methods for modulation are described. An important property of the modulated signal is its spectrum which must be put in relation to the bandwidth that the channel can provide. The book describes how the demodulator must be designed to retrieve the information as accurately as possible. Ways of calculating the probability of the demodulator misinterpreting the signal, i.e. the bit error rate, when certain properties of the channel are given beforehand, are also discussed. Fundamental methods for adaptive equalisers are also treated.

Synchronisation is treated, where the phase-locked loop is an important component and is therefore treated in a special section. The adaptive antenna is a system component which is finding an increasing number of applications, even outside military systems, and its fundamental aspects are also treated in this book.

1.4 Multiple user systems

1.4.1 METHODS FOR MULTIPLE ACCESS

When several users are using the same physical transfer channel (fibre, coaxial cable, space, etc.) for communication, ways of separating the messages are required. This is achieved by some method for multiple access. Three fundamental methods exist. In the first each user is given his own relatively narrow frequency band for communication. This method is called FDMA or frequency-division multiple access. The basic concept is simple but requires efficient filters for the separation of the messages. The next method is TDMA, ''time-division multiple access''. In this case each user has access to a limited but repeated time span for the communication. This concept requires great care for time synchronisation but does not require as efficient filters as the FDMA method. These methods are depicted in Figure 1.4, where time is along the x-axis and frequency along the y-axis. A FDMA system has a relatively narrow bandwidth along the frequency axis but extends along the whole x-axis. A TDMA

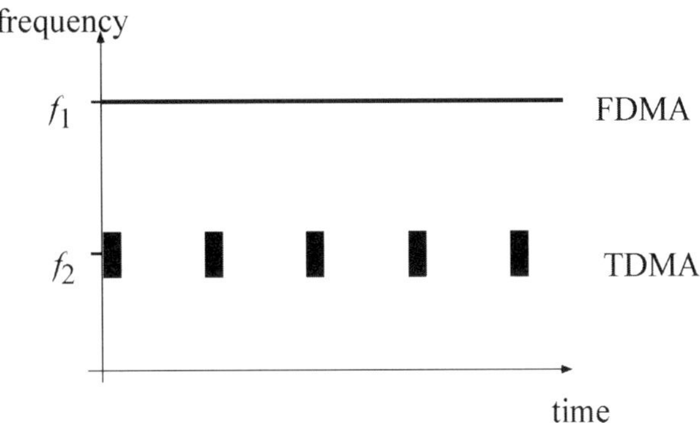

Figure 1.4 FDMA and TDMA systems.

system needs a broader bandwidth to transfer the same amount of information, but only takes up sections of the time axis.

The third method is CDMA, code-division multiple access. Each user includes a special code in his message, which is orthogonal to the other users' codes. This means that the other users will only be regarded as random noise sources in relation to the actual message. CDMA provides a flexible system but it can be complicated if the number of users and their distribution in space vary. To include CDMA in a diagram a third axis representing the coding dimension has to be added, as shown in Figure 1.5.

An acronym which ought to be explained in this context is SDMA, spatial-division multiple access. This means that the antenna is pointing the signal power in the desired direction. The same frequency, time frame or code can then be used by another user in another direction.

All real systems are hybrids of two or more of the above-mentioned multiple

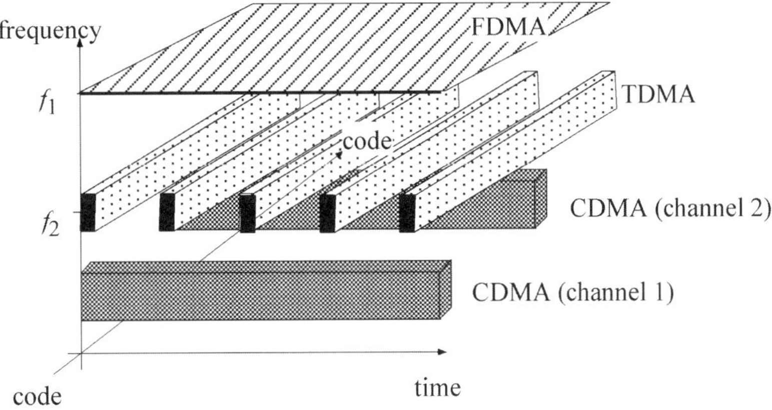

Figure 1.5 Diagram depicting FDMA, TDMA and CDMA in the multiple access space.

access methods. Thus, a TDMA system is, for instance, also an FDMA system as the TDMA frames have a limited bandwidth and other systems operate outside this frequency band. The choice of multiple-access method depends on several factors, such as the properties of the physical channel, implementation aspects and coexistence with other systems. To give the reader some insight into these issues, the multiple-access methods are described in more detail below.

The word multiplexing is used when access by the different users to the channel is controlled by a master control, i.e. as in wire-bound telephony. TDM and FDM are often used, corresponding to TDMA and FDMA.

FDMA

Each user, or message, has its own frequency band. At the transmitting end the messages are transferred to the respective frequency bands through modulation of different carrier waves. Figure 1.6 shows an example with seven different messages, each with a bandwidth of 2 W (Hz). Each carrier wave is a sinusoidal signal having, for example, the frequency f_2 (Hz) for message 2.

The signals are transferred on a common channel which can be a coaxial cable or the space between the transmitter and receiver. At the receiving end the different signals are separated by bandpass filters, which can distinguish the relevant frequency band for each message. The signals are then transformed back to the original frequency band using demodulators.

When analysing FDMA systems the influence of the channel on each message is usually evaluated individually. In a real system it is difficult to completely eliminate the influence of the other messages and these will enter the analysis as an increased noise level. For a digital communication system, speech and channel coding of each message are separately added at the transmitting end (cf. Figure 1.3). At the receiving end corresponding decoding is obviously added.

For optical fibre communication, different messages can be transferred in the same fibre using different wavelengths. The method is basically identical to FDMA, but in this case it is called WDM, wavelength division multiplex.

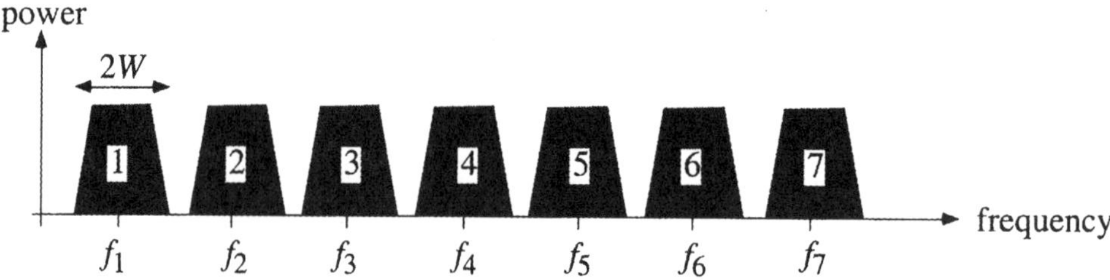

Figure 1.6   The positions of the different messages along the frequency axis. Each signal has a bandwidth of 2 W and must therefore take up at least the corresponding bandwidth on the frequency axis.

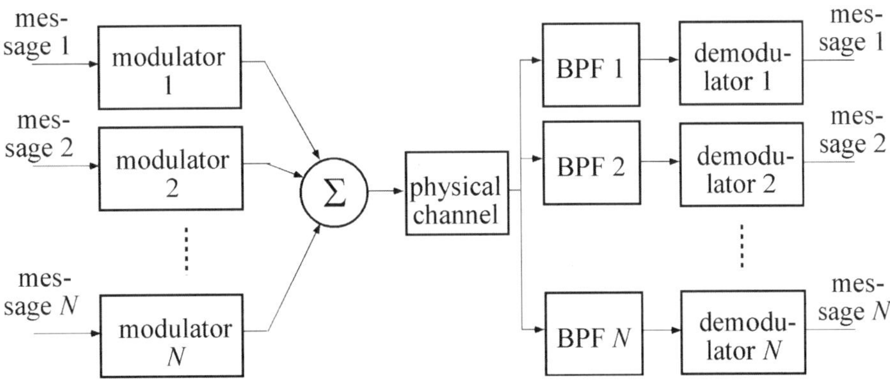

Figure 1.7 Structure of an FDMA system for N different messages. BFP is a bandpass filter.

It should be remembered that Figure 1.7 is primarily a model of the system, which is useful when understanding of, or calculations on, the properties of an FDMA system are desired. In certain cases the practical structure also looks the same, e.g. in wire telephone systems. In radio systems, on the other hand, each modulator in the figure can usually be seen as a radio transmitter and each bandpass filter with demodulator as a receiver. The addition of the signals is done in the space between the transmitter and the receiver. If the transmitters are at different locations and the receivers at the same location the model can be modified so that each transmitter has its own channel and the addition is performed at the channel output. The model is obviously influenced by the application under study.

TDMA

To give each message source access to the transfer channel during a certain time slot it is possible to let the channel be preceded and finished by a switch which connects the source at the input and guides the signal to the correct destination at the output. One or more samples from the respective source can be transferred within each time slot. Since the messages are generated in a continuous flow of samples, these must be stored in buffers which are emptied in each time slot. The transfer rate of the channel is defined by the total transfer need from all sources. The buffers at the output reset the original message rate in each branch.

In the same way as for the FDMA system, Figure 1.8 is mainly a model. In practice there is usually no switch at the channel input, as each source is well behaved and transmits information on the channel only during the allocated time slot. This is, for instance, the case for mobile telephone systems using TDMA. The channel (cf. Figure 1.8), can include features such as channel coding, modulator, physical channel, demodulator and error correction. These units can also be present in each of the branches 1 to N.

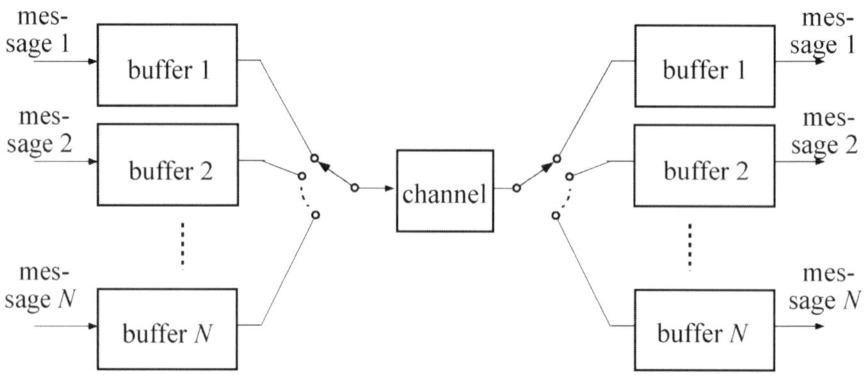

Figure 1.8   Model of a TDMA-system.

CDMA

CDMA, or code-division multiple access, is presently mainly used in wireless systems. There are several methods of CDMA. The method described below is direct sequence CDMA (DS/CDMA), which is used, for instance, in certain mobile telephone systems. Assume two signals; one message signal and one spreading code. The latter is made up of a defined sequence of the symbols $+1$ and -1. The symbol rate of the spreading code is considerably higher, 1000 times higher say, than the data rate of the message signal. The two signals are multiplied. This can be seen as a chipping of the information signal by the spreading code (cf. Figure 1.9). The rate of the spreading code is therefore called the chip rate.

The bandwidth of a signal is proportional to its data or symbol rate. The result of the multiplication is therefore that the signal spectrum is spread out and obtains a bandwidth approximately corresponding to the double chip rate. A signal with a broader bandwidth than the actual information signal is then transmitted into

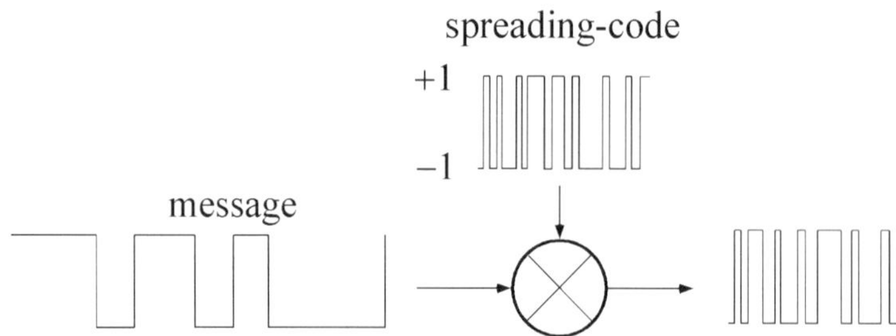

Figure 1.9   The message stream is sliced through multiplication by the spreading code. The real ratio between message and chip rate is higher than shown in the figure.

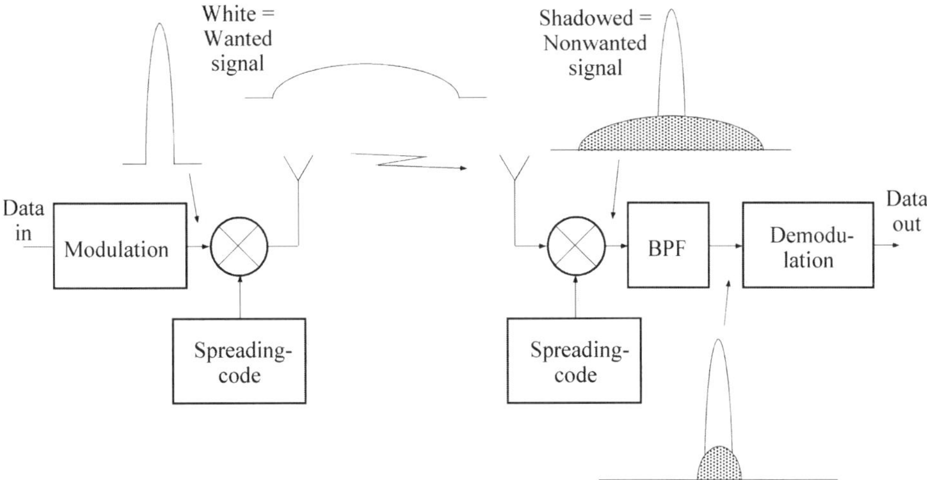

Figure 1.10   CDMA system.

space as illustrated in Figure 1.10. This type of system is therefore called a broad-band or spread-spectrum system.

At the receiving end the received signal is multiplied by an identical spreading sequence, adjusted for the time lag between transmitter and receiver. Since $-1 \times -1 = +1$ and $+1 \times +1 = +1$, the spreading code completely vanishes and the message signal resumes its original shape and bandwidth. The bandwidth of interfering signals, on the other hand, become larger than or equal to the double chip rate. Subsequent bandpass filter is adjusted to the bandwidth of the message signal and cuts off most of the power from the unwanted signals. The ratio between wanted and unwanted signal powers will then normally increase suffi-ciently not to interfere with the demodulator which regenerates the message stream. The ratio is improved in proportion to the bandwidth expansion and is called processing gain; e.g. a bandwidth expansion of 100 times gives a proces-sing gain of 20 dB.

The combination of multiplier and bandpass filter can be seen as a correlator since the filter bandwidth is narrow in relation to the double chip rate. A model for a CDMA system with several users can then look like the one shown in Figure 1.11.

The bandwidth of the information signal is approximately $1/T$, where T repre-sents the duration in seconds of the information symbols. The bandwidths of the bandpass filters are equal to the bandwidths of the information signals. The time constant of the filters is then equal to T, and the information symbols can be seen as constants while the spreading signal is integrated by the filters. The output signal from each correlator, y_k, can be divided in three terms as defined in eq.

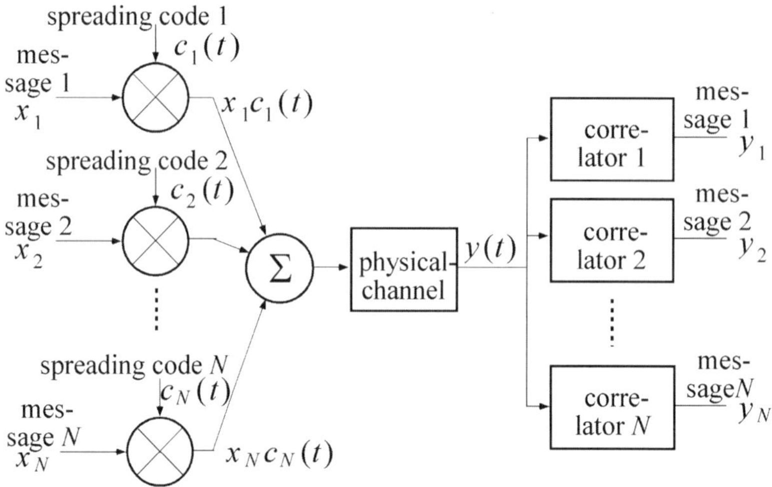

Figure 1.11 Model of a CDMA system.

(1.1). Assume that the only influence of the channel on the signal is that noise, $n(t)$, is added.

$$y_k = \frac{1}{T} \int_0^T y(t)\, c_k(t)\, dt$$

$$= x_k \underbrace{\int_0^T \frac{c_k(t)\, c_k(t)}{T}\, dt}_{=1} + \sum_{j\neq k}^{N} x_j \underbrace{\int_0^T \frac{c_j(t)\, c_k(t)}{T}\, dt}_{=\rho_{jk}} + \underbrace{\int_0^T \frac{n(t)\, c_k(t)}{T}\, dt}_{=n_0}$$

$$= x_k + \sum_{j=1,\, j\neq k}^{N} x_j\rho_{jk} + n_0 \tag{1.1}$$

The first term represents the multiplication of the information symbol by the autocorrelation of the spreading sequence, which is equal to one. Consequently, only the information symbol remains. The next term is the sum of all the other messages. Each message has been multiplied by a coefficient consisting of the cross correlation between the spreading sequence of the wanted signals and all the other spreading sequences. The last term represents a scaling of the noise which has been added in the channel.

As can be seen a factor is also included which depends on the cross correlations of the spreading sequences. This causes interference in the communication and the factor should be as small as possible. The spreading codes are consequently very important for the properties of CDMA systems. It is a matter of finding a large set of sequences with small or no mutual cross correlation. A distinct autocorrelation peak is also important, to inhibit false synchronising with an autocorrelation peak other than the wanted one. Two examples of sequences used are Walsh functions and m-sequences. The m-sequences are generated in feedback shift registers. The feedback is chosen so that the length of the sequence is the longest possible for the given shift register.

Even if codes with low or no cross correlation are chosen, the filtering that occurs in the channel will introduce correlation between the messages. Then the factors ρ_{jk} will not be negligible to the system properties.

1.4.2 CONTINUED ANALYSIS

In the analysis throughout this book, the message path from source to destination is studied. The multiple access method can then be regarded as being transparent, i.e. it does not need to be considered in the analysis. Possibly the noise level can be affected by the choice of multiple-access method. The model shown in Figure 1.2 is applied here. This does not imply that the multiple-access method is irrelevant for the properties of a real system. It strongly affects questions such as system capacity, implementation and properties in a multi-path environment. These aspects, however, are outside the scope of this book.

1.5 Appendix: the division and usage of the electromagnetic spectrum

Frequency band	Wavelength	English terminology
300–3000 Hz	1000–100 km	Voice frequency (v-f)
3–30 kHz	100–10 km	Very low frequency (VLF)
30–300 kHz	10–1 km	Low frequency (LF)
300–3000 kHz	1000–100 m	Medium frequency (MF)
3–30 MHz	100–10 m	High frequency (HF)
30–300 MHz	10–1 m	Very high frequency (VHF)
300–3000 MHz	100–10 cm	Ultra high frequency (UHF)
3–30 GHz	10–1 cm	Super high frequency (SHF)
30–300 GHz	10–1 mm	Extremely high frequency (EHF)

Usage:

- *ELF (300–3000 Hz)* provides underwater radio communication for short and long range uses, e.g. to submarines when underwater; underground communication, e.g. in mining; remote sensing under the surface of the earth; distribution through ground and water.
- *VLF (3–30 kHz)* provides stationary services over large distances, e.g. navigation systems and time signals (Omega 10–14 kHz); time and frequency reference signals. The ground wave follows the curvature of the earth.
- *LF (30–300 kHz)* provides long distance communication with ships and aircraft; long range stationary services; radio broadcasting, especially national services; radio navigation. The ground wave follows the curvature of the earth. The waves are attenuated in the ionospheric D-level, which exists during the daytime. At night a very long range can be achieved by reflection in other ionospheric levels. (The ionosphere consists of 2–3 electrically conducting layers, reflecting radio waves, at altitudes of 80 km or more above the earth's surface. The ionosphere is composed of free electrons and ions of for example hydrogen and oxygen which have been ionised by among other things ultraviolet sun rays. In order the layers are named D, E, F (F can be divided into F1 and F2).)
- *MF (300–3000 kHz)* provides radio broadcasting, especially national services; radio navigation; radio beacons for shipping; mobile and stationary services; emergency frequency for shipping (2182 kHz). The waves are attenuated in the ionospheric D-level, which exists during day time. At night a very long range can be achieved by reflection in other ionospheric levels.
- *HF (3–30 MHz)* provides stationary point to point connections; mobile services for land, sea and air transport; radio broadcasting over large distances, international services; amateur radio; military communication systems over large distances; private radio; time signals. The waves can be transmitted over long ranges as the waves are reflected in the ionosphere.
- *VHF (30–300 MHz)* provides radio broadcasting and television with a transmitting range of up to a few hundred kilometres; mobile services for land, sea and air transport, e.g. taxi, transport, police and fire brigade communications; relayed traffic for radio telephone systems; radio beacons for air traffic; emergency frequencies 121.5 and 156.8 MHz.
- *UHF (300–3000 MHz)*. Television with a transmitting range of up to a few hundred kilometres; navigation aids for air traffic; radar; mobile telephone systems (D-AMPS, GSM, etc.) and stationary point to point services; mobile radio; mobile and stationary satellite services; future personal telephone systems; cordless telephones; satellite navigation; GPS and Glonass; distribution through direct wave or reflected wave.
- *SHF (3–30 GHz)*. Stationary services (radio link systems); stationary satellite services; mobile services; mobile satellite services; remote sensing via satel-

lites; satellite television broadcasting; computer communications in so-called wireless local area networks (WLAN); usually line of sight communication. This is a frequency band where most of the radar applications are located. Long range radar is in the low part of the band, e.g. aircraft surveillance over large areas. Short range radar is in the upper part of the band.

- *EHF (30–300 GHz)*. Extreme short distance communication; satellite applications; remote sensing via satellites.

2   LINK BUDGET

An important parameter for the properties of the transmission channel is the signal to noise ratio (SNR), which is defined as the ratio between the received power of the wanted signal and the power of the received noise. To calculate SNR, both the power of the wanted signal and the power of the noise have to be known. These are obtained by setting up a link budget and by using noise models, respectively. A link budget is a calculation of the power of the wanted signal, from now on called the signal power, as a function of the various transmitter and receiver parameters.

The aim of this chapter is to introduce signal to noise calculations for the fundamental case of transmission in free space, as well as introducing the different concepts of SNR currently being used in analogue and digital transmission. For a thorough presentation the reader is referred to the special literature in wave propagation, noise in amplifiers and antenna theory.

2.1 Signal to noise ratio

The signal to noise ratio is usually expressed in decibels (dB) defined as

$$SNR = 10 \log\left(\frac{P_R}{N_R}\right) \tag{2.1}$$

where P_R is the power of the wanted signal and N_R is the noise power, both in Watts (W). For digital signals the following definition is used:

$$SNR = 10 \log\left(\frac{E_b}{N_0}\right) \tag{2.2}$$

where E_b is the bit energy (J) of the wanted signal and N_0 is the noise power density (W/Hz).

A description of how the units can be calculated is given below. We describe a radio system where the information is transmitted by an electromagnetic wave from a transmitter to a receiver located elsewhere.

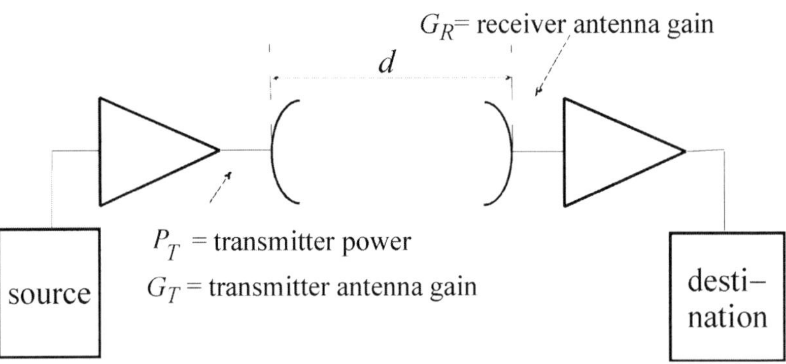

Figure 2.1   Model for signal transmission in a radio system.

2.1.1 CALCULATION OF THE SIGNAL POWER

The transmitted electromagnetic signal is distributed in space and only a fraction reaches the receiver. To calculate the received signal power the following simple model for radio transmission in free space can be used. For other kinds of transmission or storing of signals, other models are used. The signal is generated in a source and amplified in some kind of power amplifier having the output power P_T. The power is radiated by the transmitter antenna, passes through the transmitting medium, is captured by the receiving antenna, amplified in the receiver and finally reaches its destination (Figure 2.1).

The antenna can be seen as a device for converting an electric current to an electromagnetic wave or vice versa. Assume a point-source antenna. The output power of the transmitter will then be distributed equally in all directions, i.e. the antenna is an isotropic antenna. The total power is then distributed evenly, as on the surface of a sphere (cf. Figure 2.2), and the power density, in W/m^2, at

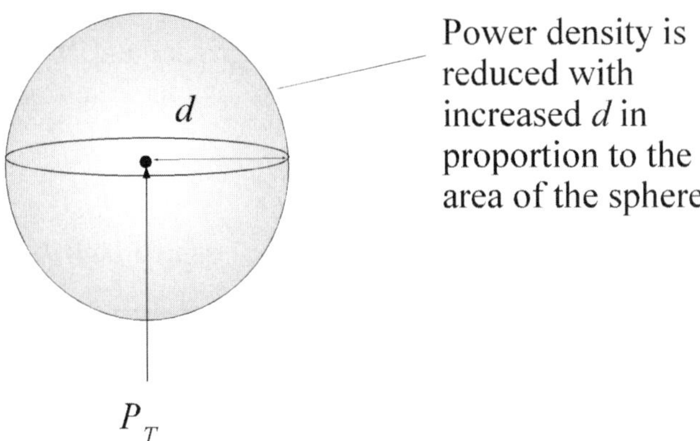

Figure 2.2   Power distribution on the surface of a sphere.

distance d becomes

$$\text{power density} = \frac{P_T}{4\pi d^2} \tag{2.3}$$

In most cases the antenna is not isotropic, but concentrates the power in a specific direction. More power can then be received in the desired direction and, at the same time, the risk of interfering with other communication is reduced (Figure 2.3).

The increased power in the main lobe can be interpreted as an amplification in the directional antenna, which is denoted antenna gain and is defined as

$$G_T = \frac{\text{max radiation intensity}}{\text{mean value of radiation intensity over the whole sphere}} \tag{2.4}$$

(for a lossless antenna, gain and directivity are the same). The radiation intensity is defined as radiated power per steradian (W/str) and is therefore independent of the distance to the antenna. The gain of the transmitting antenna is denoted G_T and the gain of the receiving antenna is denoted G_R. The antenna is a reciprocal system component, which means that its function is independent of how the power is delivered to the antenna, be it from the surrounding space to the connecting joint or vice versa. The antenna gain is the same whether the antenna is used for transmission or reception and is often given in dB relative to an isotropic antenna and is then denoted dBi. The power which is captured by a receiving antenna located within the main lobe of the transmitting antenna is higher than it would be if the transmitting antenna diagram were isotropic. The received power corresponds to a transmitted power of *EIRP* (effective isotropic radiated power) and depends on the product of the input power and the antenna gain.

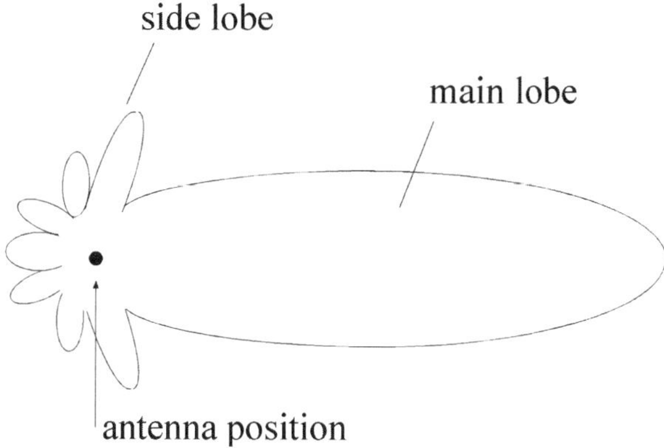

Figure 2.3 Directional antenna with the desired main lobe and unwanted, but unavoidable, side lobes.

$$EIRP = P_T G_T \tag{2.5}$$

The power given by *EIRP* corresponds to the transmitting power which would be needed if an isotropic antenna were being used instead.

Suppose free space prevails between transmitter and receiver. The power captured by the receiving antenna is then

$$P_R = \frac{P_T G_T}{4\pi d^2} A_R \tag{2.6}$$

When the free space assumption is not valid some further terms which model the increased losses are included in eq. (2.6). A_R represents the effective antenna area of the receiving antenna. The effective antenna area corresponds to the concept of cross-section in particle physics and is not a real area. A relationship between effective antenna area and antenna gain G_R, is given by

$$G_R = \frac{4\pi A_R}{\lambda^2} \tag{2.7}$$

where λ is the wavelength of the signal. (The relation between frequency, f, and wavelength, λ, is $\lambda = v/f$, where v is the propagation velocity of the wave, usually 2.997925×10^8 m/s.) Eq. (2.6) in (2.7) gives the received signal power:

$$P_R = \frac{P_T \, G_T \, G_R \, \lambda^2}{(4\pi d)^2} \tag{2.8}$$

We have thus found an expression for the received signal power given as a function of parameters, which are normally given in connection with hardware specifications. When the condition of free space is not fulfilled, more complex models for the wave propagation are needed, taking into account the topology of the environment. A simple way of taking increased distribution losses into account within the UHF field is to let the exponent of the distance parameter d assume a higher value, e.g. between 2 and 4.5. A common value in urban areas is 3.5. The situation is even more complicated if it is assumed that the transmitter and receiver are in relative motion. For those who wish to do further studies within this field, special literature dealing with wave propagation is available. Since it is often difficult to design a reliable model, measurements are always important complements to calculations of signal power in radio systems.

It is practical to sometimes express the signal power in dB, even though it represents an absolute value and not a ratio. In such a case the power is referenced to as 1 Watt or 1 milliwatt, and is denoted dBW and dBm, respectively. Thus, the power 1 W is equal to 0 dBW or 30 dBm.

In order to obtain the SNR the noise power is needed, and this will be calculated in the next section. But first a few words about the variation in signal strength that may occur if the signal propagates along many paths.

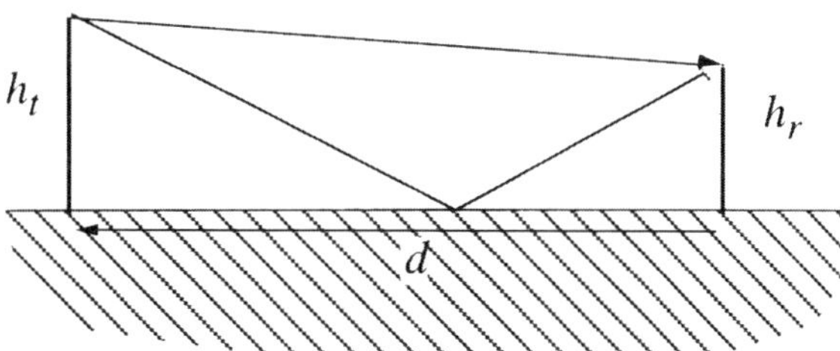

Figure 2.4   Direct and reflected signal path.

Fading

A propagating electromagnetic wave is reflected in objects, large enough
compared to the wavelength. Signals in the MF range are reflected for example
in the ionosphere. In the UHF range the signals are reflected in houses, cars, etc.
Assume a situation where a transmitter is located at height h_t above ground. The
signal propagates both through a direct wave and a wave that has been reflected in
the ground. The signal arrives at the receiver situated at distance d and height h_r
(cf. Figure 2.4).

The difference in runtime, $\Delta\tau$, along the separate paths is

$$\Delta\tau = \frac{1}{v}\left[\sqrt{(h_t + h_r)^2 + d^2} - \sqrt{(h_t - h_r)^2 + d^2} \right] \tag{2.9}$$

When the distance between transmitter and receiver is large relative to the
antenna heights an approximation is often used:

$$\Delta\tau \approx \frac{2}{v}\frac{h_t\, h_r}{\sqrt{h_t^2 + h_r^2 + d^2}} \tag{2.10}$$

The reflected signal strength depends on the reflection coefficient, which depends
on the constitution of the reflector. Usually the signal is attenuated several dB in
the reflection. Assume for a change a perfect mirror, i.e. the signal is not attenuated at the reflection. If the sent signal is $A\cos(\omega_c t)$ then the received signal is

$$A\cos(\omega_c t) + A\cos(\omega_c t + \omega_c\,\Delta\tau) = 2A\cos\left(\frac{\omega_c\,\Delta\tau}{2}\right)\cos\left(\omega_c t + \frac{\omega_c\,\Delta\tau}{2}\right) \tag{2.11}$$

As is seen the signal amplitude will depend on the factor $\cos(\omega_c\Delta\tau/2)$ which is
also zero for certain time differences $\Delta\tau$. These variations in signal strength are
called fading. When it occurs it is superponed on the attenuation in free space
propagation (see Figure 2.5).

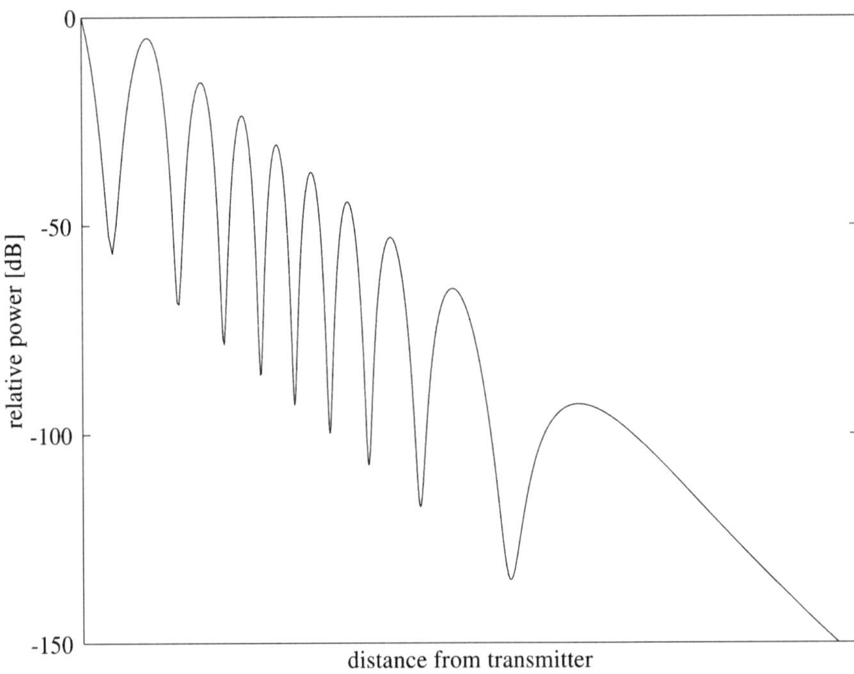

Figure 2.5   Example of a fading pattern for a two ray channel.

2.1.2 CALCULATION OF NOISE

Noise and other kinds of interference are always present at the transmission of electromagnetic signals and will limit the quality of the transmitted information. Interference is usually divided into two main categories: natural and man-made. Natural noise is generated by physical phenomena in nature, such as thermal noise, cosmic noise, solar radiation and atmospheric lightning. This kind of noise is difficult to control at the source and is therefore always present as a noise floor. Man-made noise is generated by human activities through the use of different technical apparatus, such as cars, engines, electrical motors, electric light, power transmission lines, etc. Man-made noise can be controlled by shielding and other measures at the source, or by choosing a location for the receiving equipment which is relatively free of man-made noise.

For frequencies above 50 MHz thermal noise usually dominates (under 50 MHz the atmospheric noise dominates). Thermal noise is always present and is unavoidable because it is generated by the thermal motion of the electrons in the electrical conductor materials. Thermal noise has a Gaussian amplitude distribution and can for all technical purposes be regarded as 'white', i.e. it has a constant spectrum. For frequencies larger than 10^{12} Hz thermal noise decreases. Man-made noise can vary with the application and is more difficult to define in

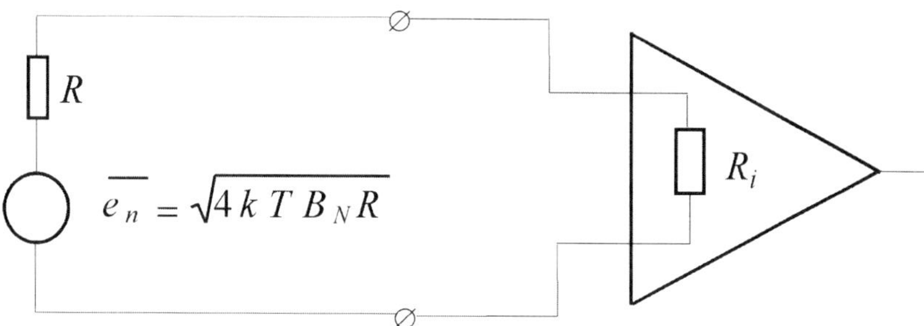

Figure 2.6 Model for thermal noise in an amplifier.

terms of a general model. The presentation below is limited to thermal noise. To calculate SNR this noise power needs to be known. For this reason a description of how to calculate the thermal noise power in an electrical system is given below. To calculate noise in electrical circuits a voltage source and a series resistance is used as a model (Figure 2.6). The source is generating the noise e_n. The amplifier is assumed noise free.

The root mean square (rms) value of the voltage source is given by

$$\overline{e_n} = \sqrt{4k\,T\,B_N R} \tag{2.12}$$

where k is Boltzman's constant (1.38×10^{-23} J/K), T is the absolute temperature (K), B_N is the noise bandwidth (Hz), R is the resistance (Ω).

As can be seen in eq. (2.12) the bandwidth of the system is included as a parameter to calculate the noise power. The noise bandwidth is not the same as the often used 3-dB bandwidth, but corresponds to the bandwidth a system with a rectangular-shaped frequency response would need in order to generate the same noise level as the real system. The noise bandwidth is defined as

$$B_N = \frac{1}{G_{\max}} \int_0^\infty G\,(f)\,df \tag{2.13}$$

where $G_{\max}$ is the maximum power gain and $G(f)$ is power gain versus frequency ($= |H(f)|^2$). The definition of noise bandwidth can be made single-sided when frequencies from 0 Hz and up are being used, or double-sided, which means that both sides of the frequency axis are included. The integration in eq. (2.13) is then taken from minus infinity to plus infinity. The single-sided definition uses the range of the spectrum available at practical measurements. In this presentation the single-sided noise bandwidth is used in order to simplify the connection to practical applications and because this definition is the most common one used in text books within this field. At single-sided analysis the power density of the noise in W/Hz is twice that of the double-sided case. The total noise power, however, remains the same in both cases since the bandwidth at double sided analysis is twice that of single-sided.

Maximum power into a load is obtained when the output resistance of the source is matched to the input resistance of the amplifier, which is fulfilled when $R = R_i$. The best possible SNR is obtained when the system is matched. When matched, the noise power at the input is

$$N_R = kTB_N \quad \text{(W)} \tag{2.14}$$

In many cases it is not the total power of the noise which is of interest but the power density of the noise in W/Hz. This is often denoted noise density. The single-sided noise density is defined as

$$N_0 = \frac{N_R}{B_N} = kT \quad \text{(W/Hz)} \tag{2.15}$$

When the noise density is measured with a spectrum analyser, a filter with a certain noise bandwidth is always used. In order to obtain the real noise density, N_0, the measured density must therefore be divided by the noise bandwidth of the analysing filter. For double-sided noise the density becomes $N_0/2$.

In amplifiers more noise than just the thermal is always added because of different noise-generating mechanisms in active components. In order to maintain a noise free amplifier in the model, the amplifier noise is represented in the way that the input resistance is assumed warmer than the ambient temperature T_a. The assumed temperature of the resistance is called total noise temperature. It corresponds to the temperature the source resistance would have to assume to generate the same noise as is measured at the output of the amplifier.

Example 2.1

Calculate the noise bandwidth B_N for a Butterworth lowpass filter of nth order and compare with the 3-dB bandwidth. What is the power of the output signal if the input signal consists of white noise having the double-sided power density $N_0/2$?

Solution   For a Butterworth lowpass filter the power transfer is (found in books on filter or circuit theory)

$$|H(\omega)|^2 = \frac{1}{1 + (\omega/\omega_0)^{2n}} \tag{2.16}$$

where $\omega = 2\pi f$, ω_0 is the upper limit of the angular frequency or the 3-dB bandwidth in rad/s and n is the order of the filter. Since the noise bandwidth, B_N, is usually expressed in Hz, f is used instead of ω, which gives the noise bandwidth (see Figure 2.7)

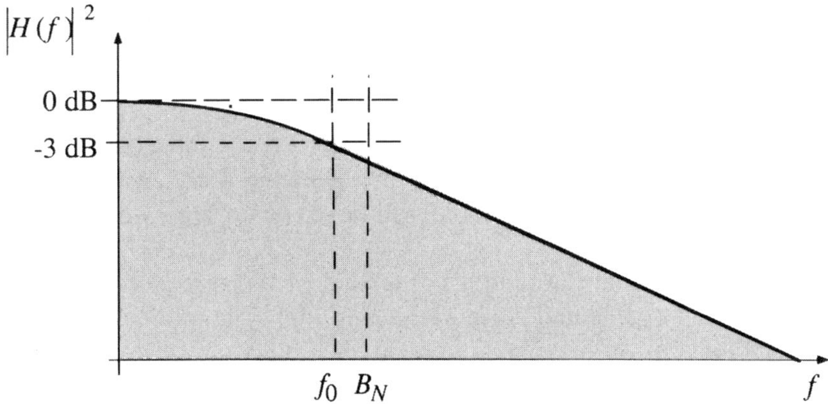

Figure 2.7   Frequency response and noise bandwidth of the filter.

$$G_{max} = |H(0)|^2 = 1$$

$$B_N = \int_0^\infty \frac{1}{1 + (f/f_0)^{2n}} \, df = \frac{\pi f_0}{2n \sin(\pi/2n)} \tag{2.17}$$

For n between one and four the ratio between the 3-dB bandwidth and the noise bandwidth then becomes as listed in Table 2.1. When the input signal consists of white noise with the double-sided power density $N_0/2$ (W/Hz) the double-sided power density of the output signal, $G_y(f)$, is

$$G_y(f) = \frac{N_0}{2} |H(f)|^2 = \frac{N_0}{2} \frac{1}{1 + (f/f_0)^{2n}} \tag{2.18}$$

The way in which the noise bandwidth has been defined means that the single-sided noise density must be used. For a second order Butterworth filter, for

Table 2.1

Filter order	Noise bandwidth, B_N
1	$\dfrac{\pi}{2} f_0 \approx 1.57 f_0$
2	$\dfrac{\pi}{2\sqrt{2}} f_0 \approx 1.11 f_0$
3	$\dfrac{\pi}{3} f_0 \approx 1.05 f_0$
4	$1.03 f_0$

instance, the noise power at the output becomes $N_0 B_N = N_0 f_0 \pi / 2\sqrt{2}$ (W).

Noise in systems

A communication system consists of several connected units which amplify the signal, but unfortunately they also generate noise. An amplifier whose input is connected to a signal source will have the noise power P_N at the output (Figure 2.8).

The noise at the output is the sum of the noise of the signal source and the internally generated noise in the amplifier. The internally generated noise can be thought of as a noise source at the input of the amplifier having the noise temperature T_e. The total noise will then be the sum of two noise sources at the input, one corresponding to the input noise and the other corresponding to the noise generated internally. The noise at the output is

$$P_N = GP_{in} + P_{internal} = GkT_sB + GkT_eB = GkB(T_s + T_e) \tag{2.19}$$

where T_s is the noise temperature of the source, which usually is 290 K if the source noise only consists of thermal noise. P_{in} is the noise power at the amplifier input. When the input noise only consists of thermal noise then $P_{in} = N_R$. The total noise at the amplifier output corresponds to a noise source at the input having the temperature $T_s + T_e$. As mentioned earlier, the parameter T_e is called equivalent noise temperature and gives the noise contributed by the amplifier.

The ratio S/N at the input and output differs by a factor F and can be written

$$\left(\frac{S}{N}\right)_{out} = \frac{1}{F}\left(\frac{S}{N}\right)_{in} \qquad \Leftrightarrow \qquad \left(\frac{GS}{P_N}\right) = \frac{1}{F}\left(\frac{S}{P_{in}}\right) \tag{2.20}$$

The factor F is called the noise factor and can be obtained from eq. (2.20) giving

$$F = \frac{P_N}{GP_{in}} = \frac{GP_{in} + P_{internal}}{GP_{in}} = 1 + \frac{T_e}{T_s} \tag{2.21}$$

In order to obtain sufficient gain, several amplifiers are usually connected in cascade (Figure 2.9).

The noise power at the output of the last amplifier becomes:

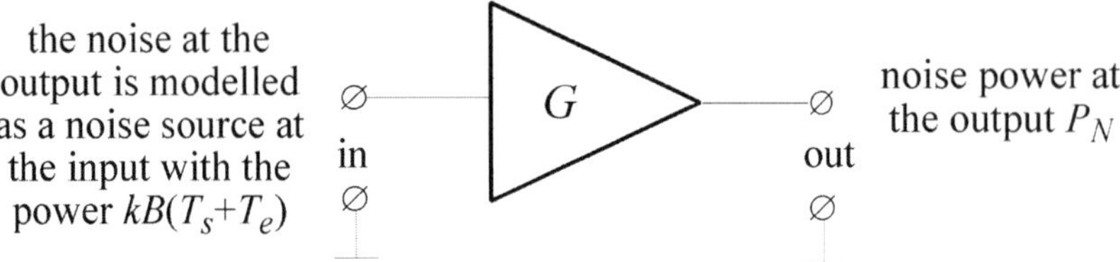

Figure 2.8   Amplifier with the power gain G.

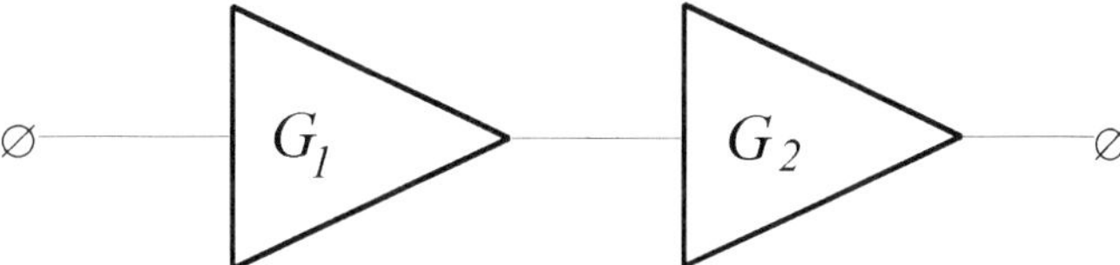

Figure 2.9 Amplifiers in cascade. Total gain $G = G_1 \cdot G_2$.

$$P_N = \overbrace{G_1 G_2 k T_s B}^{\text{source}} + \overbrace{G_1 G_2 k T_{e_1} B}^{\text{noise in stage 1}} + \overbrace{G_2 k T_{e_2} B}^{\text{noise in stage 2}} \tag{2.22}$$

The total noise factor of the system then becomes (cf. eq. (2.21))

$$F = \frac{P_N}{G_1 G_2 P_{in}} = 1 + \frac{T_{e_1}}{T_s} + \frac{T_{e_2}}{G_1 T_s} = F_1 + \frac{F_2 - 1}{G_1} \tag{2.23}$$

This means that the noise factor of stage two, F_2, will be reduced by one and divided by the gain in stage one. If this gain is large enough, the noise factor of stage one, F_1, will dominate the total noise factor of the system. The formula is sometimes called Friis formula and can obviously be generalised to more amplifier stages. The more stages which are included in the analysis, the more obvious it is that the input noise dominates. The conclusion is that a system with low noise should have high gain and low noise in its first amplifier stage. On the other hand can the amplification, in for example a radio receiver, not be too high prior to the stages that separate the wanted signal. The input stage will otherwise be subject to the risk of distorting the wanted signal if a strong signal, except for the wanted one, should be received. The gain of the input stage must be a compromise between these contradicting requirements.

When an amplifier or some other system component is delivered from a manufacturer, the noise factor has been measured at a standardised temperature, which is $T_0 = 290K$. If the signal source has another noise temperature, i.e. $T_s \neq T_0$, the noise factor has to be modified. This can be done by starting with eq. (2.21) and norm F by F_0, which has been specified at the standard temperature. Thereafter solve for F, which gives the relation

$$F = 1 + (F_0 - 1)\frac{T_0}{T_s} \tag{2.24}$$

The modification of the value of F by means of eq. (2.24) can for example be necessary when the system is connected to an antenna capturing atmospheric disturbances, which make the noise at the antenna output larger than if only the thermal noise should exist.

2.1.3 SNR IN DIGITAL TRANSMISSION

The signal to noise ratio of the receiver is, as mentioned earlier, the ratio between the power of the input signal and the noise power,

$$SNR = \frac{P_R}{N_R} \tag{2.25}$$

This signal to noise ratio is inappropriate in digital transmission. The rate at which new symbols are clocked influences the system bandwidth and since the noise effect, but not the signal effect, depends on the bandwidth one should have to specify the symbol rate to give the SNR value a meaning. A SNR measure which is independent of the bandwidth of the transmission system is therefore needed. In analogue systems, the information is transmitted as an analogue signal modulated on a sinusoidal signal. At digital transmission the information is in the form of symbols which are represented by pulses with limited duration in time. The signal used at transmission is sinusoidal and has been modulated by the pulse train and can be written $A_c(t)\cos(\omega_c t + \phi(t))$. The parameters included here are explained in detail in Chapter 5. The shape of the pulse is arbitrary, but each system has its well defined shape. The pulses are time-limited signals, i.e. energy signals. The signal is non-zero only during a time interval of D seconds, which means that the mean power over an infinite time interval is zero. As a measure of the signal strength the mean power during one symbol time is used instead. This is defined as the signal power, E, divided by the length of the pulse interval, D. A modulated carrier wave with the angular frequency ω_c achieves

$$P_R = \frac{E}{D} = \frac{1}{D} \int_0^D |s(t)|^2 dt$$

$$= \frac{1}{D} \int_0^D \frac{A_c^2(t)}{2} (1 + \cos(2\omega_c t + 2\phi(t))dt$$

$$\cong \frac{1}{D} \int_0^D \frac{A_c^2(t)}{2} dt \tag{2.26}$$

The unit is Watts. The approximation in eq. (2.26) is based on ω_c being large, which means the contribution from the cosine term is zero or very small. When the amplitude of the sinusoidal signal is constant, it can be seen from eq. (2.26) that the relation between amplitude and energy is $A_c = \sqrt{2E/D}$.

The noise is an ergodic power signal (for an ergodic signal time and ensemble means are the same). The mean power of the noise during the time interval D is obtained by letting the noise pass through an integrator having the impulse response $h(t) = 1/D$ for $-D/2 \leq t \leq D/2$ and zero for other values of t. Since noise is a stochastic signal, the phase is of no importance and the integration interval may very well be from the negative side of the t-axis. The autocorrelation of the output signal from a system with an ergodic input signal and the impulse response $h(t)$ is $R_{YY}(\tau) = \int \int R_{XX}(\tau + t_1 - t_2)h(t_1)h(t_2)\,dt_1 dt_2$. The power of the output signal is given by $R_{YY}(0)$. After integration over the time interval D the noise power or variance therefore becomes

$$N_R = \frac{1}{D^2} \int_{-D/2}^{D/2} \int_{-D/2}^{D/2} R_{nn}(t_1 - t_2)\,dt_1\,dt_2$$

$$= \frac{1}{D^2} \int_{-D}^{D} R_{nn}(\tau_1)\,(D-|\tau_1|)\,d\tau_1 \tag{2.27}$$

For the last integral in eq. (2.27) the substitutions $\tau_1 = t_1 - t_2$ and $\tau_2 = t_1 + t_2$ have been made. Then $R_{nn}(\tau_1)$ can be left out of the first integration, which gives the last line in eq. (2.27). Noise having infinite time distribution is regarded as a power signal and for white noise the autocorrelation is

$$R_{nn}(\tau) = \frac{N_0}{2}\,\delta(\tau) \tag{2.28}$$

Also note, that among most real existing noise spectra, white noise gives the lowest possible value of N_R for a certain noise density since the integration interval in eq. (2.27) in this case becomes infinitesimal. The discussion can also be done the other way around. Given that the noise is white, $R_{YY}(0)$, as specified above, consists of an integral along the line $t_1 = t_2$. The integral is taken over the square of the impulse response which means that the signal power on the output of the filter is $R_{YY}(0) = E_h N_0/2$, where E_h is the energy in the filter impulse response.

Equation (2.28) inserted in eq. (2.27) gives the noise power for white noise $N_R = N_0/(2D)$. For white noise, using the result in eq. (2.26), SNR then becomes

$$\frac{P_R}{N_R} = \frac{E}{D}\frac{2D}{N_0} = 2\frac{E}{N_0} = 2\frac{E_b \log_2 M}{N_0} \tag{2.29}$$

where E_b is the energy per bit (from **bi**nary dig**it**) if each transmitted symbol consists of M possible characters. Since D can be cancelled we get variables which are independent of the transmission rate. It can be seen from eq. (2.29) that E and N_0 may be used as a measure of signal to noise instead of P_R and N_R. Instead P_R and N_R may be useful for measurements on digital communication systems. Note, however, that the two measures differ by a factor of two, which corresponds to 3 dB. The value of N_R depends on the transmission rate. A shorter symbol time, for instance, requires a larger bandwidth and hence gives higher noise power. To compensate for the noise increase a higher signal power is required. This normalising problem disappears if E and N_0 are used as measures. Rearranging eq. (2.29) gives the SNR unit bit-energy/noise density, which is usually used in digital transmission,

$$\gamma_b = \frac{E_b}{N_0} = \frac{P_R}{2N_R\log_2 M} \tag{2.30}$$

Example 2.2

A receiver system consists of a preamplifier with $F_1 = 3$ dB and $G = 20$ dB, a cable with loss $L = 6$ dB and ambient temperature $T_{amb} = 290$ K, and a receiver with noise factor $F_3 = 13$ dB (see Figure 2.10).

(a) Calculate the noise factor of the system expressed in dB.
(b) Repeat (a) with the preamplifier between the cable and the receiver.

Solution The lossy cable does not add noise but still influences the noise properties of the system since the wanted signal is attenuated. By writing $G = 1/L$ the loss can be expressed as a gain factor, which is then less than one. The noise factor of the lossy cable is denoted by F_2. When the cable impedance has been matched to the source, the amplifier will sense the same source impedance whether the source is connected directly to the amplifier or the cable has been placed in between. The thermal noise level is therefore not changed by insertion of the cable. From eq. (2.19) the noise at the output end of the cable is then given as

$$P_N = kBT_{amb} = \frac{1}{L}kB(T_s + T_e) = \frac{1}{L}kB(T_{amb} + T_e) \tag{2.31}$$

where T_s is the source temperature, assumed equal to the ambient temperature, T_{amb}, and T_e is the equivalent noise temperature. Using the second and fourth parts in eq. (2.31), the equivalent noise temperature can be obtained:

$$T_e = T_{amb}(L - 1) \tag{2.32}$$

and, according to eq. (2.21), the noise factor becomes

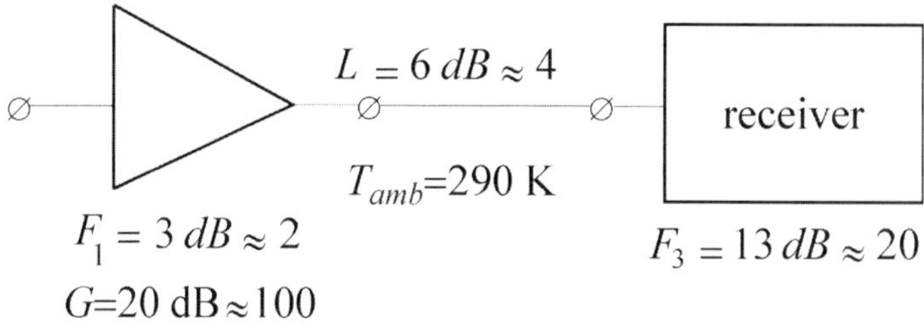

Figure 2.10 System with a cable inserted between preamplifier and receiver.

$$F_2 = 1 + \frac{T_e}{T_{amb}} = 1 + \frac{T_{amb}(L-1)}{T_{amb}} = L = 4$$

(a) According to eq. (2.23) the total noise factor of the system with the amplifier before the cable becomes

$$F = F_1 + \frac{F_2 - 1}{G_1} + \frac{F_3 - 1}{G_1 \dfrac{1}{L}} = 2 + \frac{4-1}{100} + \frac{20-1}{100\dfrac{1}{4}} = 2.79 \rightarrow 4.5 \text{ dB}$$

$$(2.33)$$

(b) and with the amplifier after the cable,

$$F = 4 + \frac{2-1}{\dfrac{1}{4}} + \frac{20-1}{100\dfrac{1}{4}} = 8.76 \rightarrow 9.4 \text{ dB} \tag{2.34}$$

In case (a), only the first stage dominates the noise factor as it has a gain of 20 dB. In (b) the noise is strongly influenced by the first two stages, since stage one has no gain. The conclusion is that there is a lot to be gained by placing a possible pre-amplifier as close to the antenna as possible.

Example 2.3

For a satellite link (cf. Figure 2.11), the following parameter values are given for the down link (i.e. from satellite to ground receiver):

satellite EIRP ($G_T = $ 28 dB and $P_T = $ 10 W)	38 dBW
carrier frequency	400 MHz
noise temperature of ground receiver, incl. antenna	1000 K
ground receiver antenna gain	0 dB
system losses	3 dB
system bandwidth	2000 Hz

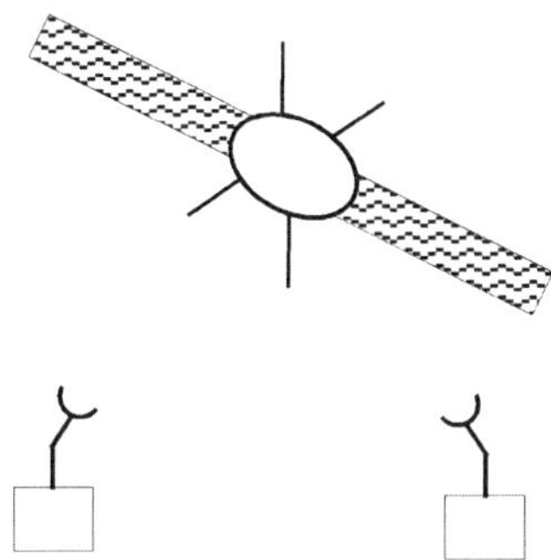

Figure 2.11   Satellite link with transmitting and receiving ground stations.

distance satellite to ground station 41000 km

Find the signal to noise ratio in a bandwidth of 2000 Hz at the output of the ground receiver.

Solution Received signal power is obtained from eq. (2.8), which can be rewritten in order to be able to use units expressed in dBW directly:

$$10\log P_R = 20\log\frac{\lambda}{4\pi d} + 10\log P_T + 10\log G_T + 10\log G_R - L$$

$$= \underbrace{20\log\frac{2.997925\times10^8}{0.4\times10^9\times4\times\pi\times41\times10^6}}_{-176.74} + 38 + 0 - 3$$

$$= -141.74 \tag{2.35}$$

where P_T is transmitted power, P_R is received power, G_T, G_R are the antenna gain, transmitter and receiver, respectively, λ is the wavelength ($\lambda = c/f$), d is the distance between the transmitter and receiver and L is other system losses. The noise power is often expressed by using the standard temperature $T_0 = 290$ K and is (sky noise is not included)

$$P_n = k(T_s + T_e)B_N = kT_0\frac{(T_s + T_e)}{T_0}B_N \tag{2.36}$$

or in dBW if the source temperature is $T_s = T_0$,

$$10\log P_n = 10\log kT_0 + 10\log\left(1 + \frac{T_e}{T_0}\right) + 10\log B_N$$

$$= -204 + 10\log\frac{1000}{290} + 10\log 2\times10^3$$

$$= -165.60 \quad (\text{dBW}) \tag{2.37}$$

Rewriting using T_0 gives the standard term $10\log kT_0 = -204$. The *S/N* then becomes

$$S/N = -141.74 + 165.50 = 23.76 \quad (\text{dB}) \tag{2.38}$$

Exercise 2.1

In a radio link the receiver antenna has the affective area 1 m^2 and the attenuation in the antenna cable 3 dB. The first stage in the receiver, the mixer, has the noise factor of 7 dB and gain 0 dB. The effective noise bandwidth is 500 MHz and the antenna noise temperature is 100 K. Compute the maximum distance between the receiver and transmitter antennas, if the last one emits 1 W of signal power and 20 dB antenna gain. A signal to noise ratio of 30 dB is required in the receiver. Compute how much the maximum distance can be increased by the insertion of a FET amplifier with the noise factor 5 dB and the gain 20 dB. Free space wave propagation is assumed.

3 INFORMATION THEORY AND SOURCE CODING

A typical problem for the communication engineer is that a certain amount of information is to be transmitted over a given channel. The transmission capacity of the channel must be larger than the amount of information to be transmitted per unit time, otherwise it is not possible to solve the problem. In those cases when transmission capacity is a commodity in short supply, it is very important to efficiently code the information to be transmitted. Information theory gives us tools to analyse these questions through the definition of some important parameters of the channels and of the information to be transmitted. The parameters which we focus on here are: the amount of self information generated at the source per unit time and the information transmission capacity of the channel. Depending on the inherent properties of the information source, different kinds of coding can reduce the need for transmission capacity. This chapter therefore includes several passages dealing with source coding. The aim of this chapter is to provide knowledge about the fundamental concepts of information theory so the reader will be able to calculate the information transmission need and relate it to the capacity of the transmission channel. The aim is also to provide some fundamental methods for the coding of information from different source types, as well as to give information about some methods for speech coding.

In order to answer the questions asked in Figure 3.1 we start from the block diagram in this figure.

We live in a changing world where we cannot predict everything about the circumstances around us. Hence, we must collect information about our environment in order to be able to function at all. Anything of interest to us can be a source of information. During telecommunication, information is transmitted in time or space. Transmission in space takes place in a channel which can be a telephone wire or a radio link, etc. Transmission in time takes place by using magnetic tapes or discs, etc. The information should reach its destination, the user of the information, as undistorted as possible. Coding of the information is needed at transmission. The information can be coded by human speech, for example, and transmitted by shouting at the receiving party. Electric transmission can be coded by analogue to digital conversion, which can be followed by more

How much information has to be transmitted per unit time?

What is the transmission capacity of the channel?

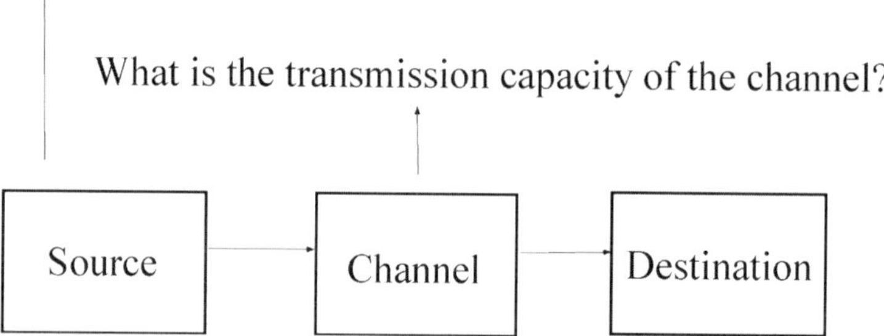

Figure 3.1 Questions addressed in information technology.

efficient coding. Independent of the way in which transmission takes place, the coding must be done by using symbols or characters. Information about a smudged shirt, for instance, can be transmitted without coding by taking the shirt to the destination, but then we no longer talk about tele- or remote communication.

A source generates a stream of information consisting of a series of random messages which can be described by a stochastic variable, X. Each message has M different possible outcomes. The sample space can be expressed as $\mathbf{X} = [x_1 \quad x_2 \quad \cdots \quad x_M]$. Each message is coded and transmitted over a channel. Depending on which questions are to be answered by the use of a specific model, the channel does not need to represent the actual channel only, such as a telephone wire, but can also include modulation, actual channel and demodulation in the channel model (cf. Figure 3.2).

The concepts that are important for the analysis of the source are: information per event I_i (where $i = 1,2,\ldots,M$), average information $H(X)$, number of events per second r and the information rate $R = r \times H(X)$, which is the same as the average information per second.

3.1 The concept of information

A message delivered to someone contains information if it reduces the uncertainty of the recipient about some circumstance. The more uncertainty that vanishes the more information is contained in the message. The messages originate from the information source, which thereby can be said to contain more or less information. Suppose, for instance, you receive a message stating that 'the sun rises today'. Then your knowledge about this condition is not improved since the probability that the sun rises is very high. If, on the other hand, you receive a

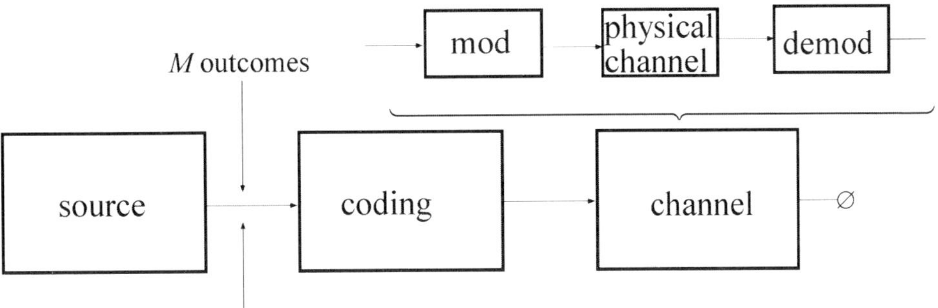

Important parameters in this point are:
Information per event, I_i (where $i = 1,2,....M$)

Expected information, entropy, $H(x)$
Information rate $R = r\,H(x)$ [expected inform./second]

Figure 3.2 Model for calculations and some important parameters for the study of the source from an information theoretical point of view.

message stating that 'there will be a hurricane today', your uncertainty about this condition is considerably reduced since hurricanes are very rare. This last message therefore contains a lot of information. A continuously changing picture requires more information acquisition, i.e. it generates more information than a static one. The complexity is also of importance for the information content. More information is needed to study seasonal variations in nature than to follow how the temperature varies with different weather conditions. An important question is: how large do the changes have to be to be significant? When temperature is studied it obviously varies by some micro degrees per second or, in other words, undergoes an infinitesimal change in an infinitesimal time. If the ambition were to reduce the uncertainty about the temperature with this precision, it can immediately be concluded that the information acquisition becomes infinite. The information acquisition must therefore be limited. This is achieved by sampling the condition of interest on certain occasions and in some way limit the values which the temperature can take. The temperature is hence seen as a stochastic variable and its acquisition is seen as an event where the sample space is made up of the possible temperatures that can be obtained.

In order to find out how much information a source generates, the information has to be quantified in some way. As has been mentioned earlier the information reduces the uncertainty of the receiver about some circumstance. The information measure should therefore be a function which decreases with increasing probability of a certain event. Another important property is that when a number of independent events are studied the amount of information generated should be the sum of the amount of information of the individual events.

3.1.1 DISCRETE SOURCE

The information per event (self information) is defined as

$$I_i = -\log_b P_i \tag{3.1}$$

where P_i is the probability of outcome i. The constant b can be chosen arbitrarily, but is usually 2, which gives the unit bit.

The mean information generated at the source is called entropy, and becomes

$$H(X) = \sum_{i=1}^{M} P_i I_i = \sum_{i=1}^{M} P_i \log_b \frac{1}{P_i} \tag{3.2}$$

When all events have equal probability, i.e. $P_i = 1/M$ for all possible i, the mean information becomes

$$H(X) = \sum_{i=1}^{M} \frac{1}{M} \log_b M = \log_b M \tag{3.3}$$

Since the entropy is limited by

$$0 \le H(X) \le \log_b M \tag{3.4}$$

maximum entropy occurs when all events have equal probability.

The entropy can be seen as a measure of the source complexity. In the sequel it is shown that if a signal is to be coded in such a way that it later could be completely restored, the entropy gives a lower limit for the maximal compression of the signal.

Example 3.1

Assume an equal number of each of the grades A, B, C, D, and F have been given in a course. How much information in bits have you received if the teacher tells you that you did not get the grade F? How much more information is needed to determine your grade?

Solution

$$P\,(\text{not } F) = \frac{4}{5} \quad \Rightarrow \quad I = -\log_2 \frac{4}{5} = -\frac{\ln(4/5)}{\ln 2} \approx 0.322 \text{ bits} \tag{3.5}$$

$$P\,(\text{certain grade}) = \frac{1}{5} \quad \Rightarrow \quad I = \log_2 5 \approx 2.32 \text{ bits} \tag{3.6}$$

To estimate the grade another $2.32 - 0.32 = 2$ bits are needed.

Example 3.2

A source has eight symbols and transmits data in blocks of three symbols at the rate of 1000 blocks per second. For synchronisation reasons the first symbol in each block is always the same. At the other two positions some of the eight symbols appear with equal probability. Find the information rate of this source.

3 symbols/block	Each symbol contains
of which the first	one of 8 letters, all
is always the same	are equally likely

then there are $8 \times 8 = 64$ different blocks $\Rightarrow$ $\qquad\qquad$ (3.7)

$$H(X) = \log_2 64 = 6 \text{ bits} \qquad\qquad (3.8)$$

1000 blocks/second $\quad \Rightarrow \quad R = 6 \times 1000 = 6 \text{ kbits/s} \qquad (3.9)$

3.1.2 CONTINUOUS SOURCE

A signal source for which the output signal can assume all values on the amplitude axis can be seen as a source where an event has a sample space which consists of a continuum of points infinitely close together, rather than consisting of discrete points. The amount of information and entropy can then be calculated by using integrals. The entropy is (differential entropy)

$$H(X) = -\int_{-\infty}^{\infty} p_X(x) \log_2(p_X(x))\, dx \qquad (3.10)$$

The entropy gives an absolute measure on the randomness of the source for discrete sources. In the continuous case the entropy relative to the assumed definition is obtained. This is of no importance in a comparison, with for example the channel capacity, provided that the same definition is used.

Assume we have a signal source with a given output power. An interesting question for the communication engineer is: what should the amplitude density of the signal look like for its information content to be as large as possible? This is calculated in the next section and the result reveals what type of signal should be used for maximum utilisation of the communication system from a theoretical point of view.

Maximum entropy

How shall $p_X(x)$ be chosen to maximise eq. (3.10)? The power limitation of the source gives a constraint. Since the signal power S of a signal with mean value zero is equal to the statistical second order moment the constraint can be expressed as

$$S = \int_{-\infty}^{\infty} x^2 p_X(x)\, dx \tag{3.11}$$

The limitation to a signal with zero mean value is logical, since a constant DC power level of the signal would not transmit any information about the source. Another constraint can be formulated because of the requirement that the surface under a statistic density function should be one:

$$1 = \int_{-\infty}^{\infty} p_X(x)\, dx \tag{3.12}$$

The problem now is to find a method for solving the above-mentioned optimisation problem. Luckily such a method, called calculus of variations, is provided by the mathematicians. It is being used for optimisation problems expressed in the form of an integral:

$$J = \int_{x_1}^{x_2} f(y(x), x)\, dx \tag{3.13}$$

J is the quantity to be maximised. The function f is known, cf. eq. (3.10), but the relation between x and y, i.e. $y(x)$, is unknown and is to be solved (in physics problems a constraint of the derivative, dy/dx, is often included). In this case the function f is under the integral sign in eq. (3.10) and $y(x)$ is the amplitude density function of the signal $p_X(x)$. The constraints are handled by using Lagrange's multiplier method. Variational calculus then gives the solution as

$$\delta J = \delta \int \left[f(y, x) + \sum_k \lambda_k \varphi_k(y, x) \right] dx = 0 \tag{3.14}$$

where λ_k are the multipliers, the constraints are $\int \varphi_k(y, x) dx = $ constant. The symbol δ is the differential, and in the variational calculus the difference must approach zero. The differential then reduces to a derivative. According to eq. (3.14) the solution of the optimisation problem treated here then becomes

$$\delta J = \frac{d}{d p_X(x)} \int_{-\infty}^{\infty} \left[-p_X(x) \log_2 p_X(x) + \lambda_1 x^2 p_X(x) + \lambda_2 p_X(x) \right] dx = 0 \tag{3.15}$$

Using the rules for derivation of integrals, eq. (3.14) can be rewritten and provides a simpler expression for the solution. In the general case we obtain

$$\frac{\partial f}{\partial y} + \sum_k \lambda_k \frac{\partial \varphi_k}{\partial y} = 0 \tag{3.16}$$

In the special case treated here the following equation is obtained:

$$-1 - \log_2 p_X(x) + \lambda_1 x^2 + \lambda_2 = 0 \tag{3.17}$$

The equation is solved and the values for the multipliers are obtained by using the constraints. The result is that the amplitude probability distribution of the signal should be

$$p_X(x) = \frac{1}{\sqrt{2\pi S}} \exp[-(x^2/2S)] \tag{3.18}$$

This result is very interesting as it tells us that a Gaussian amplitude probability distribution is needed for a signal to have the highest possible information content measured in 'bits'.

Example 3.3

What is the entropy, $H(X)$, as a function of the signal power, S, for a signal having a Gaussian amplitude probability distribution?

Solution   The entropy is obtained by inserting eq. (3.18) into eq. (3.10), which gives

$$H(X) = \frac{1}{\sqrt{2\pi S}} \int_{-\infty}^{\infty} \left[\exp[-(x^2/2S)] \left(\log_2\left(\sqrt{2\pi S}\right) + \frac{x^2}{2S}\log_2 e \right) \right] dx$$

$$= \left(\log_2\left(\sqrt{2\pi S}\right) + \frac{S}{2S}\log_2 e \right) = \left(\log_2\left(\sqrt{2\pi S}\right) + \log_2(\sqrt{e}) \right)$$

$$= \log_2\left(2\sqrt{\pi e S}\right) \tag{3.19}$$

As can be seen, the entropy increases as the 2-logarithm of the amplitude of the signal, i.e. as the square root of the power.

3.2 Channel

Information must be transmitted via a channel of some kind which in all real telecommunication cases has a limited transmission capacity. The physical properties of the channel can be described by different models. Some different channel models, the concept of transmission capacity and how the capacity of some channels is calculated are described below.

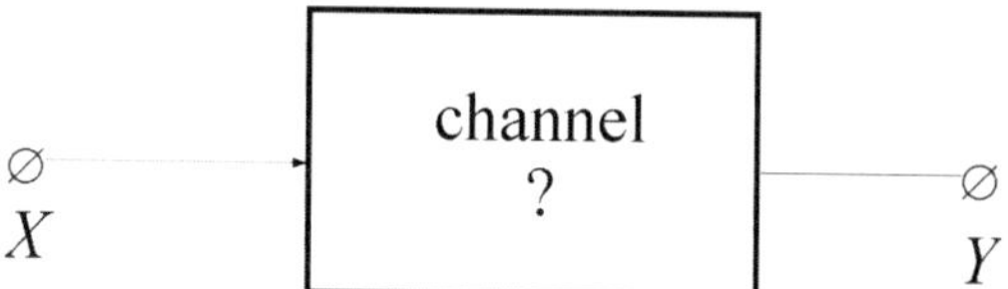

Figure 3.3 The general problem within telecommunication is that the influence of the channel on the message at a certain time is unknown. A statistical model can be used for the description.

The situation of the receiver is that only the variable Y is observed while information about X is required (Figure 3.3).

3.2.1 MUTUAL INFORMATION

The mutual information is a concept which describes how much uncertainty about the wanted variable, X, that has been reduced by observing the variable Y. For a transmission channel in which no symbol error is introduced, the reduced uncertainty equals the uncertainty in X, i.e. $H(X)$. If, on the other hand, X and Y are completely independent no uncertainty about X is reduced by observing Y. The mutual information is in this case zero.

Discrete channel

When y_j has been observed the uncertainty $I(x_i; y_j)$ has been reduced about the event x_i, where

$$I(x_i; y_j) = \log_b \frac{P_X(x_i|y_j)}{P_X(x_i)} \tag{3.20}$$

The base $b = 2$ gives the unit bit. Note that if x_i and y_j are independent the quotient of the probabilities is one and $I(x_i; y_j)$ becomes zero. The mutual information is the mean value of all events, i.e.

$$I(X, Y) = \sum_{i=1}^{M} \sum_{j=1}^{N} P_{X,Y}(x_i, y_j) I(x_i; y_j)$$

$$= \sum_{i=1}^{M} \sum_{j=1}^{N} P_{X,Y}(x_i, y_j) \log_2 \frac{P_X(x_i|y_j)}{P_X(x_i)}$$

$$= \sum_{j=1}^{N} \sum_{i=1}^{M} P_{X,Y}(x_i, y_j) \log_2 \frac{P_Y(y_j|x_i)}{P_Y(y_j)} \tag{3.21}$$

where M is the number of symbols on the channel input and N is the number of symbols on the output. For the simultaneous outcome the following holds:

$$P_{X,Y}(x_i, y_j) = P_Y(y_j|x_i)P_X(x_i) = P_X(x_i|y_j)P_Y(y_j) \tag{3.22}$$

which has been used for the last equality in eq. (3.21). From eq. (3.21) it is clear that $I(X, Y) = I(Y, X)$.

To get a better idea of what the mutual information can express, the second line of eq. (3.21) can be rewritten using the relation

$$\log\frac{P_X(x_i|y_j)}{P_X(x_i)} = \log P_X(x_i|y_j) - \log P_X(x_i) \tag{3.23}$$

Two different terms can hence be defined:

$$I(X; Y) = H(X) - H(X|Y) \tag{3.24}$$

The first term is the entropy of the source and the second term $H(X|Y)$ is called the equivocation or confusion, and identifies the information loss due to channel noise, or, in other words, due to the entropy of the channel. This information can never be retrieved by any kind of signal processing in the receiver.

$$H(X|Y) = -\sum_{i=1}^{M}\sum_{j=1}^{N} P_{X,Y}(x_i, y_j)\log P_X(x_i|y_j) \tag{3.25}$$

Since $H(X|Y)$ is always positive, eq. (3.24) is always smaller than the entropy of the source, which means that information is lost at transmission over a noisy channel.

Continuous channel

Due to the continuous amplitude distribution, the sums in eq. (3.21) change to integrals. The mutual information for channels with continuous amplitude distribution is obtained:

$$I(X, Y) = \int_{-\infty}^{\infty}\int_{-\infty}^{\infty} p_{X,Y}(x, y)\log_2\frac{p_X(x|y)}{p_X(x)}\,dy\,dx$$

$$= \int_{-\infty}^{\infty}\int_{-\infty}^{\infty} p_X(x|y)p_Y(y)\log_2\frac{p_X(x|y)}{p_X(x)}\,dy\,dx \tag{3.26}$$

The ratio in the logarithm is rewritten as a difference between two logarithms. Some known expressions can then be identified and we get

$$\underbrace{\int_{-\infty}^{\infty} \int_{-\infty}^{\infty} p_X(x|y)p_Y(y)\log_2 p_X(x|y)dy\, dx}_{-H(X|Y)} -$$

$$\int_{-\infty}^{\infty} \left[\underbrace{\int_{-\infty}^{\infty} p_X(x|y)p_Y(y)dy}_{p_X(x)} \right] \log_2 p_X(x)dx$$

$$= -H(X|Y) - \underbrace{\int_{-\infty}^{\infty} p_X(x)\log_2 p_X(x)dx}_{-H(X)}$$

$$= H(X) - H(X|Y) \tag{3.27}$$

In the same way as in eq. (3.24) the difference between entropy and equivocation is obtained for the continuous channel.

3.2.2 CHANNEL MODELS

In the following section some common mathematical models for the transmission channels are described. The models can represent either the real, i.e. the physical, channel or several parts in a communication system. Which model is used at any particular time depends of course on the properties of the real world being modelled, as well as on the query which is being answered at the moment. In order to simplify the connection to the real communication system, the term 'symbol' instead of 'event' is used in the following as input variable to the channel. The usual problem an engineer faces is to transmit information which has been encoded by the use of symbols. The transmission consists of a flow of symbols. Each symbol is collected from an alphabet with M different characters, each representing an event. If the symbols are among the integers 0–9, we have ten characters, i.e. $M = 10$.

Binary symmetric channel

The binary symmetric channel (BSC), has two characters as input variables, for instance 0 and 1. Given a specific character at the input, e.g. a zero, the probability is α to get the other character, i.e. a one, at the output of the channel. The error probability is, in other words, $P_Y(Y = 1|X = 0) = \alpha$. The stochastic variables X and Y are normally not written inside the expression for conditional probability. The fact that the channel is symmetric means that the probability for making an error is equal, independently of whether the input signal is a one or zero, $P_Y(0|1) = P_Y(1|0) = \alpha$. The probability that the symbol passes the channel unchanged is the complementary event to the one when it is changed, which

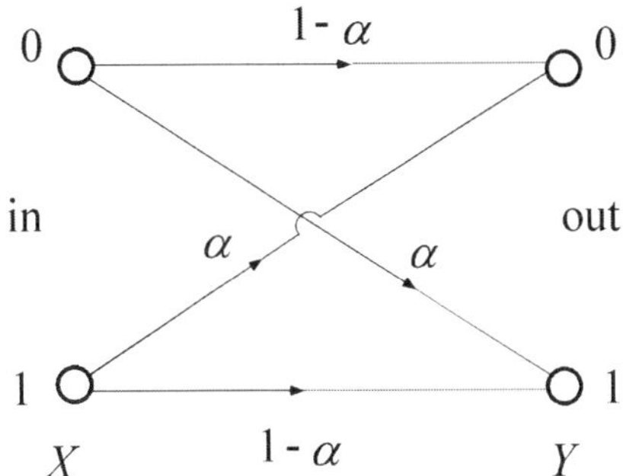

Figure 3.4   Flow chart for a binary symmetric channel.

means that the probability of a correct symbol becomes $P_Y(0|0) = P_Y(1|1) = 1 - \alpha$. The binary symmetric channel is often represented by a flow chart as shown in Figure 3.4.

The total error probability for a BSC becomes

$$P_e = P_X(0)\, P_Y(1|0) + P_X(1)\, P_Y(0|1) \tag{3.28}$$

A common variation of BSC is the binary erasure channel. This has two inputs and three outputs. The third output represents a symbol which has been erased and a decision about zero or one cannot be made.

Discrete memory-less channel

For a discrete memory-less channel (DMC), a certain output symbol among the flow of output symbols only depends on one single input symbol. A dependence on several input symbols occurs, for instance, in a channel with multipath, i.e. the signal in the actual channel reaches the receiver through different paths with different time delays. Such a channel can therefore not be modelled using DMC. For a discrete memoryless channel, the probability of a certain symbol sequence to occur is equal to the product of the individual probabilities:

$$P_Y([y_1...y_n]|[x_1...x_n]) = \prod_{k=1}^{n} P_Y(y_k|x_k) \tag{3.29}$$

Additive white Gaussian noise channel

The additive white Gaussian noise channel (AWGN) is the most widely used channel model. It describes several types of physical channels very well. An example is the signal transmission between modulator and detector, for instance in radio transmission, wire transmission, storing on magnetic media etc. (cf.

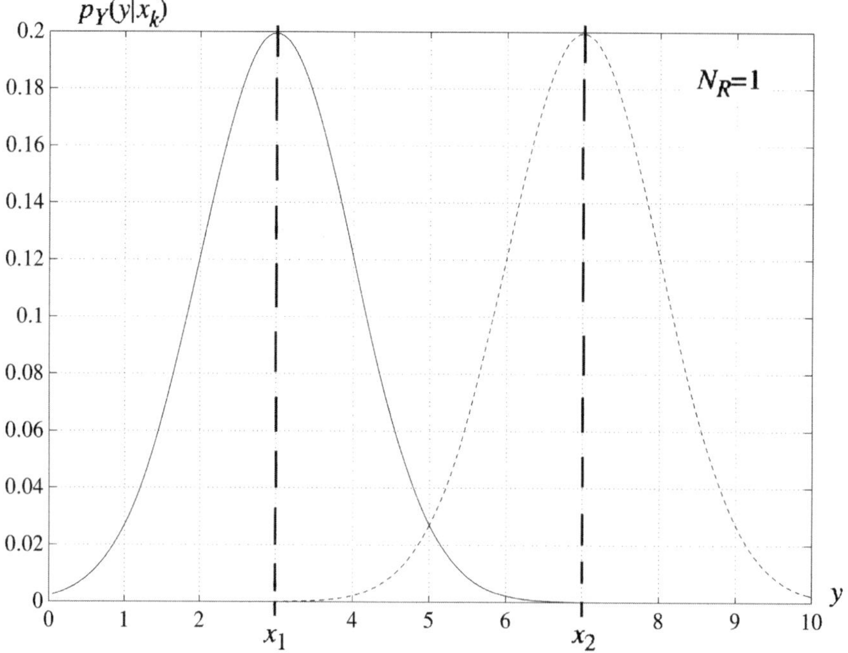

Figure 3.5   Amplitude density for a signal consisting of two information signals with the values x_1 and x_2 disturbed by Gaussian noise.

Figure 3.2). As the name indicates, the disturbance is additive, i.e. white Gaussian noise G is added to the input signal X to form the output signal

$$Y = X + G \tag{3.30}$$

The signal at the output of the channel is continuous, but at the input it can be either continuous or discrete. Given a certain input signal, or event, x_k is the probability density for the amplitude of the output signal y:

$$p_Y(y|X = x_k) = \frac{1}{\sqrt{2\pi N_R}} \exp[-(y - x_k)^2/2N_R] \tag{3.31}$$

where y is the observed and x_k is the occurred event. The model consists of a conditional probability which is described by a Gaussian or normal distribution. As an example, the density functions of y with the two signals x_1 and x_2 at the input are shown in Figure 3.5.

3.2.3 CALCULATION OF CHANNEL CAPACITY

As has been mentioned earlier, the information required about an event is reduced when the message is carried over the channel. An interesting question rises: if a given channel is utilised as efficiently as possible how much information can it transfer? The channel capacity corresponds to the amount of information which can be transmitted per symbol (bit/symbol) over a certain channel or, in other words, the maximum amount of transmitted mean information, see also eq. (3.24). The channel capacity is a function of signal to noise ratio or the error probability on the channel, and is defined as

$$\max_{P_X} I(X;Y) = C_s \tag{3.32}$$

Hence, the channel capacity is obtained by maximising the mutual information with regard to the probability distribution of the input signal. In practice this means that encoding the information in a way which makes optimal use of the channel is chosen. We are usually interested in transmitting a certain message in a certain time. The interesting parameter is then the information transmission per second, C. With a symbol rate of r symbols/second the following channel capacity in bits/second is obtained:

$$C = r\, C_s \tag{3.33}$$

The channel capacity of a binary symmetric channel and of a Gaussian white noise channel is calculated below.

Binary symmetric channel

Assume that the probability of a zero on the input is $P_X(0) = p$ and a one on the input is $P_X(1) = 1 - p$, respectively. The probability of an incorrect and a correct transmitted symbol is given in Section 3.2.2 and is

$$P_Y(0|0) = P_Y(1|1) = 1 - \alpha, \quad P_Y(0|1) = P_Y(1|0) = \alpha \tag{3.34}$$

respectively. P_X influences the entropy of the source while eq. (3.34) influences the equivocation or confusion in the channel. From eq. (3.21) the mutual information becomes

$$I(X;Y) = P_X(0)P_Y(0|0)\log_2 \frac{P_Y(0|0)}{P_Y(0)} + P_X(0)P_Y(1|0)\log_2 \frac{P_Y(1|0)}{P_Y(1)}$$

$$+ P_X(1)P_Y(1|1)\log_2 \frac{P_Y(1|1)}{P_Y(1)} + P_X(1)P_Y(0|1)\log_2 \frac{P_Y(0|1)}{P_Y(0)} \tag{3.35}$$

Equation (3.35) has a maximum when $P_X(1) = P_X(0) = 0.5$ and the error probability $\alpha = 0$. The transmission capacity is then, not unexpectedly, $C_s = 1$ bits/symbol. Note that the capacity becomes zero when the error probability $\alpha = 0.5$. Half of all symbols are then incorrect and, since the receiver does not know which ones, no information about the input signal, x, of the channel is transmitted. When

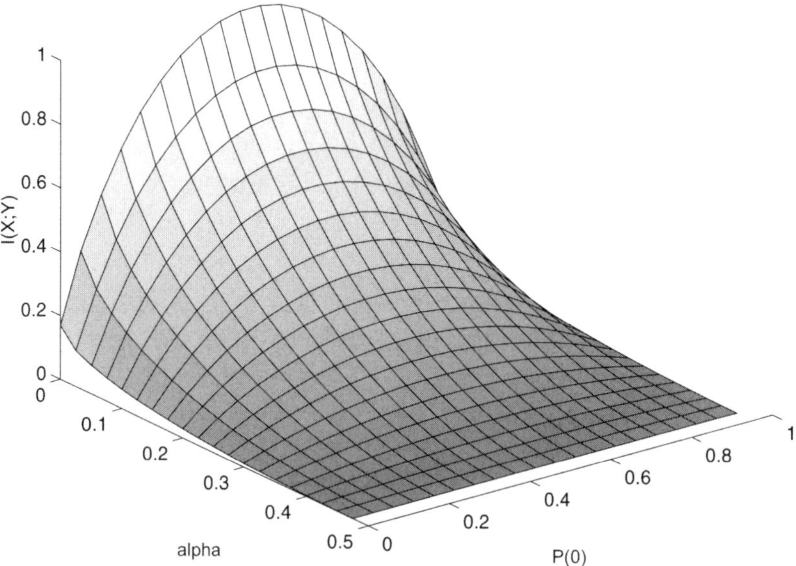

Figure 3.6   Transmission capacity of a BSC versus error probability, α, and the probability of a zero at the input, $P_X(0)$.

the error probability increases further, the transmission capacity increases again because the receiver can then choose the strategy of assuming all received signals are incorrect and in this way obtain information about the input signal (Figure 3.6).

Example 3.4

The binary erasure channel has two possible characters at the input: 0 and 1, and three at the output: 0, 1, and E. The character E corresponds to a detected but uncertain symbol (cf. Figure 3.7). The transmission probabilities are (for simplicity we write only $P(\cdot)$ and not $P_Y(\cdot)$ for the conditional probabilities below)

$$P(0|0) = 1 - \alpha \quad P(E|0) = \alpha \quad P(1|0) = 0$$
$$P(0|1) = 0 \qquad P(E|1) = \alpha \quad P(1|1) = 1 - \alpha \tag{3.36}$$

Because of the symmetry $I(X,Y)$ has its maximum when the symbols of the source have equal probabilities. Find C_S versus α.

Solution   The condition for maximum channel capacity is obtained from eq. (3.32), and from eq. (3.21) the mutual information is obtained.

A symmetric channel has maximum capacity when both symbols are equally

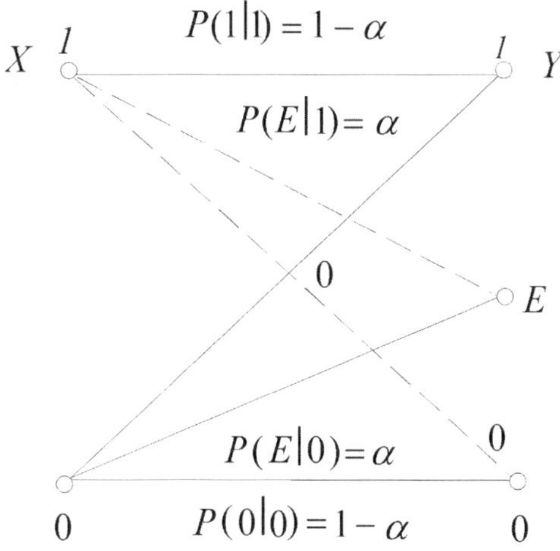

Figure 3.7 Binary symmetric erasure channel.

probable $P_X(0) = P_X(1) = 1/2$. The probability of the respective symbols at the output becomes

$$P_Y(1) = P_Y(0) = \frac{1}{2}(1 - \alpha) \tag{3.37}$$

Due to the symmetry the same probability is obtained for one and zero, but for erasure the probability becomes

$$P_Y(E) = \frac{1}{2}\alpha + \frac{1}{2}\alpha = \alpha \tag{3.38}$$

Inserted in a slightly modified eq. (3.21) the mutual information becomes

$$I(X;Y) = P_X(1)\left(P(1|1)\log_2 \frac{P(1|1)}{P_Y(1)} + P(E|1)\log_2 \frac{P(E|1)}{P_Y(E)}\right.$$

$$\left. + P(0|1)\log_2 \frac{P(0|1)}{P_Y(0)}\right) + P_X(0)\left(P(1|0)\log_2 \frac{P(1|0)}{P_Y(1)}\right.$$

$$\left. + P(E|0)\log_2 \frac{P(E|0)}{P_Y(E)} + P(0|0)\log_2 \frac{P(0|0)}{P_Y(0)}\right)$$

$$= \frac{1}{2\ln 2}\left((1 - \alpha)\ln \frac{1 - \alpha}{(1/2)(1 - \alpha)} + \alpha\ln \frac{\alpha}{\alpha} + 0 + 0\right.$$

$$\left. + \alpha\ln \frac{\alpha}{\alpha} + (1 - \alpha)\ln \frac{1 - \alpha}{(1/2)(1 - \alpha)}\right)$$

$$= \frac{1}{2\ln 2}((1 - \alpha)\ln 2 + (1 - \alpha)\ln 2) = 1 - \alpha \tag{3.39}$$

Additive white Gaussian noise channel

The transmission capacity of the AWGN channel is fundamental. The channel has a continuous output signal and its maximum capacity occurs when the input signal has a continuous amplitude distribution. As can be seen above in eq. (3.27) the mutual information is $H(X) - H(X|Y)$. The problem, of finding the density function p_X which maximises the mutual information, is then the same as that of maximising $H(X)$.

Eq. (3.27) gives the mutual information. The equation requires knowledge about $p_X(x|y)$, which has not been specified in the model. By using Bayes' theorem in the form $p_X(x|y) = p_Y(y|x)p_X(x)/p_Y(y)$ a transformation can be made, which gives the mutual information with known parameters:

$$I(X;Y) = H(X) - \underbrace{H(X|Y)}_{H(Y|X) + H(X) - H(Y)} = H(Y) - H(Y|X) \tag{3.40}$$

The transformation of $H(X|Y)$ is not shown, but it can easily be performed using Bayes' theorem in eq. (3.26). As has been stated earlier, the conditioned amplitude density at the output is

$$p_Y(y|x) = \frac{1}{\sqrt{2\pi N}}\exp[-(y - x)^2/2N] \tag{3.41}$$

given the input signal x and where N is the noise power ($N = \sigma^2 = N_0/2D$). The maximum of I is obtained when p_x has a Gaussian distribution, since this gives the largest entropy, cf. eq. (3.18). For

$$\max_{p_X(x)} I = C_s$$

p_x must, in other words, have the following density function:

$$p_X(x) = \frac{1}{\sqrt{2\pi S}}\exp(-x^2/2S) \tag{3.42}$$

where S is the signal power. As y is the sum of two Gaussian signals, $p_Y(y)$ also becomes a Gaussian density function with the variance $N + S$, which is the received power, i.e.

$$p_Y(y) = \frac{1}{\sqrt{2\pi(N + S)}}\exp[-y^2/2(N + S)] \tag{3.43}$$

The entropy of (3.43) then becomes (cf. Example 3.3 and eq. (3.27))

$$H(Y) = \frac{-1}{\sqrt{2\pi(N+S)}\ln2}$$

$$\times \int_{-\infty}^{\infty} \exp\left[-\frac{y^2}{2(N+S)}\right]\left(-\frac{y^2}{2(N+S)} - \frac{1}{2}\ln[2\pi(N+S)]\right)dy$$

$$= \frac{1}{2\ln2}(1 + \ln[2\pi(N+S)]) \tag{3.44}$$

The integration seems relatively awkward to carry out, but the result is actually found in most fairly comprehensive handbooks which include tables of integrals. The first term in the integral is proportional to the second moment and the second term is proportional to the density function integrated over the entire x-axis, i.e. one. By using eq. (3.27) and inserting the Gaussian densities, the mutual information becomes

$$H(Y|X) = \frac{-1}{(\ln2)\sqrt{2\pi S}} \frac{1}{\sqrt{2\pi N}} \int_{-\infty}^{\infty} \exp\left(-\frac{x^2}{2S}\right) \int_{-\infty}^{\infty} \exp\left[-\frac{(y-x)^2}{2N}\right]$$

$$\times \left[-\frac{(y-x)^2}{2N} - \frac{1}{2}\ln(2\pi N)\right]dy\,dx$$

$$= \frac{1 + \ln(2\pi N)}{2\ln2}$$

$$\tag{3.45}$$

This integral also seems complicated but can be found in handbooks after the substitution $y - x = n$. Eqs. (3.44) and (3.45) inserted in (3.27) give

$$C_s = H(Y) - H(Y|X)$$

$$= \frac{1}{2\ln2}(1 + \ln[2\pi(N+S)] - 1 - \ln(2\pi N))$$

$$= \frac{1}{2}\log_2\left(1 + \frac{S}{N}\right) \tag{3.46}$$

To calculate the maximum transmission rate, $C = r\,C_s$ must exist, where r is the symbol rate. Since the channel is continuous we must find a connection between the bandwidth of the channel and a discrete channel, which has a certain value of the number of symbols per second. This connection is Nyqvist's sampling theorem, which states that if the channel is strictly band limited to B (Hz), it is sufficient to use the sampling frequency $f_s = 2B$. Compare also the maximum transmission rate with the so-called Nyquist pulses in Section 5.1.2. The sample rate f_s corresponds to the symbol rate, which inserted in eq. (3.46) gives

$$C = 2BC_s = B\log_2\left(1 + \frac{S}{N}\right) \tag{3.47}$$

which is the well-known 'Shannon's formula' for maximum transmission capacity in an analogue channel having white Gaussian noise interference and a given signal power (Hartley-Shannon). This formula is frequently used within communication, both to determine the limitations of the transmission capacity of a certain channel and also to see how well the telecommunication systems use the available channel bandwidth and SNR. We shall show later that this transmission capacity can be obtained with arbitrarily small error probability for the transmitted message.

Example 3.5

Consider an AWGN channel with the signal power $S = 10^4$ (W) and noise density $N_0 = 1$ (W/Hz). The power density spectrum of the signal is assumed to have a flat distribution within the available bandwidth and zero outside. Find the maximum transmission capacity for information when $B = 1$ kHz, 10 kHz and 100 kHz as well as for infinite bandwidth.

Solution   Maximum transmission capacity C is given by Shannon's formula which can be rewritten as

$$C = B\log_2\left(1 + \frac{S}{N}\right) = B\log_2\left(1 + \frac{S}{BN_0}\right) \tag{3.48}$$

The conditions above applied to (3.48) gives

$$\frac{S}{N_0} = 10^4 \quad \Rightarrow \quad C = B\log_2\left(1 + \frac{10^4}{B}\right) \tag{3.49}$$

For the different bandwidths, we get

$$B = 1 \text{ kHz} \quad \Rightarrow \quad C = 3.5 \text{ kbits/second}$$

$$B = 10 \text{ kHz} \quad \Rightarrow \quad C = 10 \text{ kbits/second}$$

$$B = 100 \text{ kHz} \quad \Rightarrow \quad C = 14 \text{ kbits/second}$$

For increasing bandwidth the noise level also increases. An interesting question is therefore what happens if one lets $B \to \infty$. We then get

$$C = \lim_{B \to \infty} B \frac{\ln\left(1 + \dfrac{10^4}{B}\right)}{\ln 2}$$

$$= \lim_{B \to \infty} B \frac{\dfrac{10^4}{B} - \dfrac{10^8}{2B^2} + \cdots}{\ln 2} = \frac{10^4}{\ln 2}$$

$$\approx 14.4 \text{ kbits/second} \tag{3.50}$$

This means that infinite bandwidth does not lead to infinite transmission capacity as long as noise is present in the channel.

3.3 Source coding

Information is generated in the source through the outcomes of a sequence of random messages. Before transmission of the information the message must be coded in some way, which is done by source coding. The coding should be done as efficiently as possible to avoid more symbols than necessary having to be transmitted to convey a certain message. Efficient source coding leads to reduced requirements on the transmission channel. The coding can be made suitable for a large number of signals, wave form coding, or can be done with a certain type of signal in mind, such as speech coding or picture coding. The basic concepts of data compression, divided into sources with and without memory, are given below, followed by some methods for speech coding. The starting point is that we have a source which generates an information flow. Each event at the source is coded by using symbols. The number of symbols per second is given in units of Baud. The outcome 'ball in the cup', after hitting a golf ball, can for example be coded as a 1 and the outcome 'ball not in the cup' be coded as a 0 (Figure 3.8).

The coding operation can be divided into noise-free coding and noisy coding. In the first case there is a 1:1 relation between a certain outcome and the code word. For noisy coding there is no unambiguous relation between event and code word.

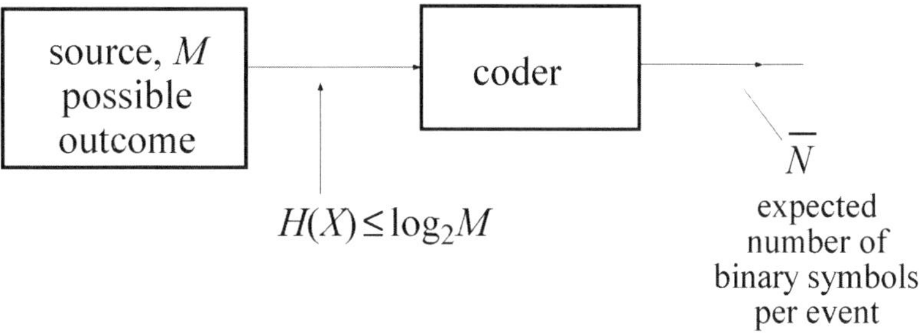

Figure 3.8 Discrete signal source and some important parameters.

Such a code would therefore not be suitable for banking transactions or storing of a computer program. The coding efficiency is increased at the cost of the introduction of errors because several outcomes give the same code word. An example is A/D conversion, which maps an amplitude-continuous signal at a limited number of discrete code words, which results in so-called quantisation noise.

3.3.1 DATA COMPRESSION

Memory-less channels

For sources without memory all events are independent, i.e. given an outcome from a certain event it is not possible to state anything about the outcome of the next event. First of all code words with a fixed length are studied. An example of this is ordinary analogue to digital conversion. Assume that the sample space at each event consists of M different outcomes. The required number of binary bits for the coding of the information then becomes

$$N = \lceil \log_2 M \rceil \tag{3.51}$$

where $\lceil x \rceil$ means the nearest integer larger than or equal to the argument x. If 5 different outcomes are coded binary, 3 binary bits, or symbols, have to be used. When each event is coded with several symbols in this way, these can be grouped together to form code words.

The efficiency of the coding is the ratio between the entropy of the signal source and the mean value, $\overline{N}$, of the number of binary bits used for the coding of each event. When coding with non-binary symbols, a transformation to binary symbols must be made. The coding efficiency is written

$$\eta = \frac{H(X)}{\overline{N}} \tag{3.52}$$

When the ith event has the probability P_i and is coded with N_i bits. The average length of the code words is then

$$\overline{N} = \sum_{i=1}^{M} P_i N_i \tag{3.53}$$

Shannon's coding theorem says that there is a coding method where

$$H(X) \leq \overline{N} < H(X) + \varepsilon \tag{3.54}$$

and where ε can be made arbitrarily small. By a correct coding, the transmission channel does not therefore need to be occupied by symbols which carry no information. To achieve this, a code with varying code-word length is often needed. How to design such a code is shown below. A noise free code must be uniquely decodable, i.e. there must be an 1:1 relationship between the message and the transmitted code word. A noisy code introduces errors in the coding. This is often called 'rate-distortion' coding, i.e. lower symbol rate is exchanged against higher distortion and the expected code word length can be less than $H(X)$. Examples are many codes for speech or image coding. Except for noiseless, other conditions which should be fulfilled are that the code must be possible to decode momentarily, i.e. the decoding can be made without having to pick up any more code words. A necessary and sufficient condition for this is the prefix condition which states that: if $\mathbf{B}_Z = (b_1\ b_2...\ b_k)$ is a code word, there must not be another code word with the identical symbols $\mathbf{B}_X = (b_1\ b_2...\ b_l)$ for which $l < k$; i.e. a proper code word can not be a prefix in any other code word. A necessary and sufficient condition on the word length for the prefix condition to be fulfilled is that Kraft's inequality holds:

$$\sum_{i=1}^{M} B^{-N_i} \le 1 \tag{3.55}$$

where B is the number of symbols in the alphabet used. For the binary case $B = 2$. An everyday occurrence in society today for which the prefix condition must be fulfilled, is a telephone number (try e.g. the country codes).

A code with variable word length can be obtained if the length of the code words, N_i, is chosen so that

$$\log_2 \frac{1}{P_i} \le N_i < \log_2 \frac{1}{P_i} + 1 \tag{3.56}$$

Multiplying the expression (3.56) by P_i and taking the sum over all i gives

$$H(X) \le \overline{N} < H(X) + 1 \tag{3.57}$$

The left hand side can be identified as the entropy, and the middle expression as the mean length of the code words. This does not give a coding where the mean length of the code words is arbitrarily close to the entropy, but one still gets fairly close. To chose the code word length according to eq. (3.56) does not give an optimal code. For example consider two outcomes, one occurring with the probability 0.0001 and the other with 0.9999. To use the code word length $\lceil -\log_2 P_i \rceil$ gives a word length of 1 and 14 bits, respectively, although the code word length obviously should be 1 bit for both outcomes. An optimal coding method is however the Huffman coding (invented by D. Huffman as an outgrowth of a homework problem during a course in information theory at Massachussets Institute of Technology, in 1952), which is described below. The mean code word length of this method also fulfils eq. (3.57).

Huffman coding

Assume a source which generates a sequence of symbols. These are to be encoded by binary code words. Each source symbol consists of one of seven characters, which appear with different probabilities. The most common character is placed on top in a column and the others follow in decreasing probability order. The source symbols with the lowest probability are paired together and one is marked with zero and the other with one. These symbols form a group and this group appears with a probability which is the sum of the probabilities of the individual symbols. Then the next to the least probable ones are grouped, and so on. If several have the same probability a random group/symbol is picked from this group. This procedure continues until all groups/symbols have been paired together. The figure will form a tree. This coding technique therefore belongs to the coding class called tree coding. The code word representing the respective character is obtained by starting with the last pairing, putting together the sequence of ones and zeros leading to the respective character (cf. Figure 3.9). The procedure can be summarised in the following table:

Letter	Probability	Self-information	Code
x_1	0.35	1.5146	00
x_2	0.30	1.7370	01
x_3	0.20	2.3219	10
x_4	0.10	3.3219	110
x_5	0.04	4.6439	1110
x_6	0.005	7.6439	11110
x_7	0.005	7.6439	11111

$$H(X) = 2.11, \quad \overline{N} = 2.21$$

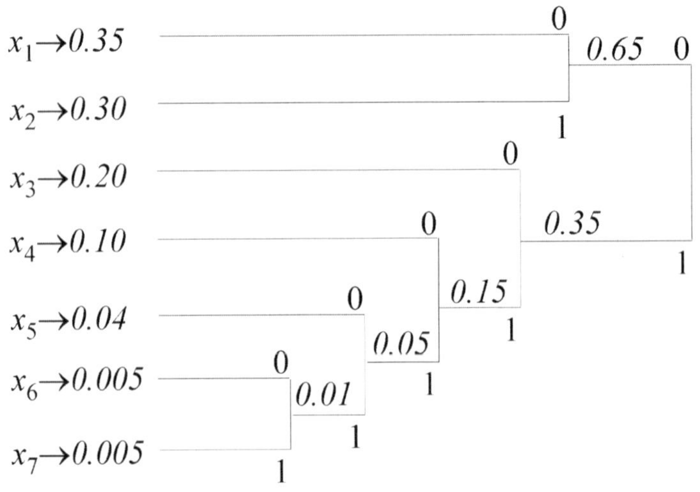

Figure 3.9 Example of a table for the synthesis of Huffman coding.

In the example above the mean code word length $\overline{N} = 2.21$ and the coding efficiency becomes 2.11/2.21 $\approx$ 0.955, i.e. about 95%.

The coding can be made more efficient by being performed in blocks, of n source symbols. The equation corresponding to (3.57) then becomes

$$n\,H(X) \le n\overline{N} < n\,H(X) + 1 \tag{3.58}$$

Dividing by n gives

$$H(X) \le \overline{N} < H(X) + \frac{1}{n} \tag{3.59}$$

The conclusion is that if a sufficient number of source symbols are coded simultaneously it is possible to get arbitrarily close to Shannon's limit.

Example 3.6

Consider a source having $M = 3$ and the symbol probabilities 0.5, 0.4 and 0.1:
(a) Perform a Huffman coding and calculate its efficiency;
(b) Repeat part (a) for a code where the source symbols have been grouped in blocks of two and two.

Solution   (a) See Figure 3.10. For symbol by symbol coding the result becomes

$$\text{symbols} \begin{cases} x_1 = 0 \\ x_2 = 10 \\ x_3 = 11 \end{cases}$$

The efficiency of the coding is

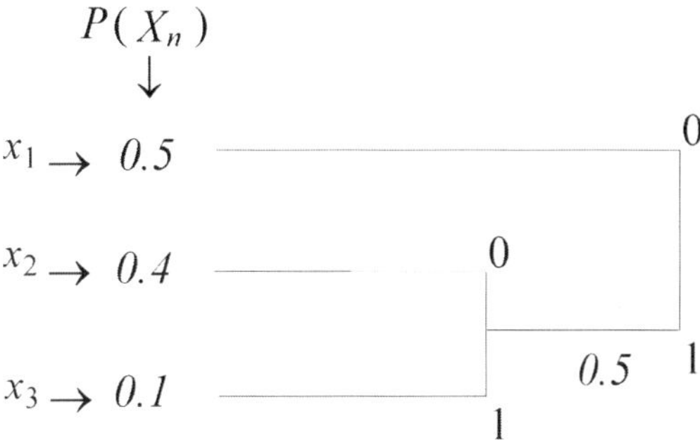

Figure 3.10   Synthesis of Huffman code symbol by symbol.

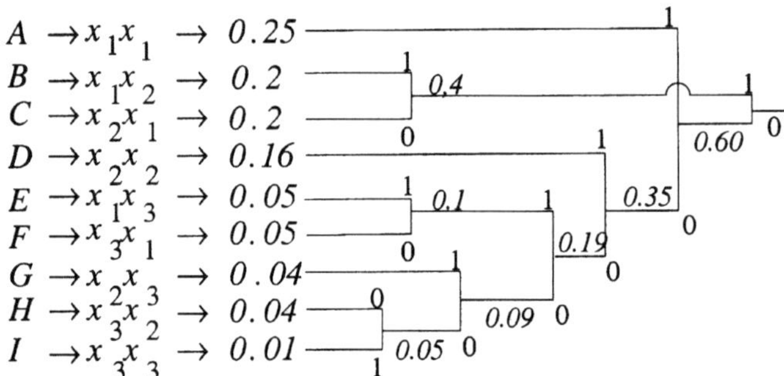

Figure 3.11 Synthesis of Huffman code where the symbols have been coded in pairs.

$$\eta = \frac{H(X)}{\overline{N}} = \frac{0.5\log_2\dfrac{1}{0.5} + 0.4\log_2\dfrac{1}{0.4} + 0.1\log_2\dfrac{1}{0.1}}{0.5 \times 1 + 0.4 \times 2 + 0.1 \times 2} = \frac{1.36}{1.5} \approx 0.907$$

$$(3.60)$$

i.e. the efficiency is 90.7%

(b) Group the source symbols in blocks of two and two (Figure 3.11). The mean value of the information per pair of source symbols becomes:

A = 01	2 × 0.25
B = 11	2 × 0.2
C = 10	2 × 0.2
D = 001	3 × 0.16
E = 00011	5 × 0.05
F = 00010	5 × 0.05
G = 00001	5 × 0.04
H = 000000	6 × 0.04
I = 000001	6 × 0.01
	$2\overline{N} = 2.78$

The coding efficiency at coding by pair then becomes

$$\frac{H(X)}{\overline{N}} = \frac{1.36}{1.39} \approx 97.8\%$$

$$(3.61)$$

The coding efficiency at coding by pair is clearly higher as is expected.

Sources with memory

For sources with a memory there is a dependence between events. Because of this dependence there is a random part in each new event which contains information, and a part which contains no new information and is therefore redundant. The concept of information is the same as before, but in order to calculate the total entropy it is necessary to go via the conditioned entropy. Assume a first order memory, which is the same as a first order Markov chain. The symbol x_i appears after the symbol x_j with a probability of

$$P_{ij} = P(x_i|x_j) \tag{3.62}$$

The conditional entropy is defined

$$H(X|x_j) = \sum_i P_{ij}\log_2 \frac{1}{P_{ij}} \tag{3.63}$$

For a stationary Markov process the total entropy is then obtained by taking the sum over all j:

$$H(X) = \sum_j P_j \, H(X|x_j) \tag{3.64}$$

Example 3.7

Calculate the entropy of a binary first order Markov source with the probabilities $P(0|1) = 0.45$, $P(1|1) = 0.55$ and $P(1|0) = 0.05$.

Solution As can be seen in the text the probability $P(0|0)$, is missing and must be calculated from the fact that the sum of branches leaving a state has to be one. A Markov chain is easily depicted by a state diagram (Figure 3.12).

According to eq. (3.64) the entropy becomes $H(X) = P(0)H(X|0) + P(1)H(X|1)$. To be able to perform this calculation the probabilities of the different symbols $P(0)$ and $P(1)$ are needed. For instance, the probability to get to state zero from a given initial state is equal to the probability of being in the initial state multiplied by the transition probability from this state to state zero. Adding the

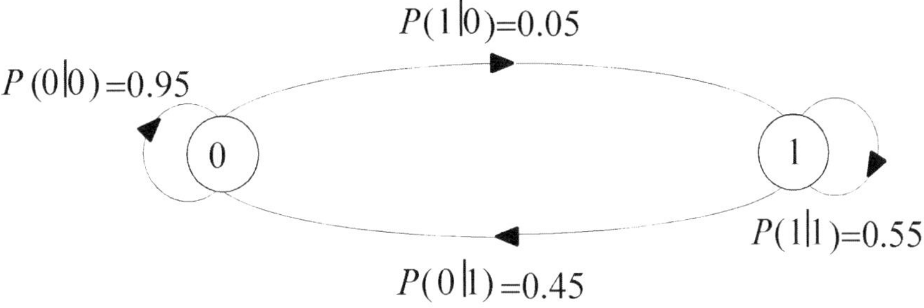

Figure 3.12   State diagram for a first order binary Markov chain.

probabilities to end up in state zero from all states gives the total probability to be in state zero. The probability of being in one of all the states is one. The conditions can be expressed by the following equation system:

$$\left.\begin{array}{l} P(0) = P(0|0)P(0) + P(0|1)P(1) \\ 1 = P(0) + P(1) \end{array}\right\} \Rightarrow P(0) = 0.9 \text{ and } P(1) = 0.1 \qquad (3.65)$$

which gives $H(X) = 0.357$ bit/symbol.

Predictive coding

For a source with a memory, the coding can be made more efficient by using the inherent dependence within the various events. A coding method using this is predictive coding, or differential pulse code modulation, DPCM as it is also often called. This is however not a modulation method as the name implies, but a method for source coding. This misleading name is mainly due to historical reasons.

The idea behind predictive coding is based on the fact that when the source outcomes are coded, for instance by A/D conversion, neighbouring symbols are correlated. This is valid for every band limited signal, e.g. speech, images, temperature series etc., provided the samples are not taken too sparse. The symbol flow is labelled $x(n)$. Because of this correlation, the variance of the difference between neighbouring samples will be

$$\text{Var}\{x(n) - x(n-1)\} \ll \text{Var}\{x(n)\} \qquad (3.66)$$

Since the variance is less than for the original signal, fewer bits are needed as only the difference is transmitted. The transmission can therefore be carried out using fewer bits than are needed if predictive coding is not used. The gain when using this method is even higher if the symbol $x(n)$ is predicted by the use of a predictor consisting of several stages, for instance, an FIR filter. The output signal from the predictor is subtracted from the original signal and the difference, i.e. the prediction error, is transmitted:

$$\varepsilon(n) = x(n) - \tilde{x}(n) \qquad (3.67)$$

The fundamental principle of predictive coding is depicted in Figure 3.13.

The decoding is done at the receiving end by the opposite operation. The error signal, $\varepsilon(n)$, forms the input signal of the decoder which is added to the output signal of the predictor. The predictor at the receiving end is identically designed to the predictor of the transmitter (see Figure 3.14). The sum of the addition is therefore $x(n)$, which is the original signal.

The transmitted message is usually quantised. This can be done by placing a quantiser, i.e. performing the operation $q[\cdot]$, at the output of the coder (cf. Figure

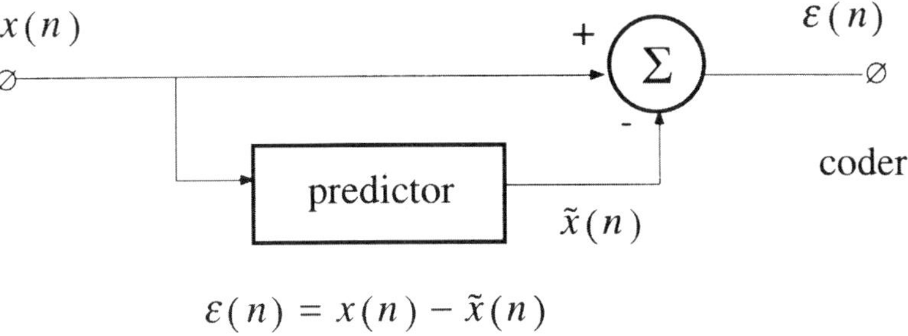

$$\varepsilon(n) = x(n) - \tilde{x}(n)$$

Figure 3.13 Predictive coder.

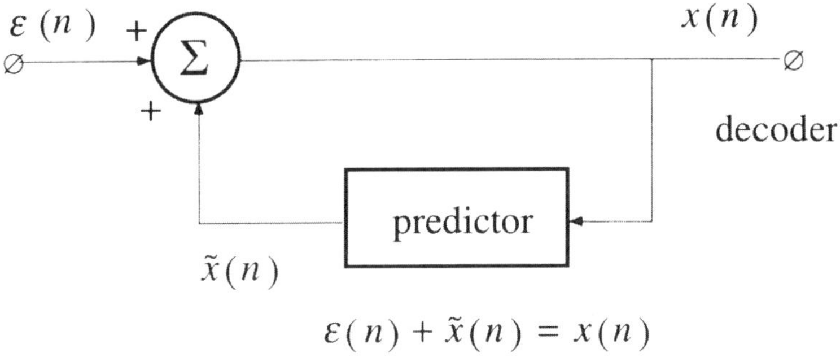

$$\varepsilon(n) + \tilde{x}(n) = x(n)$$

Figure 3.14 Decoder.

3.15). When the signal already is A/D converted, it is common to reduce the number of quantisation levels before transmission.

The decoder will then operate with $q[\varepsilon(n)]$ as input signal, which means that the predictor will not give the same $\tilde{x}(n)$ in the coder as in the decoder. This error can become quite large since the predictor may increase the error and the created signal will be considerably different from $x(n)$.

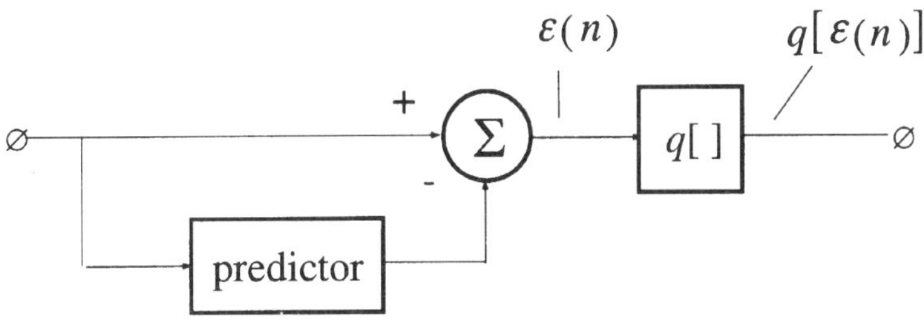

Figure 3.15 Encoder with subsequent quantisation of the error signal.

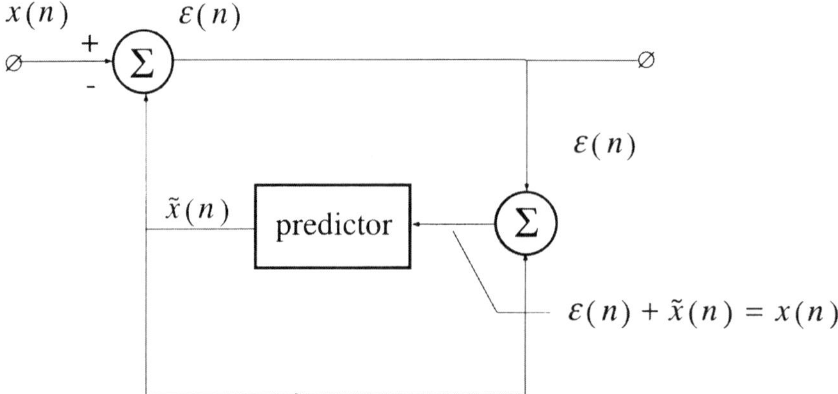

Figure 3.16　Coder with changed topology. Note that the predictor operates with the same input signal as before, so that the resulting error signal is the same as before.

A solution to this problem is to change the design, or topology, of the coder. In Figure 3.16 such a change is shown. The function of the coder is fully identical with the one in Figure 3.13, but the input signal of the predictor is created by adding the error signal to the predicted signal. This gives $x(n)$ as a result, and the input signal of the predictor becomes identical in the two cases.

When the error signal is quantised, this can be done prior to the summation just before the output of the predictor. This means that the predictors of the coder and decoder will use the same signal, i.e. $q[\varepsilon(n)] + \tilde{x}(n)$ and the prediction will therefore be the same (Figure 3.17).

Example of application of DPCM, delta modulation (DM)

A simple predictor can consist of a delay $\tilde{x}(n) = x(n - 1)$ and quantisation is done in one bit, having the amplitude $\pm\Delta$. The system of delay and feedback forms an integrator. The decoder therefore only consists of an integrator.

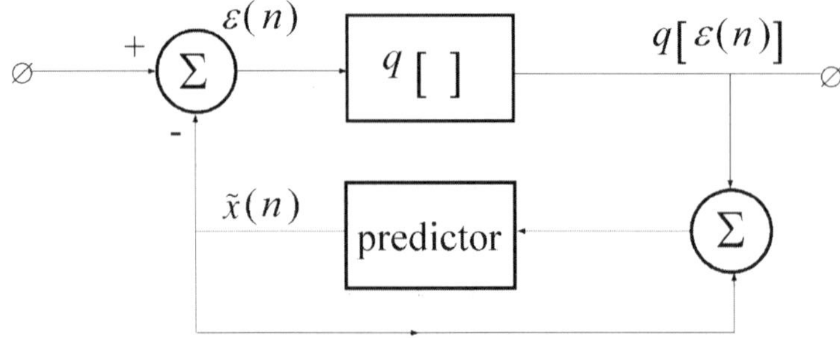

Figure 3.17　Final design of the predictive coder with quantised error signal.

To achieve a larger reduction of the required number of transmitted bits, a higher order, i.e. several stages, predictor can be used. With the input signal $x(n)$ a p-stage predictor gives the output signal

$$\tilde{x}(n) = \sum_{i=1}^{p} a_i x(n - i) \tag{3.68}$$

Compare the case $p = 1$ and $a_i = 1$ with eq. (3.66). The predictor coefficients a_i must be determined in some way to achieve as small an error signal as possible. This is usually carried out by minimising the mean value of the squared difference, i.e. the error between input signal and predicted signal (minimum mean square error, MMSE). Such a formulation gives

$$E\{\varepsilon^2(n)\} = E\left\{\left(x(n) - \sum_{i=1}^{p} a_i x(n - i)\right)^2\right\} = E\{x^2(n)\}$$

$$-2\sum_{i=1}^{p} a_i E\{x(n)x(n - i)\} + \sum_{i=1}^{p}\sum_{j=1}^{p} a_i a_j E\{x(n - i)x(n - j)\}$$

$$= \phi(0) - 2\sum_{i=1}^{p} a_i \phi(i) + \sum_{i=1}^{p}\sum_{j=1}^{p} a_i a_j \phi(i - j) \tag{3.69}$$

where $\phi(m)$ is the autocorrelation function for the signal if $x(n)$ is wide sense stationary (WSS). Minimising the error (3.69) gives a well-known equation system which is usually called the normal equations or Wiener–Hopf equations:

$$\sum_{i=1}^{p} a_i \, \phi(i - j) = \phi(j) \tag{3.70}$$

for $j = 1, \ldots, p$. The solution gives the optimal predictor coefficients for a source with Gaussian distributed signal and is further described in Chapter 7. The autocorrelation $\phi(n)$ really requires a signal of infinite length for the calculation, but it may be estimated by

$$\hat{\phi}(n) = \frac{1}{N} \sum_{i=1}^{N-n} x(i)x(i + n) \tag{3.71}$$

If $\hat{\phi}(n)$ is estimated for each new sample and new parameters a_i are calculated, adaptive DPCM (ADPCM) is obtained. The filter coefficients must then be transmitted to the receiver. For ADM $\phi(n)$ is estimated continuously and a_1 is updated for each new sample, i.e. a_1 becomes time dependent $a_1(n)$.

By inserting eq. (3.70) into eq. (3.69) and by changing the order of summation, the variance of the predictive error is obtained,

$$E\{\varepsilon^2(n)\} = \phi(0) - \sum_{i=1}^{p} a_i \phi(i) \tag{3.72}$$

The coding gain can then be defined as

$$G_P = \frac{E\{x^2(n)\}}{E\{\varepsilon^2(n)\}} = \frac{\phi(0)}{\phi(0) - \sum_{i=1}^{p} a_i \phi(i)} \tag{3.73}$$

which gives the reduction in signal power, or variance, after the coding.

For a typical speech signal sampled at the frequency 8 kHz is $\phi(1) = 0.85$. The gain with a one stage predictor is then a little more than 5 dB.

Example 3.8

Calculate the filter coefficients a_i and coding gain in dB for DPCM with a predictor having one and two taps, respectively. The input signal consists of numbers for which the autocorrelation function is $\phi(0) = 1.0$, $\phi(1) = 0.8$, $\phi(2) = 0.3$ and for $n > 2$ is $\phi(n) = 0$ (Figure 3.18).

Solution   Because of the symmetry around $n = 0$, the total autocorrelation function turns out to be as shown in Figure 3.19.

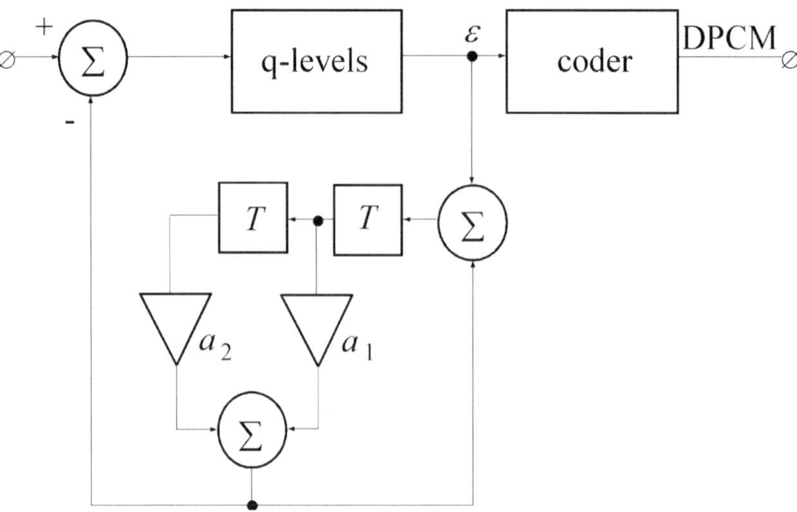

Figure 3.18   Differential PCM coder with two stage predictor.

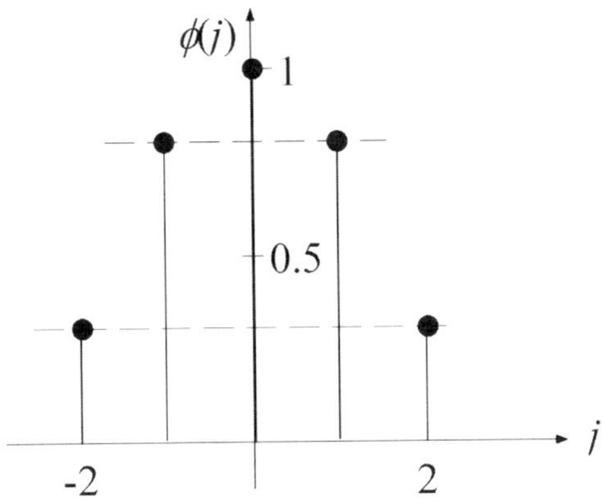

Figure 3.19 Autocorrelation function.

(a) One-stage predictor: using the normal equations one-coefficient coding gives (eq (3.70))

$$a_1 \phi(0) = \phi(1) \quad \Rightarrow \quad a_1 = \frac{\phi(1)}{\phi(0)} \tag{3.74}$$

which gives $a_1 = 0.8$ and

$$G_P = \frac{1}{1 - 0.8^2} = 2.78 \rightarrow 4.4 \ (dB) \tag{3.75}$$

(b) Two-stage predictor:

$$\begin{cases} a_1 \phi(0) + a_2 \phi(1) = \phi(1) \\ a_1 \phi(-1) + a_2 \phi(0) = \phi(2) \end{cases} \Rightarrow \begin{cases} a_1 + 0.8a_2 = 0.8 \\ 0.8a_1 + a_2 = 0.3 \end{cases} \Rightarrow \begin{matrix} a_1 = 1.56 \\ a_2 = -0.944 \end{matrix} \tag{3.76}$$

$$G_P = \frac{1}{1 - 1.56 \times 0.8 + 0.944 \times 0.6} = 26 \rightarrow 14 \ (dB) \tag{3.77}$$

Every bit of resolution in the A/D conversion means 6 dB in dynamic range for a digital signal. The one stage predictor offers no possibility to save bits. The two stage predictor with its 14 dBs of coding gain gives on the contrary a possibility to save two bits in the digital word.

The difference between delta modulation and adaptive delta modulation is illustrated in Figure 3.20. Note that for pure delta modulation the stages are always of the same size, while for adaptive delta modulation the step size varies with the variation of the signal.

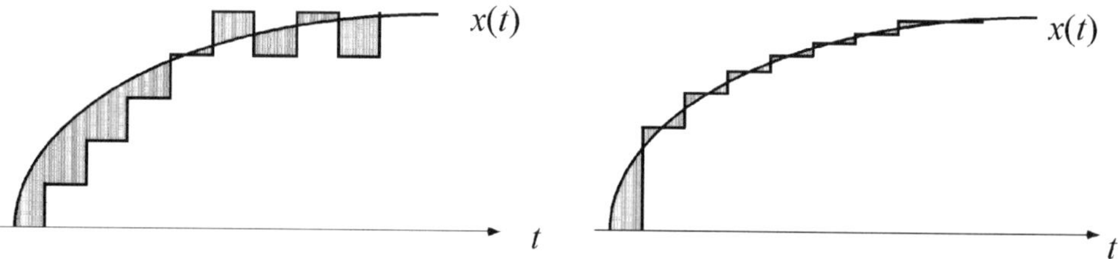

Figure 3.20   The shaded areas show the difference between the original signal and quantised signal for delta modulation (left) and adaptive delta modulation (right).

The Lempel–Ziv algorithm

For the Huffman coding a clear view is needed of what shall be coded and its a priori probabilities. If these conditions are not met the coding efficiency decreases drastically. A method which does not require such conditions is the Lempel–Ziv algorithm.

The idea behind the algorithm is based on checking if the symbol combination under study is repeated in a previous section of the symbol flow. In the coded symbol flow, only a reference to this section is then transmitted. Even in a random flow of ones and zeros there will be identical sequences to which references can be given. A Lempel–Ziv algorithm with a sliding window is described below.

A sequence of M-ary symbols $\{x_j\}_{j=-\infty}^{\infty}$ will be compressed. Assume that the sequence has been generated in a feedback shift register with one tap as shown in Figure 3.21.

The length, n, of the shift register has once and for all been determined. At the very beginning the first n symbols are stored in the shift register. These are transferred to the receiver without any compression, which requires $\lceil n\log_2 M\rceil$ binary bits. One clock pulse in the shift register means that all symbols are shifted one step to the left and that stage n is filled with a new symbol. For several shifts, new symbols must be stored in all positions that are vacant. There are three different possibilities:

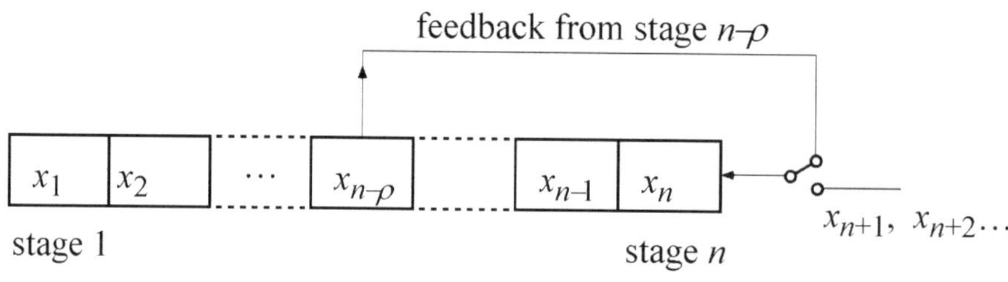

Figure 3.21   Feedback shift register as a model of the source.

- when the symbol has occurred previously and is stored in one of the positions of the shift register, the feedback is coupled to this position;
- when the symbol is stored in several positions, the feedback is chosen which generates as long a sequence as possible with the feedback in the same position during the following L clock pulses; such a sequence is called a phrase;
- when the symbol has not occurred previously it is fed in from the outside.

To describe the algorithm in more general terms, assume a symbol sequence $\mathbf{x}_1^N = (x_1 \quad x_2 \quad \cdots \quad x_N)$ for $N \geq 1$. Without losing generality it can be assumed that the symbols $\mathbf{x}_{-\infty}^0$ have been coded. For the next symbol, the algorithm starts by checking if the symbol x_{n+1} is among the symbols $\mathbf{x}_1^n$ in the shift register. If it is not there $L = 1$ is set. If x_{n+1} exists, the algorithm looks for those values of $L \geq 1$ and ρ which fulfil:

1. $\mathbf{x}_{n+1}^{n+L} = \mathbf{x}_{n-\rho}^{n-\rho+L-1}$, i.e. find a symbol sequence, L symbols long, starting at $n - \rho$ and identical to the sequence starting at $n + 1$. Note that there is no upper bound on L, so L can exceed n.
2. Maximum L, given that (1) is fulfilled.
3. Minimum ρ, given that $0 \leq \rho \leq n - 1$, and that (1) and (2) are fulfilled.

For example, if $n = 5$ the sequence $[x_1, x_2, ...] = [\,\text{abcde} \mid \text{deda...}\,]$ gives $L = 3$ and $\rho = 1$. To the decoder $[e(L) \quad \mathbf{s}]$ is transferred, where $e(L)$ is a comma-free binary coding of L and $\mathbf{s}$ is

$$
\mathbf{s} = \begin{cases} \text{binary encoding of } \rho \text{ if } \log_2 n < \lceil L\log_2 M \rceil \\[2mm] \text{binary encoding of } \mathbf{x}_{n+1}^{n+L} \text{ if } \log_2 n \geq \lceil L\log_2 M \rceil \end{cases}
$$

A comma-free binary coding of a natural number L can take the following form (a comma-free code needs no delimiter between the code words). If $l = \lceil \log_2(L + 1) \rceil$, and $b(x)$ is a binary encoding of the decimal number x, then the code word

$$
e(L) = \left[\underbrace{00\cdots 0}_{\lceil \log_2(l + 1) \rceil - 1 \text{ zeros}} \quad 1 \mid b(l) \mid b(L) \right]
$$

gives a prefix-free code, e.g. $e(5) = [\,01 \mid 11 \mid 101\,]$. The parameters $(e(L) \quad \mathbf{s})$ are transferred and the next part of the sequence is encoded.

The decoder has a similar shift register and with knowledge of $\mathbf{x}_1^n$, L and $\mathbf{s}$ it can decode $\mathbf{x}_{n+1}^{n+L}$. The decoder receives $e(L)$, decodes L and computes $\lceil L\log_2 M \rceil$. If the result is $< \log n$ the following $\lceil L\log_2 M \rceil$ bits are a binary encoding of $\mathbf{x}_{n+1}^{n+L}$ otherwise the following $\log n$ bits are a position for the feedback where a copy of $\mathbf{x}_{n+1}^{n+L}$ starts.

The efficiency and complexity of the algorithm increases with the number of positions in the shift register. It can be shown that when x_N is an infinite data string and $n \rightarrow \infty$ the coding efficiency approaches one, i.e. the number of binary bits that are needed to be transmitted per symbol becomes $H(X)$. Since a long shift register is impractical to implement, one has to compromise when the number of positions is determined.

3.3.2 SPEECH CODING

Some examples of source coding methods, which have been further developed and specialised for the coding of human voices, are given below. These are: linear predictive coding (LPC) and codebook excited linear prediction (CELP). Compared to the previously described coding methods, the coding efficiency is further increased for speech signals, while poor sound quality may result for other signals. Other examples of speech coding methods, not treated in this book, are sub-band coding and transform coding which are based on coding of the Fourier transform of the signal.

The requirements are that the sound quality obtained is good while the number of transmitted bits is reduced as much as possible and the hardware is kept within reasonable limits. The simplest form of coding is pulse code modulation (PCM), which is an ordinary sampling and A/D conversion of the speech. High sound quality is maintained at the cost of a high bit rate. To achieve a lower transmission rate than that of PCM, knowledge of the origin of human speech has to be utilised. This requires systems of a higher degree of complexity, i.e. more advanced algorithms, which in turn require more hardware or more powerful processors.

Two main groups of coders are being used for speech coding: wave-form coders and vocoders (voice coders). The group of vocoders described below use a parameter description of the speech and are therefore called parameter coders. The ordinary PCM, together with some of its variations, belongs to the group of wave form coders. Apart from linear quantisation, it is possible to utilise logarithmic μ- or a-law quantisation when using PCM. The quantisation noise is then reduced as the resolution is highest in the amplitude region where the signal is most probable. Adaptive quantisation, APCM, is a variation where the magnitude of the quantisation steps depend on how fast the signal changes. In telephone systems 8-bits logarithmic quantisation is used at 8 kHz sampling frequency, which gives a transmission rate of 64 kbit/s.

The vocoders are designed according to a parametric model of how human speech is produced. A lower sound quality characterises the vocoders together with a much lower bit rate, down to 2.4 kbit/s. This low rate is mainly used in military systems.

In the region between the vocoders and the wave-form coders are the so-called hybrid coders. They utilise a combination of the techniques mentioned above. It is among these that you can find speech coders with high sound quality and a transmission rate of around 4 kbits/s, such as CELP speech coders.

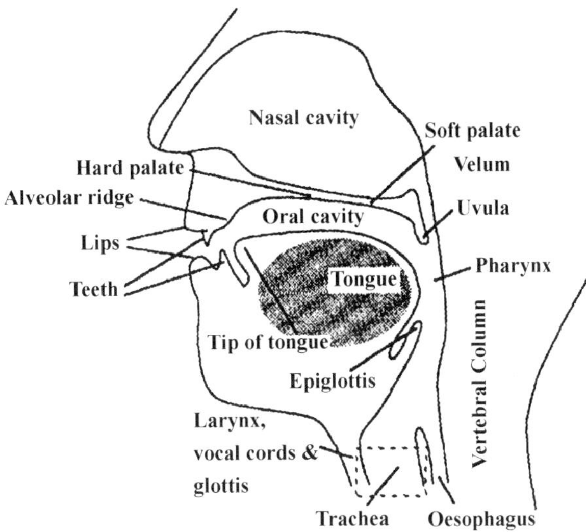

Figure 3.22 The human vocal organs. (Illustrated by John Sören Pettersson)

Speech apparatus

Speech is produced by air being forced from the lungs past the vocal cords, which then vibrate and form acoustic waves. Changing the shape of the mouth, lips and nostrils affects the acoustic waves and different sounds are produced (cf. Figure 3.22).

There are basically four different types of sound which differ according to their origins. The two most important are voiced and non-voiced sounds. Voiced sounds are made when air is forced through the vocal cords and causes them to vibrate. Short pulses of air with a certain periodicity are formed, so-called glottal pulses, which activate the oral cavity and voiced sounds are created, e.g. a, e and o. Non-voiced sounds, however, are formed when air from the lungs passes beside the vocal cords. The oral cavity is then narrowed near to the mouth where the air rushes out at high speed. This gives rise to such sounds as f, s and ch. The third type of sound is called a plosive. This is formed when the oral cavity closes, i.e. the mouth is shut. Air pressure builds up behind the opening and the air is suddenly released, producing such sounds as p and t. The final type of sound is the nasal sound such as n and ng. Such sounds are formed when the soft palate is lowered so that air is pushed up through the nasal cavity, instead of through the oral cavity, and is released through the nose.

The mouth forms a cavity for the sound waves. The resonance frequencies obtained when the mouth has a certain shape are called formants. The most important formants are the three or four which have the lowest frequencies and are the strongest. The phonetic separation of speech into vowels and consonants is based on the situation of the speech apparatus. Each situation, or in this case shape of the oral cavity, gives different formant frequencies.

To obtain a satisfactory speech coder it is necessary to create a good model of the human vocal tract. To obtain a coder with limited complexity, considerable simplifications have to be made to the model on which the vocal coder is based.

Models of the vocal tract

To model the human vocal tract it is best to distinguish between the air flow coming from the trachea into the oral cavity and the oral cavity itself, i.e. the filter. The air flow is modelled as either a pulse train or a noise source. This forms the input signal of a time variant linear system modelling the oral cavity. The pulse train and the parameters of the linear system vary with time and control the properties of the speech.

For the modelling of the oral cavity the so-called tube model is mostly used. A number of tubes, usually assumed to have zero loss, of different lengths and diameters are connected in series. The object of their different cross-section areas is to simulate the continuously changing area of the oral cavity. From this analogue model it can be shown that the oral cavity can be modelled acoustically as the time-discrete filter shown in Figure 3.23.

The transfer function of the filter modelling the influence of the oral cavity is

$$H(z) = \frac{G}{1 - \sum_{k=1}^{p} a_k z^{-k}} \tag{3.78}$$

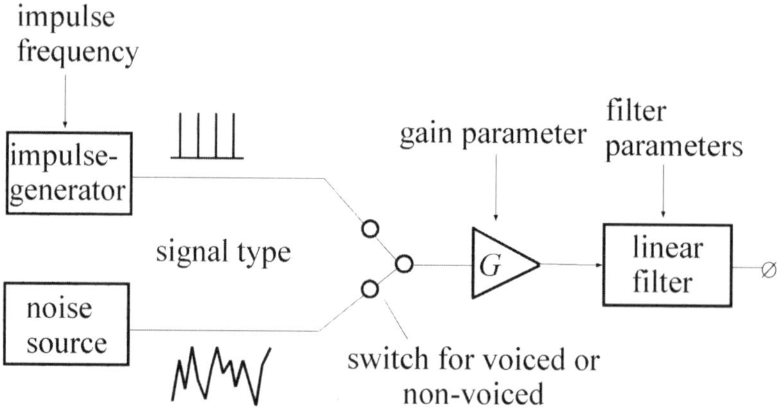

Figure 3.23   Model of the human vocal tract.

where G is a gain factor, p is the number of poles and a_k the filter coefficients. The resonance frequencies, formants, of the speech are represented by the poles of the transfer function $H(z)$. A model only based on poles has turned out to be sufficient to model the majority of existing sounds. To model nasal sounds, for instance, both poles and zeros are required. It turns out, however, that introducing more poles in $H(z)$, also improves the modelling of these sounds.

Measurements have shown that the system parameters are stationary for approximately 10–20 ms.

Speech coding with linear prediction

The idea behind a speech coder of the linear predictive coding (LPC) type, is the previously presented model for the speech mechanism in Figure 3.23 and the time-dependent, discrete time, linear filter which models the oral cavity. This filter function, which only has poles, is the natural representation of voiced speech, as mentioned above. Other kinds of speech actually also require zeros in the transfer function of the prediction filter. If the order p is high enough, however, eq. (3.78) is a good representation of all kinds of speech. To determine the coefficients, methods based on the previously described least square minimising are being used.

It can be shown that the coefficients obtained by minimising the predictive error in eq. (3.69) are identical with the filter coefficients in the model (3.78) when the model filter is excited by either an impulse or by stationary white noise. This means that the filter coefficients can be calculated in the same way as for the predictor described in previous sections. This is intuitively reasonable when the input signal $x(n)$ consists of impulses from the vocal cords. Even though its peak power is high, the mean power $E\{x^2\}$ is low, which is also the case for the power of the error signal, $\varepsilon(n)$.

For a system of the type in eq. (3.78) the output signal $y(n)$ is connected to the input signal $x(n)$ by

$$y(n) = \sum_{k=1}^{p} a_k y(n-k) + Gx(n) \tag{3.79}$$

Inserted in the least square condition the filter coefficients can be calculated from

$$\varepsilon(n) = \sum_{m} \left(y(n-m) - \sum_{k=1}^{p} a_k y(n-m-k) \right)^2 \tag{3.80}$$

In eq. (3.80) the expectation operator from eq. (3.69) has been replaced by a sum over one frame, a frame being one section of the speech. The variable m is an index for the samples included in the frame.

In a refined version of LPC coding, the filter can be excited by several pulses, typically eight, per period of the voiced sound (multipulse exciting, MPELPC). The amplitude and time position of the pulses are individually adjusted, which results in considerably improved sound quality. Different kinds of LPC are used mainly for bit rates around 9.6–16 kbits/s. The filter coefficients, voiced and voiceless sounds, pitch (i.e. impulse frequency) as well as the gain factor G are transmitted over the channel.

CELP coder

The codebook excited linear prediction (CELP) coder is a development of the LPC coder. The CELP coder is also a frame oriented technique based on the model of the speech mechanism described in Figure 3.22. The analysis of speech is organised in two basic steps: one performing the long term prediction and one performing the short term prediction. By long term prediction is meant prediction of the fundamental frequency in the speech over a longer period, which entails the vocal cord frequency that in fact changes relatively slowly. Short term prediction means prediction of changes in the frequency spectrum of the speech, i.e. the formant frequencies which are formed by changing the shape of the oral cavity, lips and nostrils. The design of a CELP coder is depicted in Figure 3.24. To

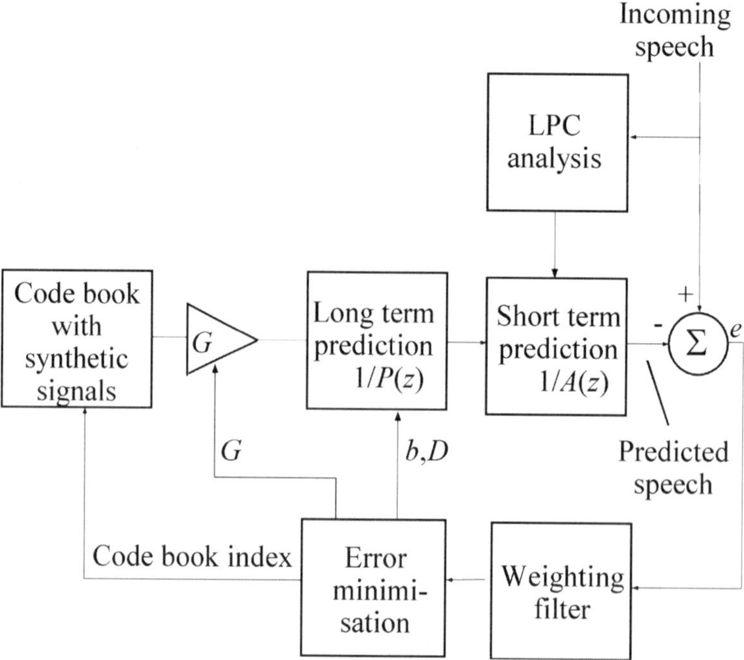

Figure 3.24 Block diagram of a CELP coder where speech coding is carried out by analysis through synthesis. G is a gain factor and e is the difference between the incoming and the predicted speech.

determine the input signal, the CELP coder utilises a technique which is called analysis by synthesis. This means that the speech synthesis in the receiver is the same as was used for the analysis in the transmitter.

The predictors are implemented as filters. The filter coefficients of the short term predictive filter, $A(z)$, are obtained by LPC analysis of the incoming speech, which is then filtered by the inverse filter in the same way as in the previous section. The filter for the long term predictor, $P(z)$, takes a very simple form:

$$P(z) = 1 - b\,z^{-D} \tag{3.81}$$

The filter $1/P(z)$ will have poles at

$$z = b^{1/D}\exp(j2\pi f_n T) = b^{1/D}\exp(j2\pi n/D), \quad n = 0, 1, 2, ..., D - 1 \tag{3.82}$$

corresponding to frequency peaks at $f_n = n/(DT)$, where $n \neq 0$ and T is the sampling period of the system. These correspond to the peaks present in the speech spectrum and are called formants. The frequency peak for $n = 1$ is typically between 50 and 500 Hz, with gives the pitch of the voice. In the signal spectrum, after the filter, a number of spikes emerge with the distance 50–500 Hz. The obtained residual, e, is used to predict the parameters b and D of the filter for the long term prediction.

In Figure 3.24 it can be seen that what in Figure 3.23 was two types of input signals, has now been replaced by one single input signal by using the code book and the long term predictor filter instead of creating voiced and non-voiced signals. The different artificial input signals are stored in a so-called code book, where each input vector is represented by an index. A code book can, for instance, be organised as a matrix or as a vector. The number of elements in an input vector (code book vector) corresponds to the number of speech samples being analysed. The code book is fixed. The elements of the artificial input signal vector are determined by using pseudo-random numbers in a Gaussian distribution having mean value zero and variance one. The reason why this organisation of the code book was chosen is that the prediction error, which is caused by both long term and short term prediction of numbers, has turned out to consist of white noise having almost Gaussian distribution with variance one. This holds for most sounds, with the exception of some sudden sounds and the changes from voiceless sounds or quiet periods to voiced sounds. The number of code book vectors can vary, but a code book coded by 8–10 bits, i.e. with 256–1024 different vectors, has proved to give satisfactory results.

The input signal to the long term predictor is now only a noise vector, 'innovation sequence', multiplied by a gain factor. During speech analysis, several predicted speech signals are formed in the transmitter. These are obtained by testing the different noise vectors stored in the code book. Searching the code book is done sequentially from start to finish. The error, i.e. the difference between the incoming speech and the predicted one, is filtered by a so-called weighting filter given by

$$W(z) = \frac{1 - \sum_{k=1}^{p} a_k z^{-k}}{1 - \sum_{k=1}^{p} a_k \alpha^k z^{-k}} \tag{3.83}$$

where p is the order of the short term prediction filter, a_k, the coefficients of the filter and α a weighting parameter. The objective of this filter is to reduce the perceived distortion by giving low weights to the noise components that fall into the part of spectrum where the speech is strong. High weights are given to the parts of the spectrum where the speech is weak and where the noise otherwise immediately will break through to the listener. The optimum input signal vector is then determined from the minimum of the square of the weighted error. This signal vector corresponds to an index in the code book (cf. Figure 3.24). If this index is transmitted to the receiver, the input signal can immediately be regenerated. By minimising the square of the error with regard to the gain, the optimum gain factor is also obtained.

CELP coders are usually used for bit rates between 4.8 and 9.6 kbit/s. The different filter parameters for the short and long term predictors and the code book index are transmitted over the channel.

3.4 Appendix

3.4.1 EXAMPLE OF IMAGE CODING, MPEG

When moving pictures are transmitted or stored on a medium, a standard called Moving Picture Experts Group (MPEG) is often used. There are different variants of the standard, such as MPEG-2, which is used in digital TV, and MPEG-4, which is designed for interactive multimedia applications. MPEG-2 allows transmission of moving pictures at 2–10 Mbit/s with a quality better than the analogue PAL standard. MPEG-4 allows transmission with low quality at between 5 and 64 kbit/s and with TV quality at up to 4 Mbit/s. To support interactive video and multimedia MPEG-4 identifies different physical objects in the image and codes them separately. The standards are complicated and below is an overview of the main ideas given. When a moving sequence is to be recorded, 25–30 still pictures per second are taken. Each of these pictures is divided into a number of image points, pixels. A recorded sequence contains a lot of statistical redundancy both in the vicinity of a certain image point and when looking at changes in the picture over time. This is utilised by the MPEG standard to achieve low redundancy in the coded data that is to be transferred. The standard is flexible and varies for different applications. MPEG uses a combination of prediction in time and transform coding of remaining spatial information, which is further described below.

Under-sampling

The eye is more sensitive to changes in the brightness than to chrominance changes. The signal is therefore divided into one luminance and two chrominance signals, Y, U and V. The chrominance signals are sampled with a lower rate than the luminance signal. Usual ratios between the sampling rate for Y/U/V are 4:1:1 or 4:2:2.

Motion compensated prediction

If all image elements in a video scenery are displaced approximately similarly, the movement between the pictures can be described by a limited number of motion parameters. Therefore the images are divided into disjoint blocks of usually 16×16 pixels whose movement are described by a single motion parameter. When a block in the original image is moved according to the motion parameter, the difference between this block and the corresponding block in the next picture is small. Since two consecutive images, after the motion compensation, have a large correlation, differential pulse code modulation (DPCM) is used on the images. For video sequences coded for storage media, a non-causal predictor is used, in a way that forthcoming images are used.

Transform coding

The DPCM-coded images that have been generated in the previous step are divided into separate blocks, $\mathbf{b}$, of $N \times N$ pixels. The transformation is done by multiplication with the transform matrix $\mathbf{A}$ to obtain the coefficients $\mathbf{c}$, where $\mathbf{c} = \mathbf{AbA}^T$. MPEG utilises a discrete cosine transform (DCT) over image blocks of 8×8 resolution points, or pixels. Only a minor part of the DCT coefficients have a variance sufficient to make it necessary to transfer them to the decoder. The human eye is more sensitive to quantisation errors in a smooth surface than in a variegated surface. Therefore, the DCT coefficients corresponding to low spatial frequencies are quantised with a larger resolution than the coefficients corresponding to high spatial frequencies. Before the transmission, the DCT coefficients are coded by a code with variable code word length (Figure 3.25).

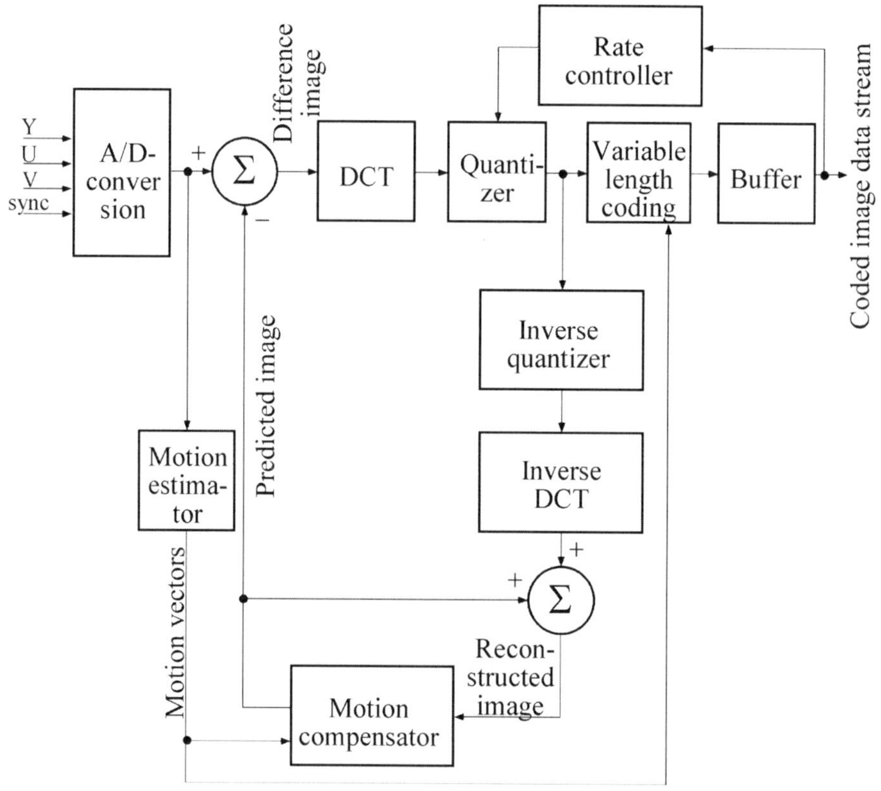

Figure 3.25 MPEG-2 encoder.

Exercise 3.1

What is the entropy for a binary source with $P(0) = \alpha$?

Exercise 3.2

A binary source has $P(0) = P(1) = 1/2$ and a first order memory so that $P(0|1) = P(1|0) = 3/4$. Compute the resulting conditional entropy.

Exercise 3.3

A DMC with two inputs and four outputs has the transition probabilities shown in the figure below. Determine the channel capacity as a function of the error probability p at a symbol rate of 2×10^3 symbols per second.

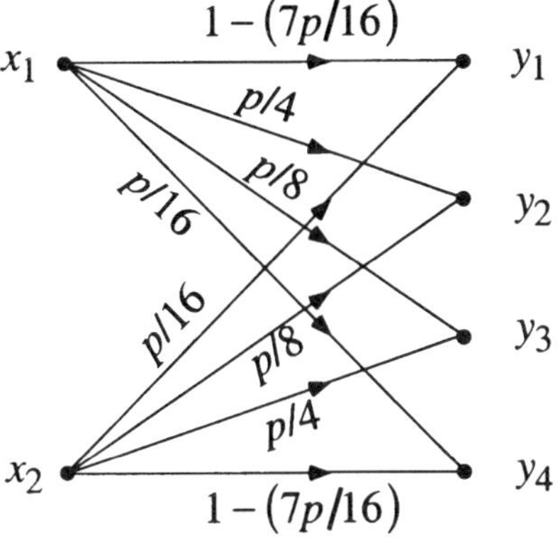

Exercise 3.4

Compute $H(X)$ for a discrete memoryless source with six symbols having the probabilities

$$P_A = 1/2, \quad P_B = 1/4, \quad P_C = 1/8, \quad P_D = P_E = 1/20, \quad P_F = 1/40$$

Determine the amount of information in the messages *ABABBA* and *FDDFDF* and compare the expected value of information in a six symbol long message.

Exercise 3.5

The samples from a signal source shall be coded in an efficient way. The samples have an autocorrelation which is $R(0) = 1$, $R(1) = 0.8$, $R(2) = 0.4$, $R(3) = 0.1$ and zero for arguments larger than three. Describe an efficient coder and give the eventual necessary parameters.

Exercise 3.6

The following statistically independent symbols shall be transmitted. Do an efficient coding and give the coding efficiency, η.

Symbol	Probability
A	3/8
B	3/16
C	3/16
D	1/8
E	1/16
F	1/32
G	1/32

Exercise 3.7

A distant civilisation uses an alphabet with four symbols $\{\Delta, *, !, \phi\}$.

(a) A fast study of the language shows that the probabilities for each symbol are (0.5, 0.3, 0.15, 0.05), respectively. Determine the entropy, $H(X)$, of the source. Design a Huffman code and compare the coding efficiency achieved with the efficiency that is achieved if two symbols are coded simultaneously.

(b) A closer study shows that the language is better described by a first order Markov model. The state diagram is shown in the figure below. Determine the symbol probability and the source entropy, $H(X)$.

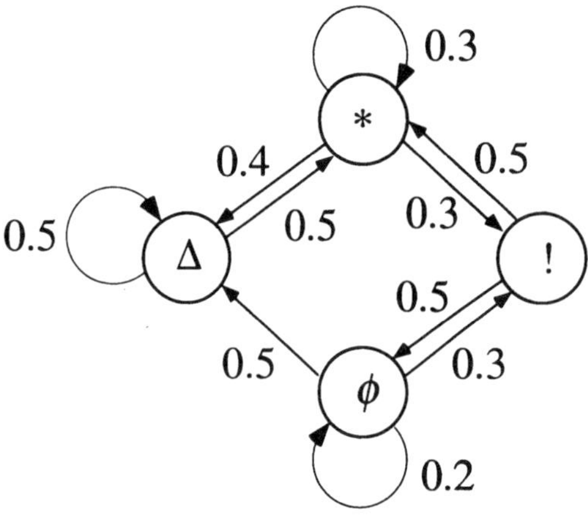

Exercise 3.8

Certain binary channels have a tendency to be either *good*, when almost no symbol error occurs, or *bad*, when the symbol error rate is approximately 1/2. A classical model for such a channel is the Gilbert–Elliot model, where the channel is described by a first order Markov process. By measurement, it has been shown that for a certain channel the transition probabilities are

$$a_{good\to good} = 0.99, \quad a_{good\to bad} = 0.01$$

$$a_{bad\to good} = 0.10, \quad a_{bad\to bad} = 0.90$$

What is the probability that the channel should by usable for data transmission?

Exercise 3.9

Find the probability density for the signal maximising the differential entropy given that the signal max and min values are limited to $x_1 \le x \le x_2$.

4 CHANNEL CODING

When a message is transmitted or stored it is influenced by interference, which can distort the message. Radio transmission may be influenced by noise, multipath propagation or by other transmitters. In different types of storage, apart from noise, there is also interference which is due to damage or contaminants present in the storage medium. There are several ways of reducing interference. However, some interference is too expensive or even impossible to eliminate. One way of solving this problem is to design the messages in such a way that the receiver can detect if an error has occurred or even possibly correct the error. This can be achieved by error-correcting coding, so-called channel coding. In such coding the number of symbols in the source-encoded message is increased in a controlled manner, which means that redundancy is introduced. Some methods for coding, decoding and methods for the calculation of the channel coding properties are described below. The chapter concludes with a very short introduction to ciphering. The reason why this is treated here is that the methods used in the description and analyse of ciphering and error correction coding have a lot in common.

4.1 Error detection and error correction

To make error correction possible the symbol errors must be detected. When an error has been detected, the correction can be obtained either by the receiver asking for repeated transmission of the incorrect codeword (Automatic Repeat reQuest (ARQ)) until a correct one has been received, or by using the structure of the error-correcting code to correct the error (forward error correction (FEC)). It is easier to detect an error than it is to correct it. FEC therefore requires more check bits and a higher transmission rate, given that a certain amount of information is to be transmitted within a certain time and with a certain minimum error probability. The reverse is also true; if the channel offers a certain possible transmission rate, ARQ permits a higher information rate than FEC, especially if the channel has a low error rate. FEC has, however, the advantage of not requiring a reply channel. The choice in each particular case therefore depends on the properties of the system or on the application in which error correcting is to be introduced. In many applications, such as radio broadcasting or compact disc

(CD), there is no reply channel. Another advantage of FEC is that the transmission is never completely blocked even if the channel quality falls to such low levels that an ARQ system would continuously ask for repeated transmission.

Typical requirements from telecom administrations on the error rate of some digital transmissions are:

telephone connection	10^{-5} errors/bit
permanent connection	10^{-6} errors/bit
high speed 2 Mbits/s	10^{-10} errors/bit
mobile telephone (after error correction)	10^{-2} errors/bit

In many applications different parts of the message are given unequally strong error protection. When a signal has been encoded in a source coder the sensitive parts of the transmitted parameters get a large error protection, for example the parameters giving the frequency of the voiced sounds in CELP coding. Less sensitive parts, for example the filter modelling the oral cavity, gives a smaller error protection. For high definition television (HDTV) the coarse structure of the image can be given a large error protection while the fine structure is then given less error protection.

There are two main methods to introduce error-correction coding (Figure 4.1). In one of them the symbol stream is divided into blocks and coded. This is consequently called block coding. In the other one a convolution operation is applied to the symbol stream. This is called convolutional coding.

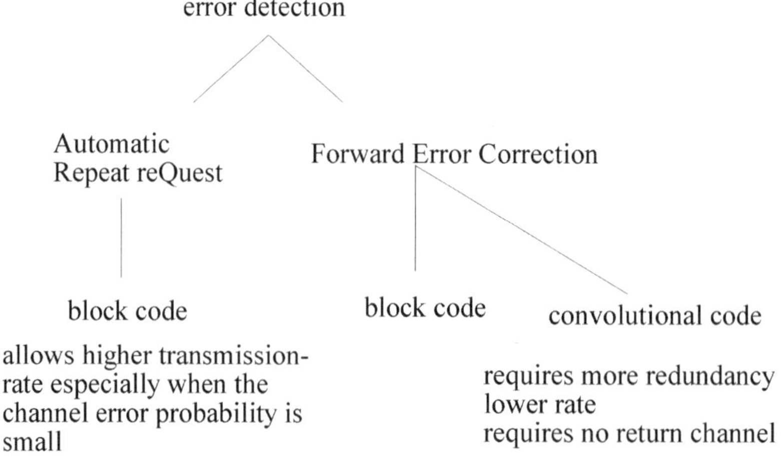

Figure 4.1   Error detection and correction.

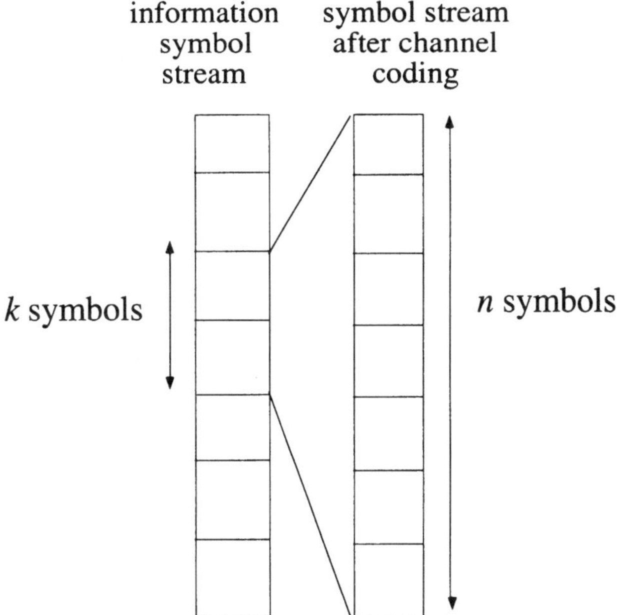

Figure 4.2 Information symbols are combined with check symbols to form a codeword.

4.1.1 FUNDAMENTAL CONCEPTS IN BLOCK CODING

Assume that the information stream is divided into blocks of length k symbols. The number of symbols in the source-coded message is increased by introducing check symbols, which increase the number of symbols per block to n. This is denoted a (n,k) code. An encoded block is called a codeword (Figure 4.2).

A simple code for binary systems is shown in Figure 4.3. The coding is carried out by arranging the message and information symbols in a rectangle and calculating check bits by modulo 2 addition of rows and columns, respectively. The check bits form parity bits and this type of code is therefore called a parity code. The check bits are transmitted together with the message bits. The receiver can correct an error and detect double errors since the bits are arranged in such a way that an even number of ones appear in each row and column. Certain multiple error patterns can also be corrected. This code is not optimal, since with the same number of check bits it is possible to correct and detect more errors.

The idea behind channel coding is that the number of possible combinations of symbols has been limited in such a way that the minimum distance, existing between two codewords, is as large as possible, as shown in Figure 4.4. The figure illustrates a case where a binary symbol, a bit, is to be transmitted with codewords of three binary symbols, i.e. with a $(3,1)$ code. The information to be transmitted is represented by a zero or a one. The words 000 and 111 are, for

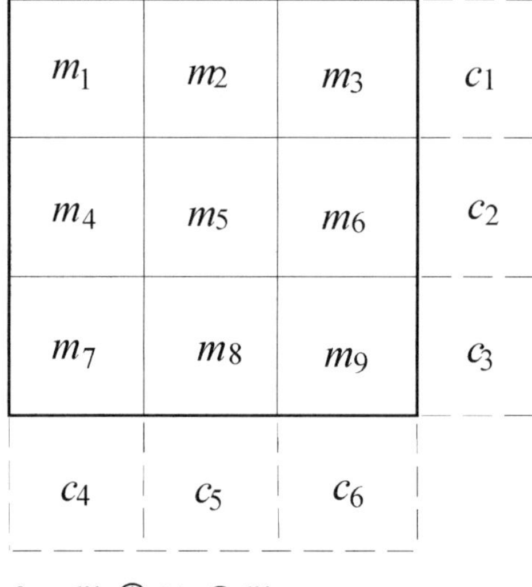

Figure 4.3 Double (15,9) parity check code as both rows and columns have special parity bits.

instance, selected as codewords (denoted repetition code) since these are separated by the maximum distance in the cube defined by the codeword symbols. Other pairs of points located diagonally in the cube offer other possible codewords, giving a code with the maximum distance between the codewords.

For the case when two information bits are coded into codewords with three binary bits, a (3,2) code, the diagonal of the cube cannot be chosen. Instead it is the diagonal in each plane that gives the maximum distance between nearest neighbours among the codewords. This leads to more closely packed points which gives less error-correcting capability than the (3,1) code, but the (3,2) code can instead transmit two information bits per codeword.

For the case when there are only three symbols in the code, the problem of finding the optimal code is easy to solve. To achieve better properties in the communication system it is desirable to make the codes longer, i.e. the number of dimensions increases. Adding the requirement that coder and decoder must be possible to implement makes the problem of finding good codes less trivial. A lot of work has been carried out on this problem and some examples of good codes are given at the end of this chapter. The theories behind these codes are, however, not presented here since it would not increase the understanding of telecommunication systems.

The code rate is defined as

$$R_c = \frac{k}{n} \tag{4.1}$$

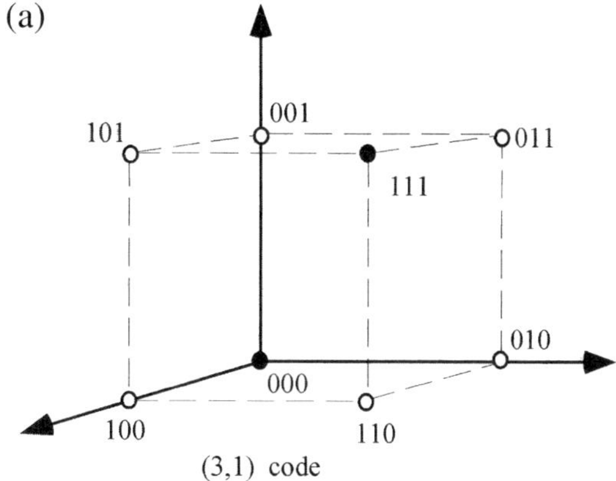

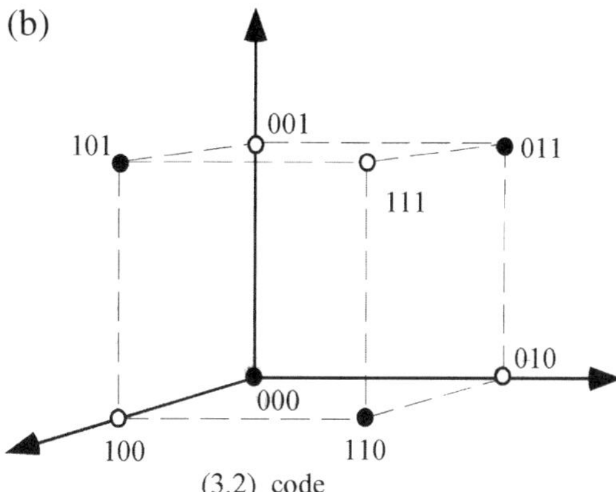

Figure 4.4 (a) For a (3,1) code the opposite corners in the cube defined by the possible code vectors are used. (b) For a (3,2) code the points along the diagonal in the same plane must be used.

If the bit rate without channel coding is r_b the bit rate after channel coding becomes

$$r = \frac{n}{k} r_b \tag{4.2}$$

Since n is greater than k, a higher symbol rate is required after the channel coding, which means a larger bandwidth of the signal to be transmitted. There are other methods which do not require a higher symbol rate since modulation and coding are regarded as a unity. This is treated in Chapter 5.

Figure 4.4 shows the Euclidean distance. Another important parameter in connection with coding is the Hamming distance $d(\mathbf{X}, \mathbf{Y})$. This is defined as the number of elements having different values of the two vectors $\mathbf{X}$ and $\mathbf{Y}$ if each element in one vector is compared to the respective element at the same position in the other vector. The vectors usually make up two codewords. For the codewords $\mathbf{X} = [011]$ and $\mathbf{Y} = [110]$, for instance, the Hamming distance is $d(\mathbf{X}, \mathbf{Y}) = 2$. Another important concept is $d_{\min}$, which is the smallest Hamming distance between two codewords among all the codewords in a certain code. Still another concept is the Hamming weight, $w(\mathbf{X})$, which gives the number of symbols not equal to zero in a codeword.

Assume one error has occurred in a transmitted codeword. The Hamming distance from the correct word is then 1. To discover the error, the smallest Hamming distance between two codewords must be larger, i.e. at least 2. In general, a code having the smallest Hamming distance d_{min} can detect e errors per codeword, where

$$e = d_{min} - 1 \tag{4.3}$$

The argument above can be further developed. Assume that the incorrect word received is 100. The three closest codewords are 000, 101 and 110 (cf. Figure 4.4b), and they have a relative Hamming distance of 2. The Hamming distance to the received word is 1 for all codewords. It is, in other words, possible to establish that an error has occurred, but it is not possible to correct it. In order to do this, the mutual distance between the codewords must be at least 3 in this case. Compare the code in Figure 4.4a, where it can easily be concluded that the transmitted codeword has been 000. Thus the distance between the codewords must be larger than double the number of errors. In general a code having the smallest Hamming distance of d_{min} can correct t errors per codeword, where

$$t = \left\lfloor \frac{d_{\min} - 1}{2} \right\rfloor \tag{4.4}$$

$\lfloor x \rfloor$ is the nearest integer which is less than or equal to x.

We return to these concepts later but there it is necessary to introduce them here.

4.1.2 PERFORMANCE OF BLOCK CODES

To be able to evaluate the improvement in transmission quality when coding is introduced and to calculate the performance of the transmission system in question, methods for calculation of the symbol error probability of the system, with and without coding, are required. Some of the methods for calculating these error probabilities are given below.

Assume that the errors are randomly distributed and mutually independent. An upper limit for the probability of having more than t incorrect symbols in a codeword of n symbols, P_{we} (word error), is obtained by using the binomial distribution

$$P_{we} \leq \sum_{i=t+1}^{n} P(i,n) = \sum_{i=t+1}^{n} \binom{n}{i} P_s^i (1 - P_s)^{n-i} \tag{4.5}$$

where $P(i,n)$ is the probability of making i errors on n symbols, and P_s is the probability of symbol error in the transmission channel.

The summation is taken from $t + 1$, since all errors shorter than that can be corrected. Since some errors of more than t bits can also be corrected, an over-estimation of the error probability will be made by eq. (4.5).

Often the required information is the probability to make an error in one symbol, P_{sc}. To do an exact computation of this, information about the mutual distance between all the codewords is necessary, resulting in a very complicated computation. Therefore is an upper limit of the symbol error probability computed here. Suppose $t + 1$ errors have occurred in the transmitted codeword. The decoder will then interpret the received word as the closest codeword, i.e. a codeword at the Hamming distance $2t + 1$ from the emitted one (cf. Figure 4.4a). Generally, imagine a sphere around every correct codeword. The spheres are not overlapping and every sphere has the radius t. A received codeword which arrives in one sphere is interpreted by the decoder to be the correct codeword connected to that sphere. If i errors occur in the transmission the received word will be displaced by the Hamming distance i from the emitted. Then the received word can land up in another sphere. If it, in addition, lands up close to a border, in the worst case the decoder can interpret the received word as a codeword located on the Hamming distance $i + t$ from the emitted. The errors are assumed to be evenly distributed over the decoded codeword. The probability that an error will appear in a given symbol is $(i + t)/n$. By utilising eq. (4.5) the probability of symbol error after decoding is

$$P_{sc} \leq \sum_{i=t+1}^{n} \frac{i+t}{n} \binom{n}{i} P_s^i (1 - P_s)^{n-i} \tag{4.6}$$

Assume that the error rate, P_s, is relatively low, which means that each codeword contains no or very few symbol errors. This means that, among the received codewords that cannot be corrected, the most probable number of incorrect symbols is $t + 1$. Eq. (4.5) can therefore be approximated as

$$P_{we} \approx \binom{n}{t+1} P_s^{t+1} \underbrace{(1 - P_s)^{n-t-1}}_{\approx 1} \approx \binom{n}{t+1} P_s^{t+1} \tag{4.7}$$

A useful approximation of P_{sc} can be obtained by the observation that for small values of P_s the first term in eq. (4.6) dominates. Together with the result in eq. (4.7) the error rate after decoding is approximated,

$$P_{sc} \approx \frac{2t + 1}{n} \binom{n}{t + 1} P_s^{t+1} \cong \frac{2t + 1}{n} P_{we} \tag{4.8}$$

To demonstrate the advantage of channel coding in a practical case, the bit error probability of binary phase shift keying (BPSK), which is further described in Chapter 6, is used. For BPSK a bit is the same as a symbol. The uncoded bit error probability, P_s, is

$$P_s = Q\left(\sqrt{2E_b/N_0}\right) \tag{4.9}$$

where $Q(\ )$ is a standard integral which is called the Q-function (see Section 6.5), E_b is the energy per information bit and N_0 is the noise power density.

For a coded system the necessary transmission rate is increased depending on the redundant symbols which have been introduced. With a constant transmitter power the bit energy is reduced. E_b/N_0 then have to be replaced by $R_c E_b/N_0$, where R_c is the code rate. Eq. (4.9) inserted in eq. (4.8) then gives the bit error probability of channel coded BPSK:

$$P_{sc} \approx \frac{2t + 1}{n} \binom{n}{t + 1} Q^{t+1}\left(\sqrt{2R_c E_b/N_0}\right) \tag{4.10}$$

To demonstrate the influence of different coding parameters, eqs. (4.6) and (4.9) have been used in Figure 4.5a–c. The results have been plotted as error probability versus signal to noise ratio in dB. The dashed and dash-dotted lines in the diagrams represent eq. (4.6). Transmission without channel coding, where the error probability is given by eq. (4.9), is used as a reference and represented by the solid line. The approximation in eq. (4.8) is represented by crosses and circles. As previously mentioned, it can be seen that the approximation is good for small values of P_s.

It is clear that the error-correcting properties improve when the block length increases provided that the code rate is constant. This is a general result which makes longer block length preferable. The errors can be differently distributed over the block and a long block gives greater possibility to make corrections when the errors appear in bursts, which now and then occur in a random sequence. The error probability decreases exponentially with increased codeword length. The problems with long blocks are more complex decoding and large delay in the system, because the entire codeword has to be acquired before decoding.

It is also clear that a single-error correcting code gives better properties when the number of information symbols (uncoded symbols) per unit time is kept unchanged. The reason for this is the large punishment for increased bandwidth.

From Figure 4.5c it is seen that the error-correcting properties are improved when the block length increases, and this is further improved by the increasing code rate. A code with small error correcting capability and high code rate is of course favoured when the channel is good anyhow; the few errors that occur can be corrected. If, on the other hand, the error probability of the transmission channel is increased the necessity of the code to be able to correct several errors per block is increased.

As can be seen from the figures above, the gain by introducing error-correcting coding can correspond to several dB in signal to noise ratio at a given error rate, or correspond to a reduction of the error rate by several orders of magnitude at a given signal to noise ratio. This occurs, however, at the cost of increased signal bandwidth and complexity of the system. The choice of code parameters is not obvious and depends on several factors. Most digital transmission or storage systems contain some kind of error correcting.

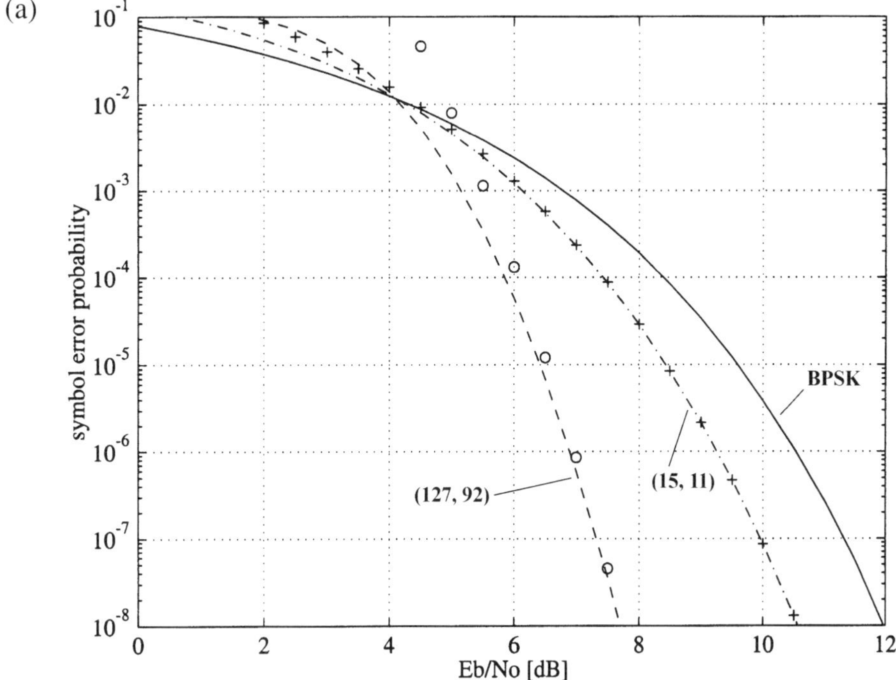

Figure 4.5 (a) Influence of block length for constant code rate. The dashed line represents a five-error-correcting (127,92) code and the dash-dotted line represents a single-error-correcting (15,11) code. The solid line represents uncoded BPSK. Rings and cross are defined in the text. (b) Influence of code rate for constant block length. The dashed line represents a single-error correcting (15,11) code and the dash-dotted line represents a three error-correcting (15,5) code. (c) Influence of varying both code rate and block length with the number of correctable symbols per block kept constant. The dashed line represents a single error-correcting (31,26) code and the dash-dotted line represents a single error-correcting (7,4) code.

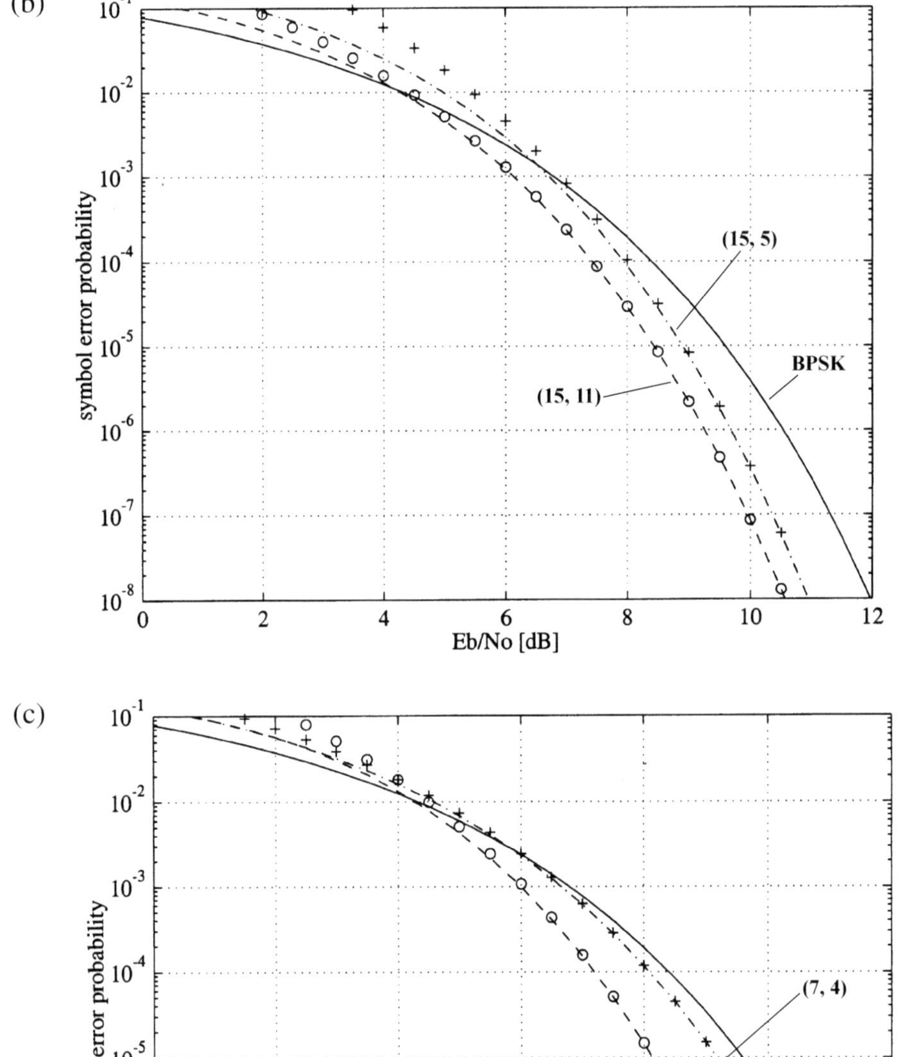

Figure 4.5 (*continued*)

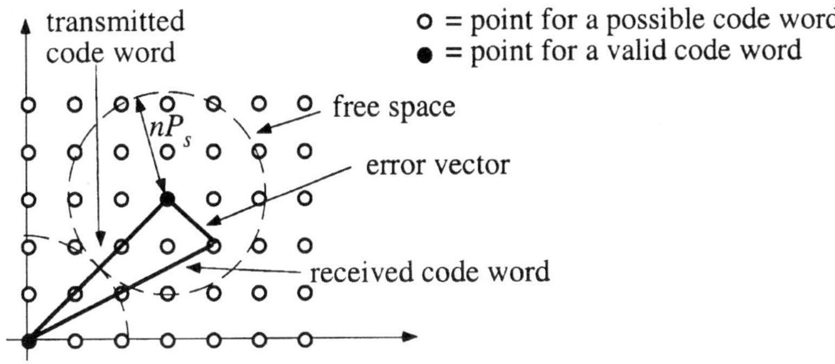

Figure 4.6   Section of a two-dimensional code space.

The channel coding theorem

According to the channel coding theorem there is a code such that the error probability in the decoded message is arbitrarily small when the code word length $n \to \infty$. How this is possible is shown in the following discussion (Figure 4.6).

As has been mentioned earlier, the channel noise introduces errors in the symbols that are transmitted over the channel. A codeword on the channel input can then be mapped on several different codewords on the channel output. To guarantee that the erroneous word is still decoded in a way that it will correspond to the right one, it has to be a protection distance between the valid codewords. Therefore, one cannot utilise all of the 2^n different codewords possible with a codeword length of n bits. Every possible codeword can be seen as a point in an n dimensional space, as shown in Figure 4.4a,b.

When the expectation of a stochastic variable is estimated by means of an infinite number of samples the estimated mean value $\hat{\mu}$ will be arbitrarily close to the true value, μ. This means that for long codewords the number of symbol errors will approach exactly nP_s. Therefore we must show that it is possible to find a code where $t \geqslant nP_s$. If such a code can be found it is guaranteed that the message can be decoded without errors.

The expectation of the number of symbol errors in a frame of n symbols is nP_s. When $n \to \infty$ the expected number of errors also approaches ∞. This means that the protection distance must also go to infinity. Then the question is whether the error correction properties approaches infinity faster than the number of symbol errors. The question is answered by first deciding the protection distance that must surround each codeword, to be able to correct up to nP_s errors per codeword. After that, the density in the code space is studied when each codeword is surrounded by the necessary protection distance.

Given a codeword of length n there are n words differing in exactly one symbol. In the same way there are $n(n-1)/2$ differing in two symbols. Generally, all words differing in i positions in a block length n are given by the binomial coefficients. The number of possible binary words within a Hamming distance of nP_s from a given codeword is therefore

$$\xi = \sum_{i=0}^{nP_s} \binom{n}{i} \tag{4.11}$$

To be able to decode correctly all codewords must be surrounded by a free space with the volume ξ. Within this area no other codeword must exist since the decoding in that case should be ambiguous when an error on $\leq nP_s$ bits occurs. Since $k = nR_c$ there are 2^{nR_c} valid codewords. The entire volume of the free space for a codeword is $\xi \times 2^{nR_c}$. The fraction, χ_n, of the entire volume, 2^n, taken by the total amount of codewords, is

$$\chi_n = \frac{2^{nR_c} \sum_{i=0}^{nP_s} \binom{n}{i}}{2^n} = 2^{n(R_c-1)} \sum_{i=0}^{nP_s} \binom{n}{i} \tag{4.12}$$

There is no closed expression for the sum in eq. (4.12). Instead an expression giving an upper limit for the free space must be computed. Since $A! \leq A^A$ it is valid that

$$\frac{(A + B)!}{A! \cdot B!} \leq \frac{(A + B)^{A+B}}{A^A \cdot B^B} \tag{4.13}$$

cf. the definition of the binomial coefficients. Provided that $P_s \leq 1/2$ it is true that the last term in the sum in eq. (4.12) is largest. Together with eq. (4.13) an upper limit is then obtained:

$$\xi = \sum_{i=0}^{nP_s} \binom{n}{i} \leq (nP_s + 1)\binom{n}{nP_s}$$

$$\leq \frac{(nP_s + 1)n^n}{(nP_s)^{nP_s}(n - nP_s)^{n-nP_s}} = \frac{(nP_s + 1)}{P_s^{nP_s}(1 - P_s)^{n-nP_s}}$$

$$= (nP_s + 1)2^{-n(P_s \log_2 P_s + (1-P_s)\log_2(1-P_s))} = (nP_s + 1)2^{-n(C-1)} \tag{4.14}$$

where C is the channel capacity for a BSC and is rendered by using eq. (3.35). Thus, around every codeword a free space that is at most the size given by eq. (4.14) is needed. An upper limit for the fraction of the entire volume taken by the free space of the codewords, then becomes

$$\chi_n \leq \frac{2^{nR_c}(nP_s + 1)2^{-n(C-1)}}{2^n} = (nP_s + 1)2^{n(R_c-C)} \tag{4.15}$$

Provided that $R_c < C$ the exponent becomes negative and decreases faster than nP_s increases. When the codeword length approaches infinity we therefore achieve

$$\lim_{n \to \infty} \chi_n = \chi_\infty = 0 \tag{4.16}$$

i.e., the entire volume of all the free space for the codewords takes a fraction of the total volume that goes to zero. It will therefore be very easy to find a code with a low enough error probability, if only n is large enough, even if one chooses the codewords randomly. When $n \to \infty$ the probability approaches one that such a code will fulfil a low enough error probability.

Note that if $R_c > C$ the upper limit of χ_∞ approaches infinity. It is thus a possibility that the density will be infinite. It can be shown that this is also the case. Then there will not be any possibility to find a code where non-ambiguous decoding can be done, even under noiseless conditions.

The conclusion is: given the channel error probability $P_s < 1/2$ it is possible, even with a random code, to make the error probability in the decoded symbols arbitrarily small if the codewords are made long enough. This is important partly as an illustration of the results in Figure 4.5a but also as a proof of Shannon's statement that the channel capacity, eq. (3.47), can be obtained with arbitrarily small error probability. The derivation has been done here for a binary channel but there is nothing to prevent an analogue channel being included in the binary channel, which is the usual case in digital communication. The conclusion is therefore valid in a typical digital communication system. It is also clear what price has to be paid for a small error probability in the communication system, i.e. large complexity and long delay.

A conclusion that seems to be paradoxical is that the S/N ratio limits the transmission capacity but does not put any limit on how small the error probability over the channel can be.

Example 4.1

Calculate the probability that a codeword has no errors, detectable errors and non-detectable errors, respectively, when a simple parity-check code having the codeword length $n = 4$ is used and the error probability per symbol is $P_s = 0.1$

Solution   The probability of a certain number of errors in a codeword has a binomial distribution, cf. eq. (4.5). The probability of zero errors then becomes

$$P(\text{no errors}) = (1 - 0.1)^4 = 0.6561 \tag{4.17}$$

A simple parity-check code can detect an odd number of errors, in this case a single or triple error. The probabilities for the respective types of error are

$$P \text{ (one error)} = (1 - 0.1)^3 \times 0.1 \times 4 = 0.2916 \tag{4.18}$$

where $(1 - 0.1)^3$ is the probability that the codeword contains three correct symbols and 0.1 is the probability of containing one error, which can be done in four different combinations with an error in the first, second, third or fourth position. The probability for three errors is

$$P(\text{three errors}) = 0.1^3 \times 0.9 \times 4 = 0.0036 \tag{4.19}$$

The undetected errors are then to be found in the rest of the sample space after 'zero errors', 'one error' and 'three errors' have been excluded. The probability of this is

$$P(\text{error}) = 1 - P(\text{no error}) - P(\text{one error}) - P(\text{three errors}) = 0.0487 \tag{4.20}$$

Example 4.2

An FEC system with BPSK modulation is contaminated by white Gaussian noise and has to achieve a bit error rate of $P_{be} \leq 10^{-4}$ with the lowest possible output power. The use of one of the three block codes shown in the table below is to be considered:

n	k	d_{min}
31	26	3
31	21	5
31	16	7

Decide which code should be used and calculate the reduction of the transmitter power in dB as compared to uncoded transmission.

Solution When using binary symbols, the symbol-error probability is the same as the bit-error probability. The error probability for uncoded transmission is given by eq. (4.9) and for coded transmission it is given by eq. (4.10).

(1) The uncoded case requires a signal to noise ratio of

$$P_s = Q\left(\sqrt{2E_b/N_0}\right) \leq 10^{-4} \Rightarrow \sqrt{2E_b/N_0} = 3.719 \Rightarrow E_b/N_0 = 6.9$$

$$\Rightarrow 8.4 \text{ (dB)} \tag{4.21}$$

(2) Encoded by the (31,26) code, where $d_{min} = 3 \Rightarrow t = 1$ which is given by eq. (4.4). The error probability becomes

$$P_{sc} \approx \frac{3}{31}\binom{31}{2}Q^2\left(\sqrt{2\frac{26}{31}E_b/N_0}\right) \leq 10^{-4} \qquad (4.22)$$

and thereby the required signal to noise ratio

$$Q(\cdot) \leq \sqrt{\frac{31}{3\times 465}10^{-4}} \approx 0.0015$$

$$\Rightarrow \sqrt{2\frac{26}{31}E_b/N_0} \approx 2.970 \Rightarrow E_b/N_0 = 5.257 \Rightarrow 7.2\ (dB) \qquad (4.23)$$

(3) Encoded by the (31,21) code, $d_{min} = 5 \Rightarrow t = 2$

$$P_{sc} \approx \frac{5}{31}\binom{31}{3}Q^3\left(\sqrt{2\frac{21}{31}E_b/N_0}\right) \leq 10^{-4} \qquad (4.24)$$

$$Q(\cdot) \leq \left(\frac{31}{5\times 4495}10^{-4}\right)^{1/3} \approx 5.167\times 10^{-3} \Rightarrow E_b/N_0 = 4.854 \Rightarrow 6.9\ (dB)$$

$$(4.25)$$

(4) Encoded by the (31,16) code, $d_{min} = 7 \Rightarrow t = 3$

$$P_{sc} \approx \frac{7}{31}\binom{31}{4}Q^4\left(\sqrt{2\frac{16}{31}\frac{E_b}{N_0}}\right) \leq 10^{-4} \Rightarrow E_b/N_0 = 5.098 \Rightarrow 7.1\ (dB)$$

$$(4.26)$$

By studying the four cases above it can be concluded that the (31,21) code should be used, as it saves 1.5 dB in output power of the transmitter as compared to uncoded transmission.

4.2 Automatic repeat request

4.2.1 FUNCTION OF ARQ SYSTEMS

Automatic repeat request (ARQ), is a system where the number of errors in the received message is reduced by detecting the errors and requesting repeated transmission of the incorrectly received blocks. ARQ requires a two-way transmission system, which means that a return channel exists where a correctly received codeword can be acknowledged; i.e. an ACK (affirmative (positive) acknowledgement) is transmitted from the receiver to the transmitter. A return channel can be obtained by using one single channel where the direction of the message flow can be reversed. This is called half duplex, and transmission can

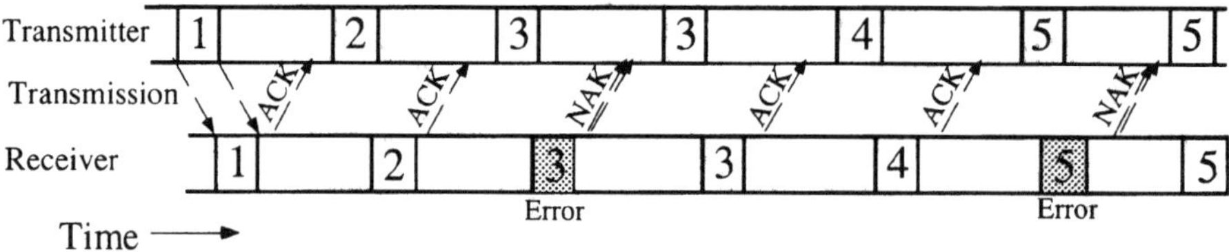

Figure 4.7 Stop and wait ARQ. The transmitter awaits an acknowledgement from the receiver regarding a correct or an incorrect codeword and then sends the next or repeats the previous codeword, respectively. Requires only a half duplex channel.

only take place in one direction at a time. Another way which uses two separate channels is called full duplex and allows simultaneous transmission both ways.

There are three main ARQ systems. The first is called stop and wait (SW), in which the transmitter waits for an ACK from the receiver before the next codeword is transmitted. When an incorrect codeword has been received the receiver responds with a negative acknowledgement, NAK (Figure 4.7).

In the second method the codewords are sent in a continuous flow and when an NAK is returned from the receiver, the transmitter backs and retransmits the codeword sequence starting from the erroneous one. Depending on the delays in the system, N codewords will be repeated. The method is therefore called Go-Back-N (GBN). It requires a full duplex channel but allows higher transmission rates than the first method (Figure 4.8).

In the third and the most rapid method only the incorrect codeword is repeated, after which the transmitter returns to the position in the message stream where it was interrupted and continues from there. This method is called selective repeat (SR) (Figure 4.9).

Detecting errors requires fewer redundancy bits than error-correcting for a certain error rate, and ARQ usually allows a higher transmission rate than FEC. The transmission rate of ARQ can, however, decrease to zero if the channel becomes too poor. For an ARQ system a lowest error rate can be guaranteed, but

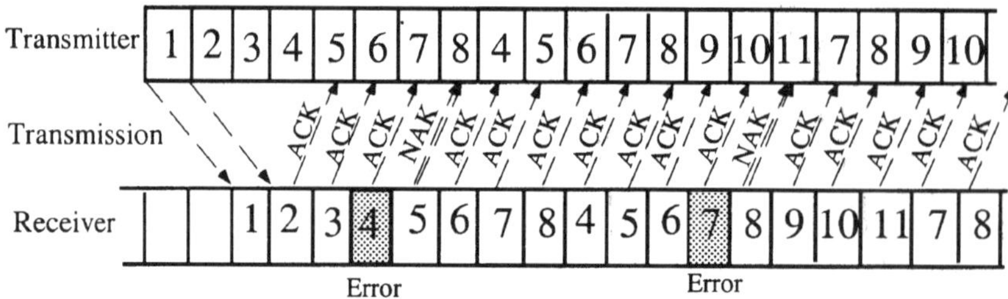

Figure 4.8 Continuous ARQ with Go-Back-N. The complete sequence is repeated after an unaccepted codeword. In this example four symbols are repeated.

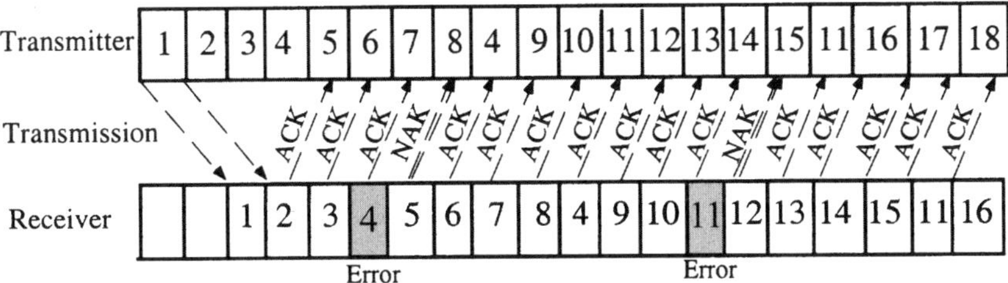

Figure 4.9 Continuous ARQ with selective repeat. Only the incorrect codeword is repeated. Requires full duplex channel.

information can be lost unless an infinite storage buffer is available on the transmitting side. When a transmission error occurs, the message flow must be stored during the time period transmission is repeated.

4.2.2 PERFORMANCE OF ARQ SYSTEMS

An important property of an ARQ system is how many codewords on average, $\bar{m}$, need to be transmitted per accepted word. Given a certain maximum transmission capacity on the channel, the possible transmission rate can then be calculated. For the calculation of $\bar{m}$ the probability of receiving an incorrect codeword must be known, and this can be obtained as follows. It is assumed that ACK and NAK signals are all received without errors and that all errors can be detected. The probability of a sequence of n bits to contain zero errors is equal to $(1 - P_s)^n$, where P_s is the symbol error probability. The probability of having at least one error in this sequence is

$$1 - (1 - P_s)^n \approx nP_s \tag{4.27}$$

To simplify the following calculations an approximation has been made in the left hand side of eq. (4.27). The approximation is valid when the symbol error probability is low and has been made using the series expansion

$$(1 - P_s)^n = 1 - nP_s + \frac{n(n-1)}{2}P_s^2 + \cdots \tag{4.28}$$

Using the approximation in eq. (4.28) the probability of having at least one error in a block becomes $P_{bl} \approx nP_s$. The probability of having zero errors in a received block, i.e. the block is transmitted only once, is

$$P(1) = (1 - P_{bl}) \tag{4.29}$$

The probability of a block being transmitted twice is equal to the probability of an incorrect block followed by a correct block, i.e.

$$P(2) = (1 - P_{bl})P_{bl} \tag{4.30}$$

The probability of $m + 1$ transmissions, i.e. m incorrect followed by a correct block or codeword is

$$P(m + 1) = (1 - P_{bl})P_{bl}^m \tag{4.31}$$

The average number of transmitted words per correctly received or accepted word, $\bar{m}$ becomes

$$\bar{m} = \sum_{m=1}^{\infty} m\, P(m) = (1 - P_{bl}) + 2(1 - P_{bl})P_{bl} + 3(1 - P_{bl})P_{bl}^2 + \cdots$$

$$= (1 - P_{bl})\left[1 + 2P_{bl} + 3P_{bl}^2 + \cdots\right] = \frac{1 - P_{bl}}{P_{bl}}\left[P_{bl} + 2P_{bl}^2 + 3P_{bl}^3 + \cdots\right]$$

$$= \frac{1}{1 - P_{bl}}$$

$$\tag{4.32}$$

For the last equality the expression for the sum of an arithmetic-geometric series has been used.

To transmit k information-carrying symbols, an average of $n\bar{m}$ encoded transmitted symbols is required. The code rate then becomes

$$R_c = \frac{k}{n\bar{m}} = \frac{k}{n}(1 - P_{bl}) \tag{4.33}$$

The above holds provided there is a continuous flow of blocks and only the incorrect block is repeated. For the other ARQ systems the code rate is lower.

Example 4.3

In an ARQ system with GBN, the message is disturbed by white Gaussian noise. The signal to noise ratio $\gamma_c = 6$ dB and the maximum code rate $r = 500$ kbaud. The distance between transmitter and receiver is 45 km. The signal is assumed to travel with the speed of light in a vacuum. Calculate the bit error rate P_{be}, the lowest number of repeated words N and the maximum value of the obtained transmission capacity r_b when a (15,11) block code having $d_{min} = 3$ is used for error detection.

Solution   First, the least number of blocks that have to be repeated, is computed (Figure 4.10). The distance of 45 km gives a delay of

$$\frac{45 \times 10^3}{2.997925 \times 10^8} \approx 150 \times 10^{-6} \text{ s} \tag{4.34}$$

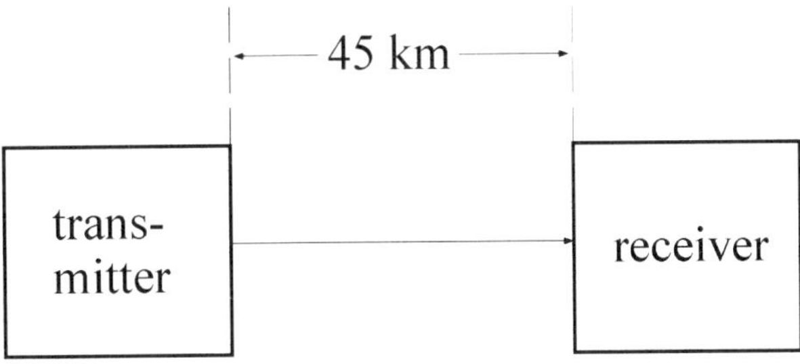

Figure 4.10 Distance between transmitter and receiver.

Since it takes 150 μs for a word to travel one way, it takes 300 μs from the time a word has been emitted until the response arrives. During this time N_d blocks have been sent out ($\lceil x \rceil$ = nearest integer $\geq x$)

$$N_d = \left\lceil \frac{300 \times 10^{-6}}{\frac{1}{r}n} \right\rceil = \frac{300 \times 10^{-6} \times 500 \times 10^3}{15} = 10 \tag{4.35}$$

where r is the symbol rate and n is the number of symbols per word. The least number of repeated words is therefore 10. The average number of transmitted words per accepted word for ARQ with GBN is derived in the same way as eq. (4.32), with the difference that N codewords are sent out for each error. This gives

$$\overline{m} = \underbrace{1 \cdot (1 - P_{bl})}_{\text{zero errors}} + \underbrace{(1 + N)(1 - P_{bl})P_{bl}}_{\text{one error}} + \underbrace{(1 + 2N)(1 - P_{bl})P_{bl}^2}_{\text{two errors}} + \cdots \tag{4.36}$$

only one block   one correct block N blocks are repeated on each erroneous block

Rewriting and utilising the formula for the sum of an arithmetic-geometric series gives

$$\overline{m} = 1 + \frac{NP_{bl}}{1 - P_{bl}} \tag{4.37}$$

As can be seen from eq. (4.35), 10 words or blocks can be sent out during 300 μs. We also have to consider that the incorrect word must be fully received, which gives the smallest value of the number of repeated symbols $N = N_d + 1 = 11$. The maximum code rate becomes

$$R_c = \frac{\text{transferred information bits}}{\text{total number of transferred bits}} \tag{4.38}$$

which gives

$$R_c = \frac{k}{n\bar{m}} = \frac{k(1 - P_{bl})}{n(1 + (N - 1)P_{bl})} \approx \frac{k(1 - nP_s)}{n(1 + (N - 1)nP_s)} \tag{4.39}$$

First, the signal to noise ratio, γ_c, has to be expressed in a linear measure instead of the given 6 dB:

$$\gamma_c = \frac{E_b}{N_0} = 3.98 \tag{4.40}$$

In eq. (4.9) the probability of an error in a symbol is given as

$$P_s = Q\left(\sqrt{2E_b/N_0}\right) \approx 2.39 \times 10^{-3} \tag{4.41}$$

The maximum transmission rate is finally obtained as

$$r_b = r\,R_c \approx r\frac{11\left(1 - 15 \times 2.39 \times 10^{-3}\right)}{15(1 + (11 - 1)15 \times 2.39 \times 10^{-3})} \approx r \times 0.520 \approx 260 \text{ kbaud} \tag{4.42}$$

The bit error rate for an ARQ system is equal to the probability that a transmission error will pass undetected, which is determined by the capability of the code to detect errors. The method to compute this is the same as the one leading to eq. (4.6). The difference is that the received word is not modified by the decoder. The share of errors in the information symbols is therefore i/n instead of $(i + t)/n$. The number of symbol errors is therefore not increased. A (15,11) block code with $d_{min} = 3$ detects $e = d_{min} - 1 = 2$ errors. The probability of a bit error in a received message is then

$$P_{be} \approx \frac{e + 1}{n}\binom{n}{e + 1}P_s^{e+1} = \binom{n - 1}{e}P_s^{e+1} = \binom{14}{2}\left(2.39 \times 10^{-3}\right)^3$$

$$= 1.24 \times 10^{-6} \tag{4.43}$$

4.3 Block codes

4.3.1 GALOIS FIELDS

Within coding theory, many mathematical operations are performed in finite number systems with certain properties, i.e. algebraic systems, which in this case is called fields. These concepts can be found in basic algebra. Most mathematical operations described in the remaining part of Chapter 4 are performed in a finite number system. As can be seen below $2 + 3$ is not always equal to 5 if the addition is performed in a finite field. For the coding theory the concept of Galois fields is of great importance, and thus a brief introduction to this concept is justified. Galois fields are denoted $GF(q)$, where q is an integer larger than one. To guarantee the existence of a Galois field, q must be a prime number or an integer power of a prime number $q = p^n$. When q is a prime number the mathematical operations become modulo q operations. A Galois field is a finite number system in the set F that fulfils the following conditions, for the elements a, b and c.

Addition:

- closed $a, b \in F \Rightarrow a + b \in F$
- associative $(a + b) + c = a + (b + c)$
- commutative $a + b = b + a$
- zero element $a + 0 = a$
- every element must have a negative element, $a + (-a) = 0$

Multiplication:

- closed $a, b \in F \Rightarrow a{\cdot}b \in F$
- associative $(ab)c = a(bc)$
- commutative $ab = ba$
- distributive $a(b + c) = ab + ac$
- every element, except 0, must have an inverse element, $b{\cdot}b^{-1} = 1$

A simple but often used field is GF(2) or binary arithmetic, where $F = \{0,1\}$ (Figure 4.11). Note that in binary arithmetic a number is both a negative and inverse element of itself, which means that $1 + 1$ and $1 - 1$ are equivalent and that $1{\cdot}1 = 1$. Plus or minus signs are therefore irrelevant in binary arithmetic, GF(2). In the following the symbol $\otimes$ is usually used for addition modulo 2.

$+$	0	1		$\cdot$	0	1
0	0	1		0	0	0
1	1	0		1	0	1

Figure 4.11 Addition and multiplication rules for GF(2), binary arithmetic.

Example 4.4

Summation and multiplication rules for GF(5) to the left and for $GF(2^2) = GF(4)$ to the right. Note that as $q = 5$ is a prime number, modulo 5 arithmetic is obtained.

GF(5) GF(4)

addition:

+	0	1	2	3	4
0	0	1	2	3	4
1	1	2	3	4	0
2	2	3	4	0	1
3	3	4	0	1	2
4	4	0	1	2	3

+	0	1	2	3
0	0	1	2	3
1	1	0	3	2
2	2	3	0	1
3	3	2	1	0

multiplication:

·	0	1	2	3	4
0	0	0	0	0	0
1	0	1	2	3	4
2	0	2	4	1	3
3	0	3	1	4	2
4	0	4	3	2	1

·	0	1	2	3
0	0	0	0	0
1	0	1	2	3
2	0	2	3	1
3	0	3	1	2

The methods described below generally hold for codes within an arbitrary GF(q). An important special case is GF(2), especially when implementations are considered. When we limit ourselves to GF(2), this is explicitly stated.

4.3.2 LINEAR BLOCK CODES

Linear block codes, or group codes, are an important class of codes for error control. In a block code the information symbols are divided into blocks, or frames, before the coding takes place. The coding operation is then performed separately on each individual block. If α_1 and α_2 are constants and if $\mathbf{X}_i$ and $\mathbf{X}_j$ are codewords, for a linear code

$$\mathbf{X}_p = \alpha_1 \mathbf{X}_i + \alpha_2 \mathbf{X}_j \tag{4.45}$$

must also be a codeword. Consequently the codeword $\mathbf{X}_0 = [0\,0\,0...0]$ must be included in a linear code since for the codeword $\mathbf{X}_i$ it is true that $\mathbf{X}_i = \mathbf{X}_i + \mathbf{X}_0$. Note that all operations are done in GF(q).

Coding

A vector formulation of the information or message symbols can be achieved if the symbols in each block are identified as the elements of a vector. For message i

the following holds:

$$\mathbf{M}_i = \begin{bmatrix} m_{i1} & m_{i2} & \cdots & m_{ik} \end{bmatrix} \tag{4.46}$$

As has been suggested earlier the codewords can be identified as a vector in an equivalent way

$$\mathbf{X}_i = \begin{bmatrix} x_{i1} & x_{i2} & \cdots & x_{in} \end{bmatrix} \tag{4.47}$$

The coding can then be achieved through the matrix multiplication

$$\mathbf{X}_i = \mathbf{M}_i\,\mathbf{G} \tag{4.48}$$

where $\mathbf{G}$ is a generator matrix, which is defined below. $\mathbf{G}$ constitutes a mapping of the information symbols onto a codeword. To simplify the implementation, a systematic code is desired, i.e. that the k information symbols are unchanged and followed by $n - k$ check symbols (Figure 4.12).

The generator matrix then obtains the following systematic form

$$\mathbf{G} = \begin{bmatrix} \mathbf{I}_k & \vdots & \mathbf{P} \end{bmatrix} = \begin{bmatrix} 1 & 0 & 0 & \cdots & 0 & P_{11} & P_{12} & \cdots & P_{1(n-k)} \\ 0 & 1 & 0 & \cdots & 0 & P_{21} & P_{22} & \cdots & P_{2(n-k)} \\ \vdots & & \ddots & & \vdots & \vdots & & & \vdots \\ 0 & 0 & 0 & \cdots & 1 & P_{k1} & P_{k2} & \cdots & P_{k(n-k)} \end{bmatrix} \tag{4.49}$$

A non-systematic code can be transformed into a systematic one by elementary row and column operations on the generator matrix. The implementation of the coding process can be performed by using shift registers. The information symbols are read into a shift register. The contents of the shift register are multiplied by the respective term in the generator matrix and are then added together to generate the respective check symbol. All mathematical operations are performed in GF(q). For a binary code the implementation thus becomes easy to do. Assume a code with four information bits, $k = 4$, and three check bits,

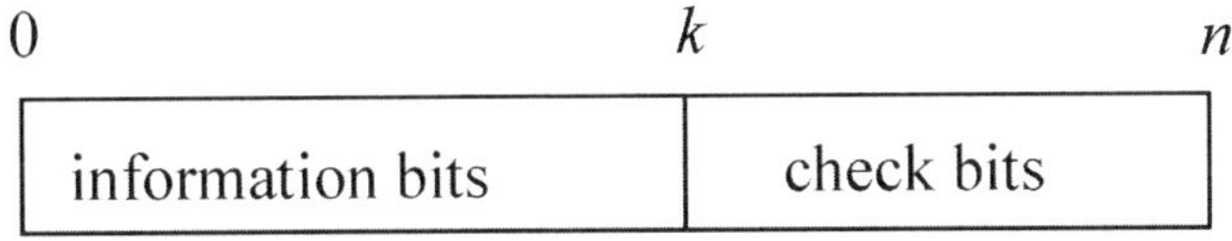

code word

Figure 4.12 Information symbols and check symbols in a codeword generated by a systematic code (for GF(2) symbol is equal to bit).

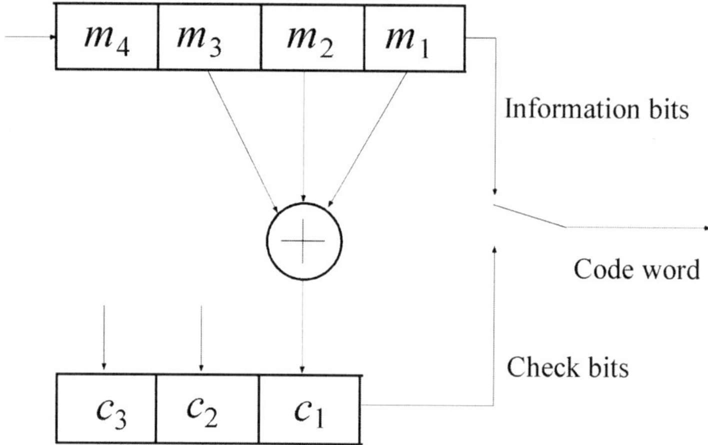

Figure 4.13   Shift register encoder for a systematic code.

which means $n = 7$. The codeword is obtained by the matrix multiplication **M·G**, which in this case becomes

$$
[m_1 \quad m_2 \quad m_3 \quad m_4]
\begin{bmatrix}
1 & 0 & 0 & 0 & 1 & 0 & 0 \\
0 & 1 & 0 & 0 & 1 & 1 & 1 \\
0 & 0 & 1 & 0 & 1 & 1 & 0 \\
0 & 0 & 0 & 1 & 0 & 1 & 1
\end{bmatrix}
\tag{4.50}
$$

Check bit 1, for example, is the product of the information bits and the fifth column of the generator matrix, i.e.

$$
c_1 = 1 \cdot m_1 \oplus 1 \cdot m_2 \oplus 1 \cdot m_3 \oplus 0 \cdot m_4 \quad (\text{mod } 2)
\tag{4.51}
$$

The operations are illustrated in Figure 4.13.

When a code is to be synthesised, or created, the question arises of how to find the matrix **P**, which gives a code that has good error-correcting properties and is easy to decode. To create a code with good correcting properties, the possible codewords have to be chosen in a suitable way. Suppose that the elements of the codeword vector indicate the parameters in an n dimensional space. In the binary case the parameters can only assume the values 0 or 1. The possible combinations of ones and zeros in the codeword vector will be in the corners of a hyper cube. For codewords with 3 bits an ordinary cube is obtained, as shown in Figure 4.4. The codewords giving the best error-correcting properties are those furthest apart. For a (3,1) code the signal points as in Figure 4.4a are chosen and for a (3,2) code as in Figure 4.4b. If all codeword points have their nearest neighbour at the same distance and the whole hyper cube is filled, we have obtained a code with the best possible error-correcting properties. Any other choice would lead to a reduced distance between two codewords. Such a code is therefore called a perfect code.

Decoding

We have now discussed how coding is carried out mathematically. To find a method for the decoding of the codewords it is assumed that a matrix exists so that

$$\mathbf{X}_i \mathbf{H}^T = \mathbf{0} \tag{4.52}$$

for each codeword $\mathbf{X}_i$, and where the right hand side is a null vector. The codewords are therefore in the null space of $\mathbf{H}$. Since this holds for every codeword $\mathbf{X}_i$, it must also hold for every message $\mathbf{M}_i$. It holds, for instance, for $\mathbf{M} = [1\ 0\ 0\ \dots\ 0]$ which gives a codeword which is a row in $\mathbf{G}$, and it also holds for $\mathbf{M} = [0\ 1\ 0\ \dots\ 0]$, which gives the next row in $\mathbf{G}$, etc. We thus get the following expression:

$$\mathbf{0} = \mathbf{G}\mathbf{H}^T = \begin{bmatrix} \mathbf{I}_k & \mathbf{P} \end{bmatrix} \begin{bmatrix} -\mathbf{P} \\ \hline \mathbf{I}_{n-k}^T \end{bmatrix} = -\mathbf{P} + \mathbf{P} \tag{4.53}$$

where $\mathbf{0}$ is a null matrix and $\mathbf{H}$ is a $(n-k \times n)$ matrix, which is called the parity-check matrix. One can recognise that for systematic codes the parity check matrix becomes

$$\mathbf{H} = \begin{bmatrix} -\mathbf{P}^T & \vdots & \mathbf{I}_{n-k} \end{bmatrix} \tag{4.54}$$

In other words, $\mathbf{H}$ consists of one part which is the identity matrix and another part which is the transpose of $-\mathbf{P}$. In binary arithmetic the minus sign is irrelevant as $1 - 1 = 1 + 1$ in this case. Assume that the received codeword can be written as the sum of the correct, emitted, codeword and an error pattern, which has been added during the transmission, i.e.

$$\mathbf{Y} = \mathbf{X} + \mathbf{E} \tag{4.55}$$

where $\mathbf{X}$ is the emitted codeword and $\mathbf{E}$ the error pattern which is the $\mathbf{0}$-vector if the received word contains no errors. The syndrome, $\mathbf{S}$, is defined as the product

$$\mathbf{S} = \mathbf{Y}\mathbf{H}^T = \underbrace{\mathbf{X}\mathbf{H}^T}_{=\mathbf{0}} + \mathbf{E}\mathbf{H}^T = \mathbf{E}\mathbf{H}^T \tag{4.56}$$

It is seen that the syndrome only depends on the error pattern and not on the codeword. A property which comes from the definition of $\mathbf{H}$ which has been done in eq. (4.52). Since $\mathbf{H}$ has been defined according to eq. (4.52), it can be seen by inserting eq. (4.55) that the syndrome only depends on the error pattern. The syndrome is a $1 \times (n - k)$ vector which is equal to the null vector if the received codeword is correct, otherwise $\mathbf{S} \neq \mathbf{0}$. There are $2^{n-k} - 1$ possible syndrome patterns different from the zero vector, which means that there is no unambiguous mapping between syndrome and error pattern. A suitable strategy for solving the problem is to let a certain syndrome pattern correspond to the most probable error, which is the error pattern containing the smallest number of errors. For a (15,11) code, for example, the number of possible error patterns is $2^{15} - 1 = 32767$, but the number of syndrome patterns is only $2^{15-11} - 1 = 15$. When the error pattern has been found it can then be subtracted from the received word in order to restore the correct word as much as possible.

A table showing which of the error patterns corresponds to each syndrome respectively, is called a standard array. The error pattern with the smallest Hamming weight, i.e. the most probable, for each syndrome is called the coset leader.

The implementation of the decoder (cf. Figure 4.14), can be done by reading the received codeword into a shift register. Syndrome calculation is then carried out through multiplication by the parity check matrix (compare the coding in Figure 4.13). The result constitutes the address to a memory, ROM, where the most probable error patterns have been stored. The current error pattern is read into a shift register, after which the subtraction from the received word can be done. For non-binary codes an error-magnitude calculation must also be

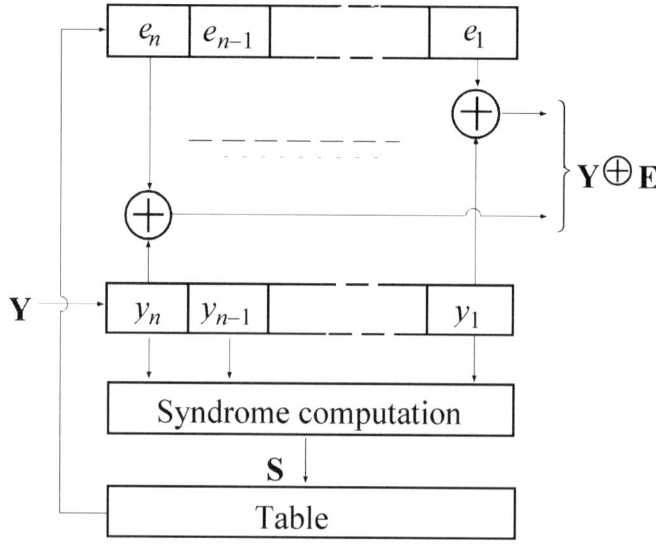

Figure 4.14   Syndrome decoder with table look up.

performed. This is not necessary for binary codes, however, since all errors in this case have the magnitude one.

Example 4.5

A repetition code is a code which repeats the message symbol a certain number of times. Determine the parity matrix $\mathbf{P}$ for a systematic (3,1) repetition code. Make a table of all possible received vectors $\mathbf{Y}$ and the corresponding syndromes. Determine the corresponding most probable maximum-likelihood error pattern, and corrected vectors $\mathbf{Y} + \hat{\mathbf{E}}$.

Solution For a (3,1) repetition code the message is repeated three times, i.e. the codeword becomes

$$\begin{bmatrix} m_1 & m_1 & m_1 \end{bmatrix} = m_1 \mathbf{G} = m_1 \begin{bmatrix} 1 & 1 & 1 \end{bmatrix} \tag{4.57}$$

Since $\mathbf{G} = \begin{bmatrix} \mathbf{I} & \mathbf{P} \end{bmatrix}$ the parity matrix is $\mathbf{P} = \begin{bmatrix} 1 & 1 \end{bmatrix}$ (in this case a vector). The parity check matrix becomes

$$\mathbf{H} = \begin{bmatrix} \mathbf{P}^T & \vdots & \mathbf{I}_2 \end{bmatrix} = \begin{bmatrix} 1 & 1 & 0 \\ 1 & 0 & 1 \end{bmatrix} \tag{4.58}$$

The syndrome is calculated by using the relation $\mathbf{S} = \mathbf{Y}\mathbf{H}^T$. The syndrome of the received vector (1 0 1), for instance, becomes

$$(101) \begin{bmatrix} 1 & 1 \\ 1 & 0 \\ 0 & 1 \end{bmatrix} = (1 \oplus 0 \oplus 0, \ 1 \oplus 0 \oplus 1) = (1, 0) \tag{4.59}$$

All possible received vectors then give the following table over syndrome and error pattern.

$\mathbf{Y}$	$\mathbf{S}$	$\hat{\mathbf{E}}$	$\mathbf{Y} + \hat{\mathbf{E}}$
000	00	000	000
001	01	001	000
010	10	010	000
011	11	100	111
100	11	100	000
101	10	010	111
110	01	001	111
111	00	000	111

The different error vectors $\hat{\mathbf{E}}$ in the table give the most probable errors, i.e. the error patterns containing the smallest number of errors (coset leaders). The listed error patterns are therefore all combinations with only one error. Note that the table above is not sufficient for making a standard array since, in this case, it should contain 2^3 different error patterns. The table connecting the syndrome to the most probable error pattern then becomes

S	$\hat{\mathbf{E}}$
01	001
10	010
11	100

Hence, $\hat{\mathbf{E}}$ is the error pattern to be added to the received word.

Hamming codes

Hamming codes are perfect single-error-correcting and double-error-detecting codes with $d_{min} = 3$ and

$$(n, k) = (2^m - 1, 2^m - 1 - m) \qquad (4.60)$$

for integers $m \geq 2$. These codes can be generated by a simple algorithm. The parity-check matrix, for example for a (7,4) code, is formed in the following way. Put the binary numbers from 1 to 7 in a table:

$$
\begin{array}{ccccc}
1 & \rightarrow & 0 & 0 & 1 \\
2 & \rightarrow & 0 & 1 & 0 \\
3 & \rightarrow & 0 & 1 & 1 \\
4 & \rightarrow & 1 & 0 & 0 \\
5 & \rightarrow & 1 & 0 & 1 \\
6 & \rightarrow & 1 & 1 & 0 \\
7 & \rightarrow & 1 & 1 & 1
\end{array}
$$

First pick those binary numbers that can form the columns in an identity matrix, then pick the others in arbitrary order. In the parity-check matrix below, the columns from left to right have been formed by the numbers 7, 6, 5, 3, 1, 2 and 4.

$$\mathbf{H} = \begin{bmatrix} 1 & 0 & 1 & 1 & 1 & 0 & 0 \\ 1 & 1 & 0 & 1 & 0 & 1 & 0 \\ 1 & 1 & 1 & 0 & 0 & 0 & 1 \end{bmatrix} \tag{4.61}$$

Then the generator matrix is formed according to eq. (4.53).

$$\mathbf{G} = \begin{bmatrix} 1 & 0 & 0 & 0 & 1 & 1 & 1 \\ 0 & 1 & 0 & 0 & 0 & 1 & 1 \\ 0 & 0 & 1 & 0 & 1 & 0 & 1 \\ 0 & 0 & 0 & 1 & 1 & 1 & 0 \end{bmatrix} \tag{4.62}$$

Example 4.6

Arrange the parity matrix $\mathbf{P}^T$ of a (15,11) Hamming code in such a way that the decimal values for each column is increasing from left to right. Make a table with the error patterns, and give check bit equations.

Solution The Hamming code is obtained by generating the **H**-matrix. Write down the binary numbers from 1 to 15:

1	0001
2	0010
3	0011
4	0100
5	0101
6	0110
7	0111
8	1000
9	1001
10	1010
11	1011
12	1100
13	1101
14	1110
15	1111

First pick out the identity matrix followed by the rest of the columns:

$$
\mathbf{H} = \begin{array}{cccccccccccccccc}
\ \ 3 & 5 & 6 & 7 & 9 & 10 & 11 & 12 & 13 & 14 & 15 & 1 & 2 & 4 & 8 \\
\end{array}
$$

$$
\mathbf{H} = \left[\begin{array}{ccccccccccc|cccc}
1 & 1 & 0 & 1 & 1 & 0 & 1 & 0 & 1 & 0 & 1 & 1 & 0 & 0 & 0 \\
1 & 0 & 1 & 1 & 0 & 1 & 1 & 0 & 0 & 1 & 1 & 0 & 1 & 0 & 0 \\
0 & 1 & 1 & 1 & 0 & 0 & 0 & 1 & 1 & 1 & 1 & 0 & 0 & 1 & 0 \\
0 & 0 & 0 & 0 & 1 & 1 & 1 & 1 & 1 & 1 & 1 & 0 & 0 & 0 & 1
\end{array}\right] \qquad (4.63)
$$

$$
\underbrace{}_{\mathbf{P}^{\mathrm{T}}} \quad \underbrace{}_{\mathbf{I}_4}
$$

In order to find the syndrome, all error patterns are multiplied by $\mathbf{H}^T$. The Hamming code is, however, single-error-correcting, which means that it is sufficient to investigate patterns with only one error, such as

$$
\mathbf{S} = [1000000000000000] \begin{bmatrix}
1 & 1 & 0 & 0 \\
1 & 0 & 1 & 0 \\
0 & 1 & 1 & 0 \\
1 & 1 & 1 & 0 \\
1 & 0 & 0 & 1 \\
0 & 1 & 0 & 1 \\
1 & 1 & 0 & 1 \\
0 & 0 & 1 & 1 \\
1 & 0 & 1 & 1 \\
0 & 1 & 1 & 1 \\
1 & 1 & 1 & 1 \\
1 & 0 & 0 & 0 \\
0 & 1 & 0 & 0 \\
0 & 0 & 1 & 0 \\
0 & 0 & 0 & 1
\end{bmatrix} = [1 \quad 1 \quad 0 \quad 0] \qquad (4.64)
$$

From eq. (4.64) it can be seen that the syndrome in this case is formed by the first row in the transposed $\mathbf{H}$ matrix. The other single-error patterns will successively be formed by the rest of the rows in $\mathbf{H}^T$. This gives the syndrome table:

S	error in bit
0000	no error
1100	1
1010	2
0110	3
1110	4
1001	5
0101	6
1101	7
0011	8
1011	9
0111	10
1111	11
1000	12
0100	13
0010	14
0001	15

Check bit equations can be obtained by recognising that the codeword of a systematic, linear code is

$$\mathbf{X} = [\mathbf{M} \ \vdots \ \mathbf{C}] = \mathbf{MG} \tag{4.65}$$

where $\mathbf{G} = [\mathbf{I} \ \vdots \ \mathbf{P}]$, $\mathbf{M}$ consists of message bits and $\mathbf{C}$ consists of check bits, which in the systematic case are the last $n - k$ bits of the codeword; here four bits. Hence, the check bits will be made up of the product of the message and the $\mathbf{P}$-matrix, which in this case becomes

$$c_1 = m_1 \oplus m_2 \oplus m_4 \oplus m_5 \oplus m_7 \oplus m_9 \oplus m_{11}$$
$$c_2 = m_1 \oplus m_3 \oplus m_4 \oplus m_6 \oplus m_7 \oplus m_{10} \oplus m_{11}$$
$$c_3 = m_2 \oplus m_3 \oplus m_4 \oplus m_8 \oplus m_9 \oplus m_{10} \oplus m_{11}$$
$$c_4 = m_5 \oplus m_6 \oplus m_7 \oplus m_8 \oplus m_9 \oplus m_{10} \oplus m_{11} \tag{4.66}$$

A shift register implementation is shown in Figure 4.15.

4.3.3 CYCLIC CODES

A code is cyclic if the cyclic shift, or rotation, of a codeword results in a new codeword. The left shift of, for example,

$$1 \quad 0 \quad 0 \quad 0 \quad 1 \quad 1 \quad 0$$

produces the following codeword:

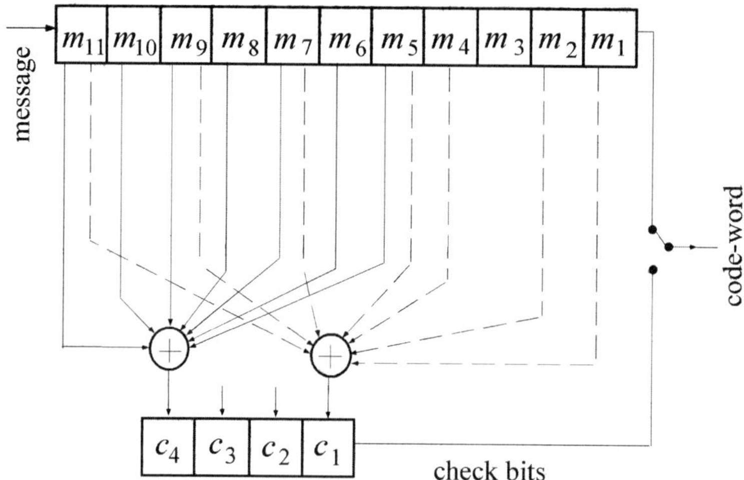

Figure 4.15   Shift register implementation of a code generator to a (15,11) Hamming code.

$$0 \quad 0 \quad 0 \quad 1 \quad 1 \quad 0 \quad 1$$

If each such shift of a codeword results in a new codeword the code is said to be cyclic. This gives the code a structure, which is suitable for the implementation. BCH codes and some Hamming codes are examples of cyclic codes. To study cyclic codes, a polynomial representation is used, rather than the matrix representation. For a word which in the previous section was represented by a vector, the vector elements are now the coefficients of the polynomial. The positions of the elements correspond to the exponents of the variable p. The left element represents the leading coefficient. A codeword $\mathbf{X} = [x_{n-1} \ldots x_0]$ is in the polynomial representation written as (the standard in cyclic coding is that the subscripts run in the reverse order, from $n-1$ to 0)

$$X(p) = x_{n-1}p^{n-1} + \cdots + x_1 p + x_0 \tag{4.67}$$

For example the vector

$$1 \quad 0 \quad 0 \quad 0 \quad 1 \quad 1 \quad 0$$

is in the polynomial representation written as

$$X(p) = p^6 + p^2 + p \tag{4.68}$$

p can be said to represent a delay operator. Compare z^{-1} from digital signal processing and q from control theory. Another common notation for p is D. By multiplication of the polynomials in GF(2), i.e. in a binary number field, the coefficients are calculated modulo 2; for instance

$$(x + 1)(x^3 + x + 1)(x^3 + x^2 + 1) = x^7 + 1$$

Multiplication by p leads to a left shift of a codeword. To obtain a cyclic shift, the coefficient in front of the highest power of x, must be shifted to the zero exponent term. In other words, $x_{n-1}p^n$ in the polynomial $pX(p)$ should be removed and x_{n-1} added, which is equivalent to adding $-(p^n - 1)x_{n-1}$. A cyclic shift can then be expressed in the following way by using the polynomial representation

$$X^{(1)}(p) = pX(p) - x_{n-1}(p^n - 1) = R_{p^n-1}[pX(p)] \tag{4.69}$$

In binary arithmetic the minus signs can be replaced by plus signs. The notation R_{p^n-1} represents the remainder after the division within a field, which means that the shift can also be written as

$$R_{p^n-1}[pX(p)] = \mathrm{rem}\left(\frac{pX(p)}{p^n - 1}\right) \tag{4.70}$$

A one step left shift, for example, becomes

$$X^{(1)}(p) = p(p^6 + p^2 + p) + p^7 + 1 = p^3 + p^2 + 1 \tag{4.71}$$

Generally a j-step cyclic left shift is

$$X^{(j)}(p) = R_{p^n-1}\left[p^j X(p)\right] \tag{4.72}$$

which for all j produce a new codeword when a codeword is rotated.

For every cyclic code there is a corresponding generator polynomial $G(p)$. For the cyclic property it is required that the generator polynomial is a factor in $p^n - 1$,

$$G(p) = p^{n-k} + g_{n-k-1} p^{n-k-1} + \cdots + g_1 p + 1 \tag{4.73}$$

To factorise $p^n - 1$ in a finite field is a non-trivial mathematical problem. If you also have the requirement that the resulting polynomial shall produce a good code, with large d_{min} and preferably perfect, the problem becomes very difficult and is the reason why tables are the common start point to find $G(p)$.

The codewords are formed through the multiplication

$$X(p) = Q(p)G(p) \tag{4.74}$$

Instead of $Q(p)$, the polynomial representing the message, $M(p)$, can be inserted in eq. (4.74). This leads, however, to a non-systematic code. If a systematic code is desired, further operations must be performed. These are described in the next section.

Systematic coding of cyclic codes

The information symbols should be the coefficients of the highest order terms of the polynomial. The codeword then looks like

$$X(p) = p^{n-k}M(p) + C(p) = Q(p)G(p) \tag{4.75}$$

where $C(p)$ is a polynomial which represents the check symbols. Dividing by the generator polynomial and rewriting gives

$$\frac{p^{n-k}M(p)}{G(p)} = Q(p) - \frac{C(p)}{G(p)} \tag{4.76}$$

$Q(p)$ can be identified as a ratio and $-C(p)$ as a remainder after the division by $G(p)$. The polynomial $C(p)$ can then be written

$$C(p) = -R_{G(p)}\left[p^{n-k}M(p)\right] \tag{4.77}$$

During coding the information symbols are left shifted $n-k$ steps. The result are divided by $G(p)$ and the remainder is added to the original shifted information symbols; see the middle section of eq. (4.75). Note that the operation $R_{G(p)}[X(p)]$ is equivalent to expressing $X(p)$ modulo $G(p)$. (A $G(p)$ which cannot be factorised in $GF(q)$ can be used as a base to generate an extended field.) The division of a polynomial $A(p)$ by $G(p)$ can be performed by using a feedback shift register as in Figure 4.16.

Assume that the initial state of the shift register has zeros in all positions. Let $A(p) = p^{n-k}M(p)$ be the input signal to the shift register. $A(p)$ is fed in with the leading coefficient first, i.e. a_{n-1} (in a polynomial with degree n, $a_n x^n$ is called the leading term and a_n is the leading coefficient.). The signal is shifted from left to right with one step per clock pulse. After $n-k$ clock pulses a_{n-1} has reached the output of the rightmost memory cell. At the output of the feedback shift register is the value a_{n-1}/g_{n-k}, which is the leading coefficient in the quotient. By the feedback loop and preceding memory cell $a_{n-2} - g_{n-k-1}a_{n-1}/g_{n-k}$ is fed to the input of the right memory cell. After the next clock pulse this constitutes

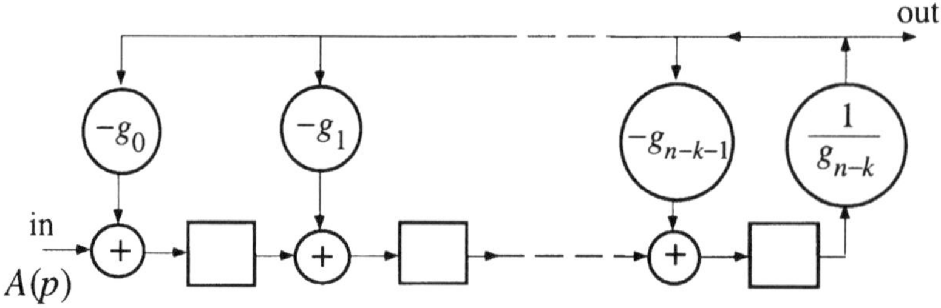

Figure 4.16   A feedback shift register.

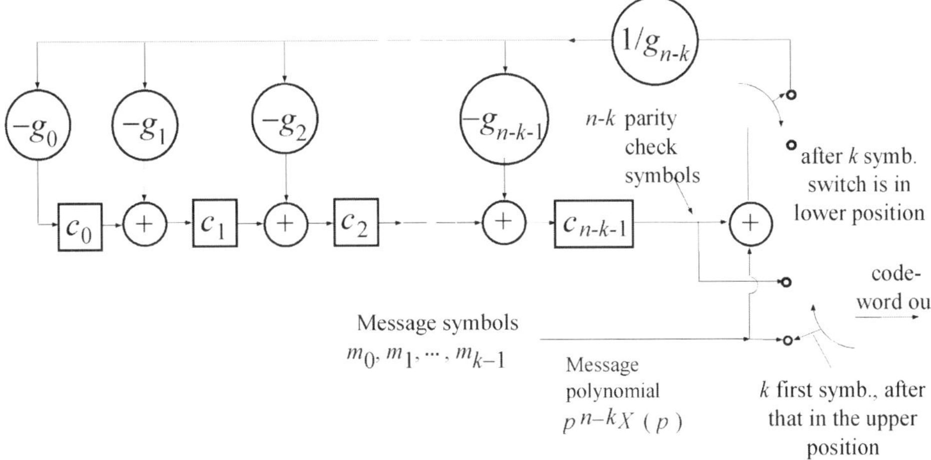

Figure 4.17 Shift register encoder for systematic cyclic codes.

the contents of the right memory cell. The contents of all memory cells are modified in the same way. The procedure is exactly the same as in the algorithm for long division which is done by pencil and paper. After n clock pulses the quotient $Q(p)$ has been serially fed to the output. The remainder $C(p)$ can be found in the shift register. If the feedback is broken the remainder can be shifted out through the output. Compare also with an IIR filter which in the frequency domain performs division by the denominator of the transfer function.

A circuit for the implementation of the coder according to eq. (4.75) is shown in Figure 4.17. The division required is done by means of the shift register. Initially the contents of the memory cells are zeros. Premultiplication of the information symbols by p^{n-k} amounts to moving the injection point to the right-most part of the shift register. After k clock pulses the wanted remainder resides in the shift register memory cells. The switches are changed and the remainder is fed out. Note that for binary codes $-g_0 = 1/g_{n-k} = 1$.

When implementing an actual circuit, one often has access only to whatever shift registers are provided by the semiconductor manufacturers, and it is thus difficult to perform the required additions between the steps in the shift register. The solution to this problem is to implement the transpose of the feedback shift register in Figure 4.16. When transposing a block diagram the signal flows are reversed, the summation points become branch points and branch points become summation points. A shift register performing division by

$$G(p) = p^8 + p^7 + p^4 + p^2 + p + 1 \tag{4.78}$$

has the direct implementation shown in Figure 4.18. After transposing and flipping over so that the input is also to the left, in this case we obtain the situation

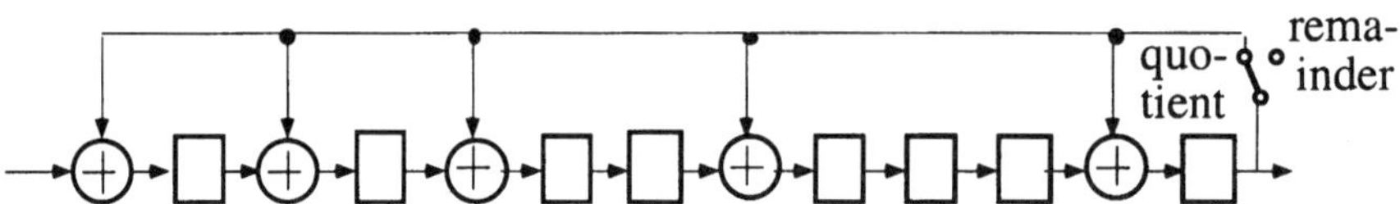

Figure 4.18   Direct implementation. The switch is in the particular positions when quotient and remainder, respectively, will be clocked out.

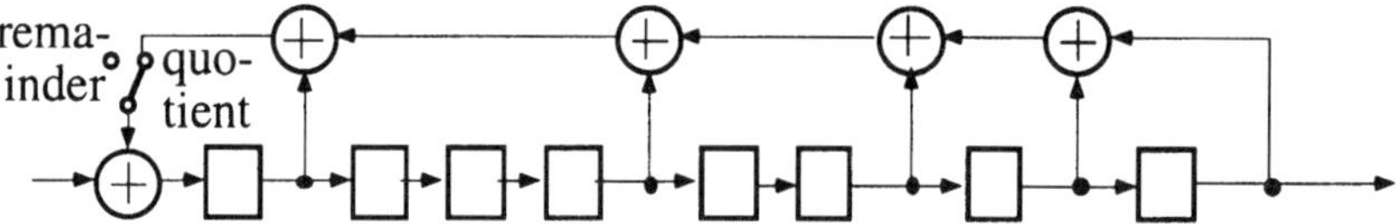

Figure 4.19   Shift register implementation by using a direct transpose.

shown in Figure 4.19 which gives a format for the implementation suitable for the available shift registers produced by the manufacturers. The implementation shown in Figure 4.19 leads to the same output sequence as in Figure 4.18. This may not be obvious at first sight, but can be verified by the reader by assuming a certain input sequence which is then clocked into the registers (this is more easily done with a shorter register).

Decoding of cyclic codes

The drawback with non-cyclic codes is that the table for the mapping between syndrome and error pattern becomes large. The table must contain $k(q^{n-k} - 1)$ symbols to correct the information symbols. The properties of the cyclic codes under cyclic shifts can be utilised to simplify the decoding, which is described below.

The received symbol vector $\mathbf{Y}$ corresponds to the polynomial $Y(p)$, which consists of an emitted correct codeword, cf. eq. (4.55), to which an error possibly has been added:

$$Y(p) = X(p) + E(p) = Q(p)G(p) + E(p) \tag{4.79}$$

The polynomial $E(p)$ is at most of the power of $n - 1$, as there are n symbols in a codeword and the errors can be spread over the whole of the received word.

Divide the received symbols by the generator polynomial. The result can be written as a ratio $T(p)$ and a remainder $R(p)$:

$$\frac{Y(p)}{G(p)} = T(p) + \frac{R(p)}{G(p)} \tag{4.80}$$

Rewriting gives

$$Y(p) = T(p)G(p) + R(p) \tag{4.81}$$

If $Y(p)$ is a codeword, $Y(p)$ is divisible by $G(p)$, i.e. there is no remainder and $R(p)$ thus becomes zero. Note that the remainder $R(p)$ can at the most be of power $n - k - 1$, and is therefore not equivalent to the polynomial $E(p)$ which can assume higher power. Eq. (4.81) inserted in eq. (4.79) gives

$$E(p) = [-Q(p) + T(p)]G(p) + R(p) \tag{4.82}$$

For a received codeword with no errors $E(p) = 0$ and $Q(p) = T(p)$, which means that $[-Q(p) + T(p)]G(p) = 0$. The conclusion is that eq. (4.82) looks like $E(p) = 0 + R(p)$. This means that when $E(p) = 0$ is true, then $R(p) = 0$ must also be true. Since $E(p)$ can be of higher power than $R(p)$, there is no 1:1 relationship between the error pattern and the remainder. The polynomial $T(p)$ is always a codeword but not necessarily identical to the one sent. The remainder $R(p)$ can therefore be equal to zero also for non-correctable errors, but for other errors $R(p)$ is different from zero and can thus be denoted a syndrome $S(p)$:

$$Y(p) = T(p)\,G(p) + S(p) \tag{4.83}$$

The syndrome is

$$S(p) = R_{G(p)}[Y(p)] \tag{4.84}$$

The syndrome now gives a probable error pattern, which can be corrected in the same way as before, for instance by table look up. In this case there is nothing gained compared to decoding of linear block codes in general. We will therefore go further and study what happens to the syndrome under a cyclic shift of the received symbols. According to what has been said before an arbitrary polynomial can be written as a quotient, $B(p)$, and a reminder, $Z(p)$, after division by the polynomial $G(p)$. For the shifted symbols it is given, from the Euclid division algorithm

$$Y^{(1)}(p) = B(p)G(p) + Z(p) \tag{4.85}$$

where $Z(p)$ according to eq. (4.84) is the same as the syndrome for $Y^{(1)}(p)$. From eqs. (4.69) and (4.83) it is obtained that a cyclic left shift of $Y(p)$ can be written

$$Y^{(1)}(p) = p\big(T(p)G(p) + S(p)\big) - y_{n-1}(p^n - 1) \tag{4.86}$$

According to the assumptions for a cyclic code $G(p)$ is a factor in $p^n - 1$, which means that the rewriting, $p^n - 1 = G(p)H(p)$, for some $H(p)$, can be done. By insertion of eq. (4.85) in the left side of eq. (4.86) and rearranging the terms between the left and the right sides we obtain

$$pS(p) = (B(p) - pT(p) + y_{n-1}H(p))G(p) + Z(p) \tag{4.87}$$

By identification it is given that

$$Z(p) = R_{G(p)}[pS(p)] \tag{4.88}$$

i.e. to obtain the syndrome for $Y^{(1)}(p)$ the original syndrome is shifted one step to the left (note this is not a cyclic shift), after which the remainder of the division by $G(p)$ is kept. This can be utilised to make the decoding more effective since the operation can easily be done with a feedback shift register, as has been described before.

The error pattern only tells where in the block the errors are located. To correct the errors, their magnitude must also be determined. This is not described in any more detail here due to lack of space. For a binary code, the errors can only take the value one. Hence, the magnitude of the errors does not need to be determined when binary codes are being used. The received block can be immediately corrected as soon as the mapping between syndrome and suitable error pattern is completed. When the cyclic codes are being used only for error detection, the method is usually called cyclic redundancy check (CRC).

The Meggitt decoder

There are many different ways to utilise the structure in the cyclic codes to simplify decoding. One method, which is rather simple to describe and which in practice can be used for reasonable long codewords, is the Meggitt decoder. In this, a large memory bank, that is required for the table lookup decoder, is exchanged for fast logic.

Suppose the symbols $y_0, y_1, \ldots, y_{n-1}$ have been received. These are read into a buffer consisting of a shift register. The idea is to do the decoding sequentially, in the way that only the symbol located most to the right in the buffer register is corrected (cf. Figure 4.20). The stored word is thereafter rotated one step and the rightmost symbol is corrected. After all the symbols successively have been shifted through the rightmost position, the whole word is corrected, provided the error pattern is correctable. Since only the right symbol is corrected in every step, it is only required that the syndrome pattern with errors in that symbol must be recognised. This means that the number of error patterns is reduced radically. For single error correcting codes it is therefore only one syndrome that has to be recognised. Mapping between syndrome and error pattern is done in a combinatorial network that can perform the necessary operations very fast.

Suppose, for example, a two error correcting $(15,7)$ BCH code with the generator polynomial $G(p) = p^8 + p^7 + p^6 + p^4 + 1$. In the initial state the content in all registers are zeros. The symbols $y_0, y_1, \ldots, y_{n-1}$ are read into the buffer. After n clock cycles the symbol y_{n-1} is in the right position of the shift register. At the same time the syndrome has been computed and resides in the shift register of the syndrome calculator. The mapping between syndrome and error pattern must

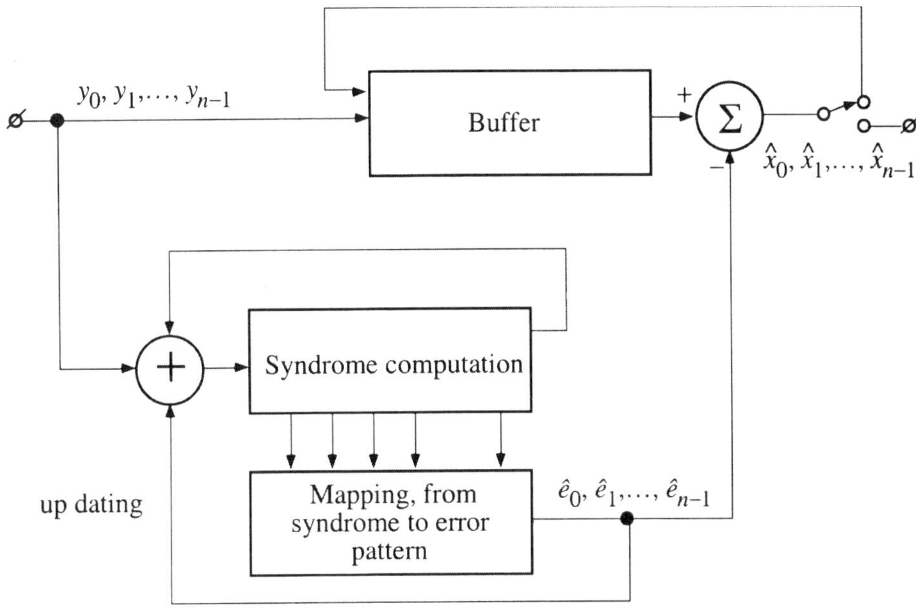

Figure 4.20 The Meggitt decoder.

then be able to identify all single and double error with one error in the position $n - 1$. There is one single error pattern with an error in position 14 and 14 double errors with one error in position 14 and the other error in some other position, i.e. a total number of 15. Some of the syndromes for this code are:

No:	Error pattern, $\mathbf{E}$	Syndrome, $\mathbf{S}$
1	100000000000000	11101000
2	110000000000000	10011100
3	101000000000000	11010010
4	100100000000000	11110101
…	…	…
15	100000000000001	11101001

The mapping is done by a combinatorial network that has $n - k$, in this case 8, inputs and one output. When the syndrome corresponds to an error pattern with a one in position $n - 1$ the estimated error is subtracted $\hat{e}_{n-1}$, which gives

$$Y_{corr}(p) = Y(p) - \hat{e}_{n-1}p^{n-1} \tag{4.89}$$

The symbol y_{n-1} is corrected if necessary. After that, a cyclic right shift of the buffer register content is done. The syndrome for the shifted word is computed according to eq. (4.88). The implementation is done by a feedback shift register of the type shown in Figure 4.16. When the received symbols have been corrected, the syndrome must be modified. The syndrome is now recomputed to correspond to the shifted $Y_{corr}(p)$. With the aid of the syndrome definition, eq. (4.84), and eq. (4.88) we obtain

$$Z_{corr}(p) = R_{G(p)}\left[Y^{(1)}(p) - \hat{e}_{n-1}p^n\right] = \underbrace{R_{G(p)}\left[Y^{(1)}(p)\right]}_{R_{G(p)}[pS(p)]} - \underbrace{R_{G(p)}\left[\hat{e}_{n-1}p^n\right]}_{\hat{e}_{n-1}}$$

$$= R_{G(p)}[pS(p)] - \hat{e}_{n-1} \qquad (4.90)$$

where it also has been used that $R_{G(p)}[p^n] = 1$ for cyclic codes. From eq. (4.90) it is clear that the syndrome is modified in the usual way according to eq. (4.88) and the possible correction is subtracted on the input of the syndrome calculator. The symbol y_{n-2} is now positioned in the right position of the buffer register and the syndrome that has been corrected according to eq. (4.90) corresponds to the error pattern for the shifted word in the buffer register. Possibly an error in y_{n-2} is corrected if correctable and new clockings are done. After n such steps, when received symbols and corresponding syndromes have been corrected and rotated, the entire received word has been corrected. If the received word is correctable the syndrome register now only contains zeros. Finally, the switch in Figure 4.20 is placed in its lower position and the corrected symbols, $\hat{x}_0, \ldots, \hat{x}_{n-1}$, are shifted out.

Example 4.7

A binary, systematic (7,3) cyclic code is generated by

$$G(p) = p^4 + p^3 + p^2 + 1 \qquad (4.91)$$

Find $Q(p)$, $C(p)$ and $X(p)$ when $\mathbf{M} = (1\ 0\ 1)$. Assume that the received codeword $Y = X^{(1)}$ (i.e. single step cyclic left shift) and verify that the syndrome $S(p) = 0$.

Solution The information bits $\mathbf{M} = (1\ 0\ 1)$ written in polynomial form become $M(p) = p^2 + 1$. Written according to eq. (4.76) the following expression results (remember that in binary arithmetic $1 - 1 = 1 + 1$):

$$Q(p) - \frac{C(p)}{G(p)} = \frac{p^4\,\overbrace{(p^2 + 1)}^{M(p)}}{p^4 + p^3 + p^2 + 1} \qquad (4.92)$$

Perform the division

$$\begin{array}{r}
p^2 + p + 1 \\
p^4 + p^3 + p^2 + 1 \overline{\smash{\big)}\, p^6 + p^4}
\end{array}$$

$$\frac{p^6 + p^5 + p^4 + p^2}{p^5 + p^2}$$

$$\frac{p^5 + p^4 + p^3 + p}{p^4 + p^3 + p^2 + p}$$

$$\frac{p^4 + p^3 + p^2 + 1}{p + 1}$$

which gives the result

$$p^2 + p + 1 + \frac{p+1}{p^4 + p^3 + p^2 + 1} \tag{4.93}$$

From eq. (4.76) we can identify $Q(p) = p^2 + p + 1$ and $C(p) = p + 1$.
The codeword becomes

$$X(p) = \overbrace{p^4 \, (p^2 + 1)}^{M(p)} + \overbrace{p + 1}^{C(p)} = p^6 + p^4 + p + 1 \Leftrightarrow \mathbf{X} = [1010011] \tag{4.94}$$

The cyclic left shift is $X^{(1)}(p) = pX(p) + X_{n-1}(p^n + 1)$. When the codeword in
eq. (4.94) is shifted we obtain:

$$\mathbf{X}^{(1)} = [0\ 1\ 0\ 0\ 1\ 1\ 1] \tag{4.95}$$

The syndrome is computed, $S(p) = \mathrm{rem}\big[Y(p)/G(p)\big]$, and the result becomes
$S(p) = 0$, which means that the shift is also a codeword.

Example 4.8

Set up a generator matrix from a given generator polynomial.

Solution A generator matrix is easily set up directly from the generator polynomial. Each
row in the generator matrix can be determined by using message vectors
consisting of zeros in all positions except one, where a one is placed. For the
first row we obtain

$$\mathbf{X} = \mathbf{MG} = \begin{bmatrix} 0 & \cdots & 0 & 0 & 1 \end{bmatrix} \mathbf{G} = \begin{bmatrix} G_{k1} & G_{k2} & G_{k3} & \cdots & G_{kn} \end{bmatrix} \tag{4.96}$$

The other rows can be determined by successively shifting the 'one' one step to
the left. For each step k is reduced by one, and finally the top row is reached.
Starting off from eq. (4.74) it can be seen that the codeword can be written

$X(p) = M(p)G(p)$ by using a polynomial. If $M(p)$ represents messages of the same type as above, $X(p)$ will simply correspond to the rows in the generator matrix. If this is performed, the following simple connection between the coefficients, g_l, of the generator polynomial and the elements of the generator matrix is obtained:

$$\mathbf{G} = \begin{bmatrix} 0 & \cdots & 0 & g_{n-k} & g_{n-k-1} & \cdots & g_1 & g_0 \\ 0 & & g_{n-k} & g_{n-k-1} & g_{n-k-2} & & g_0 & 0 \\ 0 & & g_{n-k-1} & g_{n-k-2} & g_{n-k-3} & & 0 & 0 \\ \vdots & & & & & & & \vdots \\ g_{n-k} & \cdots & & & & \cdots & 0 & 0 \end{bmatrix} \quad (4.97)$$

As can be seen, the above matrix does not give a systematic code. If this is desired, one has to start off from eq. (4.75) and calculate $C(p)$ by using eq. (4.77). For a message where the one is in position i, from the left in the message vector, the codeword written as a polynomial becomes $X(p) = p^{n-k+i-1} + C_i(p)$. For $i = 1$ we obtain the bottom row and for $i = k$ we obtain the top row. This gives a systematic generator matrix, where $c_{i,m}$ are the coefficients in $C_i(p)$

$$\mathbf{G} = \begin{bmatrix} 1 & 0 & \cdots & 0 & 0 & c_{k,n-k-1} & \cdots & c_{k,0} \\ 0 & 1 & & 0 & 0 & c_{k-1,n-k-1} & & c_{k-1,0} \\ \vdots & & \ddots & & \vdots & \vdots & & \vdots \\ 0 & 0 & & 1 & 0 & c_{2,n-k-1} & & c_{2,0} \\ 0 & 0 & \cdots & 0 & 1 & c_{1,n-k-1} & \cdots & c_{1,0} \end{bmatrix} \quad (4.98)$$

Example 4.9

Design a Meggitt decoder for a (15,11) cyclic Hamming code with the generator polynomial $G(p) = 1 + p + p^4$.

Solution A Hamming code is a single error correcting code. In this case one error is corrected in a block of 15 symbols. One single syndrome has to be identified, which is the one that has an error in the right position of the buffer register, i.e. the 15th symbol. For this error the syndrome is

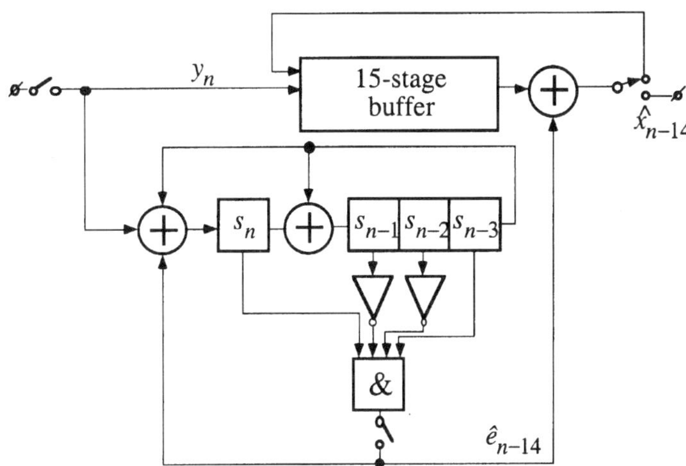

Figure 4.21 Decoder for a cyclic Hamming code with $G(p) = 1 + p + p^4$.

$$S(p) = R_{G(p)}(p^{14}) = p^3 + 1 \tag{4.99}$$

The syndrome is four bits long and from eq. (4.99) the corresponding vector $\mathbf{S} = [1\ 0\ 0\ 1]$ is obtained. The combinatorial network that creates the mapping from syndrome to error pattern therefore shall have a 'one' at the output when this syndrome appears. Such a network is shown in Figure 4.21.

4.3.4 NON-BINARY BLOCK CODES

The examples so far considered have only dealt with binary codes but, as mentioned earlier, non-binary codes are also being used. One application is to correct or detect errors in communication over channels where the errors appear in bursts, i.e. the errors tend to be grouped in such a way that several neighbouring symbols are incorrectly detected.

Suppose that the symbols used at the channel coding can assume 16 different values, while the symbols transmitted on the channel are binary. Every code symbol is then expressed by using four ncighbouring binary symbols. A single error-correcting code can then do correction if all four binary symbols should be incorrect. This makes the code more efficient for burst errors even if on average it has the same performance for error-correcting as a binary code.

In decoding a non-binary code there is a great difference compared to binary codes. The error is always a number different from zero in the field. Therefore in binary codes the value of the error is always one. In a non-binary code the error can take many values. Thus the magnitude of the error has to be determined by an operation which is not treated in this book. Some examples of non-binary codes are presented below.

Hamming code

A Hamming code can also be defined over a non-binary field. The parity check matrix $\mathbf{H}$ is designed by setting its columns as the vectors of $GF(q)^r$, whose first non-zero element equals one. There are $n = (q^r - 1)/(q - 1)$ such vectors and any pair of these is linearly independent. Take for example $GF(3)$ and $r = 3$. List the numbers with three digits in $GF(3)$:

001	012	100	111	122	210	221
002	020	101	112	200	211	222
010	021	102	120	201	212	
011	022	110	121	202	220	

Picking the numbers where the first non-zero element is a one gives

$$\mathbf{H} = \begin{bmatrix} 0 & 0 & 0 & 0 & 1 & 1 & 1 & 1 & 1 & 1 & 1 & 1 & 1 \\ 0 & 1 & 1 & 1 & 0 & 0 & 0 & 1 & 1 & 1 & 2 & 2 & 2 \\ 1 & 0 & 1 & 2 & 0 & 1 & 2 & 0 & 1 & 2 & 0 & 1 & 2 \end{bmatrix} \tag{4.100}$$

The generator matrix is obtained by solving $\mathbf{GH}^T = \mathbf{0}$. Then $\mathbf{G}$ generates a Hamming code, a perfect single error correcting code, of length n and $k = n - r$.

Reed–Solomon codes

Reed–Solomon codes (RS) are subsets of non-binary BCH codes. Mathematical operations are performed in $GF(q)$, where $q = n + 1$. The minimum Hamming distance is $d_{min} = n - k + 1$, which is the largest that can be achieved (Singleton's bound) for a linear code, with given values on n and k. The codes are cyclic and can therefore be specified by a generator polynomial, which for the RS codes looks like the following:

$$G(p) = (p - \alpha)(p - \alpha^2)...(p - \alpha^{n-k}) \tag{4.101}$$

The term α is the Nth root of 1, which means that

$$\alpha^N = 1 \quad \text{in } GF(q)$$
$$\alpha^i \neq 1 \quad \text{for } 1 \leq i \leq N - 1 \tag{4.102}$$

This can be explained in more detail by the following example. Suppose a (4,2) Reed–Solomon code is to be set up. According to the definition above the arithmetic of the code is performed in $GF(5)$. To determine the generator polynomial, an α-value which is the fourth root of 1 must be found. Such a value is $\alpha = 2$, which can be concluded if all powers of 2 are written down:

$$\alpha^0 = 1, \quad \alpha = 2, \quad \alpha^2 = 4, \quad \alpha^3 = 3, \quad \alpha^4 = 1 \tag{4.103}$$

It is also possible to use $\alpha = 3$, but not $\alpha = 4$. The generator polynomial becomes (note that the operations in GF(5) are the same as modulo 5 computation)

$$G(p) = (p - \alpha)(p - \alpha^2)$$

$$= (p - 2)(p - 4)$$

$$= (p + 3)(p + 1) = p^2 + 4p + 3 \tag{4.104}$$

The polynomial in eq. (4.104) specifies a single-error-correcting RS-code. Assume, for instance, that the information to be transferred is $\mathbf{M} = (1 \quad 3)$, which expressed as a polynomial is $M(p) = p + 3$. To calculate the codeword, the remainder $C(p)$ is required according to eq. (4.75). This is obtained through the divisions:

$$
\begin{array}{r}
p + 4 \\
\hline
p^2 + 4p + 3 \, \big) \quad p^3 + 3p \\
\underline{-(p^3 + 4p^2 + 3p)} \quad \text{rewritten in GF(5) gives } 4p^3 + p^2 + 2p \\
0 + 4p^2 + 2p \\
\underline{-(4p^2 + p + 2)} \quad \text{rewritten in GF(5) gives } p^2 + 4p + 3 \\
p + 3 \qquad\qquad \leftarrow \text{remainder}
\end{array}
$$

The remainder so obtained is $-C(p)$. Multiplication by minus one is required to obtain $C(p) = -p - 3 = 4p + 2$. After this modification, the codeword becomes

$$X(p) = p^2(p + 3) + C(p) = p^3 + 3p^2 + 4p + 2 \tag{4.105}$$

Expressed as a vector the codeword becomes $\mathbf{X} = (1 \quad 3 \quad 4 \quad 2)$. Checking that this is a codeword is easily done by calculating the syndrome $S(p)$, which should then be zero.

The usual choice is $q = 2^b$, where b is the number of binary bits in the symbols to be transmitted. A single error-correcting code will then be able to correct burst errors having a length of b binary bits.

4.3.5 MODIFYING BLOCK CODES

Codes available in tables are limited with respect to n, k and t. To obtain the properties required in the respective application, the codes may be modified according to the rules shown in Figure 4.22. The modified codes are possibly not optimal but are generally good codes and the implementation properties of the primitive code is preserved. The figure illustrates the modification of a (7,4) Hamming code, but the principle obviously holds for any block code.

When the error-correcting properties need to be improved, this can be achieved by adding further check symbols, i.e. the code is expanded. One bit is usually added in order to obtain even or odd parity of the codewords. See Figure 4.22 where the parity has been made even by having an even number of ones in the generator matrix rows. The opposite is when a code is punctured by removing check symbols. The minimum Hamming distance usually decreases by one when a check bit is removed. A code is expurgated by removing some of the code-words, e.g. by deleting a row in the generator matrix. For cyclic codes this is usually done by multiplying the generator polynomial by $(p - 1)$ (cf. Figure 4.22). The opposite is to augment the code by adding further codewords, without changing the block length. For cyclic codes this can be done by dividing the

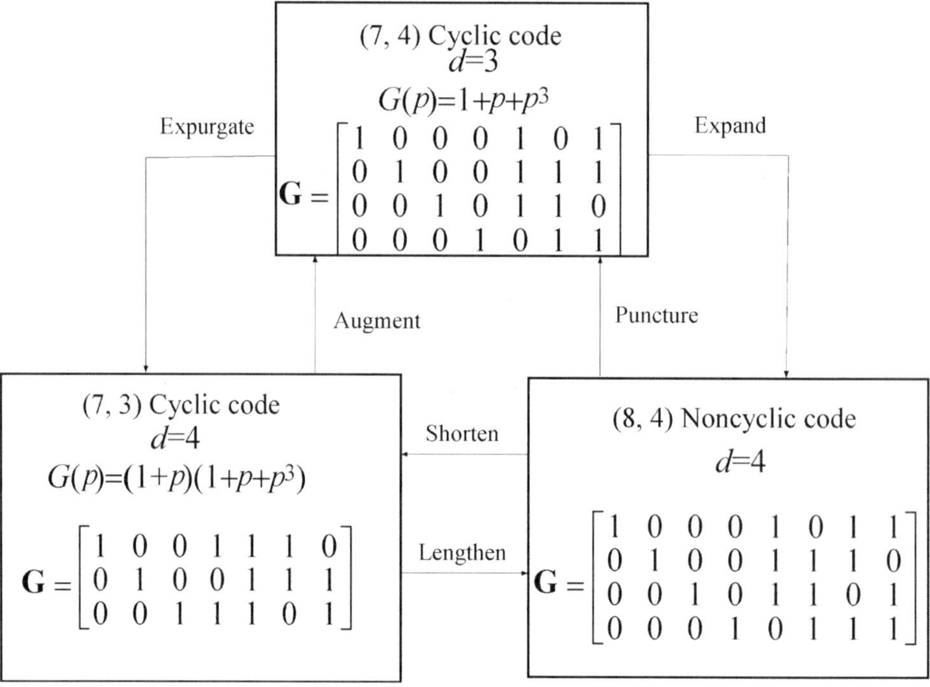

Figure 4.22   Schematic illustration of code modification.

generator polynomial by $(p - 1)$. Lengthening a code is done by adding information bits. This can usually be done without a reduction of the minimum Hamming distance. The extension is done in two steps. Start from a cyclic code which first is augmented and thereafter expanded. A code can be shortened by removing some information bits. A common way is to use only those message words which begin with a number of zeros. Extension of a code is a term subsuming the operations giving a larger block length.

Example 4.10

The Hamming code (7,4) can be expanded to a (8,4) code by adding a fourth check bit, chosen in such a way that all codewords have even parity. Perform this on the code in the table below and show that $d_{min} = 4$. Determine the equation for the check digit c_4 as a function of the message bits. Explain how the decoding in Figure 4.14 should be modified to be able to detect double errors and correct single errors.

Solution   To determine d_{min} for a linear code it is not necessary to determine the pair wise distance between all codewords. The reason for this is that for a linear code the sum of two codewords is a new codeword and since for a binary code

$$d(\mathbf{X}_1, \mathbf{X}_2) = w(\mathbf{X}_1 + \mathbf{X}_2) = w(\mathbf{X}_3) \tag{4.106}$$

it is enough to find the least Hamming weight, w_{min}, to find d_{min}. The codeword table for a (7,4) Hamming code is:

M	**C**	$w(\mathbf{X})$	**M**	**C**	$w(\mathbf{X})$
0000	000	0	1000	101	3
0001	011	3	1001	110	4
0010	110	3	1010	011	4
0011	101	4	1011	000	3
0100	111	4	1100	010	3
0101	100	3	1101	001	4
0110	001	3	1101	001	4
0111	010	4	1111	111	7

Solution: (7,4) code extended to (8,4) code by using a parity bit becomes:

M	C	$w(\mathbf{X})$
0000	0000	0
0001	0111	4
0010	1101	4
0011	1010	4
0100	1110	4
0101	1001	4
0110	0011	4
0111	0100	4
1000	1011	4
1001	1100	4
1010	0110	4
1011	0001	4
1100	0101	4
1101	0010	4
1110	1000	4
1111	1111	8

i.e. $d_{min} = 4$. The equation for the check sum, c_4, is obtained from the last column in the parity check matrix $\mathbf{P}$. This can be obtained by using the messages

$$\mathbf{M} \rightarrow \quad [1000], \ [0100], \ [0010] \text{ and } [0001] \qquad \text{thereafter study } c_4$$

$$c_4 = \quad 1 \qquad 0 \qquad 1 \qquad 1$$

which gives

$$p_{14} = 1, \ p_{24} = 0, \ p_{34} = 1 \text{ and } p_{44} = 1$$

i.e. the check sum becomes

$$c_4 = m_1 \oplus m_3 \oplus m_4 \tag{4.107}$$

To extend the decoder to become double error-correcting, the parity sum is created:

$$\alpha = \sum_{i=1}^{8} y_i \ (\text{mod } 2) \Rightarrow \begin{cases} \alpha = 0 \text{ for even numbers of ones} \\ \alpha = 1 \text{ for odd numbers of ones} \end{cases} \tag{4.108}$$

when

$\mathbf{S} = (000)$ and $\alpha = 0 \Rightarrow$ no errors	$\mathbf{S}$ are computed on the original code
$\mathbf{S} \neq (000)$ and $\alpha = 1 \Rightarrow$ odd number of ones	suppose single errors
$\mathbf{S} \neq (000)$ and $\alpha = 0 \Rightarrow$ even number of ones	suppose double errors, detected but not corrected

4.4 Convolutional codes

Convolutional codes belongs to a group called trellis codes. Binary convolutional codes are dealt with below. This connects with what already has been said about operations in GF(2). Trellis codes are discussed further in the chapter concerning modulation.

4.4.1 DESCRIPTION OF CONVOLUTIONAL CODES

When coding with a convolutional code, the information bits are processed in serial form, which is in contrast to block codes, where the flow of information symbols is processed in block form. The encoded symbols are continuously generated and the error correction properties are achieved by the introduction of a memory, or a dependence, between the symbols. When it comes to convolutional codes, only binary codes are studied here since these are the far most common. The information symbols are encoded by using a convolutional operation

$$x_j = m_{j-L}\, g_L \oplus \cdots \oplus m_{j-1}\, g_1 \oplus m_j\, g_0 = \sum_{i=0}^{L} m_{j-i}\, g_i \ (\text{mod } 2) \qquad (4.109)$$

This can be realised by letting the symbols pass through a shift register. Some symbols, which are defined by the coefficients $g_L \ldots g_0$ in the generator polynomial, are added modulo 2 to each other and form the output signal. Compare this to an FIR filter where m represents the input signal and g the impulse response. To increase the redundancy in the transmitted message several such convolutional operations are required, i.e. several filters are needed. In the general case we have n filters. With three filters the output signals from the respective filter become $x_j' x_j'' x_j'''$. The output signal from these is successively taken from each filter to form an encoded bit flow as shown in Figure 4.23. Convolutional codes can, as well as block codes, be grouped into non-systematic or systematic codes. In the systematic group is always $x_j' = m$. However, systematic convolutional codes have in general lower performance than non-systematic ones, which makes the latter more commonly used.

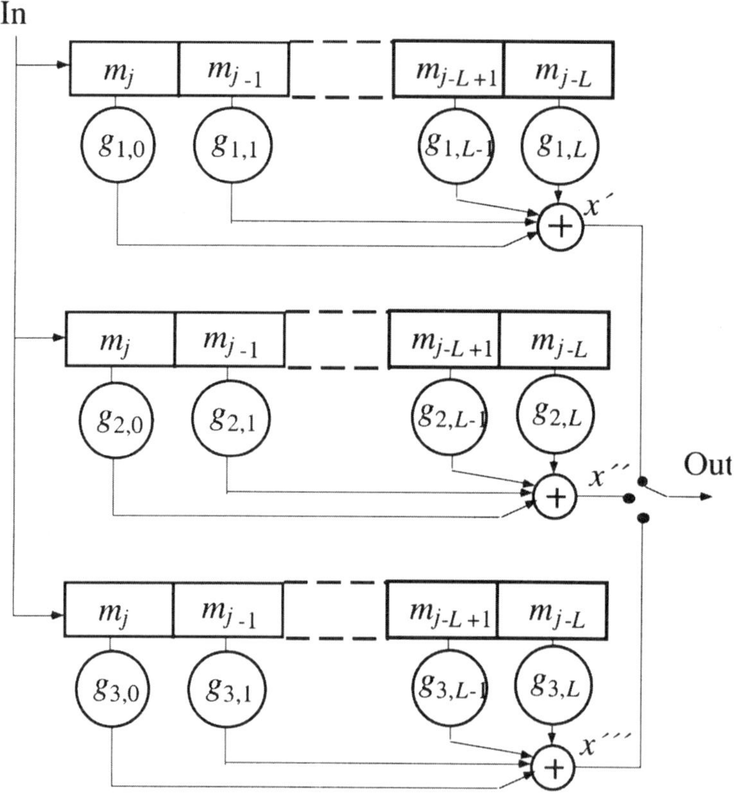

Figure 4.23 Convolutional encoder for a 1/3 code. The coder consists of three FIR filters which are successively connected to the output. Each filter has its own generator polynomial or response function.

The state of an encoder is determined by the values of m_j to m_{j-L+1}. A binary rate $1/n$ encoder, i.e. of the type shown in Figure 4.23, has therefore 2^L states. (In some text books, the states are determined by the message bits $j-1$ to $j-L$. This is an equivalent definition provided all other parameters are determined according to that definition.)

If k symbols in to the encoder lead to n symbols out, we have a rate k/n code. For codes of this type, where $k \neq 1$, the encoder consists of k sub-encoders, where each one is of the kind shown in Figure 4.23. The output signals from the corresponding filters in the respective sub-encoder are added together before the switch. As an example the coded symbol x' then becomes

$$x' = \sum_{m=1}^{k} x'_m \quad (\text{mod } 2) \tag{4.110}$$

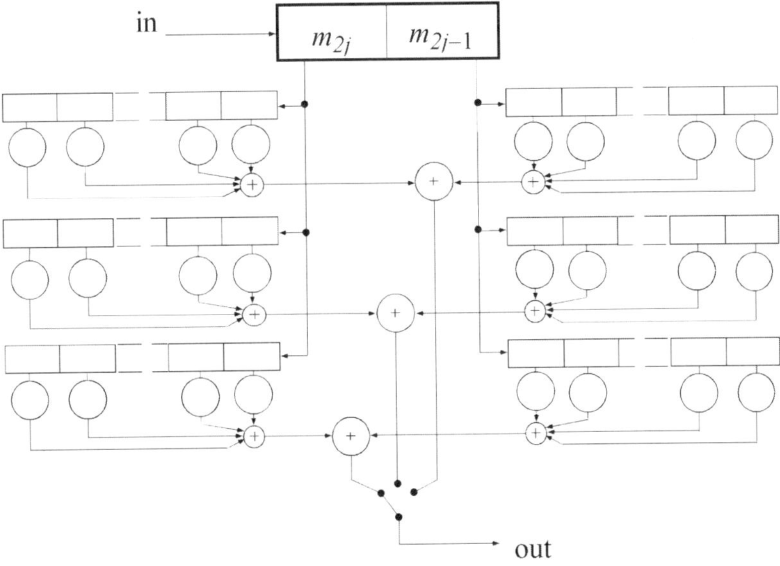

Figure 4.24 A schematic diagram showing a shift register implementation of a 2/3 encoder.

A block diagram of a rate 2/3 encoder is shown in Figure 4.24. The message symbols are encoded in pairs, and for each pair, three encoded symbols are obtained.

If each register is of length $L + 1$, then $k(L + 1)$ is called the constraint length of the code. The concept corresponds to the total length of the encoder impulse response.

Convolutional coding is closer described by means of a simple binary rate 1/2 code with the generator polynomials whose coefficients are given by

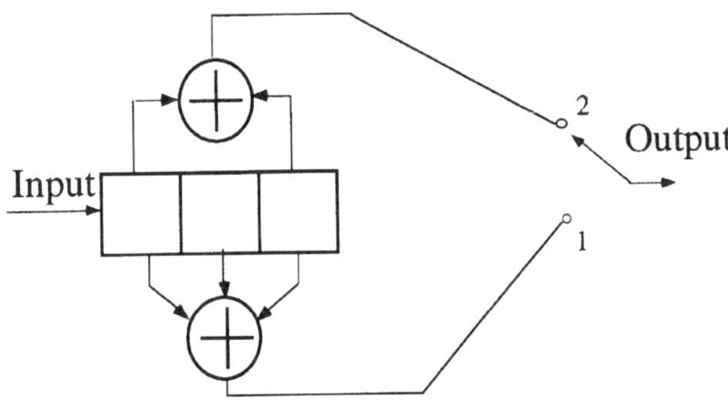

Figure 4.25 Convolutional encoder for a code with $L = 2$, $k = 1$ and $n = 2$. The two positions to the left in the shift register identify the state of the encoder.

$$\mathbf{g}_1 = \begin{bmatrix} 1 & 1 & 1 \end{bmatrix} \overset{\text{octal}}{\rightarrow} 7$$
$$\mathbf{g}_2 = \begin{bmatrix} 1 & 0 & 1 \end{bmatrix} \rightarrow 5$$

(4.111)

The definition of the generator polynomial is the same as in eq. (4.73). The shift registers which, according to Figure 4.23, are needed to implement the code, can of course be joined to a single one. This makes the realisation simpler and the resulting encoder is shown in Figure 4.25.

Graphic code representations

The output signal of the encoder can be described by using a tree-diagram as illustrated in Figure 4.26. When a stream of information symbols is being shifted into the shift register, a path from left towards right is followed along the tree. For the design of the tree diagram it is assumed that the input signal to the encoder at the start was a series of zeros so that all positions of the shift register are zeros. This is the starting point, the root, for the tree diagram and this state can be denoted a.

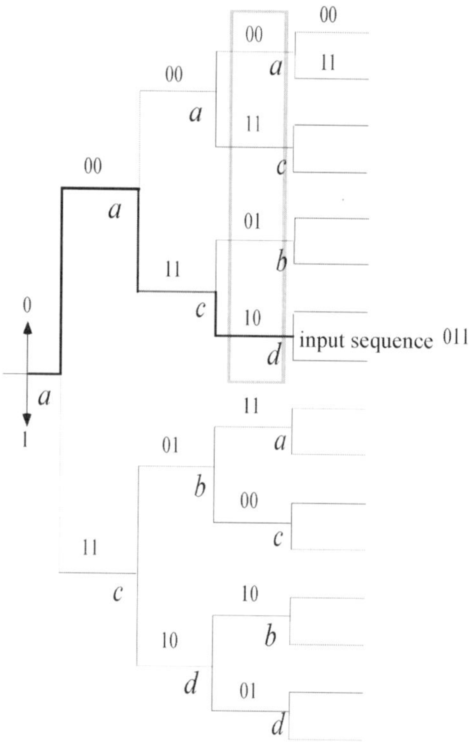

Figure 4.26 Code tree for the rate 1/2, $L = 2$ convolutional code as described in the text. The thicker line represents the input sequence 011, which gives the code sequence 001110.

Assume further that another zero is shifted into the shift register. This is indicated by choosing the upper branch of the tree. The arrows at the starting point indicate which way to go at the branching points, depending on whether the input signal is a one or a zero. The output signal from the register, i.e. the encoded symbols, is indicated in the tree above the respective branch and is in this case 00. Since nothing has changed in the content of the shift register, the state remains at a as indicated at the following branching point. Note that it is every new step in the tree that generates an output signal.

Return to the starting point and assume that a one is now shifted into the shift register. The lower branch is then followed, and 1 0 0 is read into the cells of the shift register from left to right. The output signal and the resulting state are then obviously not the same as for a zero in, which can be seen from the tree. The state at the next branching point can be denoted c.

At each bifurcation in the tree it is assumed that a new symbol is read into the left memory cell of the shift register. Depending on whether the symbol is a one or a zero, the lower or upper branch, respectively, is chosen as the next bifurcation. The output signal of the encoder together with the new state is noted. In this way one continues from left to right in the tree. A code having $L = 2$ thus has four different states.

The tree growth is exponential. However, as can be seen in Figure 4.26, the tree has a periodic structure with the sequence shown inside the shaded rectangle regularly repeated. It is therefore sufficient to analyse one period and then place a number of such periods serially in a time sequence. This forms a trellis, which is a kind of collapsed tree giving the same information as the tree but describing the code in a more efficient way. At a stationary state, after $L + 1$ steps, the trellis has reached its complete extension and the structure does not change even if a longer time interval is shown in the figure. In Figure 4.27 the branches from each branching point have been drawn in such a way that an input zero corresponds to a solid line and a one corresponds to a broken line. A certain path is followed depending on the input signal, in the same way as in the tree.

Another way of representing the coding is to look at the encoder as a finite state machine and to use a state diagram (Figure 4.28). As for the trellis, there are 2^{kL} states. Each state is represented by a node, or a box in the diagram. There are 2^k branches coming in and 2^k going out from each node. There is no time scale in this type of diagram, which provides the most compact graphical representation. Allowed transitions between the states are identified by the branches between the respective states. Since a path depending on the input signal, can be traced through the state diagram this representation gives the same information about the output signal for a given input signal as the previous representations.

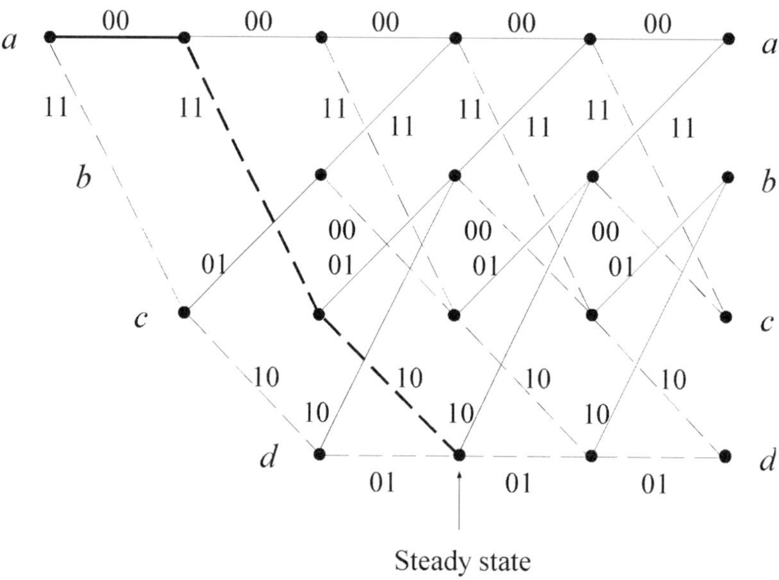

Figure 4.27 Trellis for the convolutional code described in the text. Solid line corresponds to a zero and the broken line corresponds to a one into the encoder. The input sequence 011 is indicated by thick lines. State a is the point in the top row, state b is the next row, and so on.

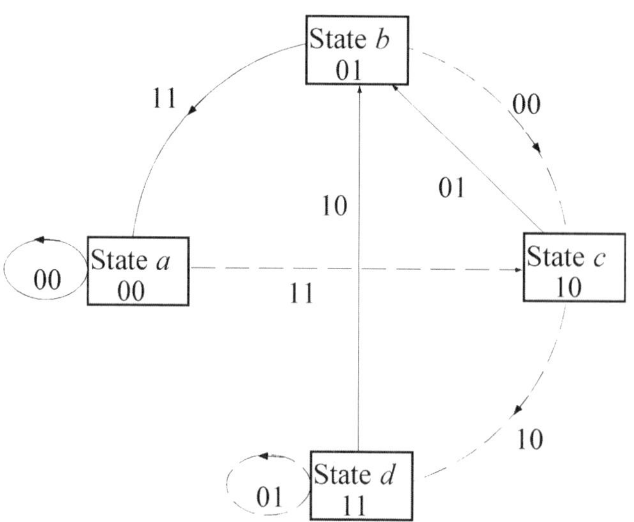

Figure 4.28 The code represented by a state diagram.

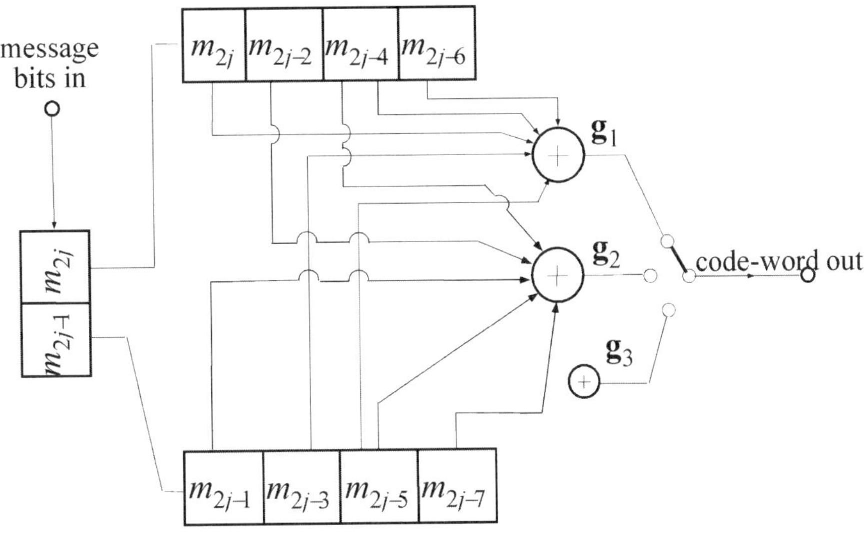

Figure 4.29 Encoder. Note that the taps corresponding to $\mathbf{g}_3$ have not been drawn.

Example 4.11

Design an encoder for a (3,2,3) convolutional code with the coefficients

$$\mathbf{g}_1 = 236_8 = (10\ 011\ 110)_2$$

$$\mathbf{g}_2 = 155_8 = (01\ 101\ 101)_2$$

$$\mathbf{g}_3 = 337_8 = (11\ 011\ 111)_2$$

Solution   A convolutional code is described by (n, k, L), where n is the codeword length, k is the information word length and $L + 1$ is the filter length. The span becomes $k(L + 1) = 2(3 + 1) = 8$. The left element in each vector is to be multiplied by m_j and the right element by m_{j-7}. The encoder is thus represented by the block diagram shown in Figure 4.29.

4.4.2 PERFORMANCE OF CONVOLUTIONAL CODES

The transfer function of the encoder

The transfer function of the encoder is used to investigate the properties of the code regarding the distance between the different coding sequences as well as regarding bit-error probability. The bit-error probability is, as mentioned earlier, a function of the distance properties of the code and decreases with increasing distance. Since the convolution in eq. (4.109) is a linear operation the convolutional codes are also linear. For linear codes it is sufficient to use one single input sequence when analysing the error-correcting properties. An input sequence consisting of a series of zeros, 00...0, is usually chosen. This then constitutes the correct sequence and the sequence one wishes to restore after the decoding. Received code sequences containing symbols other than zeros are thus incorrect. During the analysis all possible error events, or deviations from 00...0 must be investigated. Since the correct sequence means that one remains in state a, each such error event always begins and ends in state a. For a short error event one soon returns to state a, but one event can, in principle, last for any length of time. To analyse the error events, state a is divided into a start state a_s and a stop state a_t. The zero input loop disappears as it does not represent any errors. The remainder of the state diagram is unchanged. Start from the state diagram and form a flow graph. See Figure 4.30 showing a flow graph with three nodes and two branches.

The flow graph has the same structure, or topology, as the state diagram. The states are transferred into nodes. The branches remain identical between different states and nodes in the two representations, respectively (cf. Figures 4.28 and 4.31). W^β and I^γ represent the multipliers at the branches, where β is the Hamming distance from the correct output signal and γ is the number of bit errors in the information symbols, which arise along the respective branch.

In order to determine the exponents of W and I at each branch, those input and output sequences which correspond to the respective state transition are analysed and compared with the 00...0 sequence. The first branch, for instance, is the one leading from a to c. At this state transition the encoder output is the symbol sequence 11. This sequence has a Hamming distance of 2 from the correct sequence, 00. The exponent of the factor W is therefore 2. At the same time the transition corresponds to a decoded one. This is an error since the correct information symbol was zero. One bit error, or symbol error, has therefore occurred. Hence, the exponent of I is 1. The branch from a to c thus obtains a multiplication variable $W^2 I^1$. For the branch from c to b the symbol sequence 01 is obtained, which gives the Hamming distance 1 and, hence, W^1. This state transition corresponds to a decoded 0, which means that no errors were made for this particular symbol in the decoded sequence. For the branch b to e the symbol sequence 11 is obtained, which gives W^2, i.e. a Hamming distance of 2. A decoded zero means no error, which gives the factor I^0. In this case k is always equal to one or zero, since any change of state means only one information bit in, which consequently also means that there can be no more than one error per branch.

For each node a state equation can be identified consisting of the sum of all incoming branches. Each term consists of the start node multiplied by the coeffi-

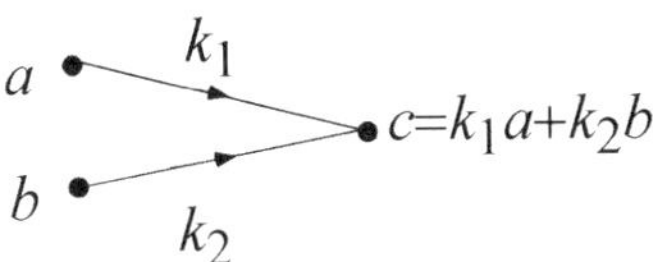

Figure 4.30 Addition and multiplication rules for a flow graph.

cients at the respective branch (cf. Figure 4.30). The system of state equations becomes in this case

$$
\begin{aligned}
X_b &= W\,X_c &&+ && W\,X_d \\
X_c &= W^2 I\,X_a &&+ && I\,X_b \\
X_d &= WI\,X_c &&+ && WI\,X_d \\
X_e &= W^2 X_b
\end{aligned}
\tag{4.112}
$$

The equation system is solved for the start and stop nodes, i.e. X_e and X_a, respectively. Their ratio identifies the transfer function, $T(W,I)$,

$$
\frac{X_e}{X_a} = T(W,I) = \frac{W^5 I}{1 - 2WI} = W^5 I + 2W^6 I^2 + 4W^7 I^3 + \cdots
$$

$$
= \sum_{d=d_{free}}^{\infty} \sum_{i=1}^{\infty} A_{d,i}\, W^d I^i = \sum_{m=5}^{\infty} 2^{m-5} W^m I^{m-4}
\tag{4.113}
$$

$A_{d,i}$ is the number of paths with the Hamming distance d and i errors. The summation over i is performed to cover error paths with the same Hamming distance but with different numbers of bit errors. From the series expansion above it can be seen that there is one path having a Hamming distance of 5 which gives one error, two paths having a distance of 6 which give two errors, and so on. The smallest value on the exponent of W gives d_{free}, in this case $d_{free} = 5$, which corresponds to the error path having the shortest distance to the correct symbol pattern. In this case it is easy to identify the paths in the state diagram, Figure 4.31. The shortest way runs through the branches in the lower end of the figure. It is also easy to realise that it is possible to circle around continuously in the closed loop paths between c and b, which coincides with long error events. If the same thing is studied in the trellis (Figure 4.27) it is realised that the correct way, i.e. the one that the transmitter has been taken, runs through the uppermost branches through the state sequence $a,a,\ldots,a$. One error in the decoding occurs because symbol errors have occurred during the transmission over the channel which makes the receiver believe that the transmitter has taken another way

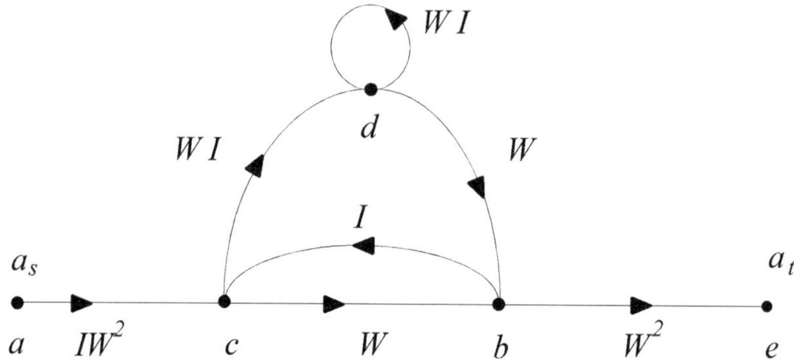

Figure 4.31 Flow graph having separate start and stop nodes.

through the trellis. The error event has ended when the transmitter and the receiver are on the same path in the trellis again.

Error probability

The error events are supposed to be statistically independent. An upper limit for the probability that an error event will commence at every new step through the trellis is obtained if the probability for every error event is summed (is called union bound and is described in more detail in Chapter 6). For the code in this case, together with eq. (4.113) we obtained

$$P_E \leq P_5 + 2P_6 + 4P_7 + \cdots + 2^{m-5}P_m + \cdots \tag{4.114}$$

where P_5 is the probability for the error event causing a decoding error with the Hamming distance 5, and so on for other P_m. The decoding error causing the distance 5 occurs if three or more of the transmitted symbols become erroneous, under a sequence length corresponding to the path from a_s to a_t. The Hamming distance from the received sequence to the incorrect path is then smaller than to the correct one. The decoder believes that the transmitter has taken the path that gives the smallest Hamming distance and an error event occurs; cf. the discussion on block coding in Section 4.1.1. The probability to do precisely the error above is equal to the probability to find sequences with at least three errors but at most five errors. Generally, the probability to find sequences with at least $\lfloor m/2 \rfloor + 1$ but at most m errors in the transmitted symbols:

$$P_m = \sum_{e=\lfloor m/2 \rfloor + 1}^{m} \binom{m}{e} P_s^e (1 - P_s)^{m-e} \leq 2^m P_s^{m/2} (1 - P_s)^{m/2} \tag{4.115}$$

cf. eq. (4.5). For error events with even Hamming distance there is a possibility that $m/2$ transmission errors occur. In that case the decoder has two sequences to chose from that have the same Hamming distance as the received one. One of them is correct and the other erroneous. For example P_6 occurs when the received sequence has the Hamming distance 3 from the correct one, which also implies a distance 3 from the erroneous one with distance 6. Suppose that the receiver then tosses a coin and therefore makes a correct choice in half of the cases. To cover this case a term is added, when m is even, which is equal to the probability that out of a length of m find $m/2$ errors. The term is multiplied by $1/2$ since the receiver is correct in half of the cases. The upper limit for P_m given in eq. (4.115) is valid for m both odd and even supposing that $P_m < 1/2$.

The upper limit described above can also be obtained by inserting $W = 2\sqrt{P_s(1 - P_s)}$ and $I = 1$ in eq. (4.113), which gives P_E in eq. (4.114). Moreover in eq. (4.113) we have a closed expression for the infinite sum.

For the communication engineer it is the probability for an error in one symbol, or bit, that is of interest. To find out that, we have to go further. If we multiply each term in eq. (4.114) with the number of bit errors done for every single error path, an expected value of the number of bit errors at each new step in the trellis, i.e. each time a decoding decision is taken, is obtained. Now the variable I is used. Differentiate eq. (4.113) with respect to I. The number of symbol errors done along each path has been counted by means of the exponent i. After the differentiation the exponent, i.e. the number of errors for the respective path, will appear as a multiplicative factor in every term:

$$\frac{dT(W,I)}{dI} = \sum_{d=d_{free}}^{\infty} \sum_{i=1}^{\infty} A_{d,i} W^d i I^{i-1} \tag{4.116}$$

The variable i is equal to the number of decoded errors along path d, i. The coefficient I has thereby completed its task and is removed by being set equal to one. By the substitution of W, in the same way as before, the wanted expectation is obtained. In each step in the trellis k bits are decoded, for a rate k/n code. The expectation therefore has to be divided by k in order to get the risk of making an error in a single decoded symbol. An upper limit for the decoded sequence is therefore

$$P_b \le \frac{1}{k} \left. \frac{dT(W,I)}{dI} \right|_{I=1,W=2\sqrt{P_s(1-P_s)}} = \frac{1}{k} \sum_{d=d_{free}}^{\infty} \sum_{i=1}^{\infty} A_{d,i} 2^d \left[P_s(1 - P_s) \right]^{d/2} i 1^{i-1}$$

$$\tag{4.117}$$

where P_s is the probability for error in a received symbol. The closed expression for the sum is of course simpler to handle and for the code used in this example is

$$P_b \le \frac{2^5 \left[P_s(1 - P_s)\right]^{5/2}}{\left(1 - 4[P_s(1 - P_s)]^{1/2}\right)^2} \tag{4.118}$$

For small P_s, a simplification can be obtained from the series expansion of the transfer function, eq. (4.113). For small enough values of P_s the most probable error dominates, i.e. d_{free}. Equation (4.113) inserted in eq. (4.117) can then be approximated:

$$P_b \approx \frac{1}{k} \sum_i A_{d_{free},i} 2^{d_{free}} \left[P_s(1 - P_s)\right]^{d_{free}/2} i \approx \frac{1}{k} \sum_i A_{d_{free},i} 2^{d_{free}} P_s^{d_{free}/2} i \tag{4.119}$$

For the code in the example above $k = 1$. There is only one error path having the shortest Hamming distance which means that i only assumes the value 1, i.e.

$$P_b \approx 2^5 P_s^{5/2} \tag{4.120}$$

For the BPSK modulation scheme, the error probability in the uncoded case is $P_s = Q(\sqrt{2E_b/N_0})$. By using eq. (4.120) we obtain for the encoded case $P_b \approx 2^5 Q^{5/2}(\sqrt{2R_c E_b/N_0})$, where the symbol error probability of the uncoded case has been used as the parameter P_s and R_c is the code rate which is $k/n = 1/2$.

The above can be used to investigate the properties of the codes with respect to different parameters; e.g. the reduction in signal to noise ratio given that the error rate is constant, i.e. coding gain, that is offered when a certain code is applied to the communication system. Such a study shows that the coding gain increases approximately linearly with the constraint length of the code. However, the complexity of the decoder increases exponentially with the constraint length, which is why a technically sound balancing must be done as usual.

The calculations above are made under the assumption that the detector always interprets the received symbol as either a one or a zero. This is called a hard decision, which can be considered logical as the emitted symbols are all binary. The received symbol can also be quantised into more than two levels, which can be used to improve the decoding. This procedure is called soft decoding and gives even better results than those shown in Figure 4.32. An upper limit of the error probability obtained from an infinite number of quantised levels can be obtained by setting $W = \exp(-E_b R_c/N_0)$ in eq. (4.117). Soft decoding is also useful in order to further improve the decoding of block codes. Soft decoding is described in connection with turbo codes.

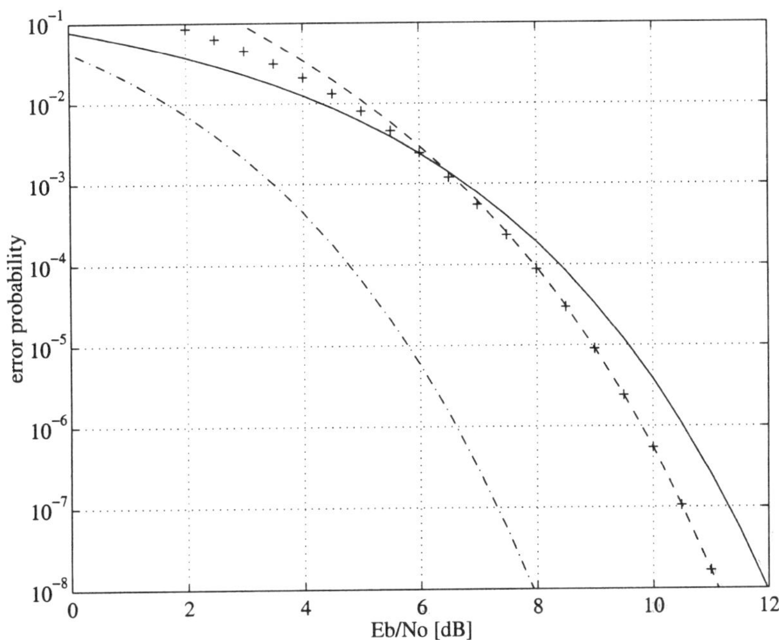

Figure 4.32 Bit error probability of a hard decoded convolutional code. The dashed line corresponds to the rate 1/2 code described in the text and calculated according to eq. (4.118). The crosses represent the same code calculated from eq. (4.120). The dash dotted line represents the error rate of a rate 1/2 code with $d_{\text{free}} = 10$. The solid line corresponds to uncoded BPSK.

4.4.3 DECODING OF CONVOLUTIONAL CODES

The two methods treated below are maximum likelihood decoding and syndrome decoding. The method described for syndrome decoding is decoding by using majority logic.

Maximum likelihood

In the maximum likelihood method one tries to find the most probable emitted symbol stream, $\hat{x}_j$, given a received stream of symbols, i.e. find

$$\left\{\hat{x}_j\right\}_{j=0}^{N-1} = \underset{i}{\mathrm{argmax}}\, P(\mathbf{X}_i|\mathbf{Y}) \tag{4.121}$$

where $P(\mathbf{X}_i|\mathbf{Y})$ is the probability of the sequence $\mathbf{X}_i$ being emitted, given that the sequence $\mathbf{Y}$ is received. The parameter N is the length of the sequence. To find this maximum, a measure or metric which is easier to calculate than the probability curve can be used. Given a BSC and the probability that there is an error in a symbol P_s, the probability that d bits differ between $\mathbf{X}_i$ and $\mathbf{Y}$ is

$$P(\mathbf{X}_i|\mathbf{Y}) = P_s^d(1 - P_s)^{N-d} \tag{4.122}$$

The natural logarithm is a monotone function and will therefore not shift the maximum point, and a metric can be defined by

$$\ln P(\mathbf{X}_i|\mathbf{Y}) = -d\ln\left[\frac{1 - P_s}{P_s}\right] + N\ln(1 - P_s) = -A\,d + B \tag{4.123}$$

where A and B are constants with respect to i. The value of eq. (4.123) is a linear function of d. The maximum value for the function is occurring for $d = 0$, i.e. when $\mathbf{X}_i$ and $\mathbf{Y}$ are identical. The maximum of eq. (4.121) can thus be found by finding the $\mathbf{X}_i$ differing from $\mathbf{Y}$ in as few bits as possible. The coefficient d is then a sufficient statistic and can therefore be used as a decision variable instead of the value of $P(\mathbf{X}_i|\mathbf{Y})$. Hence, the simple rule is: choose the sequence with the shortest Hamming distance d.

As in the code tree, the number of different symbol combinations that the receiver has to keep track of increases exponentially with the length of the sequence (Figure 4.26). This leads to tremendous administrative work to keep track of all these combinations, together with the corresponding metrics. The situation can be simplified by the use of the Viterbi algorithm, together with the trellis of the code. Since the number of possible sequences is limited by the Viterbi algorithm, the number of different symbol combinations becomes a constant with time. Assume the trellis representation of the code as in Figure 4.27. Start from the left and follow the time scale along to the right. Register the Hamming distance between the n first received symbols and the symbols indicated at each branch. The letter n indicates as before the number of coded bits per state transition. At each node, or merger, immediately to the right of the branches presently investigated, only the branch having the lowest metric value is preserved. The next set of n received symbols are compared to those branches immediately to the right. The Hamming distance at each branch is added to the value at the node from which that specific branch originates. Again the metric values, now accumulated from the start, are compared to those branches going into the same nodes. Only the path having the lowest metric value at each node is preserved. The process continues, and finally a path through the trellis having a significantly lower metric value than the others will emerge. This path is assumed to be the most probable, and corresponds to the sequence of information symbols which is the most likely, given the received symbol sequence. The method is an application of dynamic programming.

The Viterbi algorithm is thus carried out in such a way that only the path

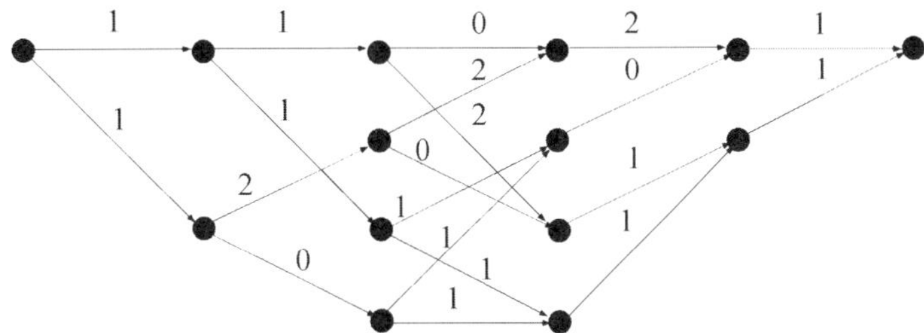

Figure 4.33 Find the correct path through the trellis by using the method used by the Viterbi algorithm. The correct path has a total metric of 3. The numbers indicate the distance for each branch.

through the trellis having the lowest d at each node is preserved, since the other paths coming in to the same node will in the sequel always have a higher metric value. In the case of two branches having the same metric value, both are equally probable and the choice is arbitrary. The preserved path is called the survivor. An example of a trellis with the corresponding metric values indicated at each branch is shown in Figure 4.33. Try to find the correct path!

The condition for a strictly unambiguous result is that the start and stop nodes are known. This is usually not the case in a telecommunication application. In practice, however, the metric will usually, after a number of symbols, differ so much between the correct path and the incorrect ones that it is possible to identify the correct path. When the symbols have been corrupted because of interference in the transfer channel, the error sequence has a finite extension, unless the transfer channel is very poor. An error sequence should therefore start from an error-free state (all other metrics are then considerably larger than the correct one) and end in an error-free state. If the Viterbi detector can handle a large enough number of symbols so the whole error sequence can be accommodated, i.e. if the decoder is deep enough, one can say that the start and stop states are known and a correct decoding can be performed.

Thus the code trellis is stored in the receiver in some way or another. In the decoding, the receiver tries to follow the same path as the transmitter has taken through the code trellis and in this way try to unravel which information stream has generated the received coded symbol stream.

In the same way as in block coding, where long blocks are preferred, convolutional codes with large constraint length are preferred from an error correcting point of view. The drawback is that the trellis will contain many states, which together with a large decoding depth gives many paths to search among. Therefore, it is reasonable to further limit the number of possible paths by some effective tree search algorithm. One way to limit the number of paths is to give up the paths where, after a given number of symbols, the metric value has reached a fixed upper limit, the so-called metric roof. The search strategy can be to search on metric, depth, or breadth first in the code tree. An algorithm searching on metric first always follows the route with the smallest metric. If this path reaches the metric roof the algorithm jumps back to some bifurcation, takes the second best path and follows it. The algorithm reaches the end point when there are no more symbols to decode. When some path reaches the end point without hitting the metric roof the algorithm probably has found the correct path and assumes the winning path as its estimate of coded symbol sequence. Another alternative is to search on depth first. All paths are searched one by one in arbitrary order until they reach a given combination of depth and metric value. The ones that have reached this roof are dropped and are not pursued. The path that reaches the endpoint first is the winning path. If no path has reached the endpoint, the metric value is increased and a new search is done. A third alternative is to search on breadth first, i.e. a given number of the best paths are saved after each symbol. When the end point is reached the correct path is probably among the paths tracked. Of course these algorithms do not give the most probable symbol sequence since the most probable path can be given up whereas it reaches the metric roof early, but should maybe later only have given a small increase in the accumulated metric value. The Viterbi algorithm for example, is a breadth first algorithm where the number of preserved paths is the same as the number of states in the code and therefore always gives the most probable sequence. Several of the algorithms implemented in practice are combinations of the methods described above.

Syndrome decoding of convolutional code

Assume a systematic convolutional code, i.e. information bits followed by check bits. For a rate 1/2 code, for instance, we have one information bit followed by one check bit, i.e. $\mathbf{X} = [x_1', x_1'', x_2', x_2''...$ The stream of encoded symbols has been generated by setting $x_j' = m_j$ and

$$x_j'' = \sum_{i=0}^{L} m_{j-i} \, g_i \ (\mathrm{mod}\ 2) \tag{4.124}$$

Define a generator matrix $\mathbf{G}$, so that $\mathbf{X} = \mathbf{MG}$, in the same way as for block codes. A parity check matrix $\mathbf{H}$ is defined, so that the product $\mathbf{XH}^{\mathrm{T}} = \{0\ 0...,$ which becomes a semi-infinite vector. The matrix $\mathbf{G}$ becomes semi-infinite and, according to the definition above, eq. (4.124) etc., it can be written as

$$
\mathbf{G} = \begin{vmatrix}
1 & g_0 & 0 & g_1 & 0 & g_2 & 0 & g_3 & 0 & & & 0 \\
 & & 1 & g_0 & 0 & g_1 & 0 & g_2 & 0 & g_L & \\
 & & & & 1 & g_0 & 0 & g_1 & 0 & g_{L-1} & \longrightarrow & \infty \\
 & & 0 & & & & 1 & g_0 & 0 & g_{L-2} & \\
 & & & & & & & & 0 & \vdots & \\
 & & & \downarrow & & & & & 0 & \vdots & \\
 & & \infty & & & & & & & &
\end{vmatrix}
$$

In order to not make the presentation too long, only the rate 1/2 encoder is treated. The method is of course valid for a code of arbitrary rate. The parity check matrix then becomes

$$
\mathbf{H} = \begin{vmatrix}
g_0 & 1 & & & 0 & \\
g_1 & 0 & g_0 & 1 & & \\
g_2 & 0 & g_1 & 0 & g_0 & 1 & \longrightarrow & \infty \\
 & & & & & & \\
g_L & 0 & & & & & \\
0 & 0 & g_L & 0 & g_{L-1} & & \\
 & 0 & & & g_L & 0 & \\
 & & \downarrow & & & \\
 & \infty & & & &
\end{vmatrix}
$$

The syndrome $\mathbf{S}$ is calculated in the usual way and, just as before, the codeword part, $\mathbf{X}\mathbf{H}^T$, becomes zero,

$$
\mathbf{S} = \mathbf{Y}\mathbf{H}^T = (\mathbf{X} + \mathbf{E})\mathbf{H}^T = \mathbf{E}\mathbf{H}^T \tag{4.125}
$$

An error event is assumed to be finite or perfectly corrected after L bits. The error vector, $\mathbf{E}$, is then finite if all bits are equal to zero, outside of the error event, are truncated. As a consequence of this, the syndrome vector $\mathbf{S}$ is also of finite length. An element of $\mathbf{S}$ then becomes

$$
s_j = y_j'' \oplus \sum_{i=0}^{L} y_{j-i}' g_i = e_j'' \oplus \sum_{i=0}^{L} e_{j-i}' g_i \tag{4.126}
$$

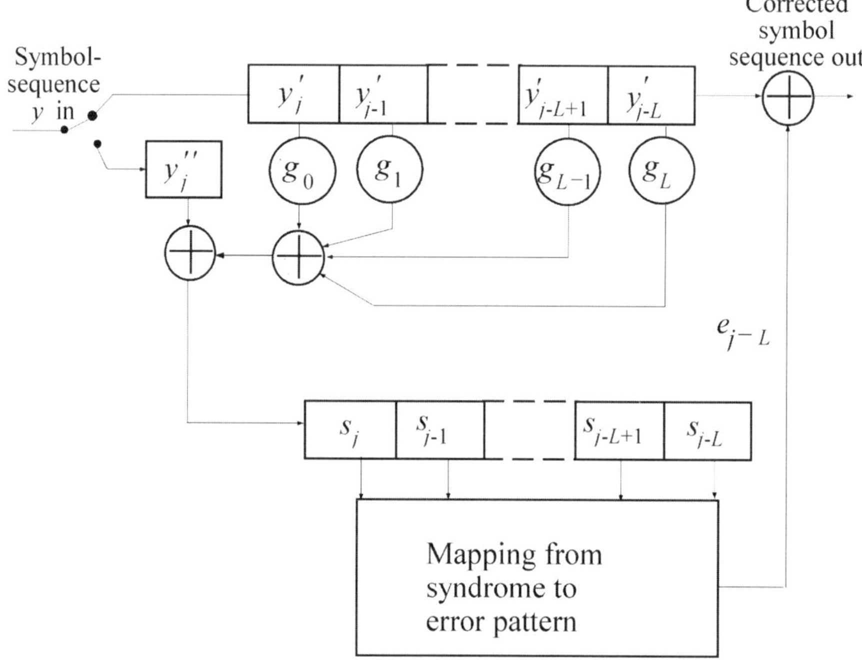

Figure 4.34   Incomplete syndrome decoder.

The L syndrome elements which form the syndrome vector **S,** are then successively calculated. The syndrome is stored in a register of length L, and a mapping from syndrome to error pattern can be performed in the same way as before (Figure 4.34; see also Figure 4.14).

Note that the computation of the syndrome is done by coding the received uncoded bit, plus errors if any, and comparing that to the check bit that has been encoded on the transmitter side.

Assume that $L = 3$. The syndrome vector then consists of four elements. Assume the latest calculated syndrome element is s_j. The complete syndrome vector then is $\mathbf{S} = \begin{bmatrix} s_j & s_{j-1} & s_{j-2} & s_{j-3} \end{bmatrix}$. The syndrome register contains syndrome elements consisting of the following sums:

$$s_j = e''_j \oplus e'_j g_0 \oplus e'_{j-1} g_1 \oplus e'_{j-2} g_2 \oplus e'_{j-3} g_3$$

$$s_{j-1} = e''_{j-1} \oplus e'_{j-1} g_0 \oplus e'_{j-2} g_1 \oplus e'_{j-3} g_2 \oplus e'_{j-4} g_3$$

$$s_{j-2} = e''_{j-2} \oplus e'_{j-2} g_0 \oplus e'_{j-3} g_1 \oplus e'_{j-4} g_2 \oplus e'_{j-5} g_3 \tag{4.127}$$

$$s_{j-3} = e''_{j-3} \oplus e'_{j-3} g_0 \oplus e'_{j-4} g_1 \oplus e'_{j-5} g_2 \oplus e'_{j-6} g_3$$

Since $L = 3$, the elements e'_{j-4}, e'_{j-5} and e'_{j-6} of the error vector will already have been corrected, which is indicated by the shaded area in eq. (4.127). The corrected elements must be removed from the syndrome. This can be done by a modification of the syndrome pattern using a feedback which erases these elements. From eq. (4.127) it can be seen that the syndrome component e'_{j-L}, in this case e'_{j-3}, must be corrected in the following way:

$$s_j^{corr} = s_j \oplus e'_{j-3} g_3$$

$$s_{j-1}^{corr} = s_{j-1} \oplus e'_{j-3} g_2$$

$$s_{j-2}^{corr} = s_{j-2} \oplus e'_{j-3} g_1$$

$$s_{j-3}^{corr} = s_{j-3} \oplus e'_{j-3} g_0 \tag{4.128}$$

The content of the shift register is then shifted one step to the right. We then obtain $s_{j-1} = s_j^{corr}$, $s_{j-2} = s_{j-1}^{corr}$, and so on. Note that g_0 does not need to be included, as the element which possibly would have been set to zero by this coefficient is shifted out anyway. The feedback is depicted in Figure 4.35.

The last section of the decoder to be treated is the mapping from syndrome to error pattern. This can be done as before by table look up. Another commonly used method is majority decoding, which works on particular codes. The method

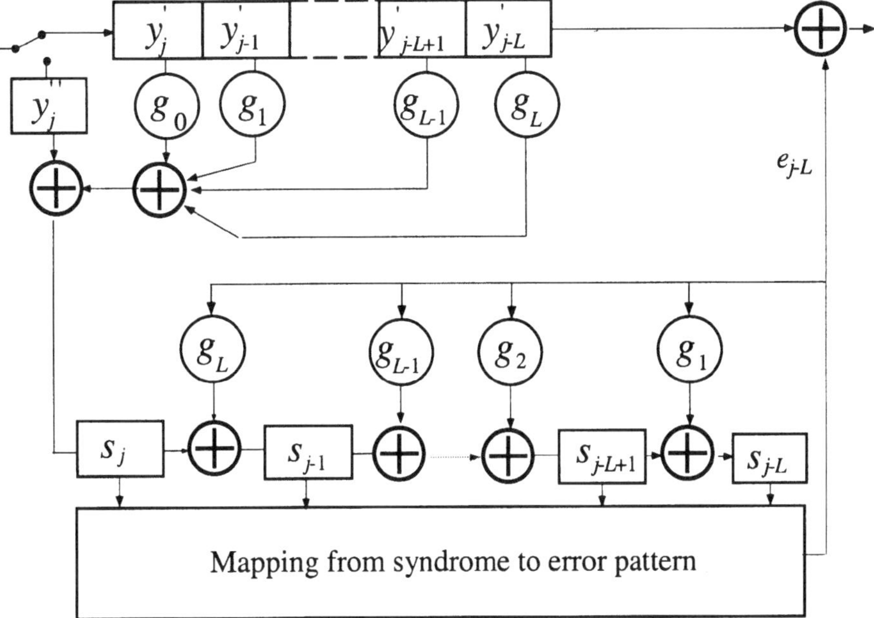

Figure 4.35 Syndrome decoder with feedback.

is illustrated by the following example. A rate 1/2 systematic convolutional code with $L = 6$ has been chosen to clarify the principle. Assume the following generator polynomials obtained from a table (in octal form 001, 145):

$$G_1(p) = 1, \quad G_2(p) = 1 + p^2 + p^5 + p^6 \tag{4.129}$$

Study the syndrome components s_0, s_2, s_5 and s_6 corresponding to the exponents of the check symbol generator polynomial. To create a running time index, j, the following modification is made. s_6 is replaced by s_j and s_5, which is a component formed one time unit before, becoming s_{j-1} etc. Starting from s_0, s_2, s_5 and s_6, the following equations can be set up by using eq. (4.126) (compare also eq. (4.127)):

$$
\begin{aligned}
s_j &= & e'_{j-6} \oplus e'_{j-5} & & \oplus e'_{j-2} & & \oplus e'_j & \oplus e''_j \\
s_{j-1} &= & e'_{j-7} \oplus e'_{j-6} & \oplus e'_{j-3} & & \oplus e'_{j-1} & & \oplus e''_{j-1} \\
s_{j-4} &= & e'_{j-10} \oplus e'_{j-9} & \oplus e''_{j-6} & \oplus e''_{j-4} & & & \oplus e''_{j-4} \\
s_{j-6} &= e'_{j-12} \oplus e'_{j-11} & \oplus e''_{j-8} & \oplus e''_{j-6} & & & & \oplus e''_{j-6}
\end{aligned}
\tag{4.130}
$$

All sums contain the element e'_{j-6}. The other elements only appear in one of the syndrome elements. If we assume one error (or only a few), only a minority of the syndrome elements are $\neq 0$. If, on the other hand, the majority of the syndrome elements are $\neq 0$, we assume that e'_{j-6} is incorrect. A simple logical operation, or so-called majority logic, can be used to determine if y'_{j-6} is to be corrected.

Example 4.12

The generator polynomials for a rate 1/3 convolutional code with span 3 (3,1,2) is defined by

$$\mathbf{g}_1 = \begin{bmatrix} 1 & 0 & 1 \end{bmatrix}, \quad \mathbf{g}_2 = \begin{bmatrix} 1 & 1 & 1 \end{bmatrix}, \quad \mathbf{g}_3 = \begin{bmatrix} 1 & 1 & 1 \end{bmatrix}$$

Construct the trellis and state diagram of the code. Give the state sequence and bit stream generated by the input sequence 1011001111. Construct the flow graph, identify the paths having the lowest Hamming weight and determine the values of d_{free}.

Solution   The state diagram gets 2^L nodes, i.e. for the (3,1,2) code $2^2 = 4$. Assume the initial state $x'_j = x''_j = x'''_j = 0$. Eq. (4.109) can be rewritten.

$$x'_j = m_{j-2} \oplus m_j, \quad x''_j = x'''_j = m_{j-2} \oplus m_{j-1} \oplus m_j \tag{4.132}$$

which results in the tree diagram shown in Figure 4.36. In the tree diagram all states of the coder can be identified and a trellis can thus be drawn (Figure 4.37).

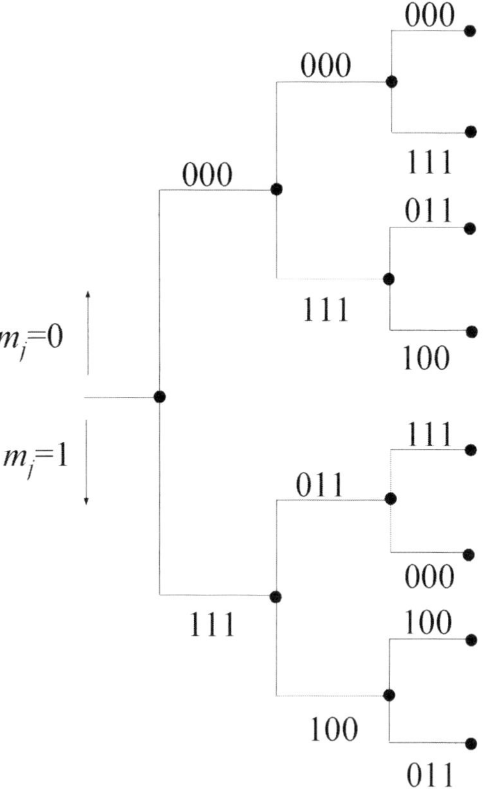

Figure 4.36   Tree diagram.

From the state diagram shown in Figure 4.38, the state and output sequences for the input signal 1011001111 become

state	a	b	c	b	d	c	a	b	d	d
output seq.	111	011	000	100	100	111	111	100	011	

The flow graph is obtained by taking a as the initial node, dividing into the beginning and end states (assume 000 as the correct word; Figure 4.39).

According to the previous definition d_{free} is equal to the Hamming distance of the path having the shortest distance between a_{start} and a_{stop}. The path $a_{start}\, b\, c\, a_{stop}$ gives $W^8 I$ and $a_{start}\, b\, d\, c\, a_{stop}$ gives $W^8 I^2$, which results in $d_{free} = 8$.

Another way of drawing the flow graph is to use a block diagram (Figure 4.40). In the case when the different paths are not as easily identified as in the

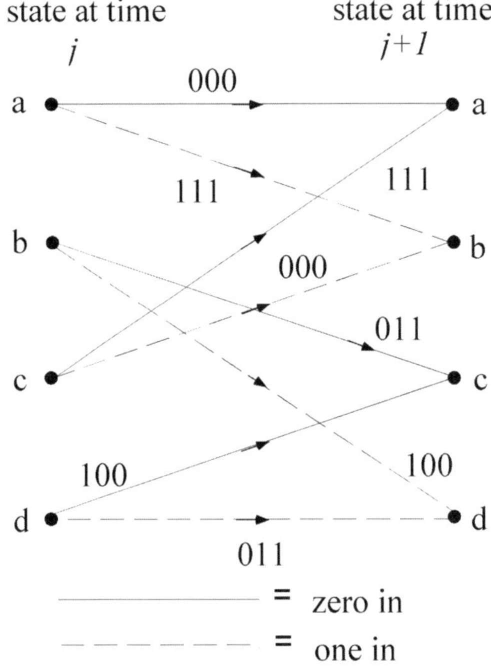

Figure 4.37   Trellis.

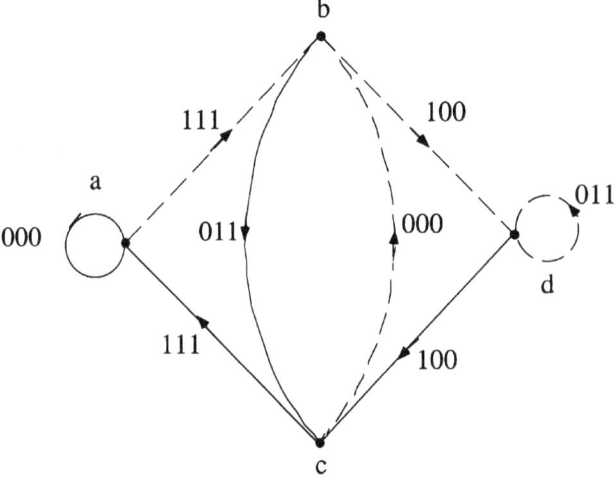

Figure 4.38   State diagram.

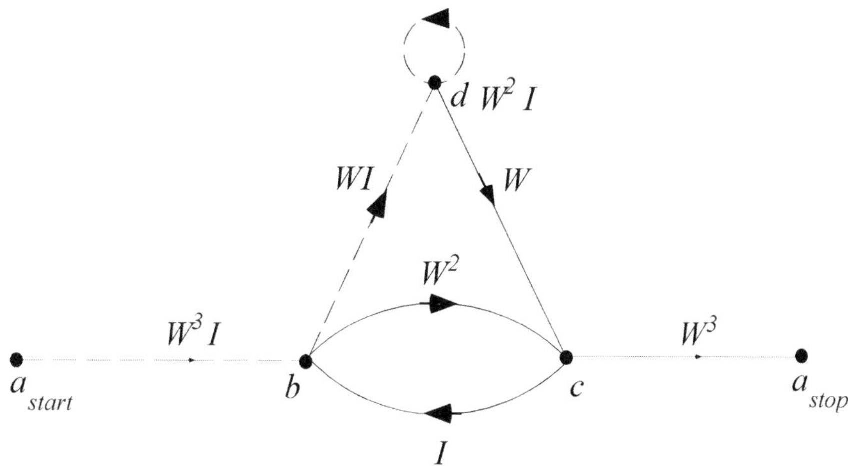

Figure 4.39   Modified flow graph. The exponent of W equals the Hamming distance from 000. For I^α, α equals 0 if the message bit is 0 and α equals 1 if the message bit is 1.

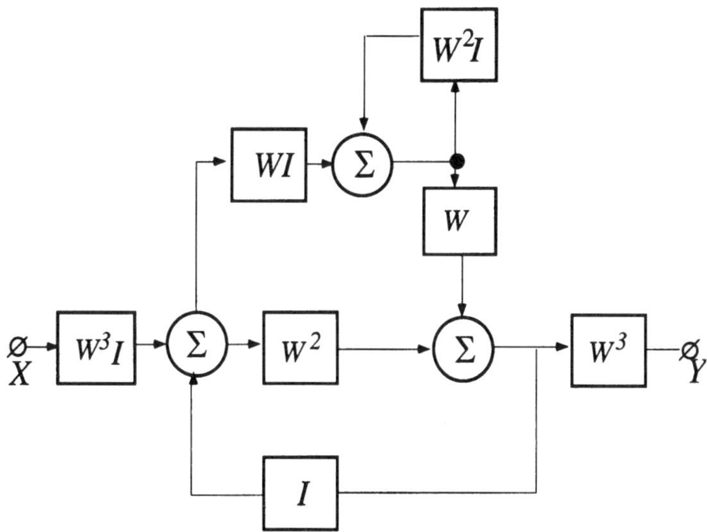

Figure 4.40   The flow graph drawn as a block diagram.

example above, the transfer function must be calculated. This is created by taking the ratio between output and input signals in the modified state diagram or block diagram:

$$\frac{Y}{X} = T(W, I) = \frac{W^8(I + I^2) - W^{10}I^2}{1 - W^2(2I + I^2) + W^4 I^2} \tag{4.133}$$

Performing the division to calculate the ratio in eq. (4.133) gives

$$
\begin{array}{r}
W^8(I+I^2)+W^{10}(I^2+3I^3+I^4)+\ldots \\[4pt]
\hline
\end{array}
$$

$$
1-W^2(2I+I^2)+W^4I^2 \Big\rvert \ W^8(I+I^2)-W^{10}I^2
$$

$$
\underline{-W^8(I+I^2)+W^{10}(2I^2+3I^3+I^4)-W^{12}(I^3+I^4)}
$$

$$
W^{10}(I^2+3I^3+I^4)-W^{12}(I^3+I^4) \tag{4.134}
$$

$$
\text{etc.}
$$

This can be expressed as an infinite series,

$$
T(W,I) = W^8(I + I^2) + W^{10}(I^2 + 3I^3 + I^4) + \cdots \tag{4.135}
$$

From eq. (4.135) it can be seen that the shortest distance is 8, which is given by W^8. There are two paths with this distance, one having one bit error and one having two. This is obtained from $(I + I^2)$.

Example 4.13

Assume a received sequence of $\mathbf{Y} = [10\ 11\ 00\ 11\ 01\ 10\ 10\ 00]$. The convolutional code used is the same as the one used as an example in Section 4.4.2. Apply the Viterbi algorithm to find the most probable message sequence, $\hat{\mathbf{M}}$. How many symbol errors are there in the sequence? For the solution of this problem it is assumed that if two paths, having identical accumulated metric, coincide at one of the nodes, the upper branch is maintained.

Solution   The check bit equations for the code are obtained from eq. (4.111):

$$
x'_j = m_j \oplus m_{j-1} \oplus m_{j-2} \quad \text{and} \quad x''_j = m_j \oplus m_{j-2} \tag{4.136}
$$

The code tree and the trellis have already been shown in Figures 4.26 and 4.27, respectively. But to simplify the decoding a section of the trellis is redrawn in Figure 4.41. The state diagram has 2^L states, which for this (2,1,2) code becomes $2^2 = 4$.

A number of trellises are positioned after each other. Then the Viterbi algorithm can be applied to obtain an estimate of the transmitted message. A known start value must be assumed, for instance a (Figure 4.42). Hence the estimated bit stream $\hat{\mathbf{M}} = [01001101]$ and 3 symbol errors have occurred in $\mathbf{Y}$, which appears from the accumulated metric that is 3.

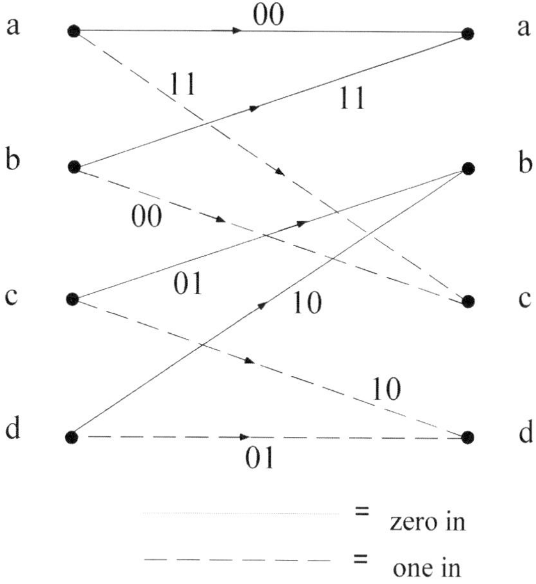

Figure 4.41   The trellis.

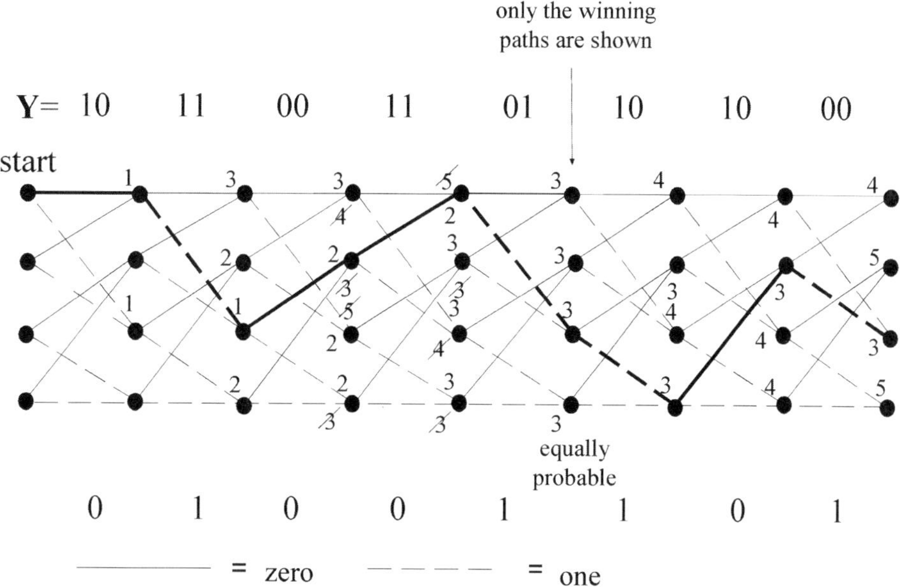

Figure 4.42   The input signal is written above the trellis and the estimated bit stream is written below. The accumulated metric along the surviving paths is written at the nodes. The path giving the smallest metric is preserved, indicated by crossing out the larger metric.

Example 4.14

Assume a systematic (2,1,4) code suitable for majority decoding, where $x_j'' = m_{j-4} \oplus m_{j-3} \oplus m_j$. Determine the parity check equations for s_{j-4} to s_j and draw a diagram of the decoder.

Solution   For a systematic code $x_j' = m_j$ holds, and in conjunction with $x_j'' = m_{j-4} \oplus m_{j-3} \oplus m_j$ the generator polynomials are obtained:

$$G_1(p) = 1, \quad G_2(p) = 1 + p^3 + p^4 \tag{4.137}$$

Consider the components s_0, s_3 and s_4 of the syndrome, which with a running time index become

$$s_j = e_{j-4}' \oplus e_{j-3}' \oplus e_j' \oplus e_j''$$

$$s_{j-1} = e_{j-5}' \oplus e_{j-4}' \oplus e_{j-1}' \oplus e_{j-1}''$$

$$s_{j-4} = e_{j-8}' \oplus e_{j-7}' \oplus e_{j-4}' \oplus e_{j-4}'' \tag{4.138}$$

It can be seen that e_{j-4}' is included in all check sums. The check sums are therefore said to be orthogonal to e_{j-4}'. If all check sums $\neq 0$, it can be assumed that e_{j-4}' is incorrect. As a consequence of the errors that have been corrected up to time j, the syndrome elements are changed through the feedback as is shown in eq. (4.128). The resulting decoder is shown in Figure 4.43.

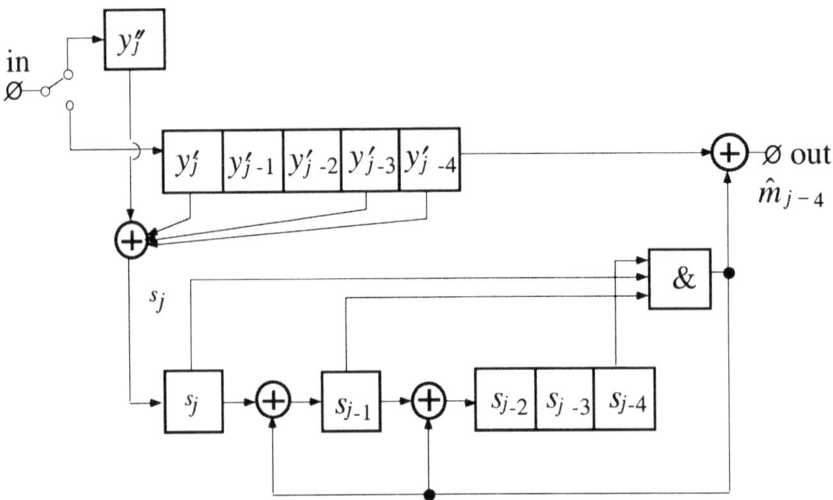

Figure 4.43   Decoder with majority logic.

4.5 Interleaving

The error-detecting and error-correcting codes work best if the errors are randomly distributed over the symbol stream. In many channels the errors have a tendency to appear in bursts, however. This is true, for instance, in the radio channel where fading leads to burst errors. Fading appears mainly in mobile communication. The signal power varies, since the signal from a transmitter travels along several paths, which gives rise to an alternating destructive or constructive interference when at least one of the terminals is moving. The symbol errors therefore tend to occur during destructive interference. The solution to the problem of burst errors is to change the order of the symbols after performing channel coding at the transmitting side. Adjacent symbols are thereby separated. When burst errors occur during transmission, several adjacent symbols become incorrect. When the receiver reorganises the symbols in correct order, the incorrect symbols become distributed over a larger time interval, and have correct symbols in between. This provides improved possibilities for the error-correcting code to detect and correct burst errors. The disadvantage of this procedure is that a time lag is introduced in the transmission. The maximum allowed time lag will then limit the depth of the interleaving. In Example 4.15 interleaving and its effects are described in more detail.

Practically the interleaving can be arranged, for example, such that at the transmiter symbols are successively written along the lines in a rectangular matrix. The reading is done along the columns. In the receiver writing and reading are done in the reverse direction, compared to the transmitter.

Example 4.15

Suppose a TDMA system transmits information in frames of 8 symbols length. Binary symbols are used. The errors appear in bursts, e.g. due to fading, so that the error rate is high in particular frames. Suppose that the probability for such a bad frame is $P_f = 0.1$. In the bad frames the symbol error probability is $P_{elf} = 0.5$ and in the good frames the symbol error probability is $P_{eln} = 0$.

(a) What is the probability for a received frame to be error free?

(b) Suppose that the information has been encoded with an error correcting code that can correct one symbol per frame. No interleaving is used. Compute the probability for an error free frame. Suppose for simplicity that each frame still contains 8 bits and that P_{elf} is unchanged.

(c) Suppose that interleaving over eight frames is used (cf. Figure 4.44). What is the probability of getting an error free frame after de-interleaving, without coding being used?

(d) Compute the probability for an error free frame if both coding and interleaving are being used.

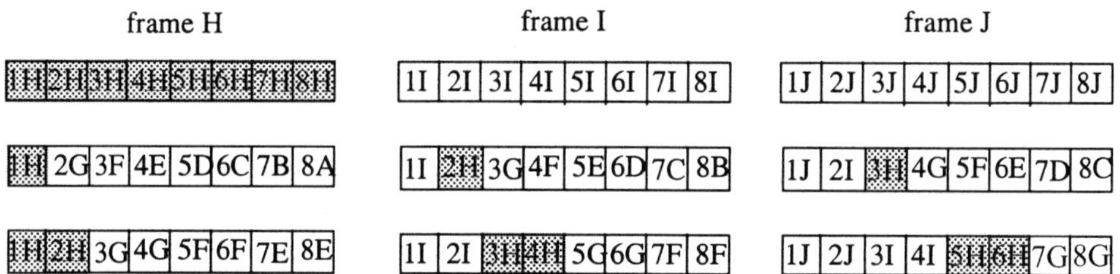

Figure 4.44 Interleaving of 8 symbols. The first three frames are shown and symbols from frame H are indicated in grey. The uppermost row shows the non-interleaved symbol stream.

(e) Suppose instead that interleaving over four frames is being used (cf. Figure 4.44). What is the probability for an error free frame?

Solution (a) Find the probability for the outcome: (good frame) ∨ (bad frame with no errors). Note that the entire frame is either a bad or good frame. The probability for a good frame is $P_n = 1 - P_f = 0.9$. The probability for a bad frame to be error free is $P_{0|f} = (1 - P_{e|f})^8 = 0.5^8 \approx 3.9 \times 10^{-3}$. The probability for an error free frame then is

$$P_0 = (1 - P_{e|n})^8 P_n + P_{0|f} \times P_f = 1^8 \times 0.9 + 3.9 \times 10^{-3} \times 0.1 \approx 0.9004$$

(4.139)

(b) The error correcting code can correct a single error. Find therefore the probability for the outcome (no errors) ∨ (one error), $P_0 + P_1$. Erroneous symbols only appear in faded frames. The probability for an erroneous symbol in such a frame is

$$P_{1|f} = \binom{8}{1}(1 - P_{e|f})^7 P_{e|f} = 8 \times 0.5^8 \approx 0.031$$

The probability for an error free frame after error correcting coding is

$$P_0 + P_1 = P_0 + P_{1|f} \times P_f \approx 0.9004 + 0.031 \times 0.1 \approx 0.904$$ (4.140)

Note that under the assumptions given in this example, i.e. burst errors, the bit error rate is only marginally reduced by the error correcting coding.

(c) When de-interleaving is done, the symbols are picked one by one from the received frames, which can be bad or good frames. The symbols picked out are put together into a new frame which therefore will contain symbols that may come from bad or good frames. A de-interleaved frame therefore consists of a random combination of symbols from bad or good frames. First, find the probability that a given symbol is correct: $P_{0|s} = (1 - P_{e|n})P_n + (1 - P_{e|f})P_f = 1 \times 0.9 + 0.5 \times 0.1 = 0.95$. The probability for an entire frame to be correct is

$$P_{0|s}^8 = 0.95^8 \approx 0.66 \tag{4.145}$$

The errors are now spread over several frames so that the probability of finding an error free frame has diminished. As a consequence the erroneous frames contain fewer errors since the total number of erroneous symbols has not been influenced.

(d) The error correcting code can correct a single error. Find the probability that a de-interleaved frame will contain no errors or one error:

$$P_{0|s}^8 + \binom{8}{1}(1 - P_{0|s})P_{0|s}^7 = 0.95^8 + 8(1 - 0.95)0.95^7 \approx 0.943 \tag{4.146}$$

The error probability has therefore been reduced from $1 - 0.904 \approx 0.1$ to $1 - 0.947 \approx 0.06$. This corresponds to an increase in power of approx. 3 dB in the transmitter, which otherwise should be required in order to reduce the error probability to 0.06.

(e) In this case two symbols are picked from four received frames and are put together into a de-interleaved frame. Therefore, study symbol pairs and find the probability that a frame will contain no errors or one error. The probability for no errors in a symbol pair is $(1 - P_{e|n})^2 P_n + (1 - P_{e|f})^2 P_f = 0.9 + 0.5^2 \times 0.1 = 0.925$. Errors only appear in faded frames. The probability for a single error in a symbol pair then is $2(1 - P_{e|f})P_{e|f}P_f = 0.05$. The probability for no errors in an entire frame is 0.925^4 and the probability for a single error in an entire frame is $4 \times 0.925^3 \times 0.05$. The total probability for an error free frame then is $0.925^4 + 4 \times 0.925^3 \times 0.05 = 0.8904$, which can be compared to the answer in (d). This shows that it is more effective to spread the symbols over many frames than a few frames.

4.6 Turbo coding

Turbo coding is a method of error correcting coding which offers very good properties at the cost of a certain delay in the system. The method is used, for instance, in third generation cellular phone systems and gives transmission properties close to Shannon's limit compared to other methods.

Turbo coding is based on the following three ideas: parallel concatenation of two systematic convolutional codes, reducing the number of paths with a distance close to d_{free} by introducing feedback in the coder, and finally performing an iterative decoding.

In Figure 4.45, an example of the layout of a rate 1/3 turbo encoder is shown. The coded bit stream consists of the symbols m_n, x_{1n} and x_{2n} in sequence, where n is a running time index. If $\mathbf{x}_n = [m_n \ x_{1n} \ x_{2n}]$, an infinitely long sequence can be written $\{\mathbf{x}_n\}_{n=-\infty}^{\infty}$. For the coded sequences to be regarded as statistically independent, interleaving is used to feed one of the two encoders with a permutation of the information sequence. The interleaving is vital for the function of the turbo

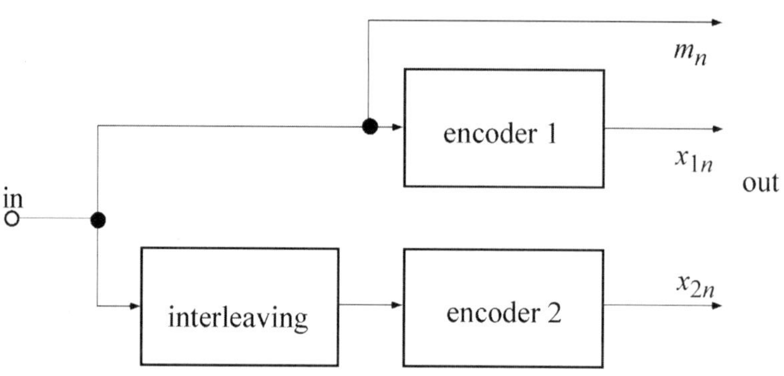

Figure 4.45   Rate 1/3 turbo encoder.

coding. To avoid statistical dependence, interleaver implementations which give too regular permutations, for instance cyclic shifts, must be excluded.

Since the coded sequences are independent, the decoder can decode the respective sequence separately and compare the results. Another way, which is described in more detail in Section 4.6.2, is to let the second decoder perform a decoding using the result of the first as a support. The result of the second decoder can be used to support a renewed decoding in the first decoder, etc. This creates a loop and iterative decoding can be performed where the errors are successively reduced. Finally, a situation is reached where the errors no longer decrease between each iteration. The decoding is then completed. The way of feeding the output signal back to the input of the decoder to improve the decoding is similar to the function of a turbo engine, which justifies the name given to the coding method. The turbo coding can be performed with block codes or convolutional codes. In the following only convolutional codes will be described; they were also the first implementations of turbo codes to be presented.

4.6.1 CODING

The coding is done by using feedback convolutional encoders. An example of such a coder is shown in Figure 4.46. The code is systematic, so the uncoded symbol makes up one part of the bit stream. The other part is generated by the shift register, in which the content in certain memory cells is fed back by connecting them to the input of the register. The reason for using feedback is that the output signal gets a more random character than that of convolutional encoders without feedback. Consider, for instance, an input signal consisting of a number of zeros followed by a 'one' and then zeros again, …0001000…. The output signal would in this case only be influenced by this single 'one' during three clock cycles if there where no feedback. The output signal of the coder would then become zero again. The corresponding input

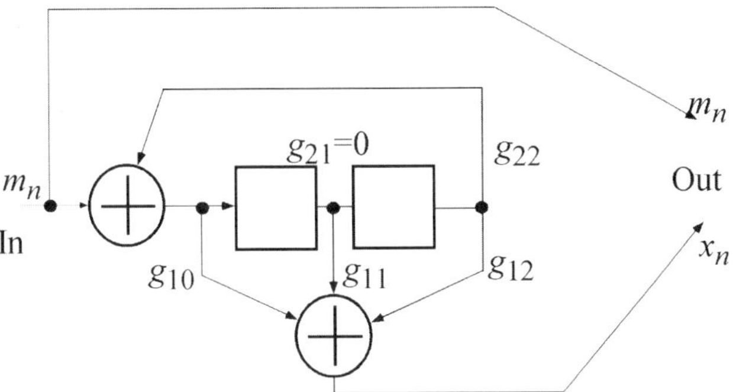

Figure 4.46 Coder for recursive systematic convolutional code. The symbol g_{ij} represents the coefficient j of polynomial i.

signal in the feedback coder gives an infinitely long response; cf. the FIR and IIR filters in signal processing. The free distance, d_{free}, becomes the same as for the non-feedback coder, but the distribution of different error paths, cf. eq. (4.113), are different. This leads to better performance at low signal to noise ratios in comparison to the non-feedback coder.

For a non-recursive convolutional code we have, for instance for a rate 1/2 code, two generator polynomials $G_1(p)$ and $G_2(p)$; see eq. (4.111). For a systematic code, one of the generator polynomials must be a one. To create a recursive coder, the generator polynomial must consist of the ratio between two polynomials, compare the transfer function for recursive linear filters. A suitable ratio can be obtained by dividing the generator polynomials of two non-recursive encoders, such as $G_1(p)$ and $G_2(p)$. For the turbo code, the polynomials

$$[1, G_1(p)/G_2(p)] \quad \text{or} \quad [1, G_2(p)/G_1(p)] \tag{4.147}$$

are used. Assume the code vectors $\mathbf{g}_1 = (7)_8$ and $\mathbf{g}_2 = (5)_8$, which give the polynomials $G_1(p) = p^2 + p + 1$ and $G_2(p) = p^2 + 1$. Form, for instance, the ratio $G_1(p)/G_2(p) = p^2 + p + 1/[p^2 + 1]$. The coefficients of the numerator determine which taps are added up and fed back to the input of the coder. An example of a coder with the generator polynomial given above is shown in Figure 4.46.

The complete encoder can now be realised by two identical or different recursive encoders (cf. Figure 4.45). Two rate 1/2 encoders give a rate 1/3 code, since the uncoded bit does not have to be transmitted for both encoders.

4.6.2 DECODING

To make the decoding as efficient as possible soft and iterative decoding is used. This is performed in such a way that the uncertainty of the decoder about the received signal is reflected in the output signal of the decoder. An AWGN channel, and not a BSC channel, is then used to derive the detector. The Bahl algorithm (also called the BCJR algorithm after the initials of the inventors family names) is described below. Another alternative is a modification of the Viterbi algorithm, to give a soft output signal. This is called the soft output Viterbi algorithm (SOVA)). It does, however, give a slightly worse result than the Bahl algorithm.

Algorithm for a sequential maximum a posteriori detector

Every state transition in a convolutional code gives rise to a number of symbols which constitute a part of the coded stream. The turbo codes are systematic convolutional codes, which means that the message bit continues unchanged to the coded stream. For a rate 1/2 coder (an extension to rate 1/3 will be done later), the symbol stream can be divided in pairs, where m_n is the message symbol and x_{1n} the output signal from the shift register, which gives $\mathbf{x}_n = [m_n, x_{1n}]$ (cf. Figures 4.26 and 4.27).

The received signal consists of

$$\mathbf{y}_n = \begin{bmatrix} y_{mn} & y_{1n} \end{bmatrix} = \begin{bmatrix} m_n + e_{mn} & x_{1n} + e_{1n} \end{bmatrix} = \mathbf{x}_n + \mathbf{e}_n \tag{4.148}$$

where $\mathbf{e}_n$ is a 1×2 vector consisting of independently Gaussian distributed samples with variance σ^2 and which represents the noise added as the symbols traverse the channel, i.e. an AWGN channel. The received signal then consists of a stream of symbols contaminated by noise. Study a block consisting of N samples of the sequence, which can be written $\{\mathbf{y}_n\}_{n=1}^N = \mathbf{y}_1^N$.

One would like to restore $\{m_n\}_{n=1}^N$ with as few errors as possible from the received sequence $\mathbf{y}_1^N$. Mathematically this can be formulated through the maximum a posteriori or the MAP receiver, which selects the most probable of the transmitted symbols given a certain received signal, $\mathbf{y}_1^N$. The probability that symbol i has been transmitted at time k when the signal sequence $\mathbf{y}_1^N$ has been received can be written $\Pr(m_k = i|\mathbf{y}_1^N)$. For a binary decision there are two alternatives to choose from, hypothesis H_1, which for instance corresponds to $m_k = +1$, or hypothesis H_2, which then corresponds to the transmitted symbol $m_k = -1$. For a binary decision at time k, H_1 or H_2 is chosen, depending on which has the largest probability.

$$\Pr(m_k = +1|\mathbf{y}_1^N) \underset{H_1}{\overset{H_2}{\lessgtr}} \Pr(m_k = -1|\mathbf{y}_1^N) \tag{4.149}$$

This can be written in the form

$$\frac{\Pr(m_k = +1|\mathbf{y}_1^N)}{\Pr(m_k = -1|\mathbf{y}_1^N)} \underset{H_1}{\overset{H_2}{\lessgtr}} 1 \tag{4.150}$$

It is more practical to use the logarithmic probability ratio. After rewriting, where $\Pr[m_k = i, \mathbf{y}_1^N] = \Pr[m_k = i|\mathbf{y}_1^N] \cdot \Pr[\mathbf{y}_1^N]$ has been used, we obtain

$$\Lambda(m_k) = \log\left(\frac{\Pr[m_k = +1, \mathbf{y}_1^N]}{\Pr[m_k = -1, \mathbf{y}_1^N]}\right) \tag{4.151}$$

Since $\log(1) = 0$, the decision limit will now be zero. Given that $\Lambda(m_k)$ is available it is possible to make a hardware decision by using a limiter, for instance an operational amplifier with very high gain. The decision then becomes $\hat{m}_k = \mathrm{sgn}[\Lambda(m_k)]$. In order to produce as good a detector as possible the problem is therefore to obtain a good estimation of $\Lambda(m_k)$.

For the presentation of the Bahl algorithm it is practical to let $\Pr[\ ]$ denote either the probability distribution or the probability density, depending on whether a discrete or continuous stochastic variable is described. The reason is that for a continuous stochastic variable X, the value x is assumed with probability $P(X = x) = 0$. In order to avoid this, the probability in a small interval x to $x + dx$ is studied. The probability then becomes $\Pr[x] = P(x \leq X \leq x + dx) \cong p(x)dx$. When the ratio between two probabilities is calculated, dx can be removed and only the densities, $p(x)$, for the respective alternative remains. $\Pr[\ ]$ can therefore also be used as a density.

The Bahl algorithm

The derivation is quite long, but is nevertheless described here as it gives a good insight into how the probability theory can be used for the detection of signals, and to give the reader a chance to understand an algorithm which gives very good results in a turbo decoder. The algorithm may seem complicated, but because of its recursive nature it is still relatively easy to implement on a computer. In the treatment, the fact that for independent events A, B and C, the probability for the joint event $\Pr\{A,B,C\} = \Pr\{A\}\cdot\Pr\{B\}\cdot\Pr\{C\}$, is utilised. This can be used to split up the sampled signal into three parts $\mathbf{y}_1^{k-1}$, $\mathbf{y}_k$ and $\mathbf{y}_{k+1}^N$. An example of a trellis for a convolutional code is shown in Figure 4.47. Assume that the trellis has S states. Two arbitrarily chosen states s and s' have been labelled in the figure, but s and s' can assume any value between 1 and S.

Assume that a complete sequence from $n = 1$ to $n = N$ has been received. Follow, for example, the path which has been marked with bold lines. As usual, start and end states are assumed to be known. Study the state s' at time $k - 1$. Given that s' and $\mathbf{y}_1^N$ are stochastic variables the probability that the transmitter was in state s' at time $k - 1$ is

$$\alpha_{k-1}(s') = \Pr\left[s', \mathbf{y}_1^{k-1}\right] \tag{4.152}$$

We are now at time $k - 1$ in state s' and want to move on within the trellis. Study the state transition at time $k - 1$ to k. Assume that the transition takes place from state s' to s. Allowed transitions are determined by the trellis. The transition probability depends on where in the trellis one starts, on the signal which was

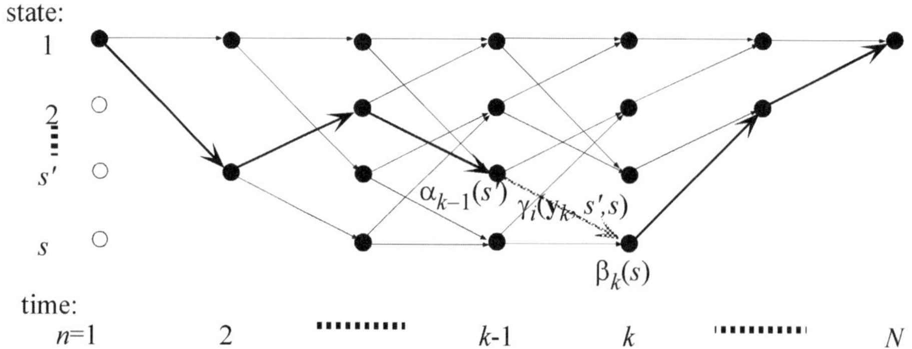

Figure 4.47　Example of a trellis for a convolutional code. The states s and s' indicate a pair of states at times k and $k - 1$, respectively among the 1 to S available states.

received at time k and also on whether it is assumed that $+1$ or -1 has been sent. Therefore $m_k = i$ is included as a parameter. Given that state s' is the starting point the transition probability becomes

$$\gamma_i(\mathbf{y}_k, s', s) = \Pr[m_k = i, s, \mathbf{y}_k | s'\} \tag{4.153}$$

Since all samples from $n = 1$ to $n = N$ are available, the estimation can be improved by also computing backwards, i.e. from N towards time k. From the end state along the marked path to s, the probability becomes

$$\beta_k(s) = \Pr[\mathbf{y}_{k+1}^N | s] \tag{4.154}$$

We now have expressions to calculate the probability for the whole sequence. Since the message symbols are mutually independent and the channel has no memory, the probability at time k for the whole sequence from 1 to N becomes equal to the product of the separate probabilities:

$$\Pr[m_k = i, s', s, \mathbf{y}_1^N] = \alpha_{k-1}(s')\gamma_i(\mathbf{y}_k, s', s)\beta_k(s) \tag{4.155}$$

In order to perform the calculations in a processor smoothly, it is an advantage to use recursive expressions for $\alpha_k(s)$ and $\beta_k(s)$. The probability that the transmitter was in state s at time k is determined by all paths leading to state s. The probability becomes the sum of all probabilities for those transitions that lead to s, i.e. $\gamma_i(\mathbf{y}_k, s', s)$, multiplied by the probability that the transmitter was in state s', which is $\alpha_{k-1}(s')$ at time $k - 1$. All transitions to state s must be included, but certain transitions have zero probability since they correspond to forbidden transitions according to the trellis. It is then possible to recursively calculate

$$\alpha_k(s) = \sum_{s'=1}^{S} \alpha_{k-1}(s')\gamma_i(\mathbf{y}_k, s', s) \tag{4.156}$$

The same reasoning holds for $\beta_k(s)$ but the calculation is performed in the reversed direction

$$\beta_k(s') = \sum_{s=1}^{S} \beta_{k+1}(s)\gamma_i(\mathbf{y}_k, s', s) \tag{4.157}$$

A certain symbol, for instance $m_k = +1$, can lead to several different states depending on the previous symbol sequence. Of interest at the detection is not the state of the encoder but what symbol is sent. This is why the probability of symbol $m_k = i$ is summed over all states. Given the number of states, S, the total probability of symbol i at time k becomes

$$\Pr[m_k = i, \mathbf{y}_1^N] = \sum_{s=1}^{S} \Pr[m_k = i, s, \mathbf{y}_1^N] \tag{4.158}$$

Insert this result in the probability ratio from eq. (4.151). This gives

$$\Lambda(m_k) = \log \frac{\displaystyle\sum_{s=1}^{S}\sum_{s'=1}^{S} \Pr[m_k = +1, s', s, \mathbf{y}_1^N]}{\displaystyle\sum_{s=1}^{S}\sum_{s'=1}^{S} \Pr[m_k = -1, s', s, \mathbf{y}_1^N]}$$

$$= \log \frac{\displaystyle\sum_{s=1}^{S}\sum_{s'=1}^{S} \alpha_{k-1}(s')\gamma_{+1}(\mathbf{y}_k, s', s)\beta_k(s)}{\displaystyle\sum_{s=1}^{S}\sum_{s'=1}^{S} \alpha_{k-1}(s')\gamma_{-1}(\mathbf{y}_k, s', s)\beta_k(s)} \tag{4.159}$$

Remember that the vector $\mathbf{y}_k$ consists of the elements y_{mk} and y_{1k}. Since the code is systematic, the probability of a certain value of y_{mk}, which is the information symbol m_k plus noise, is independent of the encoder state. The transition probability γ_i can then be factored as

$$\gamma_i(\mathbf{y}_k, s', s) = \Pr[m_k = i, s', s, \mathbf{y}_k] = \Pr[m_k = i, s', s, y_{1k}]\Pr[m_k = i, y_{mk}]$$

$$= \gamma_i(y_{1k}, s', s)\Pr[m_k = i, y_{mk}] \tag{4.160}$$

Furthermore, the simultaneous probability for $m_k = i$ and y_{1k} to occur is $\Pr[m_k = i, y_{mk}] = \Pr[y_{mk}|m_k = i]\cdot\Pr[m_k = i]$. This can be used to rewrite the logarithmic probability ratio

$$\Lambda(m_k) = \log \frac{\Pr[y_{mk}|m_k = +1]\Pr[m_k = +1]\sum_{s=1}^{S}\sum_{s'=1}^{S}\alpha_{k-1}(s')\gamma_{+1}(y_{1k}, s', s)\beta_k(s)}{\Pr[y_{mk}|m_k = -1]\Pr[m_k = -1]\sum_{s=1}^{S}\sum_{s'=1}^{S}\alpha_{k-1}(s')\gamma_{-1}(y_{1k}, s', s)\beta_k(s)}$$

$$= \log \frac{\Pr[y_{mk}|m_k = +1]}{\Pr[y_{mk}|m_k = -1]} + \log \frac{\Pr[m_k = +1]}{\Pr[m_k = -1]}$$

$$+ \log \frac{\sum_{s=1}^{S}\sum_{s'=1}^{S}\alpha_{k-1}(s')\gamma_{+1}(y_{1k}, s', s)\beta_k(s)}{\sum_{s=1}^{S}\sum_{s'=1}^{S}\alpha_{k-1}(s')\gamma_{-1}(y_{1k}, s', s)\beta_k(s)}$$

$$(4.161)$$

If the last term is denoted W_k, eq. (4.161) becomes

$$\Lambda(m_k) = \log \frac{\Pr[y_{mk}|m_k = +1]}{\Pr[y_{mk}|m_k = -1]} + \log \frac{\Pr[m_k = +1]}{\Pr[m_k = -1]} + W_k \qquad (4.162)$$

We now use the assumption that the symbols are transmitted over an AWGN channel. Since we have a continuous stochastic variable, the notation Pr[ ] must be interpreted as a probability density. Given the symbol i, the probability density of the received signal y_{mk} becomes

$$\Pr[y_{mk}|m_k = i] = \frac{1}{\sqrt{2\pi\sigma^2}}\exp\left[\frac{-(y_{mk} - i)^2}{2\sigma^2}\right] \qquad (4.163)$$

Eq. (4.163) inserted in eq. (4.162), gives

$$\frac{\Pr[y_{mk}|m_k = +1]}{\Pr[y_{mk}|m_k = -1]} = \log\left(\frac{\exp\left[\frac{-(y_{mk} - 1)^2}{2\sigma^2}\right]}{\exp\left[\frac{-(y_{mk} + 1)^2}{2\sigma^2}\right]}\right) = \frac{2}{\sigma^2}y_{mk} \qquad (4.164)$$

Finally, inserting eq. (4.164) in eq. (4.162) gives a new formulation of the probability ratio

$$\Lambda(m_k) = \frac{2}{\sigma^2}y_{mk} + \log \frac{\Pr[m_k = +1]}{\Pr[m_k = -1]} + W_k \qquad (4.165)$$

Note that the decision variable $\Lambda(m_k)$ now consists of a sum, where y_{mk} which is a noisy version of m_k, is only included in the first term and y_{1k} which is a noisy version of x_{1k}, is included in the second term. Provided that the code is systematic and that we have white noise, these two terms are statistically independent. It is therefore possible to estimate m_k based on two independent measurements, which obviously gives a better estimation than if only one measurement was used. The first term is called the trivial estimation of m_k. The second term, which is calculated by using other symbols, is called extrinsic.

So far y_{2k}, which is a noisy version of x_{2k}, has not been used. Below we illustrate how this measurement can be used and we create a number of successively improved estimations by introducing feedback in the decoder. But first the numerical problems will be treated.

Numerically more stable algorithm

Unfortunately, the formulation described above gives a numerically unstable algorithm. In order to solve this problem some reformulating must be made starting in eq. (4.165).

Since eq. (4.165) is a ratio, the denominator and numerator can be multiplied or divided by the same numerical value without changing the ratio. By dividing eq. (4.155) by

$$\frac{\Pr\left[\mathbf{y}_1^N\right]}{\Pr[\mathbf{y}_k]} = \Pr\left[\mathbf{y}_{k+1}^N\middle|\mathbf{y}_1^k\right]\Pr\left[\mathbf{y}_1^{k-1}\right] \tag{4.166}$$

$\alpha_{k-1}(s')$ can be redefined using the part $\Pr[\mathbf{y}_1^{k-1}]$, which gives

$$\tilde{\alpha}_{k-1}(s') = \frac{\alpha_{k-1}(s')}{\Pr[\mathbf{y}_1^{k-1}]} = \frac{\alpha_{k-1}(s')}{\displaystyle\sum_{s'=1}^{S}\alpha_{k-1}(s')} \tag{4.167}$$

Against each signal sequence it corresponds a certain probability that the transmitter encoder was in state s' at time $k-1$. According to eq. (4.152) the probability is $\alpha_{k-1}(s') = \Pr[s', \mathbf{y}_1^{k-1}]$. A summation over all s' gives the probability for the sequence $\mathbf{y}_1^{k-1}$, which corresponds to the last equality.

Using eq. (4.156) the recursive formulation then becomes

$$\tilde{\alpha}_k(s) = \frac{\displaystyle\sum_{s'=1}^{S}\alpha_{k-1}(s')\gamma_i(\mathbf{y}_k, s', s)}{\displaystyle\sum_{s=1}^{S}\sum_{s'=1}^{S}\alpha_{k-1}(s')\gamma_i(\mathbf{y}_k, s', s)} = \frac{\displaystyle\sum_{s'=1}^{S}\tilde{\alpha}_{k-1}(s')\gamma_i(\mathbf{y}_k, s', s)}{\displaystyle\sum_{s=1}^{S}\sum_{s'=1}^{S}\tilde{\alpha}_{k-1}(s')\gamma_i(\mathbf{y}_k, s', s)} \tag{4.168}$$

In a similar way it is possible to use the remaining part of eq. (4.166) to modify $\beta_k(s)$:

$$\tilde{\beta}_k(s) = \frac{\beta_k(s)}{\Pr[\mathbf{y}_{k+1}^N | \mathbf{y}_1^k]} \tag{4.169}$$

If the probability laws are used the denominator can be rewritten further

$$\Pr\left[\mathbf{y}_k^N | \mathbf{y}_1^{k-1}\right] = \Pr\left[\mathbf{y}_1^k\right] \frac{\Pr\left[\mathbf{y}_{k+1}^N | \mathbf{y}_1^k\right]}{\Pr\left[\mathbf{y}_1^{k-1}\right]} \tag{4.170}$$

In the same way as in the reformulating of $\alpha_k(s)$ and utilising eq. (4.151) we obtain for eq. (4.170)

$$\Pr\left[\mathbf{y}_k^N | \mathbf{y}_1^{k-1}\right] = \frac{\left[\sum_{s=1}^{S} \sum_{s'=1}^{S} \alpha_{k-1}(s')\gamma_i(\mathbf{y}_k, s', s)\right]}{\Pr[\mathbf{y}_1^{k-1}]} \Pr\left[\mathbf{y}_{k+1}^N | \mathbf{y}_1^k\right]$$

$$= \left[\sum_{s=1}^{S} \sum_{s'=1}^{S} \tilde{\alpha}_{k-1}(s')\gamma_i(\mathbf{y}_k, s', s)\right] \Pr\left[\mathbf{y}_{k+1}^N | \mathbf{y}_1^k\right] \tag{4.171}$$

Divide eq. (4.157) by eq. (4.171) and identify eq. (4.157). We can then write

$$\tilde{\beta}_k(s) = \frac{\sum_{s=1}^{S} \tilde{\beta}_{k+1}(s)\gamma_i(\mathbf{y}_k, s', s)}{\sum_{s=1}^{S} \sum_{s'=1}^{S} \tilde{\alpha}_{k-1}(s')\gamma_i(\mathbf{y}_k, s', s)} \tag{4.172}$$

By inserting values for $\tilde{\alpha}_{k-1}(s')$ and $\tilde{\beta}_k(s)$ instead of for $\alpha_{k-1}(s')$ and $\beta_k(s)$, in eq. (4.165) a numerically more stable algorithm is obtained.

The algorithm starts by using the fact that the initial state in the trellis is known as zero. Therefore the probability $\alpha_1(s' = 0) = 1$ and for the remaining states $\alpha_1(s' \neq 0) = 0$. New values are then calculated through eqs. (4.168) and (4.173), which is a modification of eq. (4.160). By using the AWGN channel and the fact that $\Pr[m_k = \pm 1] = 1/2$ we obtain

$$\gamma_i(\mathbf{y}_k, s', s) = \Pr[y_{mk} | m_k = i]\gamma_i(y_{1k}, s', s)/2$$

$$= \frac{1}{4\pi\sigma^2} \exp\left[\frac{-(y_{mk} - i)^2}{2\sigma^2} + \frac{-(y_{1k} - x_k)^2}{2\sigma^2}\right] \tag{4.173}$$

Note that the noise power, or the variance, must be known for the algorithm. This can be achieved by an estimation of the noise level. Alternatively, setting the variance to a suitable value, where the system normally operates, is often suffi-cient since the algorithm is not particularly sensitive to the choice of σ^2.

For the reverse recursion the starting position should really be $\beta_N(s = 0) = 1$, but since it is difficult to set the turbo coder in state zero, it is instead common to set $\beta_N(s) = 1/S$ for all s. The recursive calculation is then carried out using eqs. (4.172) and (4.173). The logarithmic probability ratio is then obtained from eq. (4.165).

Iterative decoding

As can be seen from eq. (4.165), the first decoder delivers the logarithm of a probability ratio which can be used to make decisions. In the first step of the iteration no more is known about the probability of the symbols than the fact that they are both equally probable, i.e. $\Pr[m_k = \pm 1] = 1/2$. This means that the second term in eq. (4.165) is zero, since $\log(1) = 0$. After the first step of the detection the logarithm of the probability ratio therefore is

$$\Lambda_1(m_k) = \frac{2}{\sigma^2} y_{mk} + W_{1k} \tag{4.174}$$

The decoding can be improved by also using the coded symbol from the second encoder. From the first detection stage we already have an estimation of the ratio of the probability that m_k is either $+1$ or -1. This estimation can be used to calculate a new value of $\log(\Pr[\hat{m}_k = +1]/\Pr[\hat{m}_k = -1])$, which now is non-zero. The value can be included as the a priori probability for the next decision. This probability is correlated to y_{mk} since it is based on a measurement of y_{mk}. No new information is therefore contributed by the variable y_{mk} which is consequently not included in the next estimation. The estimation of a given symbol depends on the preceding symbols and on the ones following. For the new estimation to be completely independent, interleaving has been introduced. The received signal sample is therefore connected to another sample in the sequence corresponding to another time, say n. Therefore it is y_{1n} which shall be decoded supported by $\Lambda_1(m_n)$. If $|k - n|$ is large enough, y_{2n} is guaranteed to be uncorrelated with $\Lambda_1(m_n)$. The estimation from the second detector then becomes

$$\Lambda_2(m_n) = \Lambda_1(m_n) + W_{2n} \approx \log(\Pr[m_n = +1]/\Pr[m_n = -1]) + W_{2n} \tag{4.175}$$

The approximation becomes better the better the evaluation of the a priori prob-abilities of the symbols from the previous detection stage. We have now obtained a better estimation of the symbol sequence $\{m_k\}_{k=1}^N$. This can now be used to repeat the first estimation to get a new one which can be called $\Lambda_3(m_k)$. Provided that the estimation of the a priori probability has been improved in stage 2, a repeated estimation in stage 1 will now give an even better estimation than the first iteration (Figure 4.48).

If this is performed in detector 1, the value that was calculated earlier must be subtracted since it is correlated to $\Lambda_2(m_k)$. A repeated calculation in detector 1 gives

$$\Lambda_3(m_k) = \frac{2}{\sigma^2} y_{mk} + \Lambda_2(m_k) - \Lambda_1(m_k) + W_{1k} \tag{4.176}$$

In principle this can go on for ever. After a number of iterations the estimated symbol probability will get closer and closer to a Gaussian distribution with variance σ_z^2. In the same way as in eq. (4.164) it is therefore possible to write the logarithmic probability ratio as

$$\Lambda_{2l-1}(m_k) = \frac{2}{\sigma^2} y_{mk} + \frac{2}{\sigma_z^2} z_k + W_{1k} \tag{4.177}$$

where $z_k = \Lambda_{2l-2}(m_k) - \Lambda_{2l-3}(m_k)$. A running index $l = 1,2,\ldots$ is used to give a general expression for estimation number $2l - 1$. In the next step the estimation

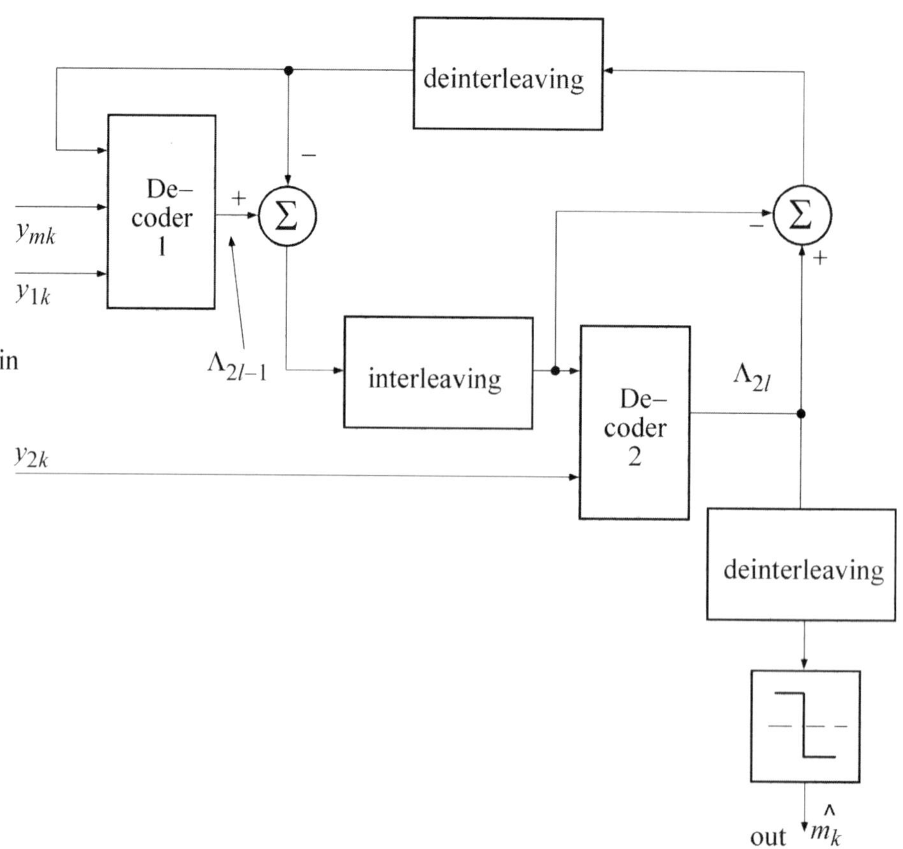

Figure 4.48   Iterative turbo decoder.

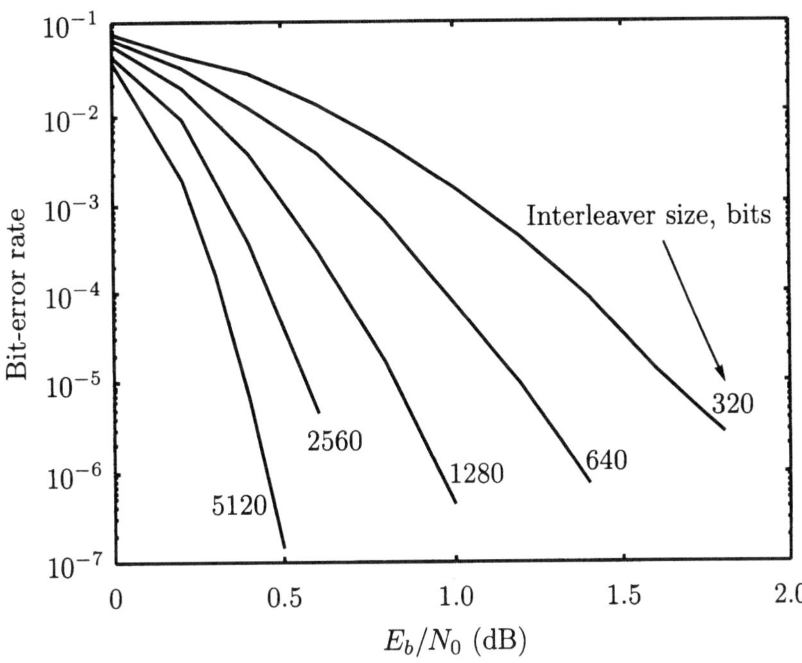

Figure 4.49 Plot of error probability for a rate 1/3 turbo code with the generator polynomials [1 17/15 17/15]; 15 iterations were performed for the decoding (illustrated by Johan Hokfelt).

$\Lambda_{2l-2}(m_k)$ cannot be used because of the correlation. The estimation therefore is

$$\Lambda_{2l}(m_n) = \frac{2}{\sigma_z^2} z_n + W_{2n} \tag{4.178}$$

where $z_n = \Lambda_{2l-1}(m_n) - \Lambda_{2l-2}(m_n)$. The next estimation is again performed in detector 1 and l increases one step. The errors will decrease for each iteration, but will level out after a number of iterations. After the desired number of iterations the final symbol decision $\hat{m}_n = \mathrm{sgn}[\Lambda_{2l}(m_n)]$ is made (Figure 4.49).

Example 4.16

To further illustrate the idea with extrinsic information, the example below can be studied, showing how this concept can be used in a block code. Assume a (7,3) code with the parity control matrix

$$\mathbf{H} = \begin{bmatrix} 1 & 0 & 1 & 1 & 0 & 0 & 0 \\ 1 & 1 & 1 & 0 & 1 & 0 & 0 \\ 1 & 1 & 0 & 0 & 0 & 1 & 0 \\ 0 & 1 & 1 & 0 & 0 & 0 & 1 \end{bmatrix} \tag{4.179}$$

For all code words it must then be true that $\mathbf{X}_i\mathbf{H}^T = \mathbf{0}$; see eq. (4.52). The multiplication gives an equation system

$$0 = x_{i1} + x_{i3} + x_{i4}$$

$$0 = x_{i1} + x_{i2} + x_{i3} + x_{i5}$$

$$0 = x_{i1} + x_{i2} + x_{i6} \tag{4.180}$$

$$0 = x_{i2} + x_{i3} + x_{i7}$$

which can be solved with respect to, for instance, x_{i1}. After minor rewriting this gives $-x_{i1} = x_{i3} + x_{i4}, -x_{i1} = x_{i5} + x_{i7}, -x_{i1} = x_{i2} + x_{i6}$. Apart from a direct estimation of x_{i1}, there are three ways of estimating the symbol x_{i1}. These estimations are based on seven totally different measurements. Provided the interference is independent between the measurements, an estimation based on all these four estimates must give a better result than if only one is used. The estimation based on the direct measurement is the trivial estimation, while the estimation based on measurements of other symbols give extrinsic information about x_{i1}. The same way of estimating symbols can be used for convolutional codes, the difference being that the number of equations is unlimited. The decoding can be performed by using only a limited selection of the equations.

4.7 Cryptography

Encryption is performed to prevent an unauthorised person from listening to or perhaps even distorting a message. The first threat can be avoided by making the message private so that no unauthorised person can understand the content (privacy) and the second threat is avoided by providing some kind of authentication. Cryptography has a long history and has played an important role in several wars.

The simplest form of cipher is a substitution cipher, where one letter is replaced by another letter in a certain pattern, for instance a displacement of the alphabet by one step so that A is replaced by B, etc. In every language different letters appear with different and known probabilities. Using statistical analysis, this type of cipher can therefore easily be deciphered by an unauthorised person. This type of cipher is called the Caesar cipher, since it was used by Julius Caesar. Cryptography therefore has many years on its account. The history of cryptography is full of heroes who invented new ciphers and others who have broken them.

The art of finding a good cipher is to find a problem which is difficult enough to solve in a reasonable time and which fulfils other conditions necessary for a cryptosystem.

Example 4.17

An example of a cipher which provides a higher degree of security than the Caesar cipher is the following. Suppose that the key is the word 'CRYPTO' and the message 'secret message keep quiet'. Since the key contains 6 letters and the message 22, we put up a 6×4 (>22) matrix. The message is introduced along the rows

C	R	Y	P	T	O
1	2	3	4	5	6
S	E	C	R	E	T
M	E	S	S	A	G
E	K	E	E	P	Q
U	I	E	T	X	X

The cipher text is then formed by reading along the columns in the order the letters of the key appear in the alphabet. Since C comes first, the 1st column is read first. The next is O which gives the 6th column, etc. The cipher text then becomes SMEUTGQXRSETEEKIEAPXCSEE. The plain text is then obtained by writing along the columns in the order determined by the key and then read along the rows.

The security of the cipher can be considerably improved by encryption of the cipher text with another key, whose length has no common factors with the length of the first key.

Fundamentals about cryptography

Modern cryptography is of course based on computer programs and is very advanced compared to the kind of ciphers described in the previous section. Those who try to force the cipher also have access to advanced computers so the security is not necessarily better just because the ciphering has become more advanced with time. The block diagram in Figure 4.50 presents a model of the assumed crypto-environment.

A plain text message, **X**, to be transmitted through a channel which may be bugged by an eavesdropper, is generated in the source. To prevent unauthorised

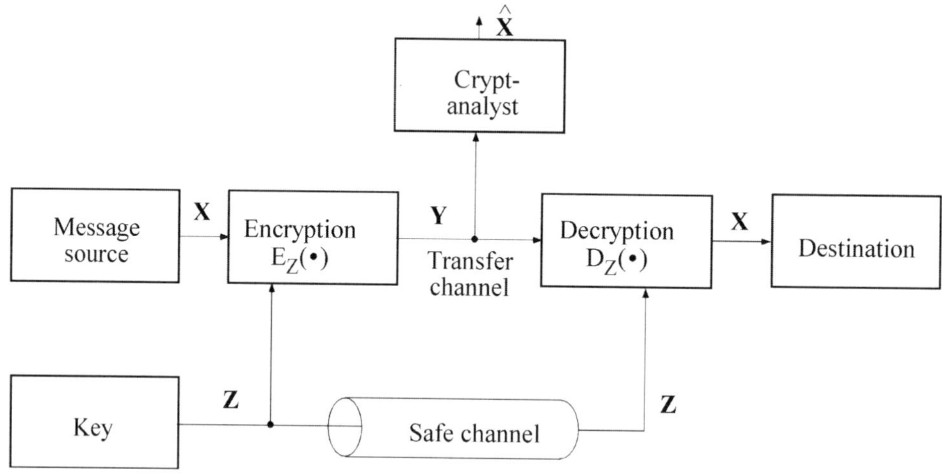

Figure 4.50   The environment of a cryptosystem.

persons taking part of the message, it is encrypted by using an invertable function $E_Z(\cdot)$, which generates the cipher text $\mathbf{Y} = E_Z(\mathbf{X})$. When the legitimate recipient at the destination receives $\mathbf{Y}$ he/she decrypts the message using

$$\mathbf{X} = D_Z\big(E_Z(\mathbf{X})\big) \tag{4.181}$$

and obtains the plain text message. The security of a cipher is assessed after the time it is expected to take for a cryptanalyst to brake the cipher, given that he has considerable computer power at his disposition. Note that the key needed to decrypt the message must be handed over through a secure and reliable channel. This is not unique for cryptosystems. The same thing holds for the key you use to lock your door. It must be kept in a safe way otherwise an intruder may get hold of it and would then have free access to your home. Another important factor for the assessment of the security of a cryptosystem is how the system works if the design is known. Only the key is unknown to the intruder. Compare to a traditional safe. Anybody can buy a safe, take it apart and investigate its design. But the number of combinations in the lock mechanism should be so large that knowledge about the construction is of no help when it comes to opening the lock.

A block cipher separates the plain text into blocks and operates on each block separately. Stream cipher, on the other hand, operate on the message symbol by symbol (see Figure 4.51).

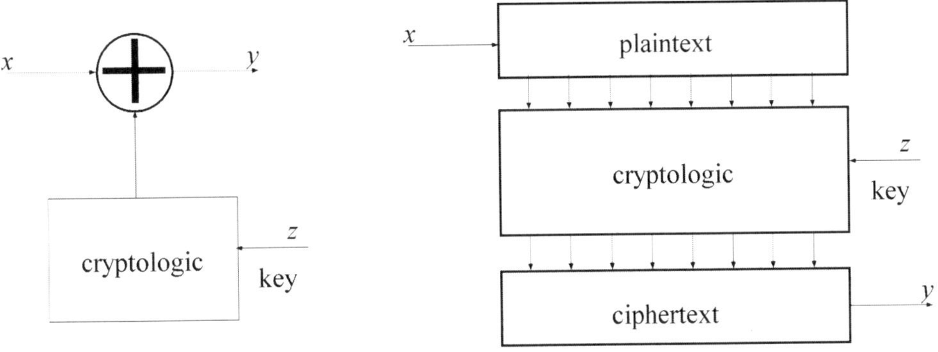

Figure 4.51   Stream cipher to the left and block cipher to the right.

Systems with open keys

In systems with open keys there are two keys: one to encrypt, lock, and one for decrypt, open, the message. If both keys are kept secret, the content of the message can obviously only be known by those in possession of the keys. If an encryption key is made public, anybody can do encryption with this key, but only those who have the decryption key can read the message. If, on the other hand, the encryption key is kept secret, and instead the decryption key is made public, then everybody can not only read the message, but will also know that the message is authentic, i.e. that it originates from the right source and has not been altered.

Example 4.18

Assume a cryptosystem consisting of open keys E_X with public access, and secret decryption keys D_X, that only each person respectively has access to

Person:	Key:
Albert	E_A
Bob	E_B
Carlo	E_C
Dai	E_D
Elias	E_E

Assume for example that the message **M** is to be sent to Carlo, then send $E_C(\mathbf{M})$. Only Carlo knows the secret decryption key D_C. Carlo can then retrieve the text message by the operation $\mathbf{M} = D_C(E_C(\mathbf{M}))$. If the operations also can be performed in reverse order, i.e. $E(D(\mathbf{M})) = \mathbf{M}$, then Bob can send the cipher text $E_C(D_B(\mathbf{M}))$ to Carlo. Bob knows his own decipher key and Carlo's encryption key. When Carlo receives the message he forms $E_B(D_C(E_C(D_B(\mathbf{M})))) = \mathbf{M}$, by using keys known by him. He is then confident that Bob sent the message, since only Bob knows D_B. This is an example of both an encryption and authenticity (digital signature). A digital signature which looks the same every time it appears can easily be forged. It must therefore vary in some predetermined way, for instance be dependent on the message.

4.7.1 THE RSA ALGORITHM

RSA (Rivest–Shamir–Adleman) is an open-key cipher which is in general use. There is one open and one secret key. The keys consist of two prime numbers, p and q, large enough to make a factorisation of their product, $p \cdot q = n$, very difficult to perform. There must also be a number $e < n$ without any common factors with the product $(p - 1)(q - 1)$.

Then compute another number, d, such that the product $ed = 1 \pmod{r}$. The base r is the least common multiple (LCM), of the numbers $[p - 1, q - 1]$. This is obtained by dividing the numbers into their primes p_i, so that no further factorisation can be made. The factorisation can be written

$$p - 1 = p_1^{\alpha_1} \cdots p_r^{\alpha_r} \quad \text{and} \quad q - 1 = p_1^{\beta_1} \cdots p_r^{\beta_r}$$

for the respective number, where $(\alpha_i, \beta_i) \in \mathbb{N}$. This gives $\text{LCM}[p - 1, q - 1] = p_1^{\mu_1} \cdots p_r^{\mu_r}$ where μ_i is the larger of the two numbers α_i and β_i for all i such that $1 \leq i \leq r$.

The number e is called the public exponent and d the private exponent. The public key is therefore the numbers n and e, while the private key is n and d. It is very hard to calculate the private key from the public one. If n could be factored in p and q, however, a cryptanalyst would find the private key d. The security of the RSA algorithm lies in the fact that factorisation is very difficult.

The encryption is performed by letting the receiver distribute his open keys n and e. The transmitter then creates his message by converting the text into digits, for instance using the ASCII code. The text is then divided into blocks, with a value $< n$, and then the public keys are used to decrypt the cipher text:

$$y = E_n(x) = x^e \pmod{n} \tag{4.182}$$

When the cipher text reaches the recipient the text is obtained through

$$x = D_n(y) = y^d \pmod{n} \tag{4.183}$$

Note that even though e is known it is very difficult to retrieve the message by taking a root $\pmod{n}$.

Example 4.19

Choose for instance $p \times q = 89 \times 113 = 10057$, and an exponent $e = 3$. For 3 there is the secret decryption exponent $d = 411$, which fulfils $ed = 1 \pmod{r}$. This gives

$$3 \times 411 = 1233 = 1 \pmod{1232} \tag{4.184}$$

Since factorisation of $88 = 2^3 \times 7^1 \times 11^0$ and of $112 = 2^4 \times 7^0 \times 11^1$ then LCM $= 2^4 \times 7^1 \times 11^1 = 1232$.

The text is divided into blocks with 4 digits and with a value < 10057. Assume that the plain text is 1419. The cipher text then becomes

$$1419^3 = 9131 \pmod{10057} \tag{4.185}$$

i.e. the number 9131 is transferred to the recipient. The plain text is then obtained by

$$9131^{411} = 1419 \pmod{10057} \tag{4.186}$$

Encryption can be performed before or after the error correcting coding (cf. Figure 1.3). The choice gives the system different properties. If error correction coding is performed first, decryption is done before error correction at the receiving side. An error which has occurred during the transfer of the message in the transmitting channel, runs the risk of being magnified after decryption. Even if the error originally could have been corrected by the error correction coding, this may no longer be the case. At ARQ it is not so important if the error increases, and when decryption is performed prior to the error correction coding an authenticity control can be done in the receiver before the message is allowed to proceed in the system. An aspect, on the other hand, supporting the idea of making the encryption before the error correction coding is applied, is that coding introduces extra redundancy in the text, which makes it easier to brake the cipher. This may be of little importance for a good cryptosystem, but in principle the cryptanalyst should know as little as possible about the plain text message. The order between making the encryption and error correction depends, in other words, on the system and the properties the system should possess.

Example 4.20

For a stream cipher it is easy to believe that a pseudo-random generator which consists of a feedback shift register is a suitable candidate for cryptologic (more about pseudo-random generators in Chapter 9). This is, however, not the case, since for a shift register with n steps it is only necessary to know $2n$ bits of the output sequence in order to establish the feedback positions and thus the rest of the sequence.

See for instance the pseudo-random generator shown in Figure 4.52. The output signal from a shift register with three taps can generally be written

$$q_{j-3} = c_2 \, q_{j-2} \oplus c_1 \, q_{j-1} \oplus c_0 \, q_j \tag{4.187}$$

where q_x is the content in the respective shift register cell and c_x are filter coefficients which can assume the values 0 or 1. If the shift register at the beginning contains 001, from left to right, the output signal successively becomes 1001110100111, etc. The first three bits provide the start content of the shift register in reverse order. The following equation system can then be set up. Assume that initially the time is $j = 0$, then

$$j = 0, \quad 1 = c_2 \cdot 1 \oplus c_1 \cdot 0 \oplus c_0 \cdot 0 \quad \text{(a)}$$

$$j = 1, \quad 1 = c_2 \cdot 0 \oplus c_1 \cdot 0 \oplus c_0 \cdot 1 \quad \text{(b)}$$

$$j = 2, \quad 1 = c_2 \cdot 0 \oplus c_1 \cdot 1 \oplus c_0 \cdot 1 \quad \text{(c)}$$

Eq. (a) tells us that c_2 must be 1, (b) that c_0 must be 1 and (c) that $c_1 \oplus c_0 = 1$, which means that $c_1 = 0$. This gives the feedback we have in the pseudo-random generator shown in Figure 4.52.

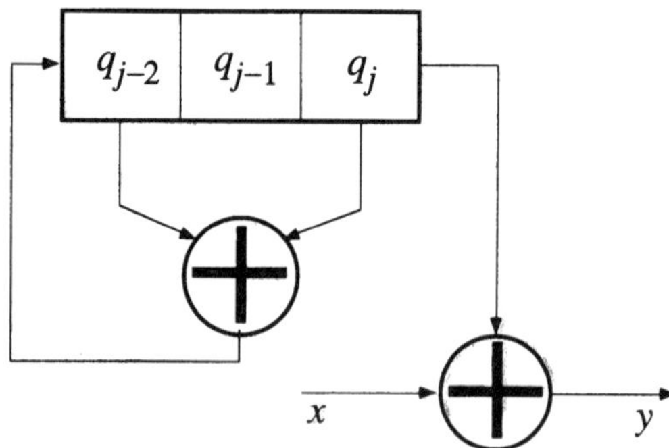

Figure 4.52 Example of maximum length pseudo-random-generator.

4.8 Appendix
BCH codes (Bose-Chaudhuri-Hocquenghem) from a table of Stenbit, 1964

n	k	t	$G(x)$
7	4	1	13
15	11	1	23
	7	2	721
	5	3	2467
31	26	1	45
	21	2	3551
	16	3	107657
	11	5	5423325
	6	7	313365047
63	57	1	103
	51	2	12471
	45	3	1701317
	39	4	166623567
	36	5	1033500423
	30	6	157464165547
	24	7	17323260404441
	18	10	1363026512351725
	16	11	6331141367235453
	10	13	472622305527250155
	7	15	5231045543503271737
127	120	1	211
	113	2	41567
	106	3	11554743
	99	4	3447023271
	92	5	624730022327
	85	6	130704476322273
	78	7	26230002166130115
	71	9	6255010713253127753
	64	10	1206534025570773100045
	57	11	335265252505705053517721
	50	13	54446512523314012421501421
	43	14	17721772213651227521220574343
	36	15	3146074666522075044764574721735
	29	21	403114461367670603667530141176155
	22	23	123376070404722522435445626637647043
	15	27	22057042445604554770523013762217604353
	8	31	7047264052751030651476224271567733130217
255	247	1	435
	239	2	267543

231	3	156720665
223	4	75626641375
215	5	23157564726421
207	6	16176560567636227
199	7	7633031270420722341
191	8	2663470176115333714567
187	9	52755313540001322236351
179	10	2262471071734043241 6300455
171	11	15416214212342356077061630637
163	12	7500415510075602551574724514601
155	13	375751300540766501572250 6464677633
147	14	16421301735371655253041653 05441011711
139	15	46140173206017556157072273 0247453567445
131	18	21571333147151015126125027744 2142024165471
123	19	12061405224206600371721032651 6141226272506267
115	21	60526665572100247263636404600276 352556313472737
107	22	22205772322066256312417300235347420 176574750154441
99	23	106566672534731742227414162015743322524 11076432303431
91	25	675026503032744417272363172473251107555076 2720724344561
87	26	1101367634147432364352316343071720462067225452 73311721317
79	27	66700035637657500020270344207366174621015326711 766541342355
71	29	2402471052064432151555417211233116320544425036255764 3221706035
63	30	10754475055163544325315217357707003666111726455267613656 702543301
55	31	7315425203501100133015275306032054325414326755010557044426035473617
47	42	253354201706264656303304137740623317512333414544604500506602 4552543173
45	43	1520205605523416113110134637642370156367002447076237303320215 7025051541
37	45	513633025506700741417744724543753042073570617432343234764435473 7403044003
29	47	302571553667307146552706401236137711534224232420117411406025475741 0403565037
21	55	12562152570603326560017731536076121032273414056530745425211531216144665 13473725
13	59	46417320050525645444265737142500660043306774454765614031746772135702613 4460500547
9	63	157260252174724632010310432553551346141623672120440745451127661155477055616 77516057

4.8.1 EXAMPLES ON APPLICATION OF ERROR CORRECTING CODING

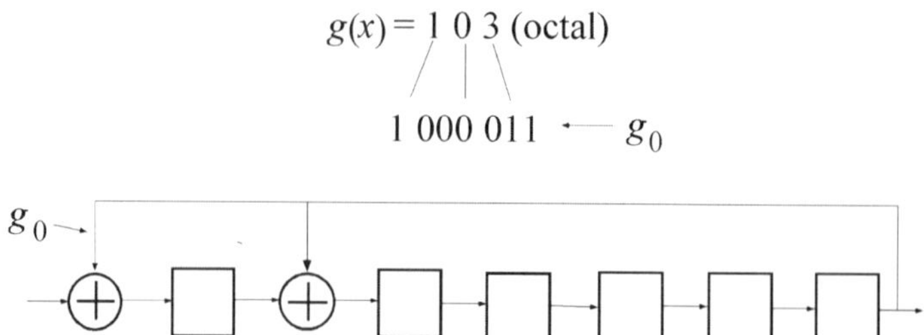

Figure 4.53   Example of how the table for the BCH code can be used to obtain a shift register encoder for a (63,57) BCH code.

Convolutional codes

Rate	Constraint length	Free distance, d_{free}	Generator polynomial
$\frac{1}{2}$	3	5	111 101
$\frac{1}{2}$	4	6	1111 1011
$\frac{1}{2}$	5	7	10111 11001
$\frac{1}{2}$	6	8	101111 110101
$\frac{1}{2}$	7	10	1001111 1101101
$\frac{1}{2}$	8	10	10011111 11100101
$\frac{1}{2}$	9	12	110101111 100011101
$\frac{1}{3}$	3	8	111 111 101
$\frac{1}{3}$	4	10	1111 1011 1101
$\frac{1}{3}$	5	12	11111 11011 10101
$\frac{1}{3}$	6	13	101111 110101 111001
$\frac{1}{3}$	7	15	1001111 1010111 1101101
$\frac{1}{3}$	8	16	11101111 10011011 10101001

Some convolutional codes for majority decoding

Rate	Generator polynomial (octal)
$\frac{1}{2}$	1 3
$\frac{1}{2}$	001 145
$\frac{1}{2}$	000001 620205
$\frac{1}{2}$	000000000001 430001202201
$\frac{1}{2}$	000000000000000001 3021000044010040005
$\frac{2}{3}$	1 3 5
$\frac{2}{3}$	00001 11401 24101
$\frac{2}{3}$	0000000000001 0014100000105 1402000100011
$\frac{3}{4}$	01 03 05 11
$\frac{3}{4}$	0000001 2100011 1400401 0024101

Compact disc

In the CD system the audio (audio = sound) signal is sampled in two channels, left and right. Every channel is sampled with 44.1 kHz and 16 bits per sample. The samples are grouped in frames of 12 samples, 6 from each channel. Each frame therefore contains 24 symbols of 8 bits. These are encoded with a (28,24) Reed–Solomon code in $GF(2^8)$. Each encoded symbol therefore constitutes 8 binary bits. Thereafter interleaving of the symbols are done over 28 frames. Another error correcting code, an outer code, is concatenated. This is a (32,28) Reed–Solomon code in $GF(2^8)$. The next step is to adjust the signal so that transitions between 0 and 1 are no closer than three bits. To prevent the bit clock from losing lock, sequences of ones or zeros longer than 11 bits are not allowed. This is solved by appending bits, so that each 8 bits symbol is represented by 14 bits, which are appended in a way to fulfil the requirements on maximum and minimum lengths. The low frequency components in the signal formed by the bit stream must be minimised in order not to disturb the servo loop following the tracks on the disc, which is done by appending another three so-called merging-bits.

Modem standards

Exercise 4.1

The generator matrix for a linear binary error correcting code is given by

Standard	Generator polynomial, $G(x)$
CRC-12	$x^{12} + x^{11} + x^3 + x^2 + x + 1$
CRC-16 (ANSI)	$x^{16} + x^{15} + x^5 + 1$
CRC-16	$x^{16} + x^{15} + x^2 + 1$
CRC-CCITT (V.41)	$x^{16} + x^{12} + x^5 + 1$
CRC-32$_B$	$x^{32} + x^{26} + x^{23} + x^{22} + x^{16} + x^{12} + x^{11} +$ $x^{10} + x^8 + x^7 + x^5 + x^4 + x^2 + x + 1$

$$
\mathbf{G} = \begin{bmatrix} 0 & 0 & 1 & 1 & 1 & 0 & 1 \\ 0 & 1 & 0 & 0 & 1 & 1 & 1 \\ 1 & 0 & 0 & 1 & 1 & 1 & 0 \end{bmatrix}
$$

- Express $\mathbf{G}$ in a systematic form
- Determine d_{min}
- Compute the parity control matrix
- Design the syndrome table

Exercise 4.2

The polynomial $G(p) = p^4 + p + 1$ is a generator polynomial to a $(15,11)$ binary Hamming code. Determine the systematic generator matrix $\mathbf{G}$ for this code.

Exercise 4.3

Plot the tree diagram, trellis, state diagram and d_{free} for the code generated by

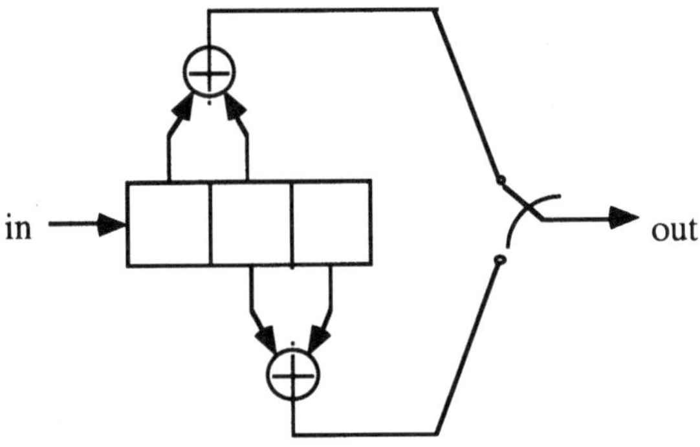

Exercise 4.4

Below is shown a rate $= 2/3$ encoder for binary symbols. Plot the code tree, trellis, and state diagram. Note that the register is shifted in two steps each time.

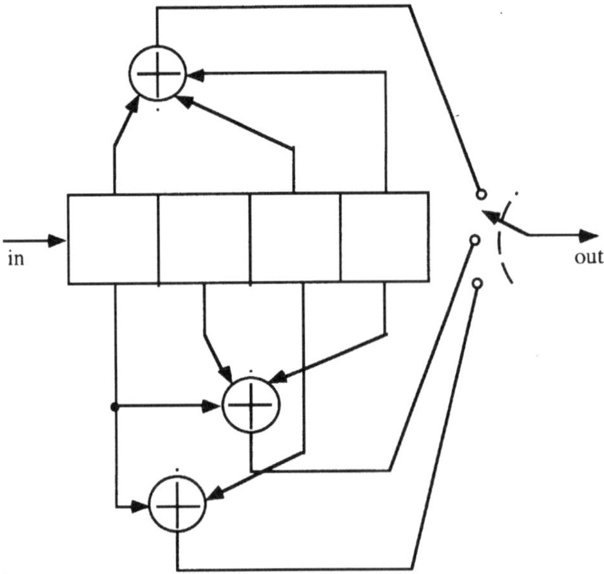

Exercise 4.5

Design the state diagram for the encoder below and determine the free distance, d_{free}. What is the constraint length of the code?

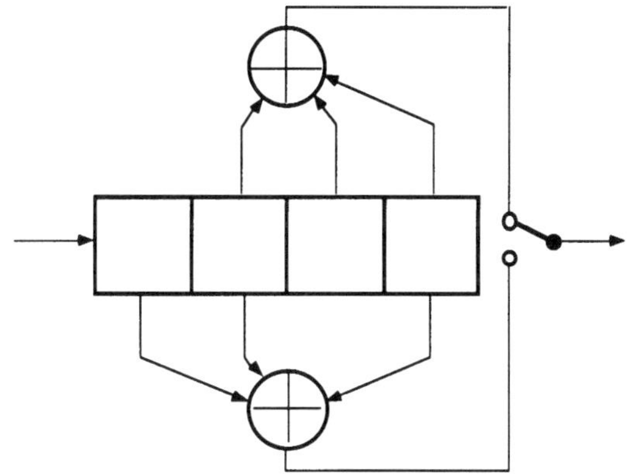

Exercise 4.6

Consider a (7,4) Hamming code defined by the polynomial

$$G(p) = p^3 + p + 1$$

The codeword 1001110, where $\mathbf{X} = [x_{n-1}...x_0]$, is transmitted over a noisy channel which alters the codeword to 1001010. Determine the syndrome $S(p)$ for the received word and compare it to the error polynomial $E(p)$.

Exercise 4.7

Determine d_{free} for the convolutional code whose generator polynomial can be obtained from

$$\mathbf{g}_1 = (73353367672)_8, \quad \mathbf{g}_2 = (53353367672)_8$$

To solve the problem, a computer has to be used.

Exercise 4.8

Assume a received word $\mathbf{y} = [1\ 1\ 0\ 1\ 1\ 1]$. The generator matrix was:

$$\mathbf{G} = \begin{bmatrix} 1 & 0 & 0 & 1 & 1 & 1 \\ 0 & 1 & 0 & 1 & 0 & 1 \\ 0 & 0 & 1 & 0 & 1 & 1 \end{bmatrix}$$

Determine the most probable word sent, if the error probability in the received signal, $P_b < 0.5$.

5 MODULATION

Modulation is an operation performed to adapt the signal shape to the channel on which the transmission will be done. The channel can be electrical wires, optical fibres, radio links, etc. Modulation can be divided into two classes; baseband modulation and carrier modulation. Baseband modulation entails transmitting the signal by using a lowpass signal. This technique is used in copper cables, optical fibres, magnetic media and other types of transmission without a carrier. Carrier modulation entails modulating the message on a carrier with a high frequency, in relation to the bandwidth of the message. In this way the signal used for the transmission is a bandpass signal. A carrier is used for radio communication and also for transmission over some types of cables, such as coaxial cables.

All transmission channels have a finite bandwidth which limits the number of transmitted symbols per unit time. (Bandwidth is defined in different ways, e.g. -3 dB level, 'null-to-null', noise bandwidth etc.) The limit sometimes depends on the physical properties of the channel or, in radio communication, often on the limited fraction of the electromagnetic spectrum that is allowed for a certain user. When storing information on magnetic media there is an upper limit for how closely packed the information can be. The user of a certain transmission channel wants to employ the channel as efficiently as possible with the above-mentioned limitations taken into account. This means that modulation techniques giving high data rates with a small signal bandwidth are desired. With efficient modulation techniques one can for example transfer 33,600 bits/s over a telephone line (V.34 modem) which has a bandwidth of 3100 Hz.

When speaking about transmission of information also the sort Baud is occurring. It denotes the signalling rate, which can be interpreted as the number of modulated symbols per second. The rate in Bauds is obtained if the number of bits/s is divided by the number of transferred bits per symbol, $\log_2 M$.

In analogue baseband transmission the message signal is transmitted unchanged. In digital baseband transmission certain types of pulse-shaped signals are used. This chapter first gives a brief presentation of digital baseband transmission, where the first part is about different wave forms with step-wise shifts and the second part about how to modify the wave form to make transmission more bandwidth efficient. A large part of this chapter is devoted to carrier transmission and is introduced by a presentation of a mathematical tool for a simple description of carrier systems. This is followed by a presentation of the common analogue systems, such as AM, FM and PM. The second half of the chapter contains methods for carrier transmission of digital signals. Among the methods mentioned, not only the ordinary standard methods ASK, PSK and FSK can be found, but also more advanced modulation techniques, such as CPM. The presentation is general so the reader can easily become familiar with other types of digital modulation not mentioned here. To be able to compare the bandwidth of the modulated signal to the bandwidth offered by the channel, the chapter also contains a section about how to calculate the spectra of modulated signals. In the next part error-correction coding and modulation are combined into a very efficient transmission method called TCM. Finally multi-carrier modulation (MCM), is discussed, which is a method to transfer high data rates over a dispersive channel.

The aim of this chapter is to present to the reader the common modulation methods for electrical and optical signal transmission, as well as to present tools for the analysis of the properties of these methods when they are used in communication systems. Methods to recover the information from the modulated signals and their properties in noisy channels are treated in Chapter 6.

5.1 Baseband modulation

A digital signal is generated either in a digital source, such as a computer, or through sampling and analogue-to-digital conversion of an analogue signal source. When the digital signal is to be transmitted on a channel having a frequency band of a lowpass character (i.e. the frequency band goes from zero up to some upper frequency limit), the signal is transformed to a form suitable for transmission by baseband modulation. Baseband modulation can be used for transmission, but it can also be seen as a preparation for carrier modulation. Some common methods for baseband modulation are shown in Figure 5.1 and are described below.

- On-off signalling, which means that a one is represented by a pulse of a length equal to the interval between the symbols, and a zero is represented by the absence of a pulse. Can also be seen as a parallel to serial conversion of a sequence of binary coded digital words.
- Non-return-to-zero (NRZ) signalling means that a one is represented by a pulse having a positive amplitude and a length equal to the interval between the symbols, and a zero is represented by a pulse with negative amplitude.
- Return-to-zero (RZ) signalling means that a one is represented by a pulse of length equal to half of the distance between the symbols, and a zero is represented by the absence of a pulse.
- Non-return-to-zero-inverted (NRZI) signalling means that a one is represented by a step in the signal level. This operation corresponds to forming a new data

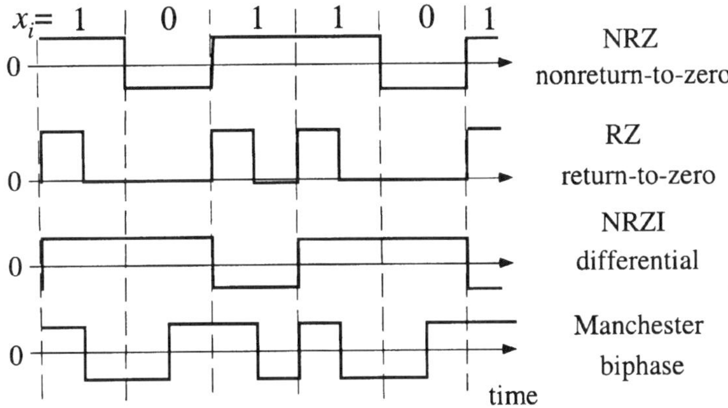

Figure 5.1   Some examples of baseband modulation. NRZ, non-return to zero; RZ, ''return to zero; NRZI, non-return to zero inverse and Manchester biphase.

sequence, z_i, where $z_i = z_{i-1} \oplus x_i$. Note that this type of transmission is insensitive to inversion of the signal along the transmission path.

- Manchester biphase (split-phase) signalling means that a one is represented by a positive pulse with a length equal to half the symbol interval followed by a negative pulse of the same length. A zero is represented in the same way but with reversed order between the pulses. Each character therefore contains a phase shift. Because of this, the spectrum contains a strong frequency component at the symbol clock frequency, which makes locking of the receiver symbol clock easier.

The shapes of the pulses can also be rounded off, which reduces the high frequency components of the spectrum and offers a more efficient use of the bandwidth of the transmission channel.

5.1.1 PRECODING

To increase the packing density on magnetic media precoding of information symbols is applied. This is achieved by letting the symbols pass a FIR filter. A common target for the combined transfer function of the channel and filter is that

$$H(D) = (1 - D)(1 + D)^n \tag{5.1}$$

where D^n is an operator which delays the symbols by n steps. The symbols will then overlap but can be decoded by the inverse transfer function or by a sequential detector of the Viterbi algorithm type. The method is called partial response maximum likelihood (PRML). One example is PR4, where $n = 1$. The resulting transfer function gives a good match to the natural response from read head, write head and magnetic media in magnetic data storing. Another advantage with the type of transfer functions specified in eq. (5.1) is that they give a small number of distinct values at the output. For the example with $n = 1$ the possible values become $\{2,0,-2\}$.

5.1.2 PULSE SHAPES FOR LOWPASS CHANNELS

The band-limited channel can be interpreted as a lowpass or a bandpass filter depending on the physical situation. A pulse passing through either of these filters will be extended in time if the pulse has a broad spectrum in relation to the bandwidth of the filter. The narrower the bandwidth the longer the pulse response. This extension of the pulses results in an overlap of successive pulses, which is an effect called intersymbol interference (ISI). In the following section pulse shapes which can be used within baseband transmission or amplitude modulated carrier transmission are discussed. In the time domain the signal can be written

$$y(t) = \sum_k x_k p(t - kT) \tag{5.2}$$

where T is the time interval between the symbols and x_k is the value of the symbol k in the symbol stream. In binary transmission x_k is zero or one. The symbols are represented by some kind of pulse shape, $p(t)$. In this section some aspects of the pulse shape are discussed. One of the requirements is that the successive symbols must not influence each other at the sampling instances, when the received signal is converted back to symbols. When no influence exists, the system is free of intersymbol interference. ISI is not wanted as it increases the error probability. For a system to be free of intersymbol interference, the pulses corresponding to the preceding and the following symbols must be zero when a certain pulse is about to be sampled. This can be stated by the following condition:

$$p(t) = \begin{cases} 1 & t = 0 \\ 0 & t = \pm T, \pm 2T... \end{cases} \tag{5.3}$$

Compare with the pulse shape function in carrier transmission, eq. (5.87). The value at $t = 0$ can in principle be an arbitrary constant. The condition (5.3) is illustrated in graphical form in Figure 5.2.

Another condition on the signal, $p(t)$, is that its spectrum must be limited to the bandwidth, B, of the channel, which means that $P(f) = 0$ when $|f| \geq B$. A possible start is then to assume a signal whose spectrum is constant within the channel bandwidth, i.e.

$$p(t) = \int_{-B}^{B} \frac{1}{2B} \exp(j2\pi ft)df = \frac{\sin(2\pi Bt)}{2\pi Bt} \tag{5.4}$$

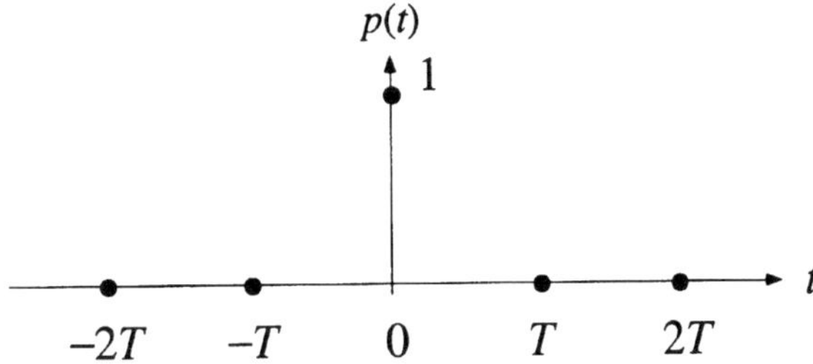

Figure 5.2   The signal must be zero at the other sampling instances.

This pulse is zero for $t = \pm 1/2B, \pm 2/2B, \pm 3/2B \ldots$ Information symbols sent out at these instances will thus be independent of the information symbol at $t = 0$. If, therefore, the symbol rate $r = 1/T = 2B$ is chosen, the pulse shape offering the highest transmission rate over a channel having the single-sided bandwidth B, is obtained.

$$p(t) = \frac{\sin(2\pi Bt)}{2\pi Bt} = \frac{\sin(\pi rt)}{\pi rt} \tag{5.5}$$

This type of pulse is called the Nyquist pulse, and three consecutive pulses are plotted in Figure 5.3.

The spectrum of the signal $\sin(r\pi t)/r\pi t$ is strictly band-limited which means that all signal power is within the frequency range $\pm 1/(2T)$ (Hz). Moreover, the power density is constant within the band limits. Such a function is usually

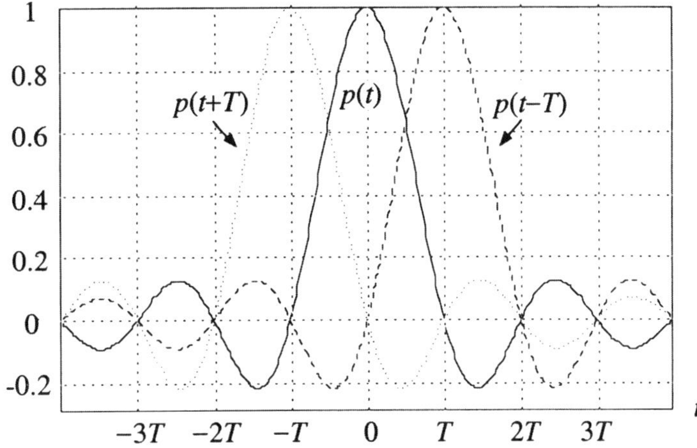

Figure 5.3   The Nyquist pulse gives the highest possible transmission rate without inter-symbol interference, which can be seen in the diagram, where three consecutive pulses have been plotted.

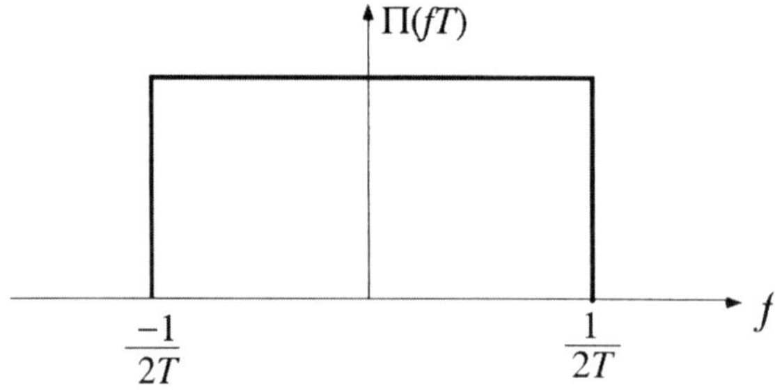

Figure 5.4   The spectrum of the Nyquist pulse.

denoted $\Pi(x)$, for which $\Pi(x) = 1$ when the argument $x \leq |1/2|$ and otherwise $\Pi(x) = 0$ (cf. Figure 5.4).

The Nyquist pulse has an infinitely long extension in time, which in practice means that it cannot be implemented exactly. It is also very sensitive to errors in the time reference, since even a small error results in considerable intersymbol interference due to slowly decreasing side lobes. It would be desirable to find a type of pulse which could reduce these disadvantages. A positive property of the Nyquist pulse is that it fulfils the requirement of having zeros which make the symbols independent. This property can be maintained if one takes the Nyquist pulse and multiplies this by another, not yet defined, function $p_\beta(t)$:

$$p(t) = p_\beta(t) \frac{\sin((\pi t)/T)}{((\pi t)/T)} \tag{5.6}$$

The spectrum of the total signal is obtained by a convolution, here symbolised by $\otimes$, between $\Pi(f)$ and the spectrum of $p_\beta(t)$:

$$P(f) = F\{p(t)\} = P_\beta(f) \otimes \left[T \cdot \Pi(f\,T)\right] \tag{5.7}$$

If the spectrum of the signal $p_\beta(t)$ is bandlimited to $\pm\beta/T$, the convolution gives a spectrum of the type shown in Figure 5.5. The spectrum has a transition

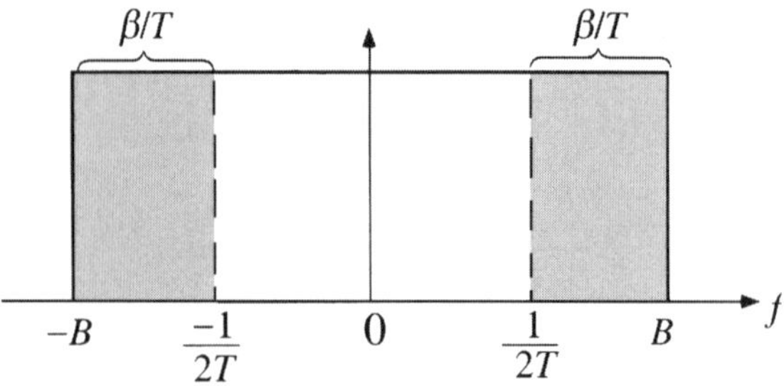

Figure 5.5   The spectrum of the signal $P(f)$ has to be confined within the bandwidth $\pm B$.

band within the shaded area where the power density decreases towards zero, but the exact appearance depends on the pulse $p_\beta(t)$ not yet defined.

The total bandwidth becomes broader than for the Nyquist pulse, and this cannot be allowed as the bandwidth, B, of the channel is determined beforehand. One solution to this problem is to reduce the transmission rate r, in relation to the maximum rate (cf. Figure 5.5). One way to define $p_\beta(t)$ is

$$p_\beta(t) = \frac{\cos(2\beta\,\pi\,t/T)}{1 - (4\beta\,t/T)^2} \tag{5.8}$$

where β is called the roll-off factor and is the same constant as the frequency intervals in Figure 5.5. With $r = 1/T$ the total pulse shape becomes

$$p(t) = \frac{\cos(2\beta\,\pi\,t/T)}{1 - (4\beta\,t/T)^2}\,\frac{\sin(\pi\,t/T)}{\pi\,t/T} \tag{5.9}$$

The spectrum of $p_\beta(t)$ is band-limited which induces a persistent band limitation of the multiplied pulse. The pulse has indeed still an infinite extension in time, but the side lobes decrease considerably faster than for the Nyquist pulse. Hence, it does not have the disadvantages of the Nyquist pulse and can easily be approximated by a time-limited signal as indicated in Figure 5.6.

The Fourier transform of $p(t)$ gives the spectrum of the pulse (eq. (5.10)), which is a cosine function with a shifted mean value greater than zero. The pulse is therefore called a raised cosine pulse.

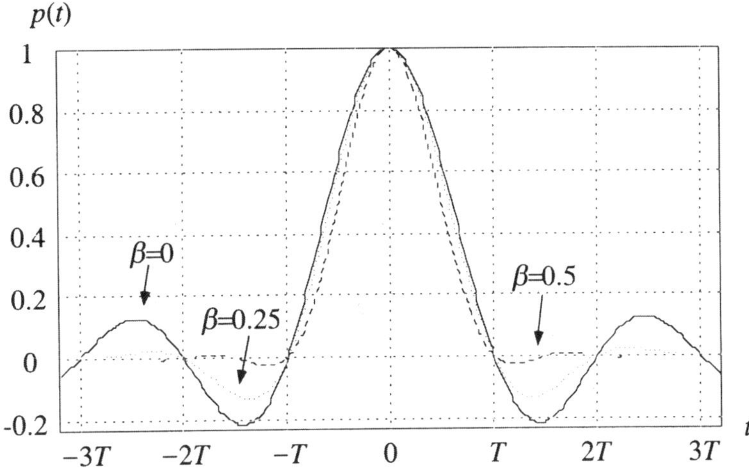

Figure 5.6 Pulse shape decreasing faster to zero than the Nyquist pulse.

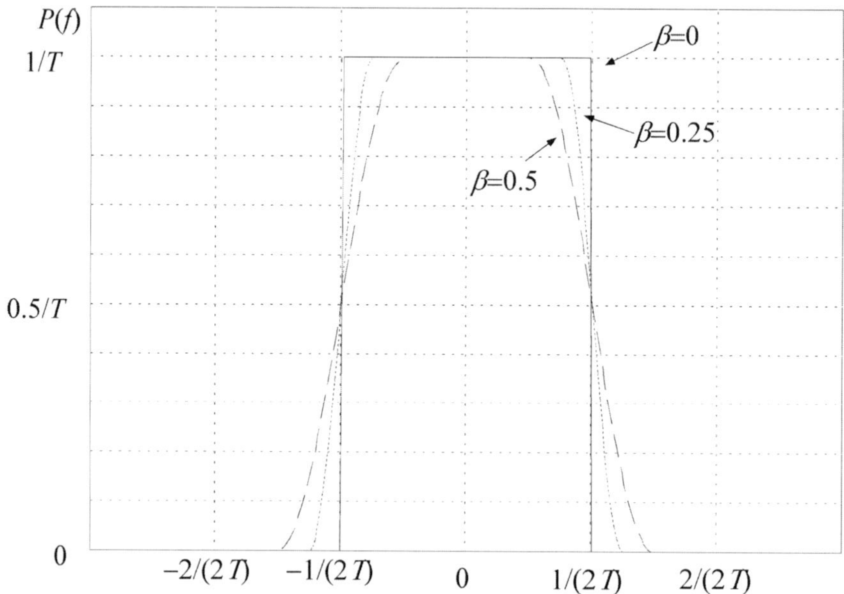

Figure 5.7   Spectrum of the raised cosine pulse.

$$P(f) = \begin{cases} T, & 0 \leq |f| \leq (1 - \beta/2T) \\[2mm] \dfrac{T}{2}\left[1 + \cos\left[\dfrac{\pi T}{\beta}\left(|f| - \dfrac{1 - \beta}{2T}\right)\right]\right], & \dfrac{(1 - \beta)}{2T} \leq |f| \leq \dfrac{(1 + \beta)}{2T} \\[3mm] 0, & \text{otherwise} \end{cases}$$

$$(5.10)$$

The shape of the spectrum for some different values of β is shown in Figure 5.7.

The raised cosine pulse offers flexibility through the parameter β, which can be adapted to different possibilities and needs, making this a frequently used pulse shape.

5.2 Calculation methods for bandpass systems and signals

When making a mathematical or computer analysis of a carrier system, it is desirable to determine the spectrum, error probability, etc. at different points in the communication system. Methods to express the signal in a way that is simple and independent of the carrier or centre frequency is then needed. A method fulfilling these requirements and moreover can be used for implementation of certain parts in bandpass communication systems is described below. By bandpass systems and signals, as a rule of thumb, we mean practical systems having

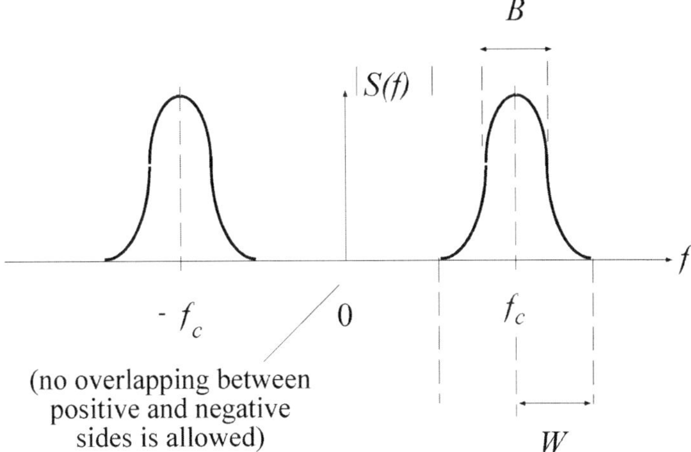

Figure 5.8 Example of spectrum for narrowband signals or amplitude response of bandpass systems.

the property $B/f_c < 0.1$. B is the signal 3 dB bandwidth and f_c is the centre, or carrier, frequency (Figure 5.8).

Formally the following must hold.

$$|S(f)| = 0 \text{ for } \begin{cases} |f| < f_c - W \\ |f| > f_c + W \end{cases} \tag{5.11}$$

for some constant $W < f_c$ and $W \geq 0$. This cannot be fulfilled for time-limited signals as these have spectra which can only be zero for some discrete frequency values. Hence, an approximation has to be made for the bandpass condition to be formally fulfilled. A bandpass signal, $s(t)$, can be written

$$s(t) = a(t) \cos[2\pi f_c t + \phi(t)] = \underbrace{a(t)\cos(\phi(t))}_{S_i \text{ in-phase component}} \cos(2\pi f_c t) - \tag{5.12}$$

$$\underbrace{-a(t)\sin(\phi(t))}_{S_q \text{ quadrature component}} \sin(2\pi f_c t)$$

where $a(t)$, $\phi(t)$, $s_i(t)$ and $s_q(t)$ are, in relation to the carrier, slowly varying lowpass signals. For these we can write

$$|S_i(f)| = |S_q(f)| = 0 \quad |f| > W \tag{5.13}$$

Signals in the time domain are labelled by lower case letters and signals in the frequency domain are labelled by upper case letters.

The signal $s_i(t)$ is in phase, i.e. synchronous, with the carrier and $s_q(t)$ is 90° out of phase, i.e. in quadrature, with the carrier. These parts are called the quadrature components of the signal and are often used when carrier modulation is discussed.

A simple way of handling bandpass signals is to use complex numbers. The actual signal is, however, real. The complex signal, $u(t)$, is defined as

$$u(t) = s_i(t) + j\,s_q(t) = a(t)\,\exp[j\,\phi(t)] \tag{5.14}$$

A bridge between the complex notation and the actual signal $s(t)$, is obtained by writing the actual signal as the real part of a complex signal, i.e.

$$s(t) = \mathrm{Re}\{u(t)\exp[j2\pi f_c t]\} = \frac{1}{2}\left(u(t)\exp[j2\pi f_c t] + u^*(t)\exp[-j2\pi f_c t]\right) \tag{5.15}$$

As been mentioned before, the included signal components are lowpass signals, which gives the result that the sum $u(t)$ is also a lowpass signal. The complex signal can be expressed by using a vector in the complex plane, as in Figure 5.9. The length of the vector, or the envelope of the modulated signal, is

$$a(t) = \sqrt{s_i^2(t) + s_q^2(t)} \tag{5.16}$$

The phase becomes

$$\phi(t) = \arctan\left(\frac{s_q(t)}{s_i(t)}\right) \tag{5.17}$$

The spectrum of a bandpass signal is obtained by taking the Fourier transform of eq. (5.15) which gives

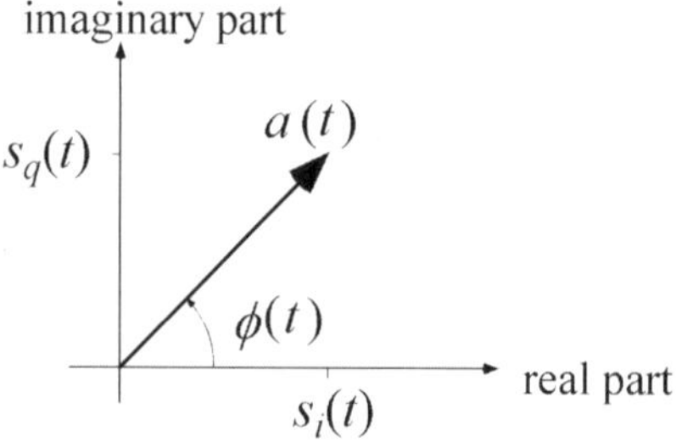

Figure 5.9   The signal represented in the complex plane.

$$S(f) = F\{s(t)\} = \frac{1}{2}\int_{-\infty}^{\infty} \left(u(t)\exp[j2\pi f_c t] + u^*(t)\exp[-j2\pi f_c t]\right)\exp[-j2\pi ft]dt$$

$$= \frac{1}{2}\underbrace{\int_{-\infty}^{\infty} u(t)\exp[-j2\pi(f-f_c)t]dt}_{U(f-f_c)} + \frac{1}{2}\underbrace{\int_{-\infty}^{\infty} u^*(t)\exp[-j2\pi(f+f_c)t]dt}_{U^*(-f-f_c)}$$

$$= \frac{1}{2}\left(U(f-f_c) + U^*(-f-f_c)\right)$$

$$= \frac{1}{2}\left(S_i(f-f_c) + S_i(f+f_c)\right) + \frac{j}{2}\left(S_q(f-f_c) - S_q(f+f_c)\right) \tag{5.18}$$

where $F\{\cdot\}$ denotes the Fourier transform. For the Fourier transform it is valid that $F\{x^*(t)\} = X^*(-\omega)$, which has been used above. For the last equality the relations $u(t) = s_i(t) + js_q(t)$ and $S^*(-f) = S(f)$, given a real signal $s(t)$, have been used. The spectrum has two components. One is the spectrum of the complex signal shifted up along the frequency axis. The other is a left shifted image obtained as a reflection of the first component around the amplitude axis. The different components of the spectrum can also be expressed by using the spectra of the in-phase and out-of-phase components, respectively, as in the last relation of eq. (5.18). An example of how the different components of the spec-

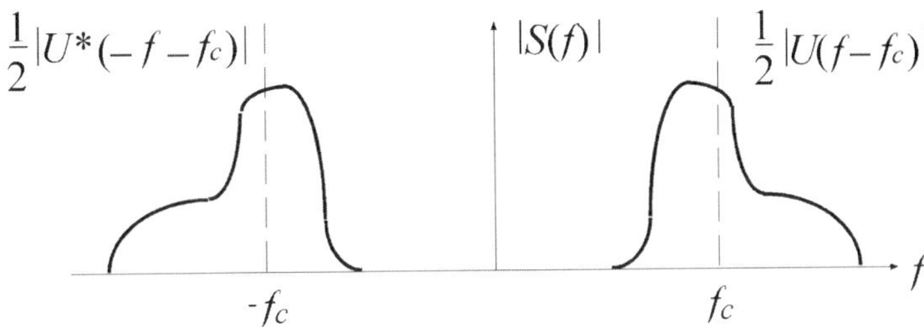

complex signal

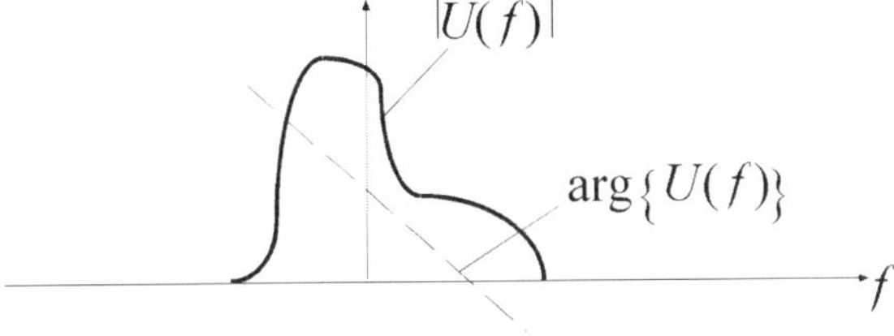

Figure 5.10 Spectra of a bandpass signal and its complex representation.

trum might look is shown in Figure 5.10. Note that the spectrum for the complex signal does not need to be symmetrical around zero.

The mathematical analysis of narrowband signals and systems is facilitated by use of the complex representation. There is no need to carry along a special function for the carrier and the results are independent of carrier frequency. If desired, the carrier is added by a multiplication of an exponential function. The intuitive understanding of different phenomena is facilitated by the natural connection between a signal and its corresponding vector. In computer simulations only the narrowband lowpass signal is represented by the use of complex numbers. The number of samples is strongly reduced compared to when the carrier is brought into the simulations. In the implementation of hardware and software it is more simple to do the signal processing on lowpass signals. Therefore the signal is converted to a low frequency and partitioned into the in phase, I-part, and quadrature parts, Q-part, where the signal processing is made on the respective part.

Example 5.1

Consider the band-pass signal

$$s(t) = v_1(t)\cos\omega_c t + v_2(t)\cos(\omega_c t + \theta) \tag{5.19}$$

where $v_1(t)$ and $v_2(t)$ are slowly varying signals. Use the vector model to obtain expressions for the in-phase component $s_i(t)$, quadrature component $s_q(t)$, envelope $a(t)$ and phase $\phi(t)$, of the signal.

Solution The larger of the signals can be regarded as a reference (cf. Figure 5.11). The in-phase component becomes

$$s_i(t) = v_1(t) + v_2(t)\cos\theta \tag{5.20}$$

The quadrature component becomes

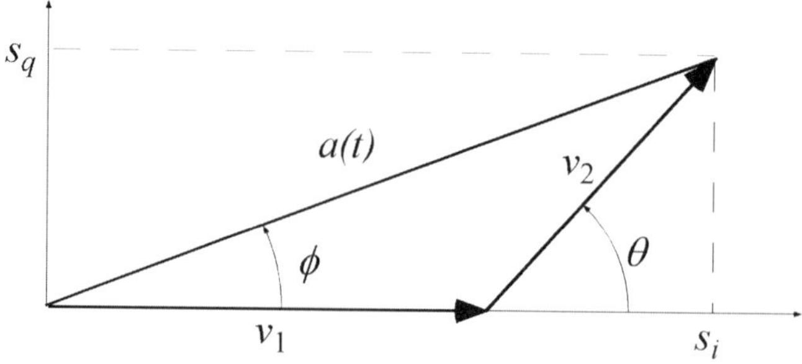

Figure 5.11 Phase diagram.

$$s_q(t) = v_2(t)\sin\theta \tag{5.21}$$

The envelope becomes

$$a(t) = \sqrt{v_1^2(t) + 2v_1(t)v_2(t)\cos\theta + v_2^2(t)} \tag{5.22}$$

The relative phase angle becomes:

$$\phi(t) = \arctan\frac{v_2(t)\sin\theta}{v_1(t) + v_2(t)\cos\theta} \tag{5.23}$$

Example 5.2

What is $u(t)$, $s_i(t)$ and $s_q(t)$ when $f_c = 1000$ and

$$S(f) = \begin{cases} 1 & 900 \leq |f| \leq 1300 \\ 0 & \text{otherwise} \end{cases} \tag{5.24}$$

Solution The spectrum of the signal can be drawn as in Figure 5.12. From eq. (5.18) it can
be seen that the spectrum of the complex or lowpass representation is obtained by
shifting the positive section of the bandpass signal spectrum to the left, a distance
corresponding to f_c. The time function of the complex signal is obtained by the
inverse Fourier transformation. The lowpass or complex representation of the
signal becomes

$$u(t) = F^{-1}\{\exp(-j2\pi f_c t)S(f); \ f \geq 0\} = \exp(-j2\pi f_c t)\int_{900}^{1300} \exp(j2\pi f)df$$

$$= \exp(-j2\pi 1000t)\frac{\exp(j2\pi 1300t) - \exp(j2\pi 900t)}{j2\pi t}$$

$$= \exp(j2\pi 100t)\frac{2\sin(2\pi 200t)}{2\pi t} \tag{5.25}$$

The in-phase component of the signal becomes

$$s_i(t) = \text{Re}\{u(t)\} = \frac{\cos(2\pi 100t)\sin(2\pi 200t)}{\pi t} \tag{5.26}$$

The quadrature component of the signal becomes

$$s_q(t) = \text{Im}\{u(t)\} = \frac{\sin(2\pi 100t)\sin(2\pi 200t)}{\pi t} \tag{5.27}$$

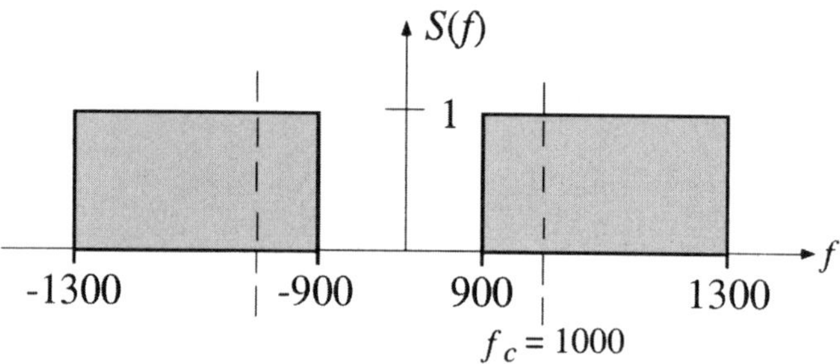

Figure 5.12 The spectrum of the signal.

When analysing how bandpass signals are affected by different system components, such as filters, an efficient and convenient notation is needed. The complex signal representation is also useful for bandpass signals passing through linear bandpass systems. In the same way as a bandpass signal can be expressed as a complex function, the impulse response of a bandpass filter can also be expressed as a complex function. For the real input signal, $x(t)$, and a real impulse response $h(t)$ a real output signal, $y(t)$, is obtained as (Figure 5.13)

$$y(t) = \int_{-\infty}^{t} x(\tau)h(t - \tau)d\tau \tag{5.28}$$

The complex representation of $x(t)$ can be labelled $u(t)$, having the real part i, and the imaginary part q. The complex representation of the impulse response $h(t)$ is denoted $g(t)$. As is shown below, the definition will be somewhat different compared to the complex representation of the signal. In the same way as in eq. (5.28) when the real and imaginary parts have been written separately, the complex output signal, $v(t)$, becomes

$$v(t) = \int_{-\infty}^{t} u(\tau)g(t - \tau)d\tau \tag{5.29}$$

$$= \int_{-\infty}^{t} u_i(\tau)g_i(t - \tau)d\tau + j\int_{-\infty}^{t} u_i(\tau)g_q(t - \tau)d\tau$$

$$+ j\int_{-\infty}^{t} u_q(\tau)g_i(t - \tau)d\tau - \int_{-\infty}^{t} u_q(\tau)g_q(t - \tau)d\tau$$

The calculation can be illustrated by a block diagram as in Figure 5.14.

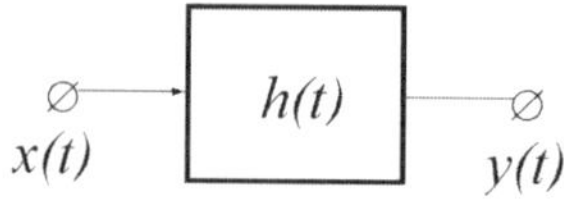

Figure 5.13 Block diagram of a real filter.

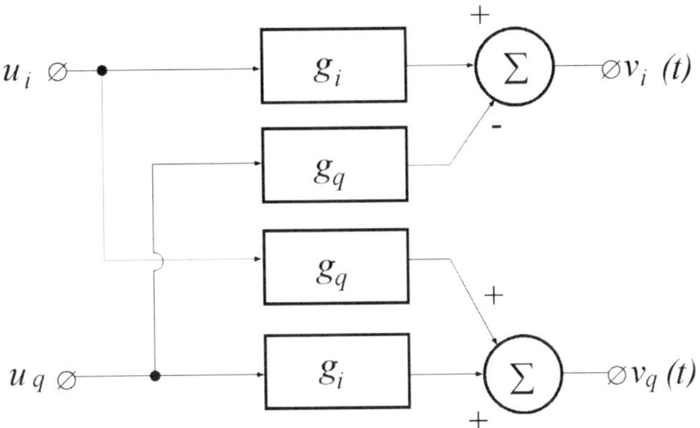

Figure 5.14 Block diagram of a complex filter. If the frequency response of the filter is symmetrical around f_c the quadrature parts of the impulse response disappear.

In the frequency domain the response from the complex filter becomes

$$V(f) = U(f) \cdot G(f) \tag{5.30}$$

The filter output signal is given in eq. (5.28). If the complex representation is used to express the output signal of the filter, the following equality is obtained:

$$y(t) = \int_{-\infty}^{t} 2\text{Re}\{u(\tau)\exp[j\omega_c\tau]\}\text{Re}\{g(t-\tau)\exp[j\omega_c(t-\tau)]\}d\tau$$

$$= \text{Re}\left\{\exp[j\omega_c t]\int_{-\infty}^{t} u(\tau)g(t-\tau)d\tau\right\} = \text{Re}\{v(t)\exp[j\omega_c t]\} \tag{5.31}$$

Since the Re{ } operation is non-linear the complex representations included must be defined differently. The usual way is that the complex representation of the filter impulse response is defined as

$$h(t) = 2\text{Re}\{g(t)\exp(j\omega_c t)\} \tag{5.32}$$

Relation to the Hilbert transform

Some readers may be familiar with the Hilbert transform, which is used, for instance, in signal processing and circuit analysis. The Hilbert transform of a function or signal, $x(t)$, is defined as

$$\tilde{x}(t) = \frac{1}{\pi}\int_{-\infty}^{\infty} \frac{x(\lambda)}{t-\lambda}\,d\lambda \tag{5.33}$$

The complex signal representation can be related to the Hilbert transform, as illustrated in the section below. Given a signal $x(t)$ with a narrow bandwidth compared to ω_c, the following Hilbert transforms are obtained:

$$H\{x(t)\cos\omega_c t\} = x(t)\sin\omega_c t \tag{5.34}$$

$$H\{x(t)\sin\omega_c t\} = -x(t)\cos\omega_c t \tag{5.35}$$

The Hilbert transform of a signal is denoted $H\{s(t)\} = \tilde{s}(t)$. An analytical signal is a signal which is expressed by an analytical function and has both real and imaginary parts (a function which is analytic in an area Ω is differentiable in a neighbourhood of every point in Ω). Form the analytical signal by using eq. (5.12) and the Hilbert transforms (5.34) and (5.35).

$$s(t) + j\tilde{s}(t) = a(t)\cos\phi(t)\cos\omega_c t - a(t)\sin\phi(t)\sin\omega_c t$$

$$+ ja(t)\cos\phi(t)\sin\omega_c t + ja(t)\sin\phi(t)\cos\omega_c t$$

$$= a(t)\big(\cos\phi(t) + j\sin\phi(t)\big)\cos\omega_c t + ja(t)\big(\cos\phi(t) + j\sin\phi(t)\big)\sin\omega_c t$$

$$= \underbrace{a(t)\exp[j\phi(t)]}_{u(t)} \exp(j\omega_c t) \tag{5.36}$$

i.e. the result is the complex representation of the signal $s(t)$. The Hilbert transform gives the imaginary part out of the real part of an analytical signal, so the result is not surprising. The complex representation of the bandpass component of the signal is obtained by multiplying eq. (5.36) by $\exp(-j\omega_c t)$, which gives

$$\big(s(t) + j\tilde{s}(t)\big)\exp(-j\omega_c t) = u(t) = s_i(t) + js_q(t) \tag{5.37}$$

Expanding the left hand side of eq. (5.37) and identifying the real and imaginary parts gives for the real part

$$s_i(t) = s(t)\cos\omega_c t + \tilde{s}(t)\sin\omega_c t \tag{5.38}$$

and for the imaginary part

$$s_q(t) = \tilde{s}(t)\cos\omega_c t - s(t)\sin\omega_c t \tag{5.39}$$

This means that the corresponding lowpass components of the bandpass signal can be obtained by the Hilbert transform multiplied by a sine or a cosine term. The in-phase component is given by eq. (5.38) and the quadrature component by eq. (5.39). By inserting these results in eq. (5.16) it can be seen that the amplitude of a bandpass signal can be obtained by using the Hilbert transform, i.e.

$$a(t) = \sqrt{s^2(t) + \tilde{s}^2(t)} \tag{5.40}$$

The argument is useful for the calculation of the instantaneous frequency of a signal $s(t)$, which is

$$f(t) = \frac{1}{2\pi}\frac{d[\phi(t)]}{dt} = \frac{1}{2\pi}\frac{d[\arctan(\tilde{s}(t)/s(t))]}{dt} \tag{5.41}$$

The instantaneous frequency is useful in many situations, but is only shown here for completeness and is not used further in this book. By letting the Hilbert transform be the output signal from a filter, a Fourier transformation of the corresponding impulse response results in the frequency response being

$$H(f) = -j\ \mathrm{sgn}(f) \tag{5.42}$$

The Hilbert transformation of a signal is, in other words, identical to letting the signal pass a filter which shifts the phase of the signal by $-90°$ and after that the spectral components located on the negative side of the frequency axis are shifted by $180°$.

Sampling of bandpass signals

The implementation of a receiver is often performed by an analogue HF or microwave section followed by a digital signal processor. The continuous time signal is sampled at the transition between the analogue and digital components. The spectrum of an ideally sampled signal can be seen as an infinite sum of shifted copies of the sampled signal spectrum, S_a:

$$S(f) = \frac{1}{T_s} \sum_{r=-\infty}^{\infty} S_a\left(j\frac{2\pi f}{T_s} + j\frac{2\pi r}{T_s}\right) \tag{5.43}$$

where $1/T_s = f_s$ is the sampling frequency. Equation (5.43) is illustrated in Figure 5.15.

From the equation and figure below it can be seen that, as long as $f_s > 2W$, where W is the bandwidth of the continuous signal, the different spectra will not overlap and the signal is correctly represented. This is in agreement with the Nyquist sampling theorem, stating that for a lowpass signal the sampling frequency must be at least twice the highest of the included frequency components.

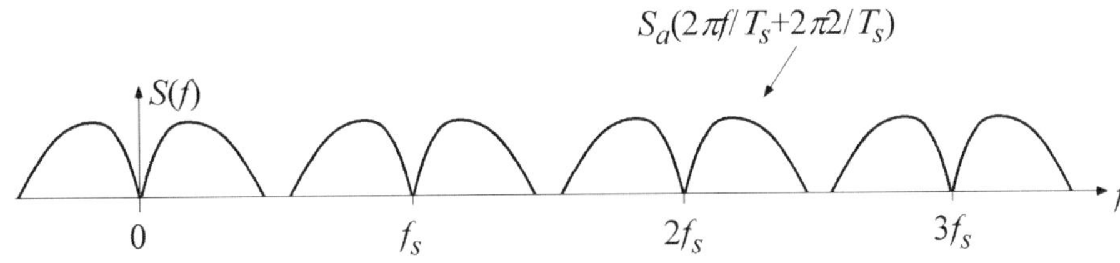

Figure 5.15 The spectrum of the sampled signal. As an example the spectrum component having $r = 2$ (eq. (5.43)) has been indicated.

One way of achieving the sampling of a bandpass signal is to down-convert the signal to the baseband and generate I and Q components, which then are sampled separately. If the bandwidth of the bandpass signal is $2W$, the down converted signals get the bandwidth W. A sampling frequency of $f_s > 2W$ is sufficient. Since both channels have to be sampled, the same number of samples are produced as when a signal with bandwidth $2W$ is directly sampled.

Another way of solving the sampling problem without down-conversion is to use so-called down sampling. Assume, for example, that the continuous time bandpass signal is the spectrum bubble in Figure 5.15 centred around the frequency $2f_s$. If such a signal is sampled by a frequency of f_s, the spectrum of the sampled signal will look like the signal in the figure. The discrete time signal processing can be performed in the same way as if the signal were a lowpass signal with bandwidth W. The sampling can be seen as if a short section is cut out of the continuous signal and averaged during the time duration of the sampling pulse. In a real system there is a lower limit for how short this duration can be. Because of this, the amplitude of the sampled signal decreases the higher the intermediate frequency. Hence, the width of the sampled pulse sets an upper limit for the centre frequency. The centre frequency of the continuous signal must be an integer multiple of the sampling frequency, $f_c = mf_s = m/T_s$. Given that the sampling has been synchronised with the carrier, the sampled bandpass signal can be written

$$s(nT_s) = s_i(nT_s)\cos(2\pi nm) - s_q(nT_s)\sin(2\pi nm) \tag{5.44}$$

This means that only the in-phase component of the signal will be sampled. To obtain the Q-component the signal may be shifted a time period corresponding to a phase shift of the carrier of $90°$. This can be achieved either by delaying the analogue signal, or by performing the sampling with a corresponding delay. The time delay of the narrowband quadrature components is negligible.

5.3 Analogue carrier modulation

During radio transmission the signal spectrum must be transferred to a suitable part of the electromagnetic spectrum. Wave propagation aspects, the size of the antenna system and international agreements on which part of the spectrum a certain type of communication can utilise, are some factors determining the frequency domain. A desire to use the electromagnetic spectrum efficiently also demands that the communication channels are closely packed, which means that the signal spectrum has to be as narrow as possible. Similar require-ments are imposed in the transmission of several signals on a single coaxial-cable by means of carrier technology.

A shift of the signal spectrum along the frequency axis can be obtained through an operation called carrier modulation, which is described below. Some important properties of the modulated signal are the bandwidth together with its properties during demodulation in the receiver. The latter is discussed in Chapter 6.

5.3.1 ANALOGUE AMPLITUDE MODULATION

The following sections describe different ways to map the information on a sinusoidal signal, which then constitutes a narrowband signal suitable to send over the available channel. The information can be transferred through the slowly varying envelope or phase of the signal in eq. (5.12). In analogue modulation, the information is analogue and in digital modulation the information is in digital form, but the modulated signal is analogue in both cases. First we give a presentation of the concepts of amplitude modulation. The first large main group to be described is amplitude modulation. In this group are three relatives with different properties described, namely AM-DSBC, AM-DSBSC and SSB. The meanings of these acronyms are described below.

Amplitude modulation

Assume the signal described by eq. (5.12). Let the information to be transmitted only influence the amplitude of the sinusoidal signal, or carrier. When only the amplitude varies we have amplitude modulation, or AM.

$$s(t) = a_c \, a(t) \, \cos(\omega_c t) \tag{5.45}$$

where

$$a(t) = 1 + \mu \, x(t) \tag{5.46}$$

a_c is the amplitude of the carrier, ω_c is the angular frequency of the carrier, $\omega_c = 2\pi f_c$, μ is the modulation index ($0 < \mu \leq 1$) often given in $100\mu\%$ and $x(t)$ is the message. The information signal $x(t)$ must fulfil $|x(t)| \leq 1$ and thereby $E\{x^2(t)\} \leq 1$, i.e. the signal power $S_x \leq 1$. Usually the properties of AM are investigated by using a sinusoidal information signal, $x(t) = a_m \cos(\omega_m t)$, where the modulation frequency $\omega_m \ll \omega_c$. The modulated signal in eq. (5.45) is for sinusoidal modulation shown in graphical form in Figure 5.16.

The shape of the information signal is found in the amplitude of the modulated signal as can be seen in Figure 5.16. Therefore the AM signal can be easily demodulated by using an electronic circuit, which tracks the contour, or envelope, of the signal. Such a circuit is shown in Figure 5.17.

Should the conditions for $|x(t)|$ and μ not be fulfilled, over-modulation occurs, as illustrated in Figure 5.18, and the signal becomes distorted in the envelope detector. The signal can, however, be retrieved by more complex detectors.

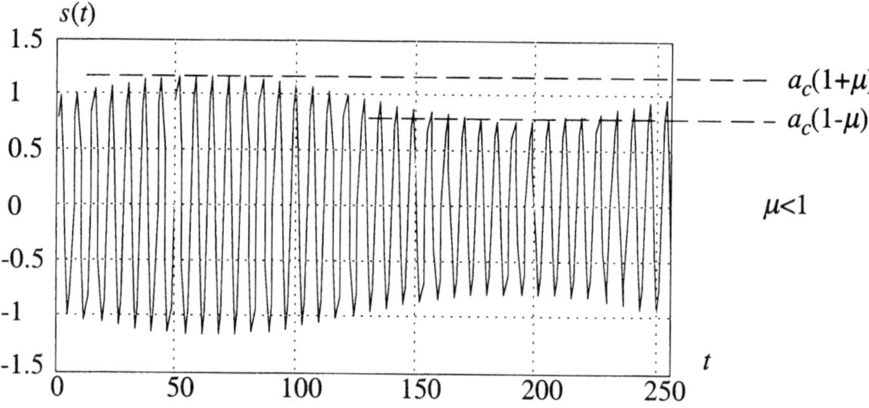

Figure 5.16   AM signal with sinusoidal modulation.

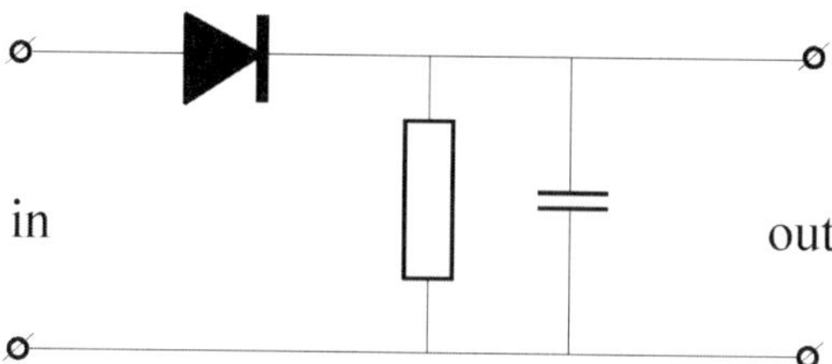

Figure 5.17   Envelope detector used as AM demodulator.

To find out the required channel bandwidth to transmit an AM signal, the signal spectrum is studied. The Fourier transform of eq. (5.45) gives the spectra of the signal ($\otimes$ stands for convolution):

$$S(f) = A(f) \otimes \frac{a_c}{2} \delta(|f|-f_c) = \frac{a_c}{2} \delta(|f|-f_c) + \frac{\mu}{2} a_c X(|f|-f_c) \tag{5.47}$$

where $X(f)$ is the spectrum of the information signal, $A(f) = F\{a(t)\}$, and $\delta(|f|-f_c)$ represent the spectra of the cosine part in eq. (5.45), i.e. the carrier. It can be seen that the spectrum consists of a peak surrounded by two side bands, one higher and one lower than the carrier frequency. An example of what the spectrum may look like is shown in Figure 5.19.

When the signal is analysed with a spectrum analyser the positive and negative sides of the frequency axis are overlapping. The carrier parts are thus added and the value a_c is read. Since both side bands and the carrier are included in the spectrum, this form of AM is sometimes called amplitude modulation with double sideband and carrier, abbreviated AM-DSBC. The total power of the signal $s(t)$ becomes

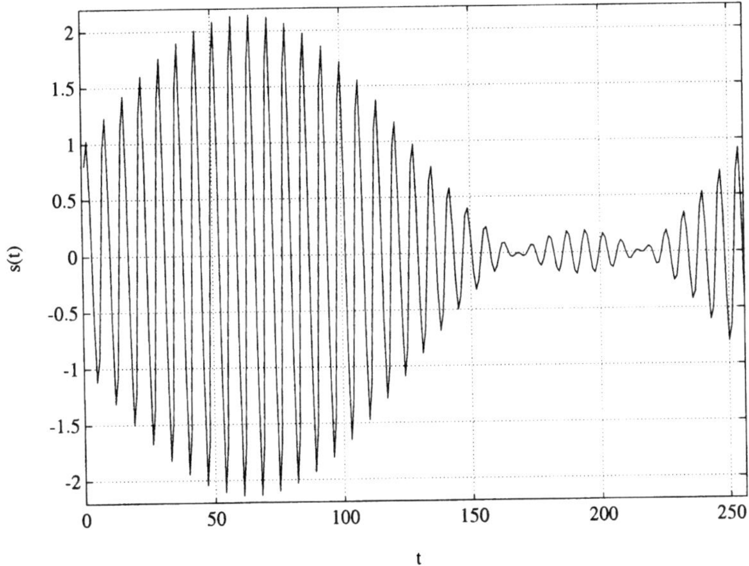

Figure 5.18   AM signal at over-modulation.

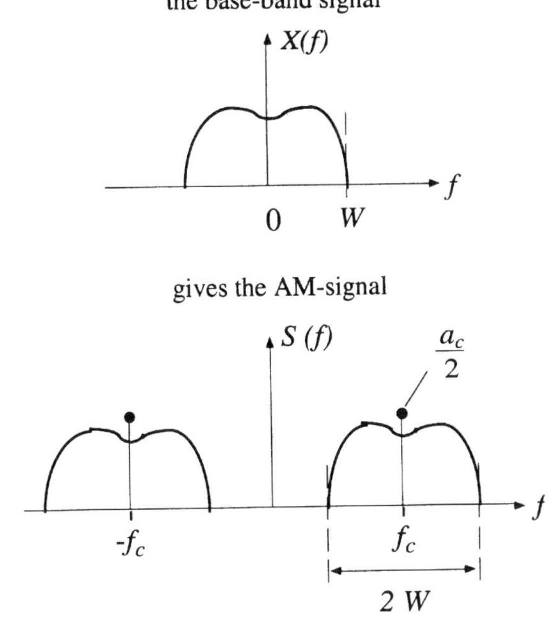

Figure 5.19   Spectrum of the message signal together with the amplitude modulated signal with double side bands and carrier.

$$E\{s^2(t)\} = E\left\{ a_c^2\left[1 + 2\mu\, x(t) + \mu^2 x^2(t)\right] \underbrace{\cos^2 \omega_c t}_{\left(\frac{1}{2} + \frac{1}{2}\cos(2\omega_c t)\right)} \right\}$$

$$= \frac{a_c^2}{2} E\{1 + \mu^2 x^2(t)\} = \left(\frac{1}{2} + \frac{\mu^2}{2} S_x\right) a_c^2 \tag{5.48}$$

This is valid on the condition that $x(t)$ and $\cos(2\omega_c t)$ are independent and that $E\{x(t)\} = 0$. By using eq. (5.48), the ratio between useful power, i.e. the power of the transmitted information signal, $\mu^2 S_x/2$, and the total signal power, can be written

$$\frac{a_c^2 \dfrac{\mu^2}{2} S_x}{a_c^2 \dfrac{1}{2}(1 + \mu^2 S_x)} = \frac{1}{\dfrac{1}{\mu^2 S_x} + 1} \leq \frac{1}{2} \tag{5.49}$$

since $\mu \leq 1$ and $S_x \leq 1$. From eq. (5.49) it can be seen that a maximum of 50% of the available transmitter power lies in the information carrying parts of the signal. This means that the available power is poorly utilised. The strength of AM-DSBC is the simple demodulator shown in Figure 5.17.

Double sideband suppressed carrier

A form of modulation which uses the available transmitter power more efficiently is obtained if $a(t)$ in eq. (5.46) is modified by removing the '1' and setting $\mu = 1$. We then get the signal

$$s(t) = a_c x(t)\cos(\omega_c t) \tag{5.50}$$

The resulting signal with sinusoidal modulation is depicted in Figure 5.20. The spectrum of this form of modulation is obtained through the Fourier transformation of eq. (5.50):

$$S(f) = X(f) \otimes \frac{a_c}{2} \delta(|f| - f_c) \tag{5.51}$$

As can be seen, the carrier component is missing but the two side bands still remain. This is called 'double sideband with suppressed carrier', AM-DSBSC. An example is shown in Figure 5.21. The power of the signal becomes

$$E\{s^2(t)\} = E\left\{ x^2(t) a_c^2\left[\frac{1}{2} - \frac{1}{2}\cos(2\omega_c t)\right]\right\} = \frac{a_c^2}{2} E\{x^2(t)\} = \frac{a_c^2}{2} S_x \tag{5.52}$$

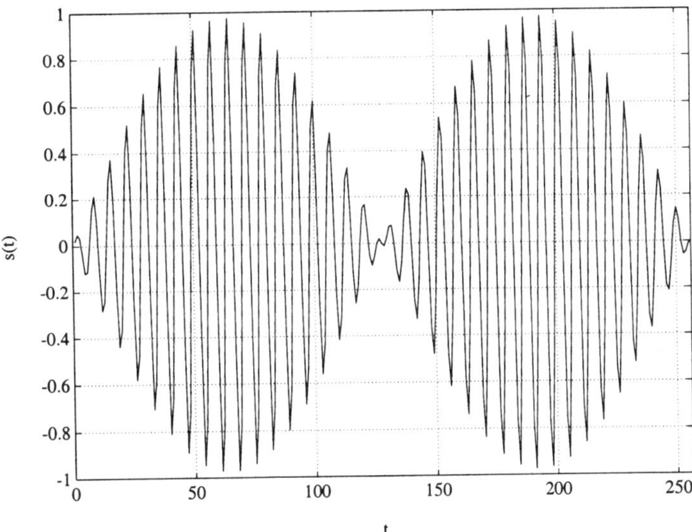

Figure 5.20 DSBSC signal with sinus modulation. It can be seen from the graph that envelope detection is not possible.

Hence, all of the power is utilised for the information transmission. The spectrum contains no components for which the power is independent of the message. A comparison of eqs. (5.49) and (5.52) shows that a radio transmitter with a given maximum mean power can transmit twice the power in the information-carrying parts of the signal when using AM-DSBSC than when using AM-DSBC. If a certain maximum peak amplitude of the transmitter is assumed, the gain is even higher as peak amplitudes are added linearly even if the signals are independent. This means that for a peak amplitude limited transmitter, the power becomes a factor of four lower for AM-DSBC than for AM-DSBSC.

To demodulate an AM-DSBSC it is required that the receiver restores the carrier in the received signal by means of a synchronous detector (cf. Figure 5.22). It is of great importance that the restored carrier has correct frequency and phase. The demodulated signal will otherwise be disturbed by a sinusoidal tone

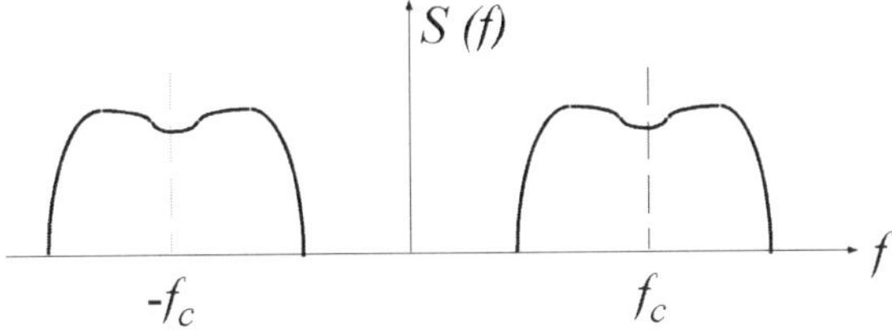

Figure 5.21 Spectrum of AM with suppressed carrier, AM-DSBSC. Note that the spectrum does not contain the carrier.

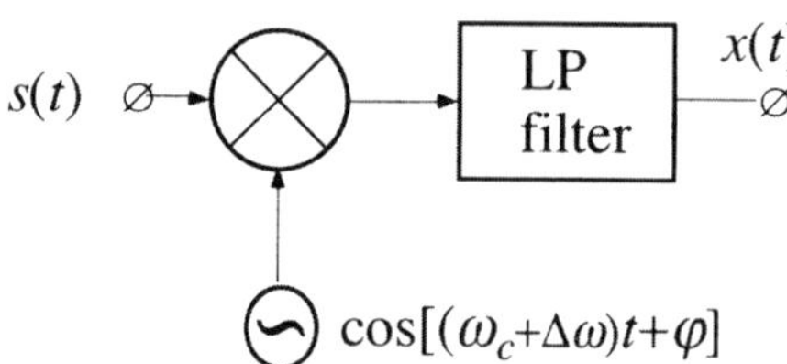

Figure 5.22   Demodulator for AM-DSBSC. The LP filter takes away the signal component at the double carrier frequency.

corresponding to the frequency difference, $\Delta\omega$, between the transmitter and the receiver carrier, respectively. A phase error, φ, gives rise to a decrease in the signal amplitude of the demodulated signal $x(t)$, possibly down to zero.

Single sideband

Since both side bands contain identical information it is sufficient to transmit only one of them. A type of modulation which therefore utilises the transmission channel very efficiently is single sideband modulation, SSB. The single sideband is created by the operation

$$s_{SSB}(t) = \frac{a_c}{2}x(t)\cos(\omega_c t) \mp \frac{a_c}{2}\tilde{x}(t)\sin(\omega_c t) \tag{5.53}$$

where $\tilde{x}(t) = H\{x(t)\}$. If the upper sideband is desired the minus sign is used and for the lower sideband the plus sign is used. This can be shown by using eqs. (5.18) and (5.42). To generate a SSB signal, the Hilbert transform of the signal is thus needed. Another way is to generate a DSBSC signal where one of the side-bands is being filtered away. This requires filters with sharp transitions between passband and stopband.

The demodulator design for AM-SSB is identical to the one for AM-DSBSC (cf. Figure 5.22). The phase of the restored carrier is not critical. Since there is only one side band, a frequency error will effect the pitch of the demodulated signal by a shift of $\Delta\omega$. SSB is usually used in speech communication and a shift in the voice pitch is not of any major importance.

Realisation of modulator

The AM signal is generated in some kind of modulator, which can be realised in different ways. One way is to use a transconductance amplifier, where the gain is current controlled. The current is made proportional to the amplitude of the information signal. By controlling the gain with this current the amplitude of the carrier is made proportional to the information signal. Another method is to use some non-linear element as in Figure 5.23.

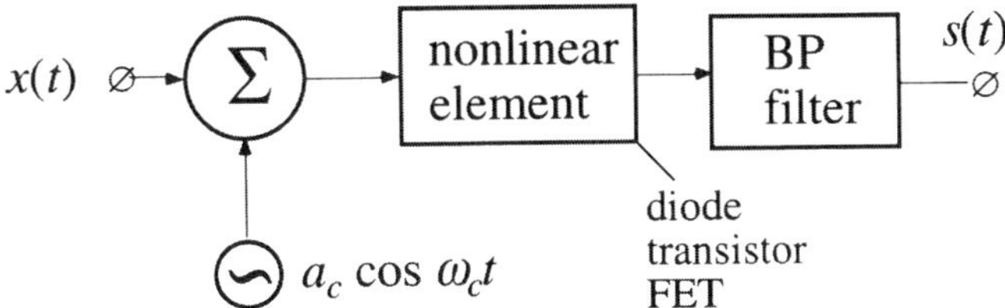

Figure 5.23 AM modulator using a non-linear element.

The non-linear element could be one of many different components, such as diodes, transistors or electron tubes. The ratio between input and output signals can then be represented by a polynomial of high degree, for which the values of the coefficients depend on the non-linearity of the component. The FET transistor is such a component which is well suited for use in modulation applications. Its non-linearity is mainly of the second degree, which gives the following relation between the input signal v_{in} and the output signal v_{ut}:

$$v_{ut} = a_1 v_{in} + a_2 v_{in}^2 \tag{5.54}$$

If the input signal, consisting of the sum of the information signal, $x(t)$, and the carrier, is inserted in eq. (5.54), the output signal will contain the following components:

$$v_{ut} = a_1 x(t) + a_2 x^2(t) + a_2 a_c^2 \underbrace{\cos^2(\omega_c t)}_{\frac{1}{2} + \frac{1}{2}\cos(2\omega_c t)}$$

$$+ \quad \underbrace{a_c[a_1 + 2a_2 x(t)]\cos \omega_c t}_{\text{this signal has the desired form and is filtered out}} \tag{5.55}$$

The last term in eq. (5.55) contains the signal components required for amplitude modulation and is filtered out by a bandpass filter.

An example of a modulator for AM-DSBSC is the double balanced diode bridge shown in Figure 5.24. If the local oscillator (LO) signal has a large enough amplitude, the diodes will be biased in a way that they will pairwise change between conducting and not conducting, in the same pace as the switching of LO signal polarity. The bridge will then mainly act as two switches, which couple the signal in alternate directions through the coil on the high frequency side. Thus the operation can be regarded as a multiplication of the LF signal by a square wave. A square wave contains a large number of frequency components and the desired one is filtered out by a linear filter.

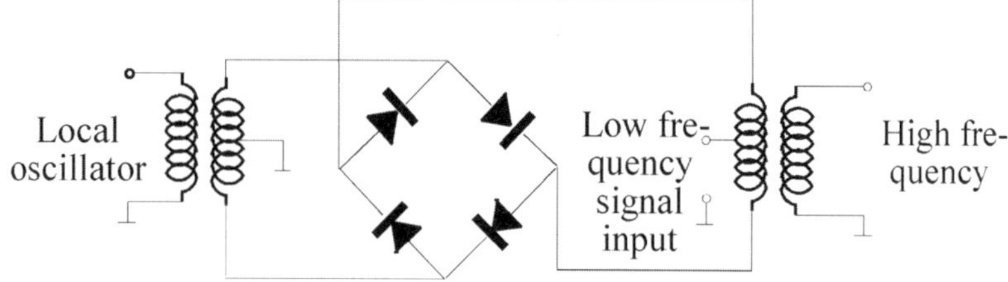

Figure 5.24 Double balanced diode bridge.

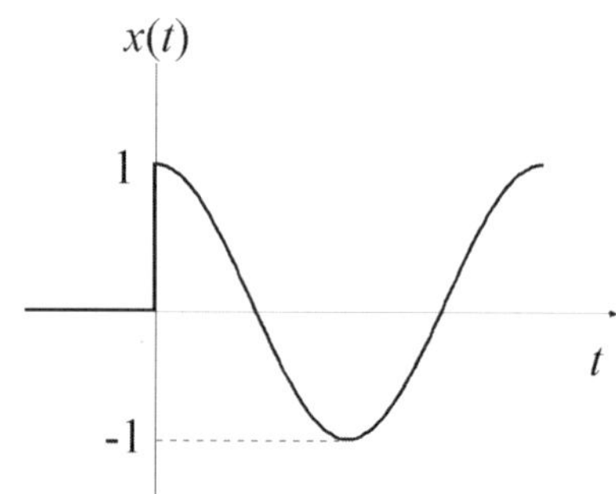

Figure 5.25 Sketch of the message signal.

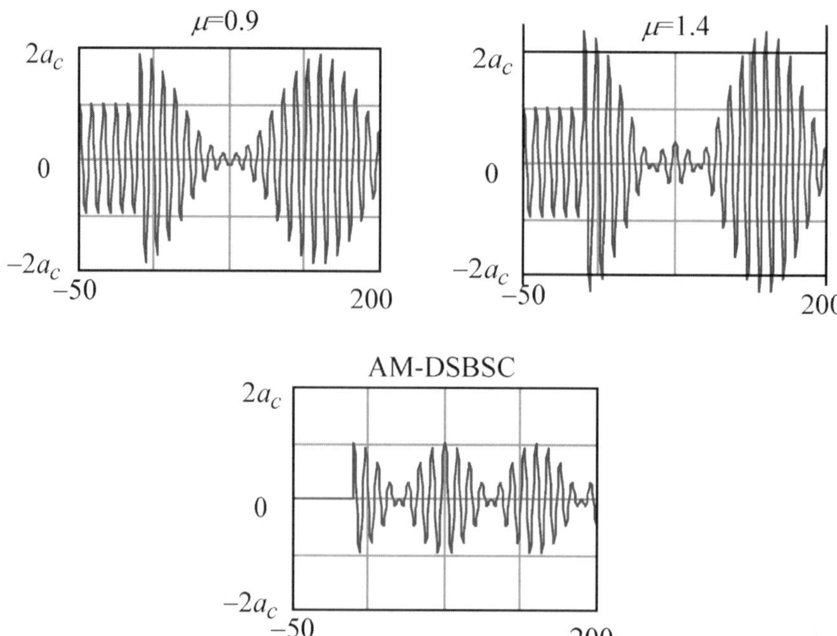

Figure 5.26 Illustration of the modulated signal for AM with carrier and different degrees of modulation together with AM without carrier.

Example 5.3

Assume the message signal $x(t) = \cos(2\pi f_m t)$ for $t \geq 0$ and $x(t) = 0$ for $t < 0$, and that $f_m \ll f_c$. Sketch $s(t)$ and its envelope for AM when $\mu < 1$ and $\mu > 1$, respectively, and for AM-DSBSC.

Solution   See Figures 5.25 and 5.26. The modulated signal is

$$s(t) = a_c[1 + \mu x(t)]\cos \omega_c t \tag{5.56}$$

Example 5.4

Suppose the signal $s(t) = a_c[1 + \mu x(t)]\cos(\omega_c t + \Phi)$ is demodulated in a synchronous detector of the type shown in Figure 5.22, where $\Delta\omega = \varphi = 0$. Because of circumstances in the transmission channel phase of the the received signal is varying. The information $x(t)$ is an ergodic stochastic signal with the average value zero, and Φ is a statistically independent random phase angle evenly distributed over 2π rad. Calculate the average power of the demodulated signal.

Solution   After the multiplier the signal is

$$x_{\mathrm{II}} = a_c^2[1 + \mu\, x(t)]\cos(\omega_c t + \Phi)\,\cos(\omega_c t)$$

$$= \frac{a_c^2}{2}[1 + \mu\, x(t)]\left(\underbrace{\cos(\Phi)}_{\mathrm{LF-part}} + \underbrace{\cos(2\omega_c t + \Phi)}_{\mathrm{HF-part}} \right) \tag{5.57}$$

The high frequency part of the signal is cancelled in the lowpass filter. The average power for the demodulated signal is $E[x_{\mathrm{II}}^2|_{\text{only the LF-part}}]$, which becomes

$$E\left\{ \frac{a_c^2}{2}[1 + \mu\, x(t)]^2\cos^2(\Phi) \right\}$$

$$= \frac{a_c^2}{2} E\{1 + 2\mu x(t) + \mu^2 x^2(t)\} E\left\{ \frac{1}{2} + \frac{1}{2}\cos(\Phi) \right\}$$

$$= \frac{a_c^2}{4}\left(1 + \mu^2\sigma_x^2\right) \tag{5.58}$$

The fact that the expectation values of $x(t)$ and the $\cos^2$ term are zero and $1/2$, respectively, is used to obtain eq. (5.58). Furthermore, since the two stochastic variables are independent, the following relation holds:

$$E\{X_1 \cdot X_2\} = E\{X_1\} \cdot E\{X_2\} \tag{5.59}$$

where X_1 and X_2 are the independent stochastic variables.

Example 5.5

Make a block diagram of a DSB modulator with a non-linear element, for which the ratio between the amplitudes of the input and output signals is $v_{out} = a_1 v_{in} + a_3 v_{in}^3$. What is the condition on the carrier frequency f_c in relation to the bandwidth, W, of the baseband signal?

Solution Let v_{in} be the sum of a sinusoidal signal with frequency ω_0 and the baseband signal, $x(t)$

$$v_{in} = x(t) + \cos(\omega_0 t) \tag{5.60}$$

The output signal is then

$$v_{out} = a_1\big(x(t) + \cos(\omega_0 t)\big)$$

$$+ \, a_3\Big(x^3(t) + 3x^2(t)\cos(\omega_0 t) + 3x(t)\cos^2(\omega_0 t) + \cos^3(\omega_0 t)\Big)$$

$$= \underbrace{\left(a_1 + \tfrac{3}{2}a_3\right)x(t)}_{\text{I}} + \underbrace{a_3 x^3(t)}_{\text{II}} + \left(\underbrace{a_1 + \tfrac{3}{4}a_3}_{\text{III}} + \underbrace{3a_3 x^2(t)}_{\text{IV}}\right)\cos(\omega_0 t)$$

$$+ \underbrace{\tfrac{3}{2}a_3 x(t)\cos(2\omega_0 t)}_{\text{V}} + \underbrace{\tfrac{1}{4}a_3\cos(3\omega_0 t)}_{\text{VI}}$$

$$\tag{5.61}$$

The six terms in the last equality of eq. (5.61), each have their counterparts in the spectrum in Figure 5.27.

The spectral component V corresponds to the desired signal. Choose an oscillator having half of the desired carrier frequency, f_c, and filter out section V of the spectrum. From Figure 5.27 it can also be seen that f_c must be high enough for the signal bandwidth $2W$ to fit into a frequency window where no other spectral components can be found. This makes it possible for the desired signal to be filtered out by a linear bandpass filter having a frequency response approximately corresponding to the dashed lines (Figure 5.28).

This example also illustrates the importance of other parts in the system, such as the transmission channel, being linear so that no undesired frequency components appear, which can cause interference in the communication system itself or other systems.

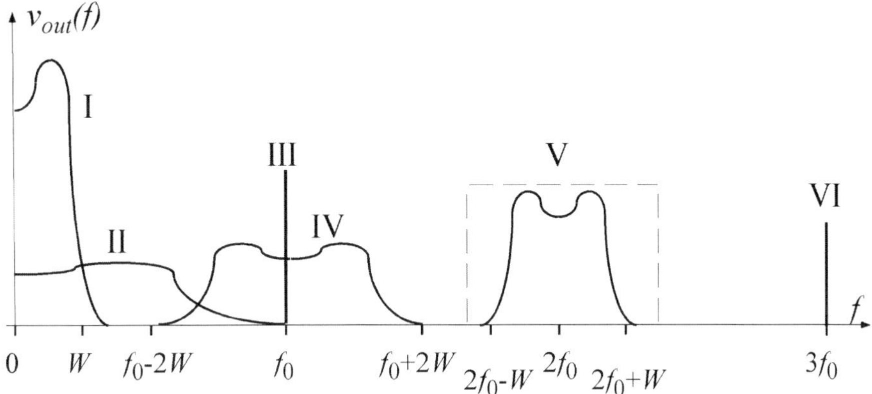

Figure 5.27   Signal spectrum at the output of the non-linear element.

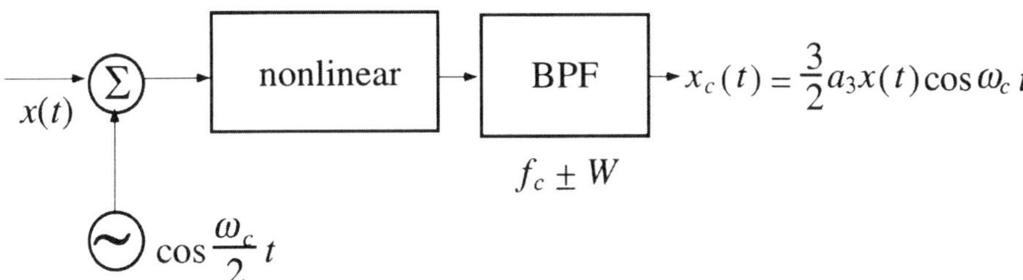

Figure 5.28   The resulting modulator.

Example 5.6

A carrier is SSB modulated by the baseband signal

$$x(t) = \cos\omega_m t - \frac{1}{3}\cos 3\omega_m t \tag{5.62}$$

The signal $x(t)$ approximates a square wave by means of the first two terms in the Fourier series. Outline the amplitude, $a(t)$, of the modulated SSB signal when the amplitude of the carrier is a_c, and compare with $x(t)$.

Solution   It can be seen from eq. (5.53), that by using the Hilbert transform an SSB signal can be written as (Figure 5.29)

$$s(t) = \frac{1}{2}a_c\left[x(t)\cos\omega_c t \mp \tilde{x}(t)\sin\omega_c t\right]$$

where $x(t)$ is the message signal, $\tilde{x}(t)$ is the Hilbert transform of the message signal and $\mp$ corresponds to upper and lower sidebands, respectively.

$$\tilde{x}(t) = \frac{1}{\pi} \int_{-\infty}^{\infty} \frac{x(\lambda)}{t - \lambda} \, d\lambda$$

$$= H\left\{ \cos(\omega_m t) - \frac{1}{3}\cos(3\omega_m t) \right\}$$

$$= \sin(\omega_m t) - \frac{1}{3}\sin(3\omega_m t) \tag{5.63}$$

According to eq. (5.40) the envelope of a bandpass signal is obtained by the Hilbert transform, which gives

$$a(t) = \sqrt{s^2(t) + \tilde{s}^2(t)} = \sqrt{x^2(t) + \tilde{x}^2(t)} \tag{5.64}$$

and the SSB signal in this example has the envelope

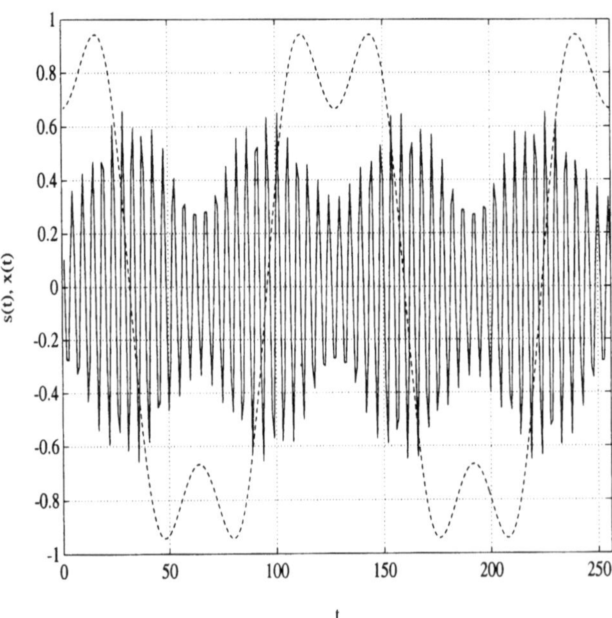

Figure 5.29   Solid line represents the modulated SSB signal. Dotted line represents the baseband signal $x(t)$.

$$a(t) = \frac{a_c}{2}\sqrt{\left(\cos(\omega_m t) - \frac{1}{3}\cos(3\omega_m t)\right)^2 + \left(\sin(\omega_m t) - \frac{1}{3}\sin(3\omega_m t)\right)^2}$$

$$= \frac{a_c}{2}\sqrt{1 + \frac{1}{9} - \frac{2}{3}\cos(2\omega_m t)} = \frac{a_c}{6}\sqrt{10 - 6\cos(2\omega_m t)} \tag{5.65}$$

Note that the envelope of an SSB signal, modulated by a sinusoidal signal $x(t) = \cos(\omega_m t)$, is constant. The spectrum only contains one peak, which has been shifted a distance ω_m from the angular frequency of the carrier.

5.3.2 ANALOGUE PHASE AND FREQUENCY MODULATION

The next large group of analogue modulation methods to be treated is phase and frequency modulation, which is usually summarised under the concept of angle modulation. In these methods the amplitude is left constant and the information is included in the slowly varying phase component (cf. eq. (5.12)). The advantage of using a constant amplitude is, for instance, that the power amplifier of the transmitter is used more efficiently since it always operates at full power. Consider the signal

$$s(t) = a_c \cos\underbrace{\left[\omega_c t + \phi(t)\right]}_{\theta_c(t)} = a_c \mathrm{Re}\left\{e^{j\theta_c(t)}\right\} \tag{5.66}$$

$\phi(t)$ contains the information to be transmitted. Since the relation between the output signal, $s(t)$, and the information confined in the phase angle, ϕ, is non-linear, this is called non-linear modulation. In phase modulation, PM, the phase, ϕ, must be proportional to the information, i.e. to the input signal $x(t)$:

$$\phi(t) = \phi_\Delta x(t) \quad \Leftrightarrow \quad \theta_c(t) = \omega_c t + \phi_\Delta x(t) \tag{5.67}$$

where ϕ_Δ is a constant called the phase deviation.

In frequency modulation (FM), the frequency of the modulated signal must be proportional to the information, i.e. to the input signal $x(t)$. Since the frequency of the carrier then varies, we must consider the instantaneous frequency, ω_i, where

$$\omega_i = \frac{d}{dt}\theta_c(t) = \omega_c + \frac{d}{dt}\phi(t) = 2\pi(f_c + f_\Delta x(t)) \tag{5.68}$$

and f_Δ is the frequency deviation. As can be seen from eq. (5.68), the frequency is proportional to the time derivative of the phase, which for $\phi(t)$ gives

$$\phi(t) = 2\pi f_\Delta \int_{t_0}^{t} x(\lambda)d\lambda + \phi(t_0) \tag{5.69}$$

As distinct from PM, in this case the phase is proportional to the integral of the input signal.

Demodulation of FM and PM signals is done by some electronic circuit (or some computer program) sensitive to variations in the phase of the received signal. The requirement is that the phase variations shall be converted to amplitude variations in a linear way, which will make up the demodulated baseband signal. Many circuits, useful in practice, are available for this. One of these is the phase locked loop, described in Chapter 10. The signal controlling the VCO is within certain limits linear to the input signal phase variations. By circuit arrangements, the loop can easily be made insensitive to amplitude variations.

To study some of the properties of PM and FM, we assume that the baseband signal $x(t)$, i.e. the message, consists of a sine signal for PM and a cosine signal for FM, both having the frequency ω_m.

$$x(t) = \begin{cases} a_m \sin\omega_m t & \text{for PM} \\ a_m \cos\omega_m t & \text{for FM} \end{cases} \tag{5.70}$$

Because of this choice of input signals, we obtain the phase angle $\phi(t) = \beta\sin\omega_m t$ for both FM and PM when eq. (5.70) is inserted in eqs. (5.67) and (5.69), respectively, where

$$\beta = \begin{cases} \phi_\Delta a_m & \text{for PM} \\ \dfrac{f_\Delta}{f_m} a_m & \text{for FM} \end{cases} \tag{5.71}$$

The number β is a parameter called the modulation index, given that $a_m = 1$.

In order to investigate the resulting spectrum when the modulating signal has a sinusoidal shape, one can start with the complex representation, where carrier and lowpass parts are separated. Rewriting eq. (5.66) gives $a_c \exp(j\theta_c(t)) = a_c \exp(j\omega_c t) \exp(j\phi(t))$. The function $\phi(t)$ is defined as above with sinusoidal modulation. Because periodic modulation results in a periodic signal, the signal can be expressed by a Fourier series. For the lowpass part we obtain

$$a_c \exp[j\phi(t)] = a_c \exp[j\beta\sin(\omega_m t)] = a_c \sum_{n=-\infty}^{\infty} J_n(\beta)\exp(j\,n\omega_m t) \tag{5.72}$$

where $J_n(x)$ is the nth order Bessel function of the first kind. The Bessel functions of order 0 to 9 have been plotted in Figure 5.30. The modulated signal, $s(t)$, can be expressed by a Fourier series:

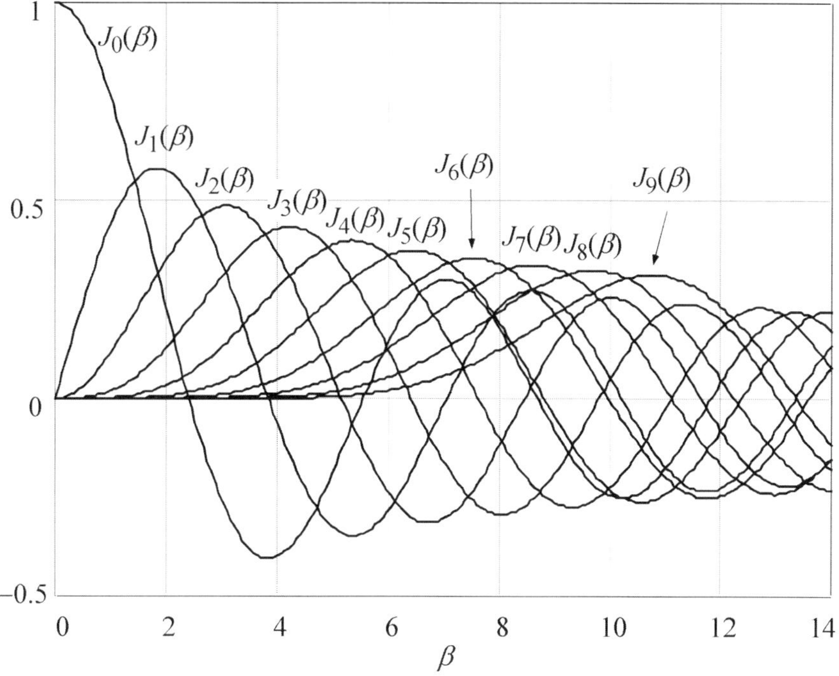

Figure 5.30 Bessel functions.

$$s(t) = \mathrm{Re}\left\{a_c \sum_{n=-\infty}^{\infty} J_n(\beta)\, \exp(j\, n\omega_m t)\, \exp(j\omega_c t)\right\}$$

$$= a_c \sum_{n=-\infty}^{\infty} J_n(\beta)\cos(\omega_c + n\omega_m)t \tag{5.73}$$

For a periodic signal the spectrum consists of a number of discrete components, where the coefficients of the Fourier series represent the amplitudes of the respective components. As can be concluded from eq. (5.73), the amplitudes of the spectral components are proportional to the Bessel function, $J_n(\beta)$. The spectrum of a sinusoidally modulated FM signal depends on the modulation index, β, i.e. the ratio between frequency deviation and modulation frequency. Some examples of the absolute value of the spectrum are shown in Figure 5.31.

As can be seen, the spectrum is considerably wider than the double bandwidth of the baseband signal as is obtained in AM-DSBC. This can seem to be a disadvantage. But as will be shown in next chapter, there are other conclusive advantages with the FM.

An interesting remark is that the power spectrum of an angle modulated signal

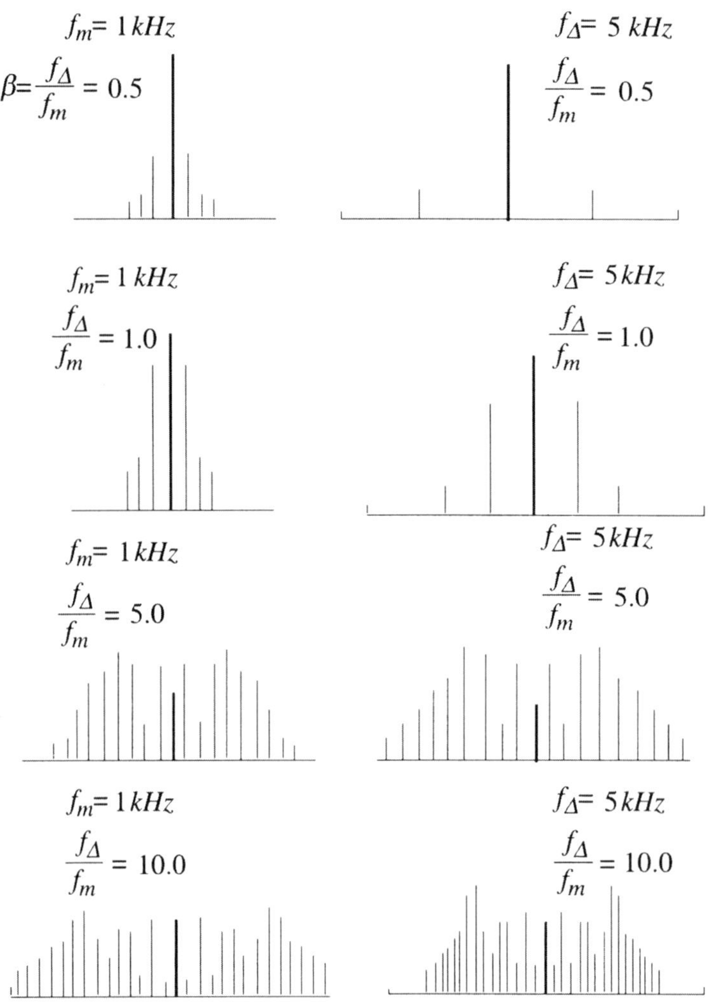

Figure 5.31 Spectra of a sinusoidally modulated FM signal as a function of the modulation index. The modulation frequency has been kept constant in the left hand column and the frequency deviation has been kept constant in the right hand column.

has, in the limit as the modulation index goes to infinity, the shape of the amplitude probability density of the modulating signal.

Narrowband FM

If the FM modulation parameters have been chosen so that $|\phi(t)| \ll 1$ (rad), the modulation type is denoted narrowband FM (cf. eq. (5.69)). To analyse the properties of narrowband FM we can use eq. (5.66) and in the same way as in eq. (5.12) divide the signal into its quadrature components. The low frequency components are collected in parts which are in phase $s_i(t)$ and quadrature $s_q(t)$ relative to the carrier

$$s(t) = s_i(t)\cos\omega_c t - s_q(t)\sin\omega_c t \tag{5.74}$$

By series expansion of the trigonometric functions the in phase and quadrature components are obtained:

$$s_i(t) = a_c\cos(\phi(t)) = a_c\left[1 - \frac{1}{2!}\phi^2(t) + \dots\right]$$

$$s_q(t) = a_c\sin(\phi(t)) = a_c\left[\phi(t) - \frac{1}{3!}\phi^3(t) + \dots\right] \tag{5.75}$$

Since $|\phi(t)| \ll 1$, $s_i \approx a_c$ and $s_q \approx a_c\phi(t)$, i.e. eq. (5.74) becomes

$$s(t) \approx a_c\cos\omega_c t - a_c\phi(t)\sin\omega_c t \tag{5.76}$$

The Fourier transform of eq. (5.76) becomes for $\omega > 0$,

$$S(\omega) = \pi a_c\delta(\omega - \omega_c) + \frac{j}{2}a_c\Phi(\omega - \omega_c) \tag{5.77}$$

where

$$\Phi(\omega) = F\{\phi(t)\} = \begin{cases} \phi_\Delta X(\omega) & \text{for PM} \\ \dfrac{-j f_\Delta X(\omega)}{\omega} & \text{for FM} \end{cases} \tag{5.78}$$

In PM the spectrum of the modulated signal consists of the spectrum of the information signal, transposed around the carrier frequency together with a frequency component from the carrier. In FM modulation the spectrum of the information signal is phase shifted and weighted by $1/\omega$. Compare how integration in the time domain influences the representation of the signal in the frequency domain ($F\{x(t)\} = X(\omega)$):

$$\phi \propto \int x(t)dt \rightarrow \frac{1}{j\omega}X(\omega) \tag{5.79}$$

When we have a sinusoidal information signal, the spectrum will in both cases consist mainly of three peaks; one corresponding to the carrier and two to the information signal, where the lower component in FM is phase reversed relative to AM.

Bandwidth

From above it can be concluded that the bandwidths of the PM and FM signals are often wider than twice the bandwidth of the information signal. A rule of thumb for the required bandwidth of the transmission channel can be obtained. Consider the number of side bands which have to be included in order that the power of higher order side bands are lower than a certain value ε. For $\beta = 2$, for instance, it becomes obvious that only the first four Bessel functions, of the orders 0, 1, 2 and 3, give spectral components with amplitudes larger than 1/10 of the carrier amplitude. The Bessel function J_0 corresponds to the amplitude of the carrier, and the other functions give in turn the different side bands. This gives three side bands on each side of the carrier, i.e. a total of $2 \times 3 + 1 = 7$ spectral components. The extension of the spectrum along the frequency axis corresponds to the bandwidth of the signal. Outside this bandwidth limit, the components are considered to be negligible. A graph, showing the number of side bands on each side of the carrier that have to be included in the bandwidth for all sidebands outside this limit to have amplitudes smaller than 1/10 and 1/100 of the carrier, is shown in Figure 5.32.

Assume the number of significant side bands is M, i.e. the amplitude is lower than ε for all side bands beyond. The required bandwidth in Hz then becomes ($\omega = 2\pi f$):

$$B_T = 2Mf_m \tag{5.80}$$

For $\beta \geq 2$ the line $M = \beta + 2$ will end up approximately in the middle between all the significant side bands. For $\beta < 2$, $M = \beta + 1$ is a suitable midpoint. Inserted in eq. (5.80), this gives for $\beta \geq 2$ and $a_m = 1$

$$B_T \approx 2(\beta + 2)f_m = 2\left(\frac{f_\Delta}{f_m} + 2\right)f_m \approx 2(f_\Delta + 2W) \tag{5.81}$$

where f_m has been replaced by the bandwidth of the modulated signal. For $\beta < 2$

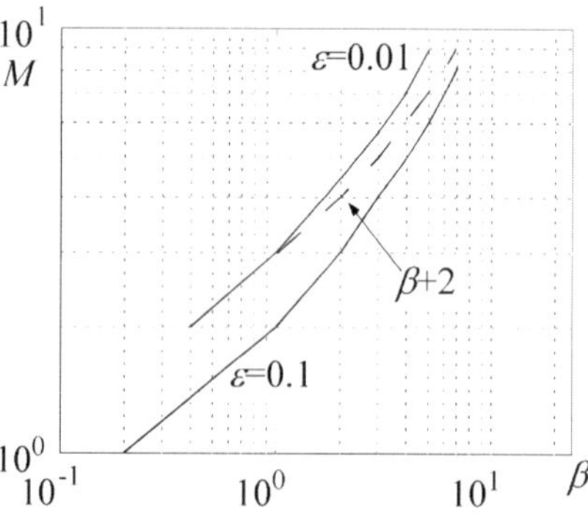

Figure 5.32 Number of significant side bands versus modulation index.

we obtain

$$B_T \approx 2(\beta + 1)f_m \approx 2(f_\Delta + W) \tag{5.82}$$

The relation above is called Carson's rule. For PM we have the corresponding expression,

$$B_T \approx 2(\phi_\Delta + 1)W \tag{5.83}$$

Example 5.7

An FM signal is modulated by a sinusoidal signal where $\beta = 1.0$ and $f_m = 200$ Hz. The FM signal is the input signal of an ideal bandpass filter (BPF) with bandwidth $B = 500$ Hz centred around f_c. Sketch the spectrum, phase diagram and envelope of the output signal.

Solution As can be seen in eq. (5.73) the amplitudes of the spectral components are determined by Bessel functions. Using a calculator or a list of tabulated Bessel functions we find that the amplitude of the component at f_c is $0.765a_c$ and the amplitude of the components at $f_c \pm 200$ is $0.440a_c$. These three components are the only ones passing through the bandpass filter. The resulting spectrum is shown in Figure 5.33.

The phase of the spectral component at f_c is used as a reference phase. The other spectral components can then be regarded as two vectors rotating in opposite directions. The phase diagram is shown in Figure 5.34.

The signal becomes

$$s(t) = a_c\big(0.77\cos(2\pi f_c t) + 0.44\cos(2\pi(f_c + 200)t) - 0.44\cos(2\pi(f_c - 200)t)\big) \tag{5.84}$$

The minus sign preceding the last term, inside the parentheses, appears due to the

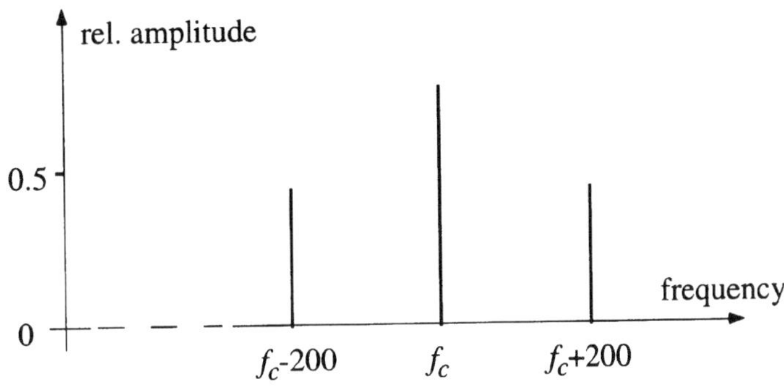

Figure 5.33 Spectrum after the bandpass filter.

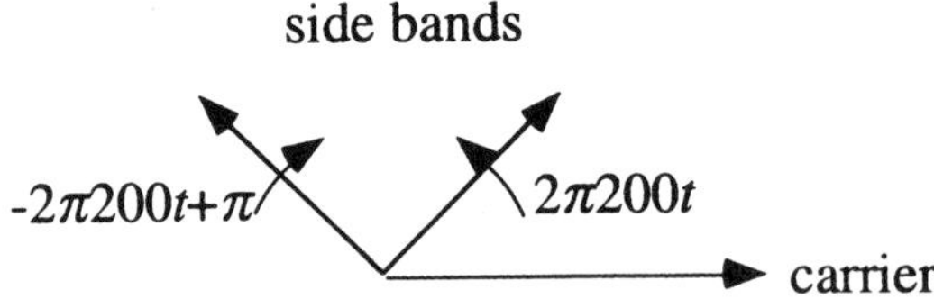

Figure 5.34 Phasor diagram.

fact that certain frequency components will be inverted, since for the Bessel functions it is valid that $J_{-n} = (-1)^n J_n$. To calculate the envelope of the signal the Hilbert transforms in eqs. (5.34) and (5.35) are utilised. This gives the result

$$a(t) = \sqrt{s^2(t) + (H\{s(t)\})^2} = \sqrt{a_c^2(0.98 - 0.39\cos(2\pi 400t))} \tag{5.85}$$

Note that the signal no longer has a constant envelope, i.e. the bandwidth limitation results in an amplitude modulation of the signal. A similar investigation of the instantaneous phase, $\phi(t) = \arctan(H\{s(t)\}/s(t))$, would show that the phase is not influenced by a bandwidth limitation as long as the limitation is symmetric around f_c.

5.4 Digital carrier modulation

In most modern teletransmission systems the signal to be transmitted is in the form of digital words. Various kinds of modulation, with different properties, have been developed for transmission of this kind of signal. The basic methods are presented here with the aid of the narrowband concept, discussed at the beginning of this chapter. In digital modulation no demodulation, with the same meaning as in analogue modulation, is done. This is because there is no need to recover the signal itself, only the symbol stream shall be recovered. The correct symbol, from a finite number of possible symbols, must be detected out of the received signal. A small part of the detection is treated in this chapter but the major part comes in Chapter 6. In digital modulation, a mapping of a discrete information sequence is made on the amplitude, phase or frequency of a continuous signal, $s(t)$. Start from the bandpass signal in eq. (5.12):

$$s(t) = a(t)\cos(\omega_c t + \phi(t)) = \mathrm{Re}\left\{ \underbrace{a(t)\exp(j\phi(t))}_{u(t)} \exp(j2\pi f_c t) \right\} \tag{5.86}$$

where $a(t)$ is the amplitude of the signal, ω_c is the angular frequency of the carrier, $\omega_c t + \phi(t)$ is the instantaneous phase and $\omega_i(t) = \omega_c + d\phi(t)/dt$ is the instantaneous frequency. Depending on which type of modulation is used, the flow of information symbols will somehow control $a(t)$ or $\phi(t)$. This is discussed below, and the description starts with relatively simple, but still commonly used, modulation schemes. This is followed by more advanced modulation schemes, such as CPM, for which the most important property is high spectral efficiency because a lot of information can be transmitted within a relatively narrow bandwidth.

5.4.1 DIGITAL AMPLITUDE MODULATION

Modulation in one dimension

In digital amplitude modulation methods like pulse amplitude modulation (PAM), or amplitude shift keying (ASK), the phase term $\phi(t)$ is constant and can be set to zero. The amplitude $a(t)$ varies as a function of the information so that $a(t) = x_k p(t)$. The index k is a running index, indicating the symbol number k in the symbol stream. For each time interval T, a new modulation symbol enters. Each character in the symbol alphabet corresponds to a certain amplitude value. The characters can be either $x_k \in \{\pm 1, \pm 3 \pm \cdots \pm (M - 1)\}$, in which case M must be an even number, or $x_k \in \{0, 1, 2, ..., M - 1\}$, in which case M can be an arbitrary integer. For example in binary modulation the digital symbol one can correspond to $x_k = +1$ and zero to $x_k = -1$. If the symbols from the preceding coding are binary, a mapping to the modulation symbol alphabet must be done. The symbol rate is then reduced to $T = T_0/\lceil \log_2 M \rceil$ if T_0 is the information symbol rate. The function $p(t)$ is a pulse shape which, among other things, influences the modulated signal spectrum and will be treated later. At the moment it is enough to assume that the pulse shape $p(t) \neq 0$ only for $0 \leq t < T$. The modulated signal then becomes a sum of non-overlapping pulses:

$$s(t) = \mathrm{Re}\left\{ \underbrace{\sum_{k=-\infty}^{\infty} x_k p(t - kT)}_{u(t)} \exp(j2\pi f_c t) \right\} = \cos(2\pi f_c t) \sum_{k=-\infty}^{\infty} x_k p(t - kT)$$

$$(5.87)$$

If $p(t)$ is a unit pulse the lowpass component of the signal, $u(t)$, can be plotted in the complex plane as in Figure 5.35. The dots indicate the amplitude values which can be assumed by the signal.

As $u(t)$ is real, the representation of the signal in the complex plane will consist of points on the real axis only. Such a modulation is called one-dimensional.

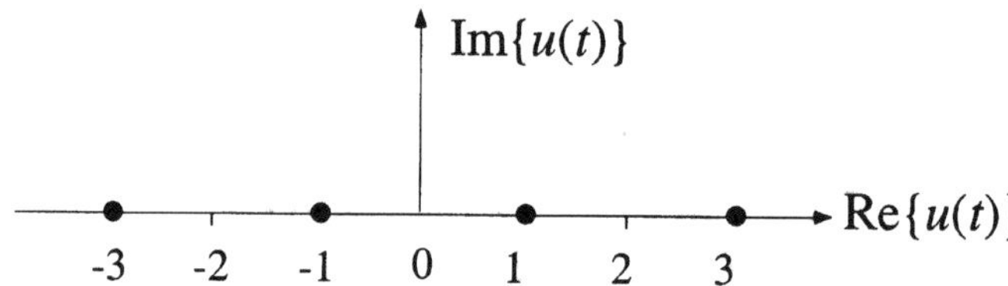

Figure 5.35 The possible signal values indicated in the complex plane.

Modulation in two dimensions

The next modulation scheme is quadrature amplitude modulation (QAM). Here, the incoming symbol stream is divided into two streams, x_{2k} and x_{2k+1}. The symbols having even index modulate along the real axis, and the symbols having odd index modulate along the imaginary axis, which gives

$$s(t) = \mathrm{Re}\left\{ \sum_{k=-\infty}^{\infty} (x_{2k} + j{\cdot}x_{2k+1})p(t - kT)\exp(j2\pi f_c t) \right\} \tag{5.88}$$

For example, suppose that the incoming symbol stream consists of binary bits divided in frames of 4 bits each. The two most significant will correspond to x_{2k} and the two least significant ones to x_{2k+1}. It is then possible to map 16 characters unambiguously. In this case the modulation type is called 16 QAM. A plot in the complex plane is shown in Figure 5.36. Since four binary bits correspond to one symbol in the modulation the pulse interval T can be four times the length of the time between two binary symbols, T_0.

Thus, one can use only the real or both the real and imaginary axes. The choice will influence the signal bandwidth and its properties on a noisy transmission channel. This will be treated later on.

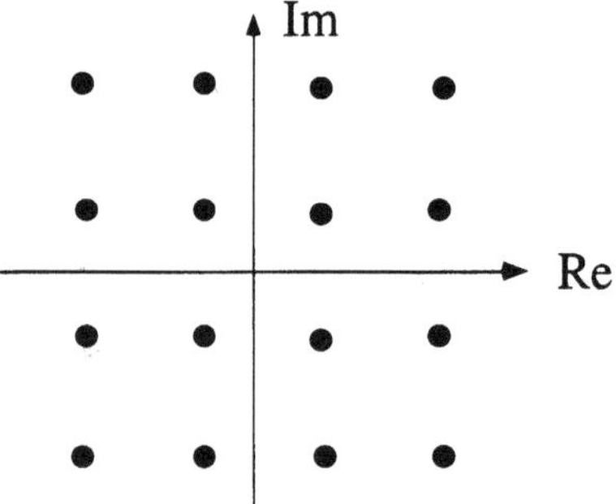

Figure 5.36 Plot of the signal configuration for 16 QAM.

5.4.2 DIGITAL PHASE AND FREQUENCY MODULATION

The next class of digital modulation to be treated is phase and frequency modulation, also commonly denoted as angular modulation. The information symbols are then mapped into the phase argument $\phi(t)$. Since there is no linear relation between information and modulated signal, these methods are called non-linear modulation methods.

Phase modulation

For M-ary digital phase modulation or M-ary phase shift keying (MPSK) the signal is

$$s(t) = A_c \sum_k \cos(\omega_c t + \phi_k) p(t - kT) \tag{5.89}$$

A_c is constant in angle modulation. The pulse shape $p(t) \neq 0$ only for $0 \leq t < T$. The signal therefore depends on one modulation symbol at a time, which means that the signal shape is only affected by the present symbol, not by previous ones, i.e. the modulated signal has no memory.

The information is contained in the phase, which is

$$\phi_k = 2\pi \frac{x_k - 1}{M}, \quad x_k \in \{1, 2, 3, ..., M\} \tag{5.90}$$

where x_k as usual is the modulation symbol. For $M = 8$ we obtain 8PSK, which is plotted in the complex plane in Figure 5.37. Note that all signal values have the same distance from the centre. For a rectangular pulse shape the envelope of the signal is therefore constant and the phase shifts occur instantaneously, according to eq. (5.89).

An important special case is when x_k assumes the values 1 and 2. This gives two phases $0°$ and $180°$, which is called binary phase shift keying (BPSK). This can also be regarded as amplitude modulation with the symbols ± 1.

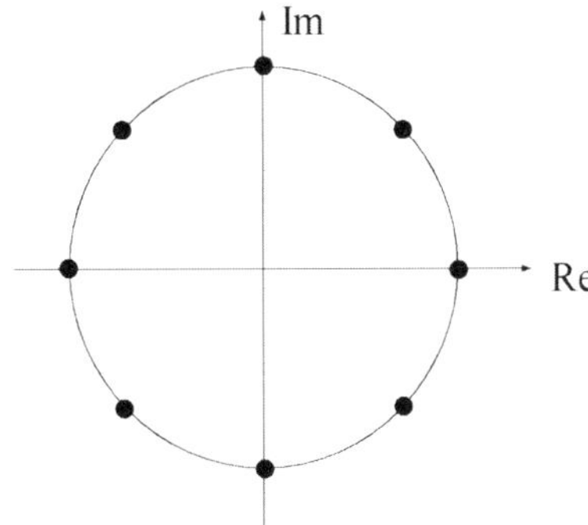

Figure 5.37 Signal configuration of 8PSK.

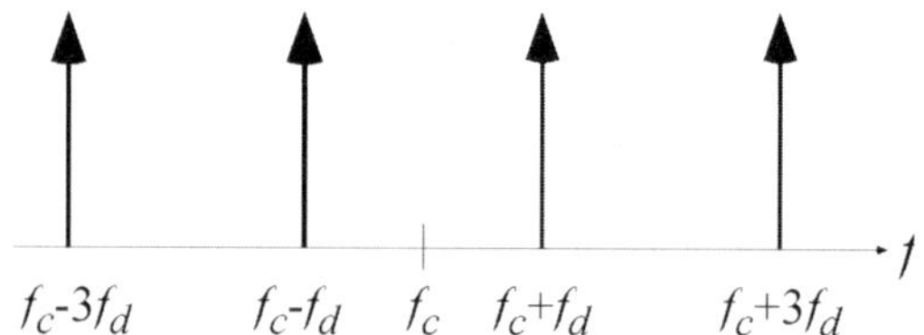

Figure 5.38 Plot of 4 FSK in the frequency plane.

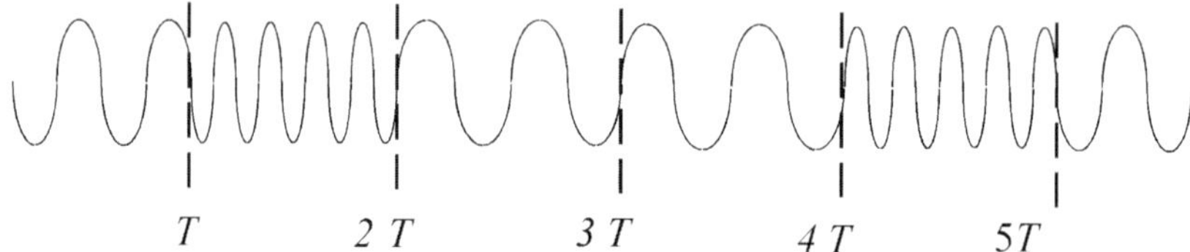

Figure 5.39 FSK signal with continuous carrier, i.e. no phase transitions in the carrier between the symbols.

Frequency modulation

Let the frequency depend on the information instead. Another modulation scheme called frequency shift keying (FSK) is then obtained. For *M* different characters it is denoted *M*FSK. The signal becomes

$$s(t) = A_c \sum_k \cos(\omega_c t + \omega_d x_k t)p(t - kT) \tag{5.91}$$

where the modulation symbols $x_k \in \{ \pm 1, \pm 3 \pm ... \pm (M - 1)\}$ and $f_d = \omega_d/2\pi$ represents the frequency deviation. This type of signal is most conveniently illustrated in the frequency plane where we get a number of peaks, one for each symbol, as shown in Figure 5.38. As will be clear later, the peaks have non-zero widths along the frequency axis.

Hence, in FSK each symbol has its own frequency assigned around the carrier frequency.

In the modulation methods described above, the phase can make a jump for each new symbol. This gives a broader spectrum than necessary. An important special case is obtained if the frequency deviation is chosen in such a way that $2\omega_d T = 2\pi N$, where *N* is an integer, which gives a continuous carrier as shown in Figure 5.39.

Realisation of digital angle modulation

The realisation of digital angle modulation, or mapping from symbol sequence to

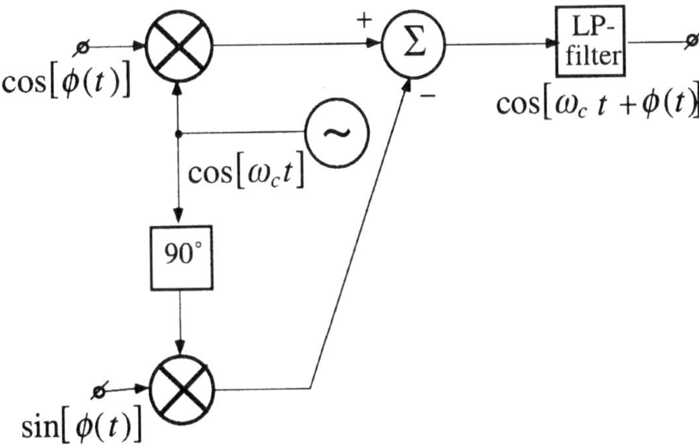

Figure 5.40 Modulator for digital angle modulation.

modulated signal, is performed by starting from an in phase and quadrature signal in the baseband. These are separately modulated by a high frequency signal and are added after the quadrature signal has been inverted (cf. Figure 5.40). Compare with the formation of the high frequency signal from the complex baseband signal illustrated in Figure 5.14.

The multipliers are usually realised by double balanced diode mixers as in Figure 5.24.

5.4.3 NON-LINEAR METHODS WITH MEMORY

Definition of CPFSK

A modulation scheme with continuous phase is presented below. The signal is not independent between the symbol intervals as in previous methods, but the form of the signal is influenced by several consecutive symbols, i.e. the modulated signal contains a memory. The method is called continuous phase frequency shift keying (CPFSK). Its foremost property is narrow bandwidth in relation to the information rate and with preserved good properties on a noisy channel. The signal is written

$$s(t) = A_c \cos[\omega_c t + \phi(t, \mathbf{x})] = \mathrm{Re}\left\{ \underbrace{A_c \exp(j\phi(t, \mathbf{x}))}_{u(t)} \exp(j\omega_c t) \right\} \qquad (5.92)$$

The signal shape depends on a sequence of modulation symbols, which can be expressed by a vector $\mathbf{x} = [x_0 \ldots x_K]$. For a frequency shift of $p_D(t)$ the phase of the modulated signal becomes

$$\phi(t, \mathbf{x}) = 2\pi h \sum_k x_k \int_{-\infty}^{t} p_D(\tau - kT)d\tau \qquad (5.93)$$

where h is called the modulation index. The modulation symbols assume the values $x_k \in \{\pm 1, \pm 3 \pm \ldots \pm (M-1)\}$. There are a couple of differences, compared to the previously described angle modulations which are worth noting. The phase depends on a sum of previous and present symbols instead of only the present symbol. The integral of the pulse shape influences the phase instead of the pulse shape coming in as a multiplier of the signal.

The frequency shift, or the instantaneous phase, becomes

$$\frac{d\phi}{dt} = \omega_d = 2\pi h \sum_k x_k p_D(t - kT) \qquad (5.94)$$

where $D = LT$ and L is an integer larger than zero. The pulse is extended over L symbol intervals and can thus include several symbol intervals.

Phase tree

Assume a rectangular shaped frequency pulse $p_D(t)$, shown in Figure 5.41.

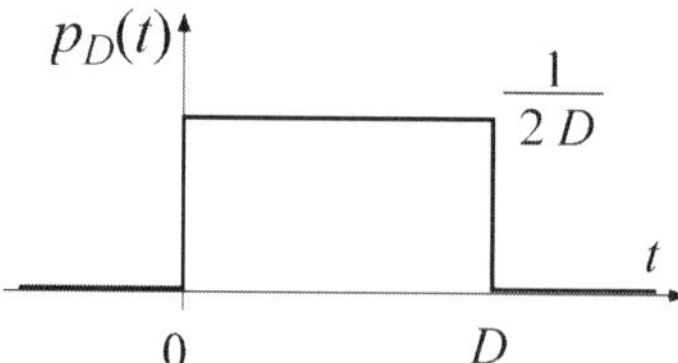

Figure 5.41   Rectangular frequency pulse.

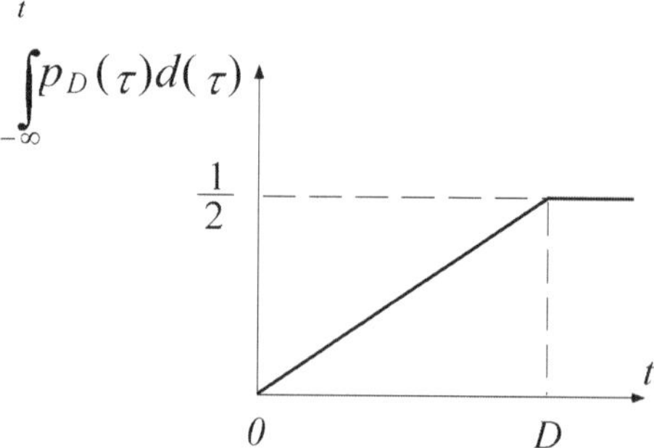

Figure 5.42   Phase for rectangular pulse.

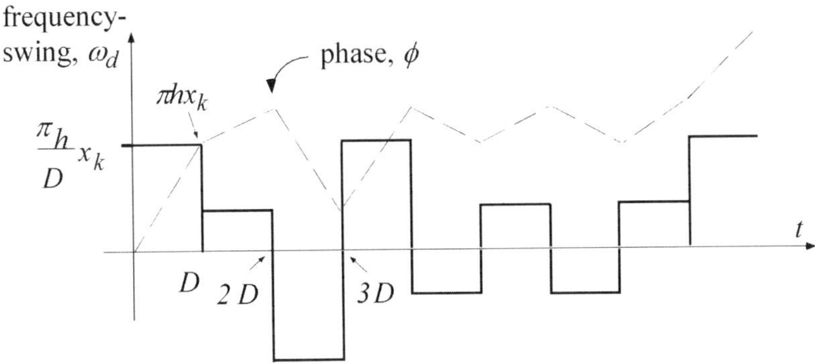

Figure 5.43 Frequency swing and phase for the symbol sequence $+3$, $+1$, -3, $+3$, -1 etc.

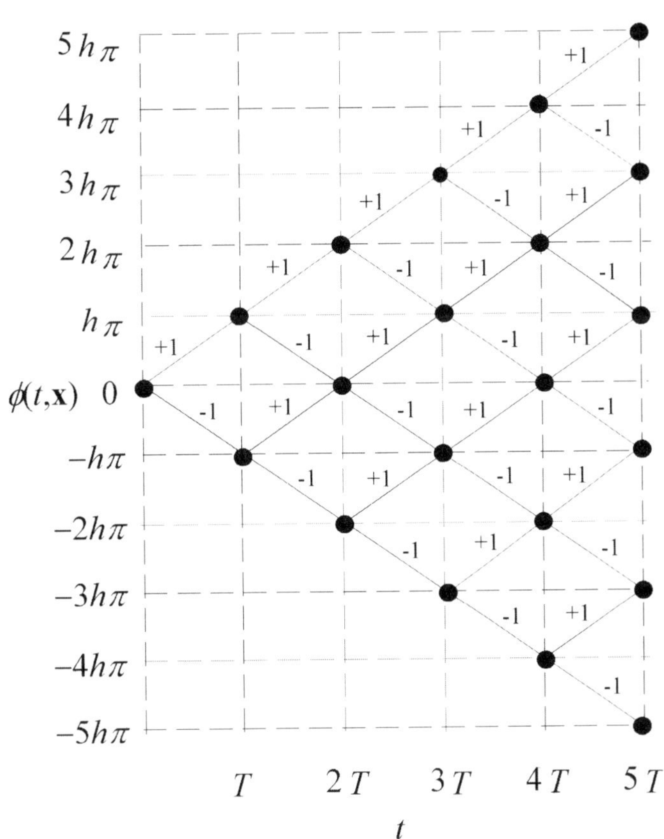

Figure 5.44 Phase tree of binary CPFSK with rectangular pulse shape.

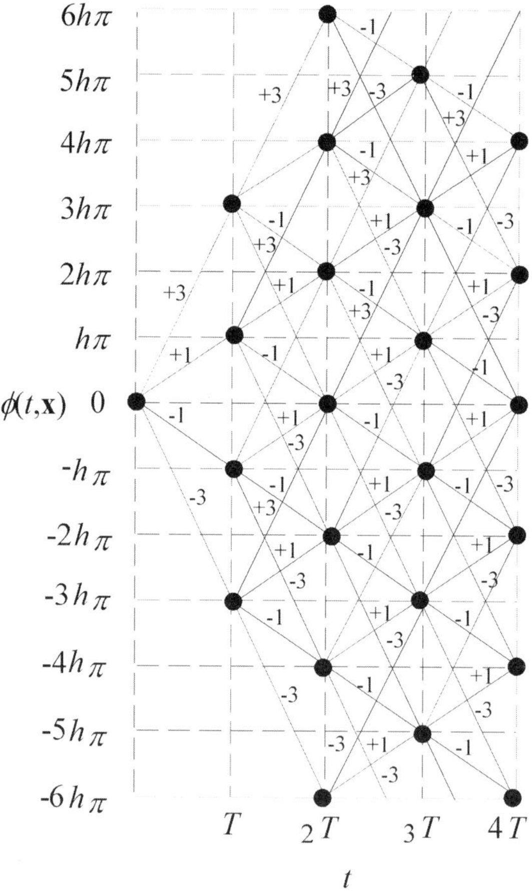

Figure 5.45   Phase tree for $M = 4$.

All kinds of frequency pulses for CPFSK are specified so that the total area is always 1/2. After integration, the phase for a rectangular pulse will then look like Figure 5.42.

Assume a symbol sequence as shown in Figure 5.43, i.e. $+3$, $+1$, -3, $+3$ etc. From eq. (5.94) it is clear that in this case the frequency shift will correspond to one symbol at a time. The phase, $\phi(t,\mathbf{x})$, makes no jumps, as is shown by the dotted line in Figure 5.43. The absence of transitions is beneficial for obtaining a narrow signal bandwidth, and comes from the integration of the instantaneous phase during the symbol interval. The phase starts at a position depending on previous symbols and increases or decreases depending on present symbol.

If, instead, the phase of all possible combinations of symbol sequences is plotted, starting from a given point, the figure will describe a phase tree (cf. Figures 5.44 and 5.45).

Assume that the phase angle is zero at the start. The symbol $+1$ then results in the phase angle $h\pi$ after a completed frequency pulse (see eq. (5.93)). If the symbol to be transmitted were -1 instead, the phase would change to $-h\pi$. The next symbol also changes the phase by $+h\pi$ or $-h\pi$, but the starting point depends on the first symbol. Each symbol sequence combination results in its own path through the phase tree. The phase tree works equivalently for other symbol alphabets than the binary one. The number of paths to and from each phase is then greater than two. For $M = 4$ the phase changes $\pm h\pi$ or $\pm 3h\pi$ depending on the symbol. The corresponding phase tree is shown in Figure 5.45.

If $M = 2$ and $h = 1/2$ we obtain a special case of CPFSK called minimum shift keying (MSK). Inserted in eq. (5.94), this gives the frequency deviation $f_d = 1/4T$, which is the smallest frequency deviation required for the signals, corresponding to different modulation symbols, to be orthogonal over one symbol interval.

Partial response

When the frequency pulse $p_D(t) = 0$ outside $0 \le t < T$ the modulation is called full response modulation, since the complete response to a certain symbol is within one symbol interval. If the extension of the pulse is longer, so that $p_D(t) \neq 0$ for $t > T$, the modulation is called partial response. The response to a certain symbol is then only partly within one symbol interval. The longer the pulse chosen, the faster the signal spectrum will decline with the distance from the carrier frequency, ω_c. From eq. (5.94) it is clear that when the frequency pulses overlap the frequency deviation is the sum of all overlapping frequency pulses at the current time. Examples of frequency and phase modulation of a pulse, whose length is equal to two symbol intervals, are shown in Figure 5.46. The frequency pulse for the first symbol extends from 0 to $2T$. At time T the next symbol arrives and its frequency pulse will persist between T and $3T$. During the time interval T to $2T$, the phase change of the signal is therefore influenced by the sum of the pulses corresponding to the symbols one and two. In this way the phase change will thereafter be influenced by two symbols at a time, in the general case by L frequency pulses, where L is the length of the frequency pulse in the number of symbols.

It can be seen that the phase plot is rounded off in Figure 5.46 as compared to the plot in Figure 5.43. This is due to the rounded pulse shape in Figure 5.46 compared to the rectangular shape in Figure 5.43. One type of rounded pulse shape is the raised cosine, further examples are given in the Appendix at the end of the chapter.

When a pulse is specified, its length and shape are given. A rectangular shaped pulse with length 3 is labelled 3REC. As previously mentioned, the length of the frequency pulse is $D = LT$, where L is the number of symbols in the pulse.

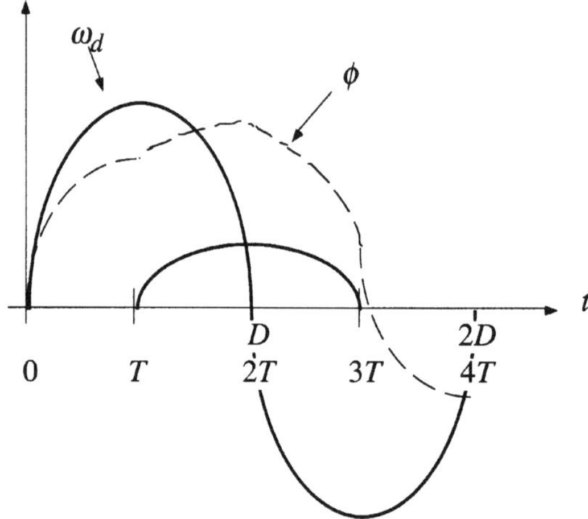

Figure 5.46 Partial response with rounded pulse shape. The figure shows the symbol sequence $+3$ $+1$ -3. The solid lines represent the frequency pulses and the dashed line represents the phase of the signal.

The phase tree for different combinations of input signals is shown in Figure 5.47. The numbers within brackets are disregarded for the time being; they will be explained later. Assume that the following conditions hold for the preceding symbols; the phase at time $t = 0$ is $\phi(0) = 0$, and that the preceding symbols are $x_{n-1} = +1$ and $x_{n-2} = +1$. Just as in full response the phase at the starting point is important, but in partial response the preceding symbols also have to be taken into account. If $L = 3$ the two preceding symbols have to be considered. As $x_n = +1$, the phase shift $\phi(T)$ will, according to eq. (5.93), assume the value given by the upper dot at $t = T$. For $x_n = -1$, the phase is given by the lower dot. For each new symbol interval, new phases are assumed, which are determined by the previous phase and the last three symbols. The upper path at each bifurcation corresponds to the last symbol value of $+1$ and the lower path corresponds to a value of -1.

Trellis

It can be seen that the phase tree grows exponentially. Due to the periodicity of the trigonometric functions, it is sufficient to calculate the phase modulo 2π. If the modulation index, cf. eq. (5.93), is a rational number, i.e. the ratio between two integers m and p gives $h = m/p$, the phase tree will be periodic and can be reduced to a trellis. One period is indicated by the shaded rectangle. The periodicity will become apparent only if Figure 5.47 is extended towards higher t-values. The concepts of phase tree and trellis have been discussed in Chapter 4 in connection with convolutional coding. The trellises cannot only be used for

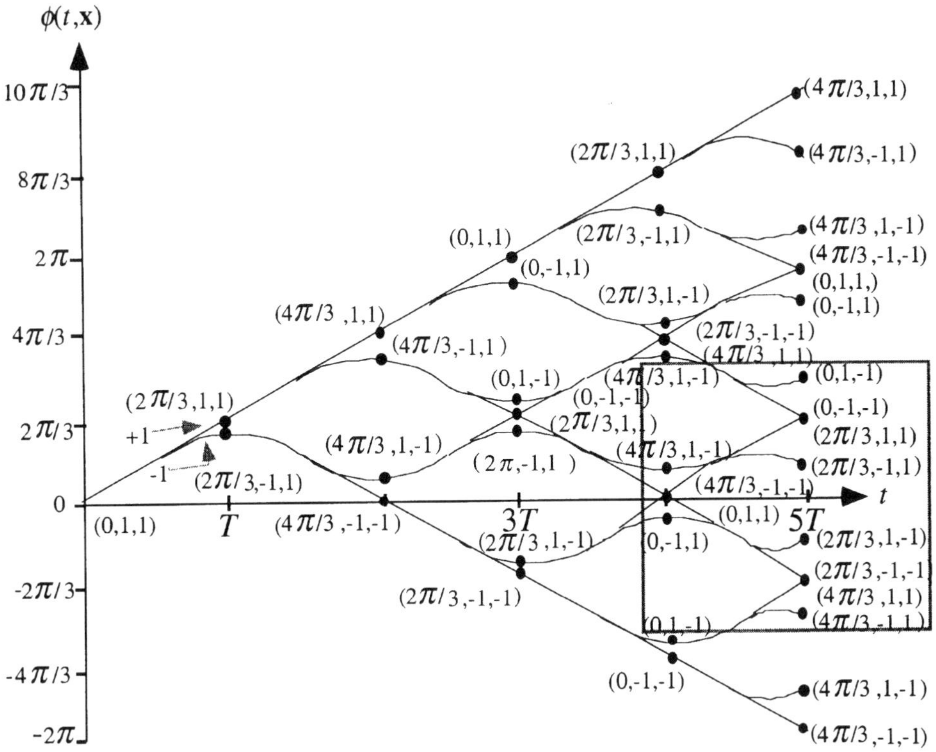

Figure 5.47   Phase tree of binary 3RC with $h = 2/3$.

the investigation of the modulation properties, but as is the case with convolutional coding the trellis is also used when the symbol stream is to be retrieved in the receiver. The number of phase states of the trellis is of great importance for the complexity of the receiver. The number of states versus modulation index and length of the frequency pulse are calculated below.

The sum in eq. (5.93) can be grouped into two parts. The first part consists of the previous symbols, whose frequency pulses have resumed zero level, compare $t > D$ in Figure 5.41. The second part consists of those symbols whose pulses have not yet resumed zero level. At times nT, the phase $\phi(t, \mathbf{x})$ will be able to assume the following values:

$$\phi_n(nT, \mathbf{x}) = \pi \frac{m}{p} \left[\sum_{k=-\infty}^{n-L} x_k + 2 \sum_{k=n-L+1}^{n} \int_{kT}^{nT} p_D(\tau - kT)\, x_k \, d\tau \right] \qquad (5.95)$$

For the first sum the frequency pulses have been integrated to their final value, which by definition is 1/2. The factor 2 has therefore been reduced in the first sum. This sum will then assume integer values ± 1, ± 2, ± 3, etc. The second sum consists of symbols which, due to the length of the symbol window, have not yet assumed their final values. For full response, i.e. for $L = 1$, it can be seen that the phases at times nT will be

$$\phi_n(nT, \mathbf{x}) = \left[\pi m/p \sum_{k=-\infty}^{n-1} x_k \right]_{\mathrm{mod}\, 2\pi}$$

since the integral in this case always reaches its final value within one symbol interval. To make the discussion easier for an arbitrary value of L, the sum is given a special symbol θ:

$$\theta_n(nT, \mathbf{x}) = \left[\pi \frac{m}{p} \sum_{k=-\infty}^{n-L} x_k \right]_{\mathrm{mod}\, 2\pi} \tag{5.96}$$

For even values of m, the following possible phases will be assumed:

$$\theta_n \in \left\{ 0, \frac{\pi m}{p}, \frac{2\pi m}{p}, \ldots \frac{(p-1)\pi m}{p}, \frac{p\pi m}{p} \right\}$$

even number times
π, i.e. phase is 0 again

After p different angles, one passage around the unit circle has been completed and the angle is again zero. The phase pattern is then repeated periodically. For odd values of m the following values are assumed:

$$\theta_n \in \left\{ 0, \frac{\pi m}{p}, \ldots, \frac{p\pi m}{p}, \ldots, \frac{(2p-1)\pi m}{p}, \frac{2p\pi m}{p} \right\}$$

odd m times π,
i.e. phase $\pm\,\pi$

even number
times π

For $L = 1$ and even values of m, p angles will thus be assumed and when m is an odd number, $2p$ angles will be assumed. For $L > 1$ the contributions from the last sum in eq. (5.95) are included. This sum consists of $L - 1$ terms, each one of which can assume one of M different symbol values. The number of combinations of the last summation therefore becomes M^{L-1}. Since some of these combinations can result in identical angles, an upper limit for the number of possible angles, S_f, is obtained. The result is

$$S_f \leq S_t = pM^{L-1} \qquad \text{for } m \text{ even}$$
$$S_f \leq S_t = 2pM^{L-1} \qquad \text{for } m \text{ odd}$$

(5.97)

The phase tree in Figure 5.47 can therefore be described by a phase trellis, having a limited number of states. Instead let each combination of angles θ_n and incomplete symbol integrations in eq. (5.95) define different states. Each state can then be described by

$$\sigma_n = (\theta_n, \; x_{n-1}, x_{n-2}, \ldots, x_{n-L+1})$$

(5.98)

A state trellis is obtained if each state is indicated as a point and the various state transitions are illustrated by lines as shown in Figure 5.48 to the right. The number of states in this state trellis becomes S_t. The phase trellis is primarily applied when the actual signal is being analysed, while the state trellis is used for decoding of the modulation and is implemented in the Viterbi decoder (cf. Section 4.4.3).

Consider the previously described MSK modulation where $h = 1/2$ and $L = 1$. Insert these values in the phase tree in Figure 5.44 resulting in a phase trellis. The two types of trellises are shown in Figure 5.48. To describe the state trellis only one symbol interval is needed, while the description of the phase trellis requires two. This has been indicated by the shaded rectangle. The indicated area is repeated periodically when a more extended section of the trellis is drawn.

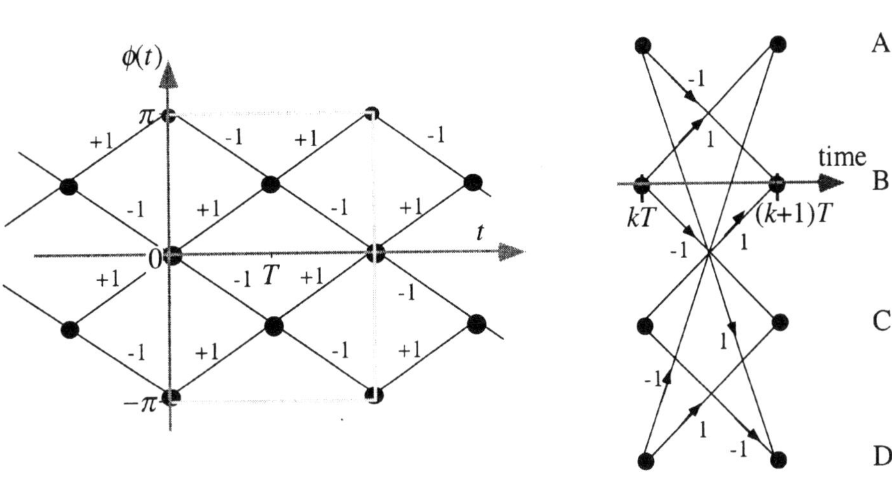

Figure 5.48   Phase trellis to the left and state trellis to the right.

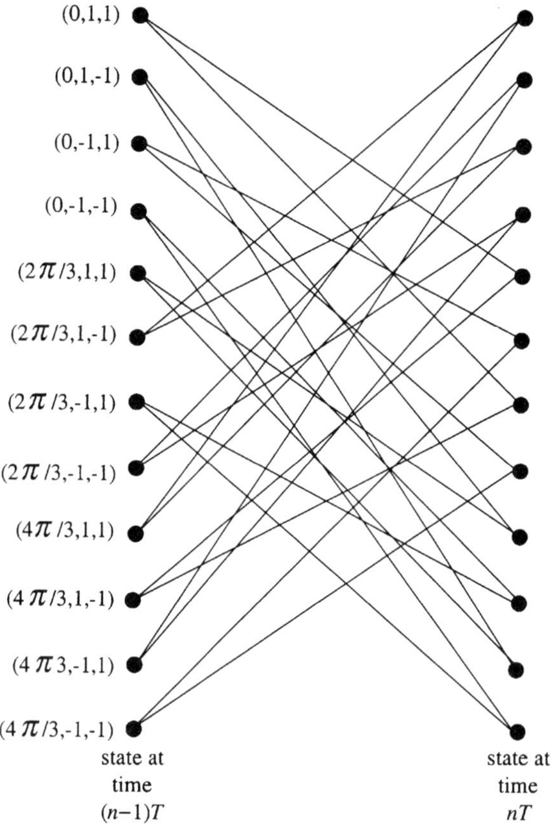

Figure 5.49   State trellis of binary 3RC with $h = 2/3$.

It is quite straightforward to construct the MSK trellis. A state trellis with a higher degree of complexity is for example obtained by using the modulation described by the phase tree in Figure 5.47. The transition from one state to another depends on the current symbol x_n, the phase θ_n as well as on the symbols x_{n-1} and x_{n-2}. The latter specify the states and are given by $(\theta_n, x_{n-1}, x_{n-2})$ to the left in the trellis. The trellis can then be connected to the phase tree where the corresponding information has been given. Since $h = 2/3$, the alphabet is binary and the length of the frequency pulse is $L = 3$, the number of states becomes $3 \times 2^2 = 12$. There are only 10 phase states in the corresponding phase trellis since some of the phases coincide (Figure 5.49).

The response of this modulation method to a certain symbol sequence can be described by a sufficient number of trellises connected in series.

It is clear from the above that CPFSK requires larger complexity at transmitter and receiver than other digital modulation schemes which have been described here. This is particularly true for the receiver where the Viterbi algorithm (accumulated Euclidean distance is used as metric), or similar, should be used to track the signal path through the trellis and in this way decode the modulated symbols. In return to the complexity, as has been said at the beginning of this section, a method is obtained that gives a high information rate with a relatively narrow bandwidth and offers good properties in a noisy channel. The method is for example used in the GSM mobile telephony system.

5.4.4 SPECTRUM IN DIGITAL MODULATION

To investigate the bandwidth occupied by a signal its spectrum has to be determined. In this section methods for studying the spectral properties of the modulated signal are discussed. The results can, among other things, be used to determine how the signals interfere with transmission channels on adjacent frequencies, so-called adjacent channel interference. Another use of the spectrum is to compute the information density, within the signal bandwidth B, by $\log_2 M/B$ (bits/Hz). Since the modulation can be regarded as a translation of the complex signal along the frequency axis, it is sufficient to calculate the spectrum of the complex signal, $u(t)$. This also gives a uniform method for analysing both baseband and carrier transmission (Figure 5.50).

The information signal can be seen as a stochastic process. Consequently the modulated signal must be treated as a stochastic signal. The signal spectrum is therefore the Fourier transform of the autocorrelation function for the signal

$$U(f) = F\{R_{uu}(t)\} \tag{5.99}$$

Irrespective of modulation scheme, the first step in the computation is usually to find the autocorrelation function of the modulation under study. In the following sections, methods for finding the autocorrelation function of some different types of modulated signals are therefore discussed. The power spectrum is then

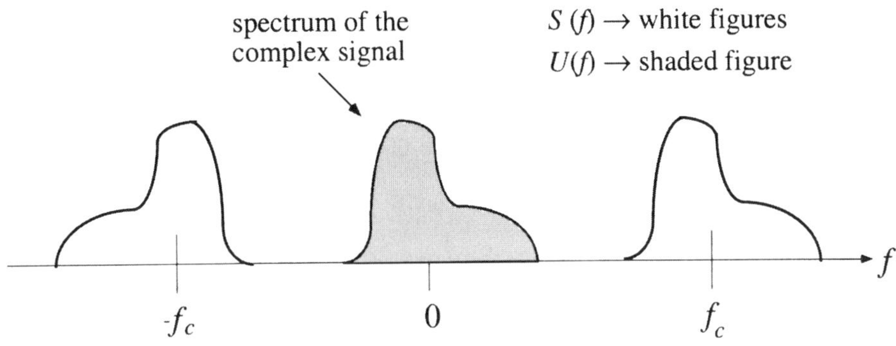

Figure 5.50   The spectrum of the modulated signal can be seen as a parallel translation of the complex baseband signal.

obtained by Fourier transformation, which is usually an easier operation once the autocorrelation function has been identified. The first part of this section is devoted to a study of how the spectrum for linear modulation methods, e.g. PAM or QAM, can be computed. In the next part a computation method for signals which can be described by a trellis, e.g. CPFSK and MSK, is treated.

Linear modulation

For a linearly modulated signal, such as an AM signal, the complex signal is (cf. eq. (5.87))

$$u(t) = \sum_{k=-\infty}^{\infty} x_k p(t - kT) \tag{5.100}$$

where $p(t) \neq 0$ only for $0 \leq t < D$. To simplify the expression below, the interval is written instead in the form $t \in [0, D)$. According to the definition given in the signal processing literature the autocorrelation function becomes

$$R_{uu}(t + \tau, t) = E\left\{ \sum_{k=-\infty}^{\infty} \sum_{l=-\infty}^{\infty} x_k x_l^* p(t - kT) p^*(t + \tau - lT) \right\} \tag{5.101}$$

Assume M symbols where $x_k \in \{ \pm 1, \pm 3, \ldots, \pm(M - 1)\}$ and all values have equal probability.

Equation (5.101) can be divided into four regions as illustrated in Figure 5.51:

- $l = k$ and $t \in [0, D)$. The pulse functions overlap completely or partly when $\tau \in [-D, D)$. $x_k x_l^*$ is the product of a complex number and its conjugate.
- $l = k$ and $t \notin [0, D)$. The pulse functions do not overlap within this area. The product of the pulse functions is then zero. In this case the expectation value is therefore zero. The variable τ may assume any arbitrary value
- $l \neq k$ and $t \in [kT, (k + L)T)$. The pulse functions overlap completely or partly for the values of τ for which $\tau \in [(l - k - L)T, (l - k + L)T)$. $x_k x_l^*$ is the product of two complex numbers. Since the symbols have a probability density which is symmetric around zero, the expectation value is zero in this case.

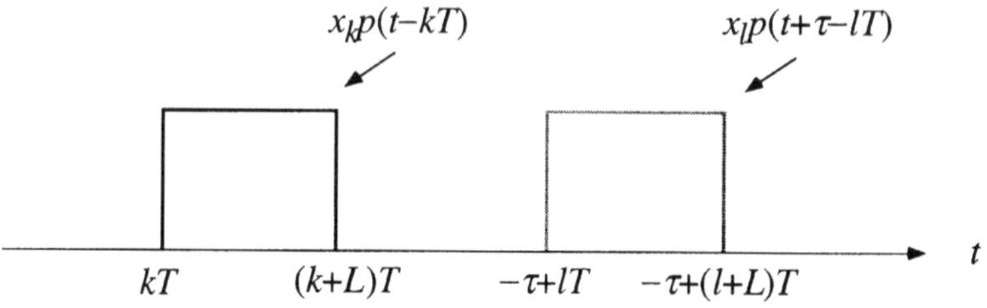

Figure 5.51 Possible overlap of the pulses for different k, l, τ and t.

- $l \neq k$ and $t \notin [kT, (k + L)T]$. The pulse functions do not overlap within this area. The product of the pulse functions is then zero. In this case the expectation value is therefore also zero.

The expectation value in eq. (5.101) then reduces to

$$R_{uu}(t + \tau, t) = E\left\{ \sum_{k=-\infty}^{\infty} |x_k|^2 p(t - kT)p^*(t + \tau - kT) \right\} \tag{5.102}$$

Since the symbols in the symbol sequence are statistically independent, the summation can be taken outside the expectation value operator, which gives

$$R_{uu}(t + \tau, t) = \sum_{k=-\infty}^{\infty} E\left\{ |x_k|^2 p(t - kT)p^*(t + \tau - kT) \right\} \tag{5.103}$$

The probability that the symbol x_k will assume the value α is equal to $1/M$. The expectation value is now rewritten by means of the standard definition. The autocorrelation function, eq. (5.101), thus becomes

$$R_{uu}(t + \tau, t) = \sum_{k=-\infty}^{\infty} \frac{1}{M} \sum_{x_k=-M+1,\, x_k \text{ odd}}^{M-1} |x_k|^2 p(t - kT)p^*(t + \tau - kT) \tag{5.104}$$

for $-D \leq \tau \leq D$. In the summation, x_k only assumes those values included in the symbol alphabet, i.e. in most cases x_k is an odd number. As $p(t - kT)p^*(t + \tau - kT)$ is a deterministic signal it can be taken outside the expectation value. We are then only left with the second order moment of the symbols, which can be written as

$$\frac{1}{M} \sum |x_k^2| = \sigma_x^2 \tag{5.105}$$

since all symbols are equally likely. As is clear from eq. (5.104) the autocorrelation function is not only depending on the time difference, τ, between the processes but also on the absolute time, t. The autocorrelation function becomes time dependent and periodic, i.e. a cyclostationary signal. The Fourier transform therefore becomes time dependent. The time dependence is of no interest, however, for the study of the signal spectrum of communication systems, only the mean value is of interest. We therefore do an averaging of the autocorrelation function by integration, where the argument giving the periodic time variation disappears. A periodic signal must be regarded as a power signal and the averaging is done over one complete period. As the signal is periodic with period D, any value of k can be chosen and it is convenient to choose zero. Equation (5.105) inserted in eq. (5.104) gives

$$R_{uu}(\tau) = \frac{1}{D} \int_0^D R_{uu}(t + \tau, t)dt = \frac{\sigma_x^2}{D} \int_0^{D-\tau} p(t)p^*(t + \tau)dt = \frac{\sigma_x^2}{D} R_{pp}(\tau)$$

$$\tag{5.106}$$

The power spectrum of the signal is the Fourier transform of the autocorrelation function, i.e.

$$U(f) = \frac{\sigma_x^2}{D} F\{R_{pp}(\tau)\} = \frac{\sigma_x^2}{D} |F\{p(t)\}|^2 \tag{5.107}$$

Equation (5.107) has been derived under the assumption that the average value of all symbols is zero and that all symbols are independent of all the others. When these assumptions are no longer valid, the autocorrelation of the symbols, $E\{x_k x_l^*\} = R_x(n)$ where $n = k - l$, will be non-zero. Each value of $R_x(n)$ for different n will give rise to a new spectrum translated in time by nD. Since the sequence is stationary and the Fourier transform linear, the shifted spectra are added together. By using the theorem for the Fourier transform of a time-shifted signal, the total spectrum can be written

$$U(f) = \frac{1}{D} F\{R_{pp}(\tau)\} \sum_{n=-\infty}^{\infty} R_x(n) \exp(-j2\pi f n T) \tag{5.108}$$

The autocorrelation function can be divided into two parts. The first part depends on the mean value of the symbols μ_x, which is constant and independent of n. The second part depends on the variance σ_n^2, which can be different for different n. The autocorrelation function thus becomes $R_x(n) = \mu_x^2 + \sigma_n^2$. For the first part, which depends on the average value, it is possible to use Poisson's summation rule in the summation of the series in eq. (5.108), and rewrite the sum as another sum:

$$\sum_{n=-\infty}^{\infty} \exp(-j2\pi f n T) = \frac{1}{T} \sum_{m=-\infty}^{\infty} \delta\left(f - \frac{m}{T}\right) \tag{5.109}$$

It is also possible to consider the sum in eq. (5.109) as a discrete Fourier transform of an infinite series of ones as a way of explaining this sum. Inserting eq. (5.109) in eq. (5.108) and using $R_x(n) = \mu_x^2 + \sigma_n^2$ we obtain

$$U(f) = \frac{1}{LT^2} F\{R_{pp}(\tau)\} \left[\mu_x^2 \sum_{m=-\infty}^{\infty} \delta\left(f - \frac{m}{T}\right) + T \sum_{n=-\infty}^{\infty} \sigma_n^2 \exp(-j2\pi f n T) \right] \tag{5.110}$$

The first sum within the square bracket tells us that the spectrum has peaks at the frequencies $|m/T|$ and is therefore usually called the discrete part. The second sum is called the continuous part of the spectrum and contributes with a spectrum part that varies as a cosine function because $\sigma_n = \sigma_{-n}$. This sum leads to oscillations in the spectrum and to repetitions at frequencies separated by $1/T$ (Hz). Such repetition of the spectrum does not add any new information to the spectrum and is therefore not wanted. The sum of the two terms are weighted by the spectrum of the pulse shape and will thereby make up the complete spectrum.

As can be concluded, the spectral peaks are cancelled if the mean value μ_x of the symbols is zero. Sometimes such a peak can be desired to possibly lock an oscillator to the received carrier but mostly it only will consume power. It is of advantage if the symbols are mutually independent since $\sigma_n = 0$ for $n \neq 0$, which makes the oscillation of the spectrum disappear. Finally, the modulated signal will have a narrow spectrum when the used pulse shape has a spectrum which falls rapidly with increasing value on $|f|$. An application of the method for computing spectra described above is given in Example 5.8.

Example 5.8

Determine the spectrum of a digitally AM modulated signal with rectangular pulse shaping, i.e.

$$p_D(t) = \begin{cases} 1 & 0 \leq t < T \\ 0 & \text{otherwise} \end{cases} \tag{5.111}$$

Assume further that the symbols $x_k \in \{0, 1, 2, ...M - 1\}$ and that they are mutually statistically independent. For binary signalling the symbols become 0 and 1, which gives on-off keying (OOK).

Solution Use the result in eq. (5.110) and first calculate $F\{R_{pp}(\tau)\} = |P_D(f)|^2$. The autocorrelation function of the rectangular pulse must be divided into two intervals (Figure 5.52):

$$R_{pp}(\tau) = \int_{-\tau}^{T} p_D(t)p_D(t + \tau)dt \quad \text{for } -T \leq \tau < 0 \tag{5.112a}$$

$$R_{pp}(\tau) = \int_{0}^{T-\tau} p_D(t)p_D(t + \tau)dt \quad \text{for } 0 \leq \tau \leq T \tag{5.112b}$$

which gives

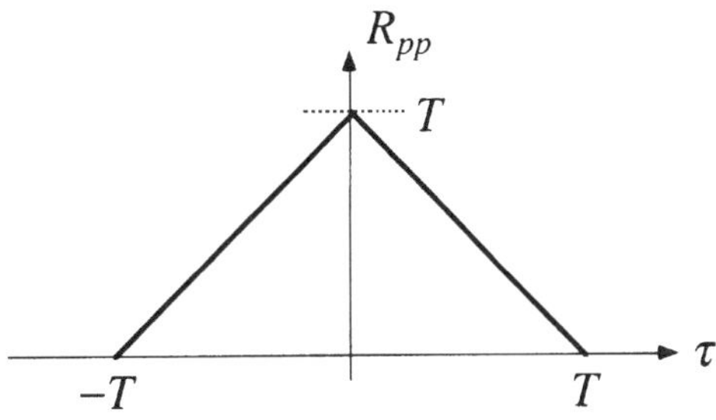

Figure 5.52 The autocorrelation function of the rectangular pulse.

$$R_{pp}(\tau) = T + \tau \quad \text{for } -T \le \tau \le 0$$

$$R_{pp}(\tau) = T - \tau \quad \text{for } 0 \le \tau \le T \tag{5.113}$$

The Fourier transform of eq. (5.113) is

$$F\{R_{pp}(t)\} = T^2 \left[\frac{\sin\left(\frac{2\pi fT}{2}\right)}{\frac{2\pi fT}{2}} \right]^2 = T^2 \text{sinc}^2(fT) \tag{5.114}$$

$(\text{sinc}(x) = \sin(\pi x)/\pi x)$. Then calculate the autocorrelation function, $R_x(n)$, of the information symbols, which will contain two parts. One part depends only on the statistical properties of the symbol sequence. Since the symbols x_k are mutually independent, this part consists of the impulse $\sigma_0^2 \cdot \delta(n)$, where σ_0^2 is the variance of the information symbols. The other part also depends on the non-zero average value of the pulse shape. Since all symbols have equal probability the average value is

$$\mu_x = \frac{M - 1}{2} \tag{5.115}$$

This results in a constant, which is independent of n. When $n = 0$ the variance of the information symbols is

$$\sigma_0^2 = \overline{x_k^2} - \mu_x^2 = \frac{1}{M} \sum_{n=1}^{M-1} n^2 - \left(\frac{M-1}{2}\right)^2$$

$$= \frac{(M-1)(2M-1)}{6} - \frac{(M-1)^2}{4} = \frac{M^2 - 1}{12} \tag{5.116}$$

To obtain eq. (5.116) we have used the fact that

$$1^2 + 2^2 + \cdots + (M-1)^2 = \frac{M(M-1)(2M-1)}{6} \tag{5.117}$$

which can be found in mathematical handbooks. The spectrum of the complex lowpass representation of the signal is then

$$U(f) = \frac{(M-1)^2}{4} \text{sinc}^2(fT) \sum_{m=-\infty}^{\infty} \delta\left(f - \frac{m}{T}\right)$$

$$+ \text{sinc}^2(fT) \cdot T \sum_{n=-\infty}^{\infty} \frac{M^2 - 1}{12} \delta(n) \exp(-j2\pi fnT)$$

$$= \frac{(M-1)^2}{4} \delta(f) + \frac{M^2 - 1}{12} T \, \text{sinc}^2(fT) \tag{5.118}$$

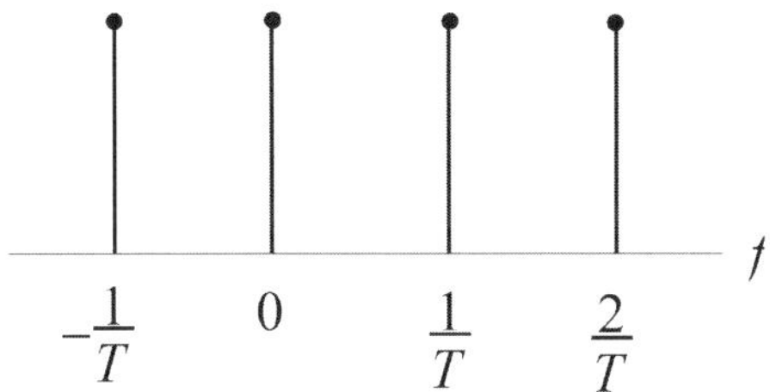

Figure 5.53 Plot of impulse train.

The part in eq. (5.118) being expressed as a sum of Dirac functions has been plotted in Figure 5.53 (cf. eq. (5.109)).

The resulting impulse sequence is multiplied in eq. (5.118) by a sinc function, whose zeros are located at $f = n/T$ for $n \neq 0$, which is exactly on the impulses in eq. (5.118). We are then left with only the impulse for which $n = 0$. The second part in eq. (5.118), which is a sum of exponential functions, has a non-zero value only for $n = 0$ because of the variance given in eq. (5.116). In other words, several terms disappear and the simplified expression, giving the spectrum of a digitally AM modulated signal with rectangular pulse, is given by the last line in eq. (5.118) (Figure 5.54).

Note that the pulse shape of a BPSK signal is the same as in this example and that the variance of the signal is $\sigma_0^2 \cdot \delta(n) = 1$. The average value μ_x is zero, however. From eq. (5.110) it can therefore be seen that the only difference between a BPSK-modulated signal and an OOK-modulated signal is a scaling factor and that the spectral peak at f_c is not present in the BPSK signal.

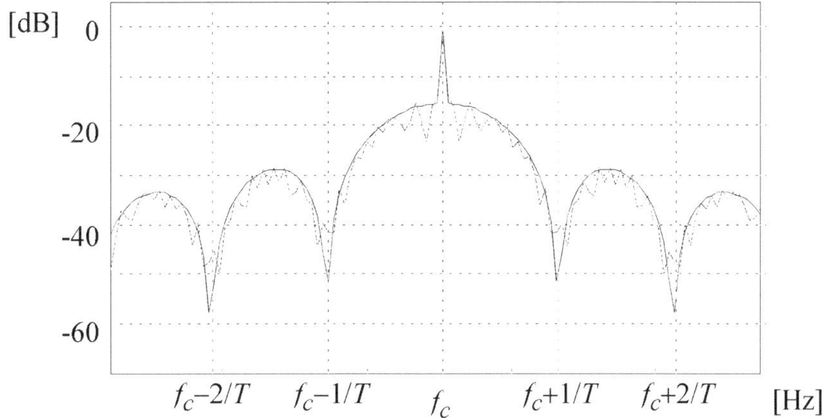

Figure 5.54 The spectrum around the carrier frequency f_c, calculated by eq. (5.118) is given by the solid line. The dotted line shows a simulated spectrum of an OOK-modulated signal.

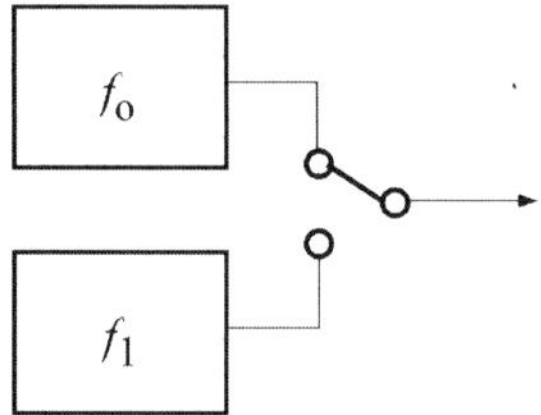

Figure 5.55   Generation of the FSK signal.

Example 5.9

Determine the spectrum of an FSK signal. The symbol 'one' is represented by a sinusoidal signal with frequency $f_1 = f_c + r_b/2$, and 'zero' is represented by $f_0 = f_c - r_b/2$, where f_c is the carrier frequency and r_b is the symbol rate. The signals are generated in two independent oscillators and the modulation is done by connecting one of the two oscillators respectively to the output as in Figure 5.55.

Solution   Since the oscillators are mutually independent the signal can be seen as two independent OOK signals. The spectrum of the FSK signal becomes the sum of these two OOK signals of frequencies f_0 and f_1, respectively. The result from the previous example can then be used. Equation (5.118) gives for $M = 2$

$$U_{FSK}(f) = \frac{1}{4}\delta(f - r_b/2) + \frac{1}{4}T\,\mathrm{sinc}^2([f - r_b/2]T)$$

$$+ \frac{1}{4}\delta(f + r_b/2) + \frac{1}{4}T\,\mathrm{sinc}^2([f + r_b/2]T) \tag{5.119}$$

The spectrum consists of two peaks centred around f_0 and f_1. The spectrum of a simulated FSK modulated signal is shown in Figure 5.56. To simulate a real signal the centre frequency has been located at $f_c = 315$ MHz.

Spectra for signals described by trellis

In this section a method for the calculation of spectrum for modulated signals which can be represented by a trellis is presented. In the same way as for linearly modulated signals, the autocorrelation function is calculated first and the spectrum is then obtained by Fourier transformation of the autocorrelation function. To obtain a correct result the signal model used must take into account whether or not the phase is constant at the symbol transitions. When the carrier phase is independent on the symbol transitions, the problem is simplified since the signal can be seen as the sum of a number of modulated signals. When the phase is not independent at the symbol transitions, the problem becomes more complex. We

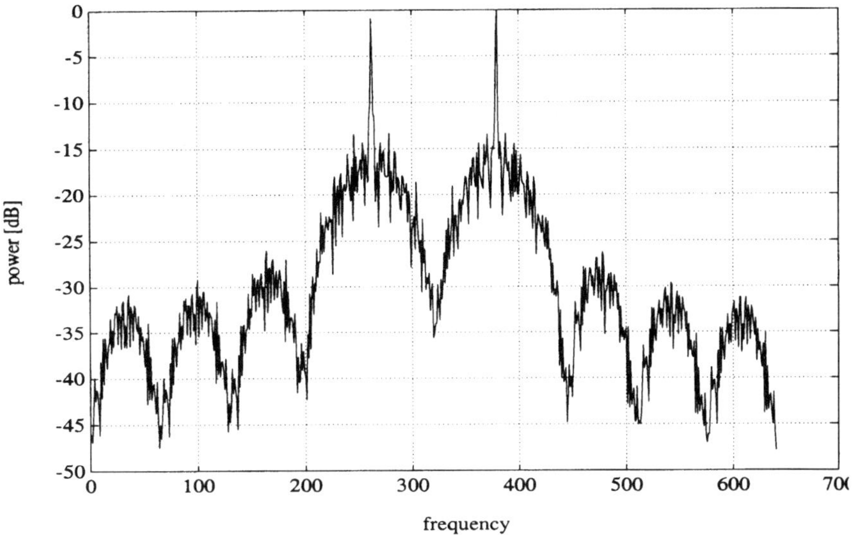

Figure 5.56 Example of the spectrum of an FSK-modulated signal.

can then use the trellis representation of the signal. In the same way as for linearly modulated signals it is difficult to find closed expressions for the spectrum. For a few simple, but nonetheless important, signals, the spectrum can be analytically calculated. For other cases we must refer to computer simulations or approximations.

For an infinitely long signal there are an infinite number of combinations of symbol and state sequences. If the signal can be expressed by a phase trellis, the calculations are simplified since only one period of the trellis needs to be considered. Each path through the trellis, resulting from different combinations of input signals within this period, must be taken into account. The autocorrelation function can be written

$$R_{uu}(t + \tau, t) = \sum_i R_i(t + \tau, t) \cdot P_i \tag{5.120}$$

where P_i is the probability of path i. All paths usually have the same probability. For the complex signal we obtain

$$R_i(t + \tau, t) = E\{u_i(t + \tau) \cdot u_i^*(t)\} \tag{5.121}$$

where $u_i(t)$ is the signal along path i. The application of this method is illustrated in Example 5.10, where the spectrum of the MSK signal is calculated.

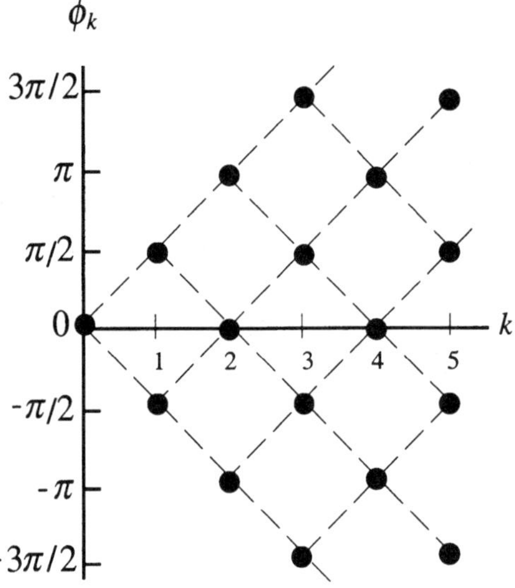

Figure 5.57   Phase tree of binary MSK.

Example 5.10

Calculate the spectrum of a binary MSK-modulated signal (cf. Section 5.4.3). The symbols are assumed to be independent, i.e. $R_x(n) = 0$ for $n \neq 0$.

Solution   The complex lowpass equivalent of an MSK modulated signal is

$$u(t) = \exp\left[j\left(\frac{\pi t}{2T} x_i + m \frac{\pi}{2} \right) \right] \tag{5.122}$$

for $0 \leq t \leq T$ and where m is a constant which prevents abrupt phase transitions of the signal

$$m = \sum_{n=0}^{i-1} x_n = 0, \pm 1, \pm 2 \text{ etc.} \tag{5.123}$$

The phase tree is shown in Figure 5.57. It is sufficient to investigate $m = 0, \pm 1, 2$ since $\exp(j\phi)$ is periodic in the interval $-\pi \leq \phi \leq \pi$ (cf. Figure 5.58).

The modulation has four phase states:

$$\theta_s \in \left\{ 0 \quad \frac{\pi}{2} \quad \frac{2\pi}{2} \quad \frac{3\pi}{2} \right\} \tag{5.124}$$

The signals at each state transition are given by the following table.

Path	Symbol	Signal, $u(t)$	t
A → B	+1	$\exp(j\frac{\pi t}{2T})$	$0 \leq t \leq T$
A → C	−1	$\exp(-j\frac{\pi t}{2T})$	$0 \leq t \leq T$
B → D	+1	$\exp(j(\frac{\pi}{2} + \frac{\pi t}{2T}))$	$0 \leq t \leq T$
B → A	−1	$\exp(j(\frac{\pi}{2} - \frac{\pi t}{2T}))$	$0 \leq t \leq T$
C → A	+1	$\exp(j(-\frac{\pi}{2} + \frac{\pi t}{2T}))$	$0 \leq t \leq T$
C → D	−1	$\exp(j(-\frac{\pi}{2} - \frac{\pi t}{2T}))$	$0 \leq t \leq T$
D → C	+1	$\exp(j(\pi + \frac{\pi t}{2T}))$	$0 \leq t \leq T$
D → B	−1	$\exp(j(\pi - \frac{\pi t}{2T}))$	$0 \leq t \leq T$

Eight paths are possible: $1 \Rightarrow A \rightarrow B \rightarrow D$, $2 \Rightarrow A \rightarrow B \rightarrow A$, $3 \Rightarrow A \rightarrow C \rightarrow D$, $4 \Rightarrow A \rightarrow C \rightarrow A$, $5 \Rightarrow D \rightarrow B \rightarrow D$, $6 \Rightarrow D \rightarrow B \rightarrow A$, $7 \Rightarrow D \rightarrow C \rightarrow D$, $8 \Rightarrow D \rightarrow C \rightarrow A$. The autocorrelation function becomes, cf. eq. (5.120):

$$R_{uu}(t + \tau, t) = \frac{1}{8} \sum_{i=1}^{8} R_i \tag{5.125}$$

For the calculation of the individual correlation functions, we can directly make a simplification by establishing that the paths starting in A or D only differ by a constant phase $\pm \pi$. Since the autocorrelation is independent of the signal phase,

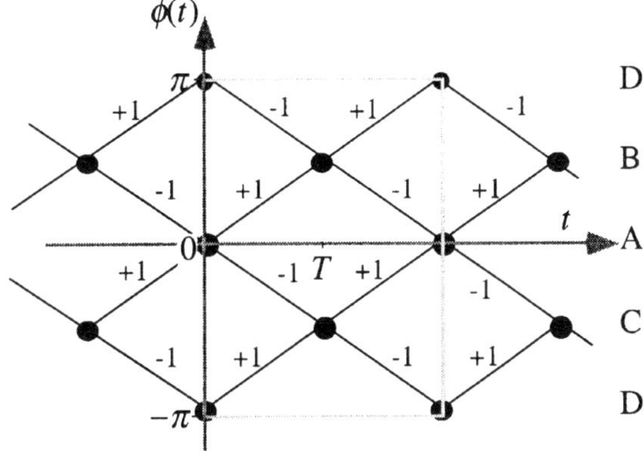

Figure 5.58 Phase trellis of binary MSK.

all paths do not need to be calculated and can be represented by the paths starting in A. The average of the periodic part of the autocorrelation function is taken over one period, i.e. from $-T$ to T, and gives

$$R_{uu}(\tau) = \frac{1}{T} \int_{-T}^{T} R_{uu}(t + \tau, t)\,dt \tag{5.126}$$

Note that the normalisation is not done over the integration interval but over a symbol interval. The calculation of the autocorrelation is straightforward, but a simplification can be made by using the fact that for a complex signal the auto-correlation function is

$$R_{uu} = R_{II} + j\,R_{QI} \tag{5.127}$$

where I is the in phase and Q the quadrature part. The calculation is thus performed by calculating the autocorrelation of the real part and the cross correlation between the imaginary and real parts. In Figure 5.59 the real parts are represented by solid lines and the imaginary parts by dotted lines. By comparing the graphs of paths two and four it can be seen that the cross correlation between the real and imaginary parts in path 2 is the same, apart from a minus sign, as the corresponding correlation for path 4. This also holds for paths one and three. The sum of the cross correlation parts is therefore zero.

For all paths the autocorrelation function becomes

$$R_{uu}(\tau) = \begin{cases} \left(T - \dfrac{\tau}{2}\right)\cos\dfrac{\pi\,\tau}{2T} + \dfrac{T}{\pi}\sin\dfrac{\pi\,\tau}{2T}, & 0 \le \tau \le 2T \\[2em] \left(T + \dfrac{\tau}{2}\right)\cos\dfrac{\pi\,\tau}{2T} - \dfrac{T}{\pi}\sin\dfrac{\pi\,\tau}{2T}, & -2T \le \tau \le 0 \end{cases} \tag{5.128}$$

Together with eq. (5.125) this gives the autocorrelation function of an MSK-modulated signal with independent random symbols. The spectrum is obtained by Fourier transformation of eq. (5.128), which gives

$$F\{R_{uu}(\tau)\} = \frac{\pi^2 \cos^2(T\omega)}{T^3\left(\left(\frac{\pi}{2T}\right)^2 - \omega^2\right)^2} = \frac{16T\cos^2(2\pi f T)}{\pi^2\left(1 - (4fT)^2\right)^2} \tag{5.129}$$

where the angular frequency ω has been replaced by $2\pi f$. Equation (5.129) represents the spectrum of an MSK-modulated signal and has been plotted in Figure 5.60. The amplitude one of the modulated signal has been assumed for the calculations. The symbol energy was thus $E_b = T/2$ (cf. eq. (2.26)). The symbol energy can then be introduced as a parameter in the spectrum

$$U(f) = F\{R_{uu}(\tau)\} = \frac{32E_b\cos^2(2\pi f T)}{\pi^2\left(1 - (4fT)^2\right)^2} \tag{5.130}$$

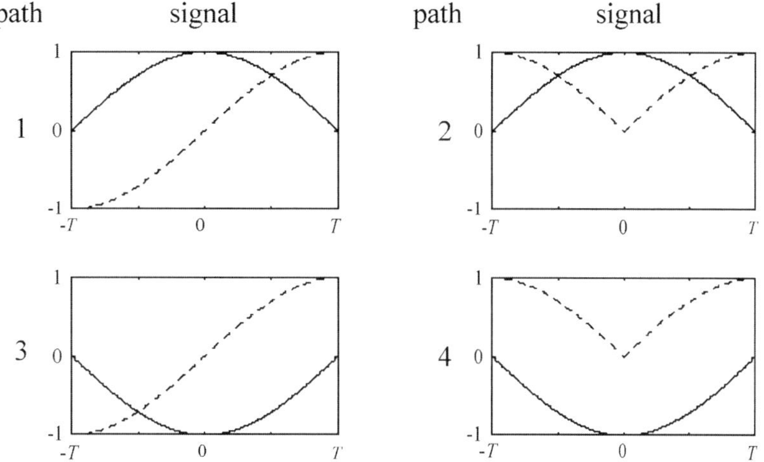

Figure 5.59 Plot of an MSK signal. The solid line represents the real part and dashed line the imaginary part.

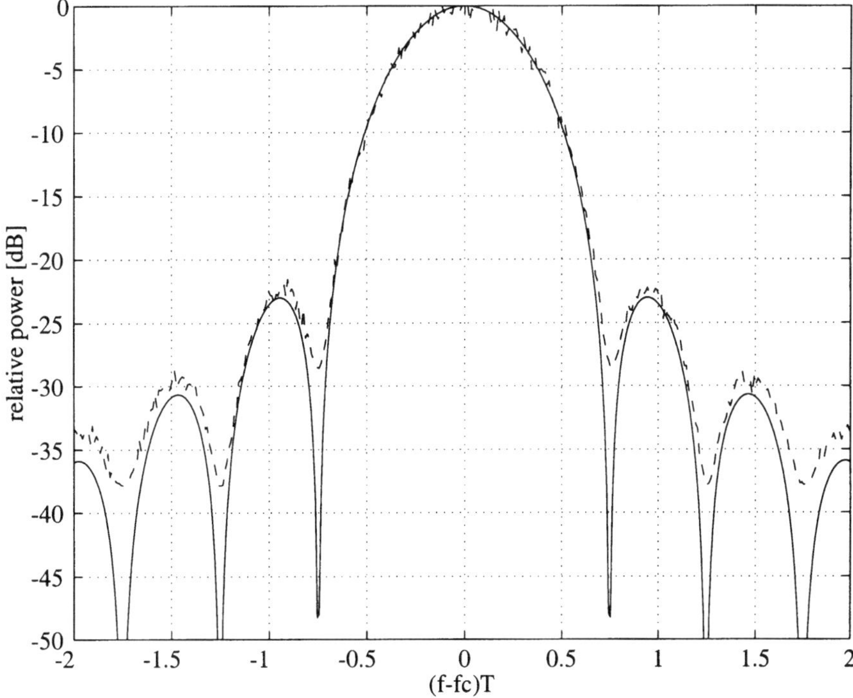

Figure 5.60 Power spectral density of an MSK modulated signal. The solid line is a plot of eq. (5.130) and the dashed line represents a simulation result.

Compared to spectrum for BPSK the sidelobes for the MSK fall off more rapidly and the zeros are located closer to the peek of the spectrum.

In this section a method for calculation of the power spectral density of non-linearly modulated signals has been presented. The method provides an analytical expression of the kind as in eq. (5.130). A second alternative is, of course, to simulate a section of the signal and then perform a Fourier transformation on a computer. The simulation requires, however, relatively long signal sections for the spectrum to converge towards the actual one (cf. the simulation performed in Figure 5.60). A third method, suitable for computation of spectra for modulation systems with high complexity, is presented in the appendix.

Optimal spectrum

An interesting question is whether the shape of the spectrum, within the bandwidth provided by the channel, is of any importance. Suppose that all signal power available has been concentrated to a spike somewhere within the frequency band allowed by the channel. In the time domain this corresponds to the modulated signal being a sinusoidal wave form, i.e. constant amplitude and phase. It is obvious that such a signal cannot transmit any information.

Also, suppose that a part of the channel frequency band is heavily interfered by some unwanted signal. The total signal to noise ratio might be better if this frequency band is cut off and all wanted signal power is assembled in the remaining frequency band. Depending on the relation between the signal bandwidth which has been cut away and the increase in SNR it is conceivable that the transmission capacity of the channel will increase.

By the reasoning above it is clear that the modulated signal spectrum will affect the system transmission capacity. The answer is given by the water pouring principle. This gives the shape of the modulated signal spectrum to fulfil Shannon's formula.

Suppose an AWGN channel with a single sided bandwidth B and noise spectrum $N(f)$. The optimal signal spectrum, $S^o(f)$, then must fulfil

$$S^o(f) = \begin{cases} K - N(f) & f \in B \\ 0 & f \notin B \end{cases} \tag{5.131}$$

If S is the available total signal power, the constant K shall be chosen such that

$$S = \int_0^\infty S^o(f)\, df \tag{5.132}$$

This means that the modulated signal spectrum shall correspond to the depth of the water filled into a jar with a bottom profile corresponding to the noise spectrum (cf. Figure 5.61).

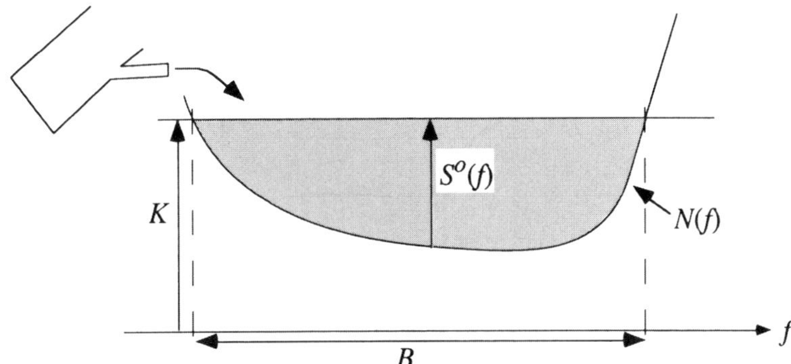

Figure 5.61 Optimal power density spectrum through water pouring.

5.4.5 COMBINED MODULATION AND ERROR CORRECTION CODING

In certain telecommunication systems, such as computer networks operating through local telephone lines, the introduction of error-correction coding is required. At the same time the bandwidth of the modulated signal must not increase and the transmitted amount of information must not decrease. The solution to this dilemma is a compression of the signal points (cf. Figure 5.36). This however also means an increased error probability for the detected symbols. A part of the gain by introducing error-correction coding then disappears. The problem is considerably reduced by considering the channel coding and modulation as a unit. The redundancy is invested at the signal points where the risk of symbol error is largest. Compression of the signal points means that redundancy can then be introduced without having to increase the symbol rate and the penalty of increased signal bandwidth does not have to be paid. Both block codes and convolutional codes can be utilised. When convolutional codes are used the coding method is called trellis-coded modulation (TCM). This is described in a practical way below by using set partitioning or the so-called Ungerboeck code.

The starting point is a bandwidth efficient type of modulation where the signal points are compressed, in relation to the uncoded case. This is achieved by, for instance, changing from 4PSK to 8PSK modulation. The flow of information symbols is arranged in a number of parallel binary bit streams. When the incoming bit stream is binary m pieces of consecutive symbols are grouped together and will constitute elements in a vector. Each element will make up a bit stream, where the rate has been reduced by a factor $1/m$ compared to the incoming bit stream. The information bits are divided into two parts where one part, consisting of $\tilde{m}$ bits, is coded by a $\tilde{m}/(\tilde{m} + 1)$ convolutional code and the other part, $m - \tilde{m}$

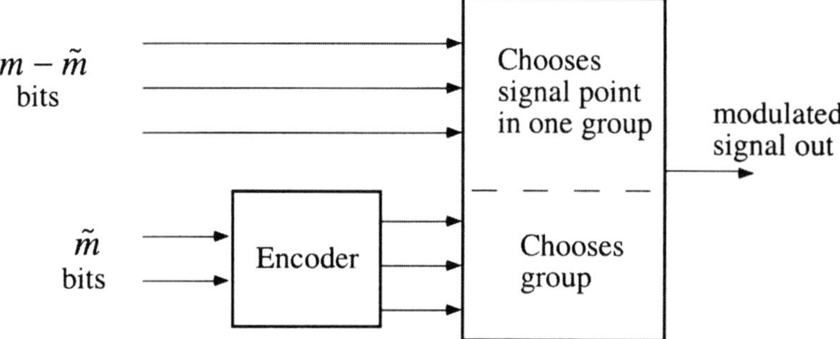

Figure 5.62 Partitioning of information bits.

bits, is sent directly to the modulator without being coded, as shown in Figure 5.62.

The signal points in the complex plane, which are taken by the modulated signal, are partitioned into groups. For a certain message, the coded bits determine from which group the signal point is to be chosen and the uncoded bits determine which one of the individual signal points in that group will represent the particular message.

Partitioning of signal points

The separation is obtained by successively partitioning the signal points in two subgroups in each step. This is illustrated in Figure 5.63. The partitioning is carried out in such a way that the distance between the signal points increases for each step. As a measure of the distance the Euclidean distance between two signal points in the complex plane is used:

$$d_E = \left| s_i - s_j \right|^2 = \sqrt{\mathrm{Re}^2\{s_i - s_j\} + \mathrm{Im}^2\{s_i - s_j\}} \tag{5.133}$$

Equation (5.133) is motivated and generalised in Chapter 6, where it is shown that an increase of the Euclidean distance leads to a reduced error probability. The smallest Euclidean distance at partition n is d_n. The partition continues until every signal subgroup contains one single signal point, or until $d_{\tilde{m}+1}$ is equal to or larger than the free distance in the convolutional code, which then provides an acceptable value of $\tilde{m}$. Each partition gives rise to a new level with groups of signal points. The objective of set partitioning is, for each step, to increase the Euclidean distances as much as possible between the signal points within each group, respectively.

At modulation of a symbol stream the uncoded bits choose signal from the subset of signal points having the largest distance. Bits protected by error-correction coding choose signal points from the ones having shorter

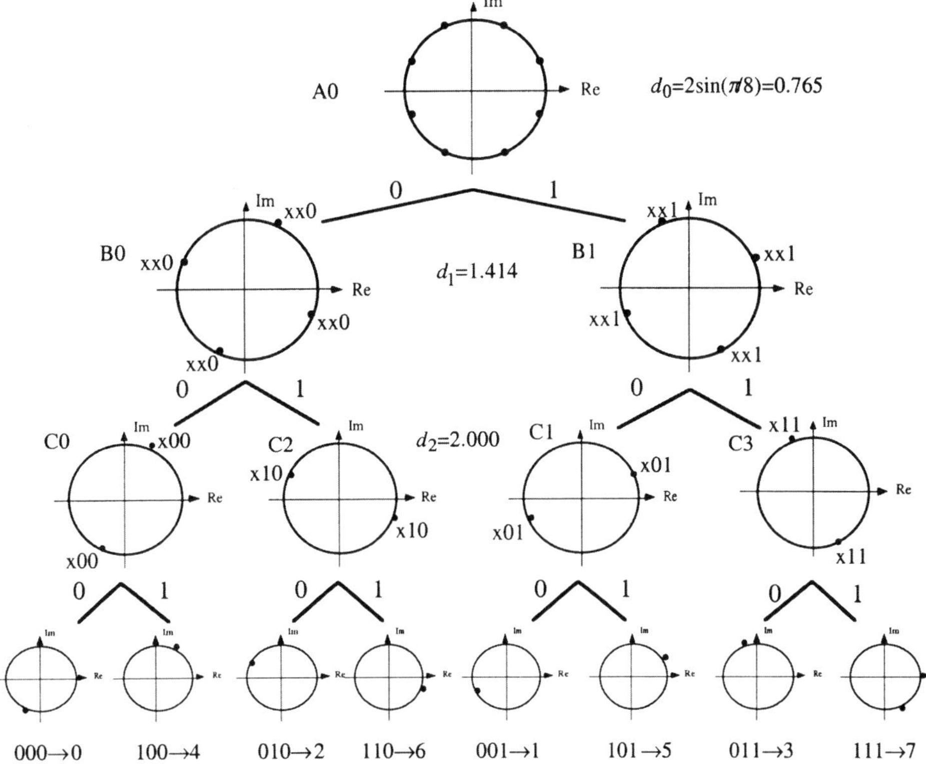

Figure 5.63 Partition of the signal points into subgroups A0, B0, B1 etc. Mapping of the binary bits and its decimal numbers are shown at the bottom.

distance. The free Euclidean distance for a TCM is

$$d_{min} = \min\left[d_{\tilde{m}+1}, d_{free}(\tilde{m})\right] \tag{5.134}$$

An error in the decoding makes the decoder choose a way other than the encoder has taken, through the trellis describing the convolutional code. One path through the trellis can be seen as a chain where every link is a state transition. An error event starts where the true and the estimated paths split and ends where they merge. The shortest error event is the one having the shortest possible accumulated distance between split up and merging. The Euclidean distance for this path is

$$d_{free}^2(\tilde{m}) = \sum_{k=1}^{K} \left(d_x^2\right)_k \tag{5.135}$$

where K is the length of the shortest error event, measured in the number of state transitions. The terms $(d_x^2)_k$ is the Euclidean distance between the correct and the incorrect paths for each state transition, respectively, in the shortest error event.

TCM for PSK modulation

The method can be illustrated by providing TCM with a rate 1/2 convolutional code to a system having 4PSK modulation. At the beginning the information is then divided into two parallel binary bit streams. One of these is coded by the rate 1/2 coder, which gives two bits out for every bit in. What originally consisted of two bit streams then increases to three. All together we then have $2^3 = 8$ different symbols, or binary words in this case, which requires 8PSK modulation. Hence, the bandwidth does not need to increase since the symbol rate is unchanged. In order to use the redundancy in an optimal way the signal points are partitioned into groups in the way described above and illustrated in Figure 5.63.

The uncoded bit stream can then determine a signal point at level C, where the Euclidean distance is largest, $d_2 = 2$. The other two levels A and B are determined by the coded bits. The channel noise may make the detector choose another signal point than the one sent. The most common error is that the nearest neighbours among the signal points are chosen; see level A. There is the convolutional code protecting though. At level C where the uncoded bits are determined the distance between the signal points is larger. It is therefore less common that an incorrect signal point is chosen there. In that way the error correction has been invested where it is most beneficial.

Consider a rate 1/2 convolutional code with two states, s_1 and s_2, as in Figure 5.64. From each state four paths are possible. The uncoded bit stream has two alternatives, 0 or 1. The paths therefore lead in pairs to the states at the next time instant. The paths originating from (or merging into) the same state are selected from the same subset. The signals from subset B0 are thus selected at a transition from s_1, and from B1 at a transition from s_2. The parallel paths are tied to the signals from C0 or C2 and C1 or C3, respectively. Parallel paths are not given any protection by the error-correction coding. This is the reason why the signal points having the longest distance are selected to represent these transitions, i.e. level C in Figure 5.63. For a good code all signals should be equally probable.

The path corresponding to the error event with the smallest possible Euclidean distance is clear by a visual inspection of the trellis. Suppose that $s_1s_1s_1$ is the correct state sequence. The error event with the shortest distance from the correct path thereby is $s_1s_2s_1$. From s_1 to s_2 the paths 6 or 2 lead. From s_1 to s_1 the paths 0 or 4 lead. The choice between 6(2) or 0(4) is done on level B where the Euclidean distance is d_1. The choice between the paths $s_1 \rightarrow s_1$, i.e. 4(0), and $s_2 \rightarrow s_1$, i.e. 1(5), is done on level A where the distance is d_0. The total free distance then becomes

$$d_{free} = \sqrt{d_0^2 + d_1^2} = \sqrt{0.765^2 + 1.414^2} = 1.608 \tag{5.136}$$

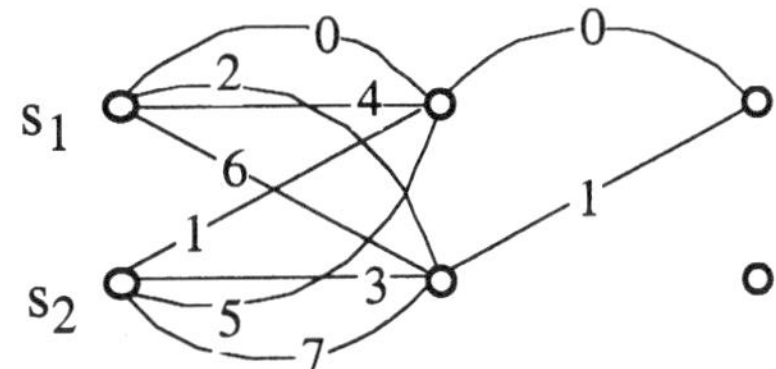

Figure 5.64 Trellis for an Ungerboeck code with two states and code rate 1/2. The numbers at every transition correspond to the numbers indicating the signal points at the bottom of Figure 5.63.

Since $d_2 = 2.000$ the smallest Euclidean distance is $d_{min} = 1.608$, according to eq. (5.134). The gain caused by the coding, compared to uncoded 4PSK which has the shortest distance between signal points of $d_{ref} = 1.414$, becomes

$$G_c = 10\log\left(\frac{d_{min}^2}{d_{ref}^2}\right) = 20\log\left(\frac{1.608}{1.414}\right) = 1.1 \quad (\text{dB}) \tag{5.137}$$

which is a moderate coding gain. Considerably improved gain is obtained if a convolutional code with four states is used (Figure 5.65). In this case d_{min} is limited by the Euclidean distance at the C-level, which is 2.000. This gives a gain of 3 dB from the coding.

An encoder for this code then has the following representation (see Figure 5.66).

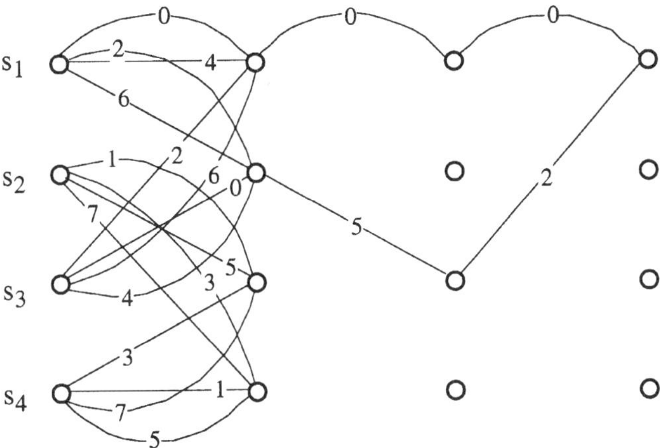

Figure 5.65 Four state Ungerboeck code.

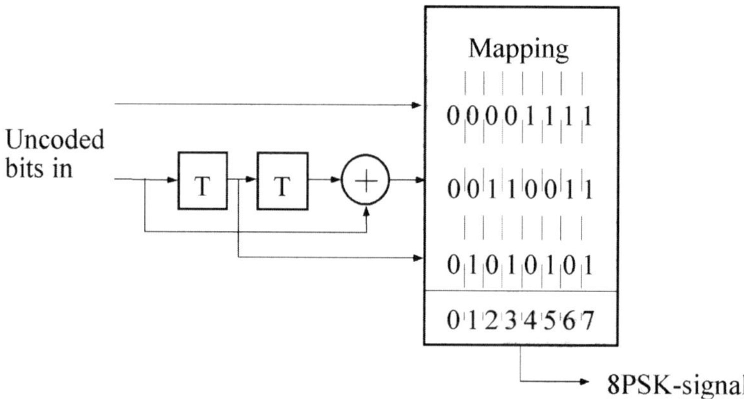

Figure 5.66   Encoder for rate 1/2 Ungerboeck code with four states and 8PSK modulation.

Example 5.11

A trellis-coded signal is obtained by coding one bit with a rate 1/2 convolutional coder while the other three bits are uncoded. Do the partition of the signal points for a 32QAM signal. What is the coding gain (Figure 5.67)?

Solution   Partition of the signal space is achieved by letting the coded bit select a signal group. The code rate is 1/2, which means that the signal points are partitioned into 4 groups as shown in Figure 5.68.

The trellis for this system is shown in Figure 5.69. Each transition is really 8 parallel transitions, corresponding to the 8 possible values of the uncoded bits. The shortest distance between two points within one signal group is d_2. This is equivalent to the distance between two parallel paths in the trellis.

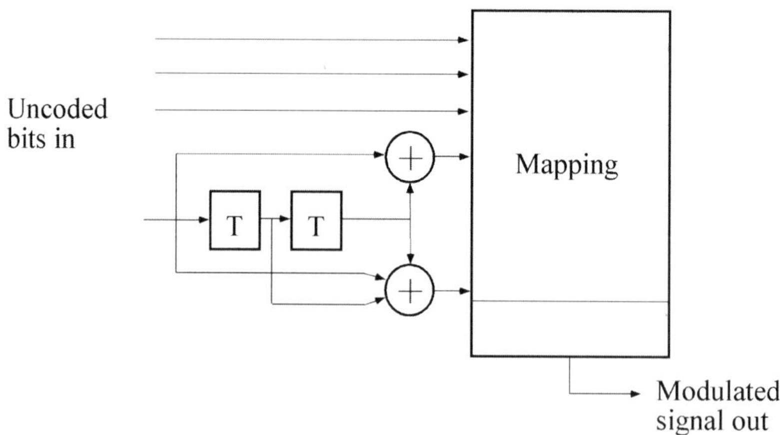

Figure 5.67   Coder.

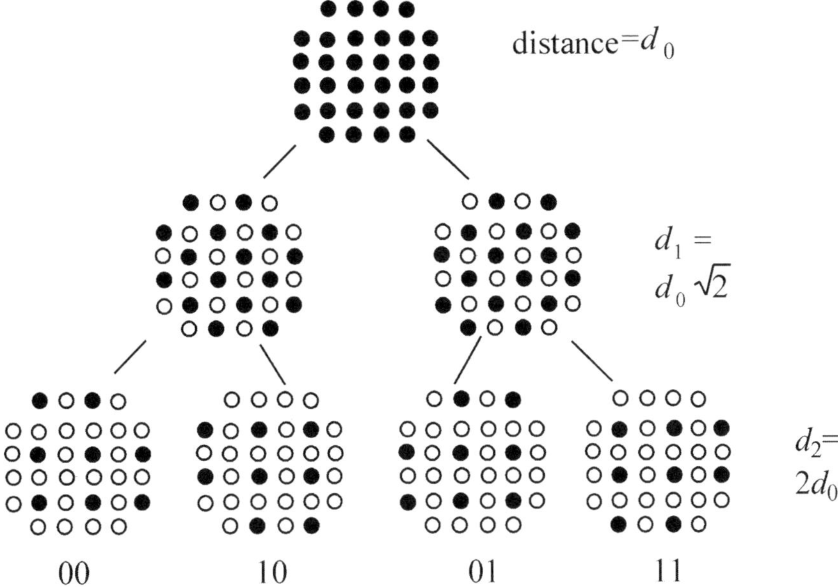

Figure 5.68 Partition of the signal set. Mapping of coded bits is shown in the bottom.

The shortest distance between two paths beginning and ending in the same state is obtained by studying the trellis in Figure 5.69. The shortest error paths are obtained from the state passages

$$s_1 \rightarrow s_1 \rightarrow s_1 \rightarrow s_1 \quad (1)$$

$$s_1 \rightarrow s_2 \rightarrow s_3 \rightarrow s_1 \quad (2)$$

which are drawn in Figure 5.70. The output signals of the encoder along the two paths are indicated in Figure 5.69.

The shortest paths correspond to one incorrect decoded bit. The Euclidean distance between the corresponding code sequences can be calculated as follows: path (1) gives the code sequence 00 00 00 and path (2) gives the code sequence 11 10 11. The shortest distance between a symbol in signal group 00 and one in

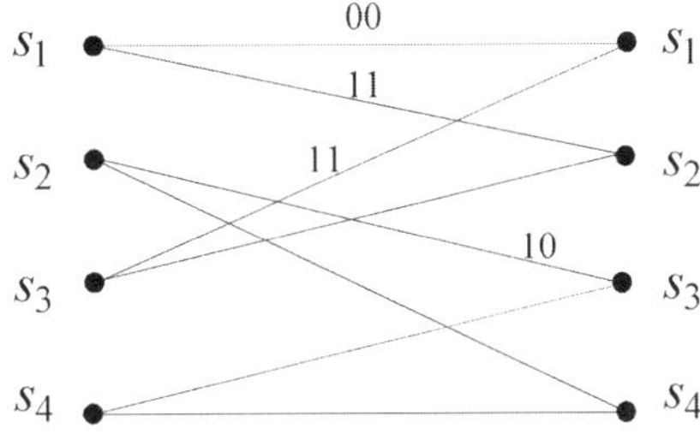

Figure 5.69 Trellis of the code. Coded bits are shown for certain transitions.

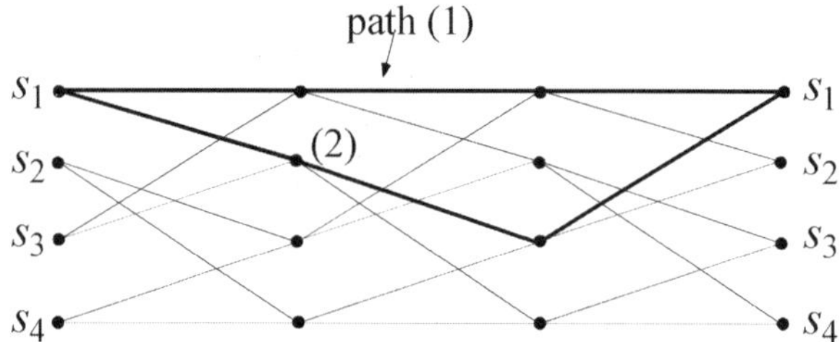

Figure 5.70   Signal paths with the shortest distance.

group 11 is d_0 and between a symbol in group 00 and one in group 10, d_1. The total distance becomes $d_{free} = \sqrt{(2d_0^2 + d_1^2)}$. The shortest distance, d_{min}, of the modulation is the smallest of the values $d_2 = 2d_0$ and $\sqrt{(2d_0^2 + d_1^2)} = 2d_0$. The minimum distance therefore becomes $2d_0$ which gives a coding gain of 3 dB.

5.4.6 TRANSMISSION USING MULTIPLE CARRIERS

Short powerful interference pulses can cause several symbol errors in high speed transmission. This is independent of whether these sources of interference are due to interfering signal sources or destructive interference of the signal owing to multi-path transmission. By partitioning of the symbol stream in several parallel streams the data rate in each stream is reduced, which reduces the sensitivity to interference. The data streams are made mutually orthogonal by some modulation or transformation process, for example the Walsh transform. This method is called multi-carrier modulation (MCM).

Orthogonal frequency division multiplex (OFDM)

Orthogonal frequency division multiplex (OFDM), is one way of creating mutually orthogonal signals, by setting up multiple carriers at a suitable frequency separation and modulate each symbol stream separately. For a given transmission requirement, the data rate per carrier can thereby be reduced by increasing the number of carriers. Since the carriers are mutually orthogonal the symbol streams do not interfere with each other. The price is increased complexity in the system. Through suitable coding and interleaving fading can be mitigated as described in Chapter 4 (Figure 5.71).

One way of making the signals independent to each other is of course to separate the bit streams at the receiving end by using a bandpass filter. The requirement on narrow transition bands of the filters involved is high. At the same time any overlap of the spectra for the different signals cannot be allowed. Another way is to select the frequency separation between each signal in such a

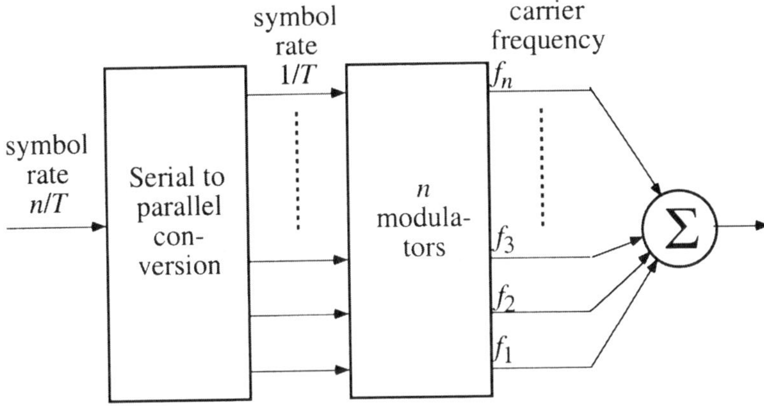

Figure 5.71   Transmitter for multi-carrier modulation.

way that orthogonality is achieved over a symbol interval. Start from the signal in eq. (5.12) and introduce the variable Δf to represent a frequency difference. The signal can then be written

$$s(t, f_c, \Delta f) = a(t)\cos[2\pi(f_c + \Delta f)t + \phi(t)]$$

To be able to draw any simple conclusions it must be assumed that a and ϕ are constant over a symbol interval, which is indeed the case for several modulation types, such as BPSK and QPSK. When $f_1 = f_c$ and $f_2 = f_c + \Delta f$, the inner product between the two first signals becomes

$$\rho = \int_0^T s(t, f_c, 0)\cdot s(t, f_c, \Delta f)\, dt \propto \frac{1}{2}\left(\frac{\sin(2\pi\Delta f\, T)}{2\pi\Delta f} - \frac{\sin[2\pi(2f_c + \Delta f)T]}{2\pi(2f_c + \Delta f)} \right)$$

$$(5.138)$$

It can be seen that a condition for orthogonality, i.e. $\rho = 0$, is that $\Delta f = n/T$ and $f_c = m/T$ where m and n are integers. If f_c is large compared to Δf the deviation from zero will be small. The conclusion is that if the frequency separation $\Delta f = 1/T$ is chosen between each carrier, the different symbol streams become orthogonal.

Implementation

An efficient way of achieving this is by letting each symbol stream constitute a frequency component in a Fourier series. The time-domain signal is obtained by using an inverse Fourier transform which is conveniently implemented by an inverse fast Fourier transform (IFFT) (cf. Figure 5.72). If, at a given symbol interval, the symbols on each stream is x_p, respectively, the complex representa-

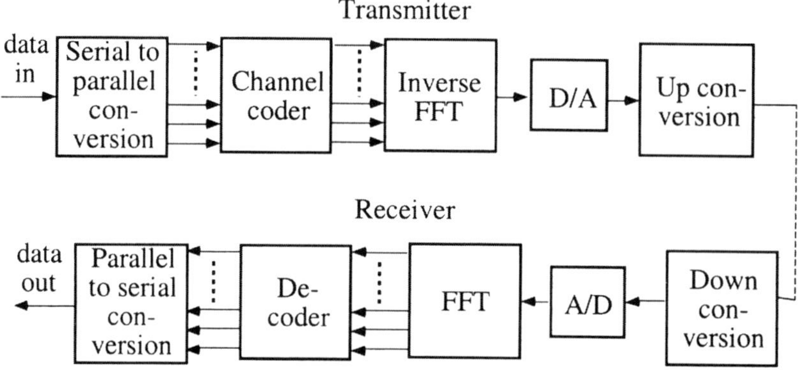

Figure 5.72 Multi-carrier system implemented using FFT.

tion of the modulated OFDM signal is

$$u_{OFDM}(k) = \sum_{p=1}^{n} x_p \exp(j2\pi kp/n)$$

$$= x_1 \exp(j2\pi k/n) + x_2 \exp(j2\pi k2/n) + \cdots + x_n \exp(j2\pi k) \quad (5.139)$$

Index k runs from 1 to n in each symbol interval, which gives eq. (5.139) as a sum of the symbols x_p modulated on carriers with the frequency separation $2\pi/n$. This corresponds to QPSK modulation if both the real and imaginary parts of the Fourier transform are used, i.e. the symbols x_p is complex. Note that the spectrum must be symmetrical around $f = 0$ for the time domain signal to be real. At the receiving end the separation of the symbol stream is appropriately done by FFT.

The strength of this modulation method is among other things its advantageous properties for transmission on channels with multi-path transmission, which is especially required for mobile reception. To further improve this property a guard period, T_g, can be introduced between each symbol. The reflected signals should then have time to fade away before the detection of the next symbol group is started. The total time between new symbols then becomes $T_{tot} = T + T_g$. Linear amplifiers are required in MCM since the ratio between the maximum and average values of the signal can be large.

5.5 Appendix

5.5.1 VARIOUS FREQUENCY PULSES

Below is a list of some common types of frequency pulses (cf. eq. (5.94) and Figure 5.41).

LRC (raised cosine with pulse length $L \times T$):

$$p_D(t) = \begin{cases} \dfrac{1}{2LT}\left[1 - \cos\left(\dfrac{2\pi t}{LT}\right)\right], & 0 \le t \le LT \\[2ex] 0, & \text{otherwise} \end{cases}$$

TFM (tamed frequency modulation)

$$p_D(t) = \frac{1}{8}\left[g_0(t - T) + 2g_0(t) + g_0(t + T)\right], \quad 0 \le t \le LT$$

where

$$g_0(t) \approx \left[\frac{\sin(\pi t/T)}{\pi t/T} - \frac{\pi^2}{24}\frac{2\sin(\pi t/T) - (2\pi t/T)\cos(\pi t/T) - (\pi t/T)^2\sin(\pi t/T)}{(\pi t/T)^3}\right]$$

LSRC (spectrum raised cosine with pulse length $L \times T$)

$$p_D(t) = \frac{1}{LT}\frac{\sin\left(\dfrac{2\pi t}{LT}\right)}{\dfrac{2\pi t}{LT}}\frac{\cos\left(\beta\dfrac{2\pi t}{LT}\right)}{1 - \left(\dfrac{4\beta t}{LT}\right)^2}, \quad 0 \le \beta \le 1, \ \ 0 \le t \le LT$$

GMSK (Gaussian minimum shift keying)

$$p_D(t) = \frac{1}{2T}\left[Q\left(2\pi B_b\frac{t - (T/2)}{\sqrt{\ln 2}}\right) - Q\left(2\pi B_b\frac{t + (T/2)}{\sqrt{\ln 2}}\right)\right], \quad -\infty \le t \le \infty$$

where

$$Q(t) = \frac{1}{\sqrt{2\pi}}\int_t^\infty \exp(-\tau^2/2)\,d\tau$$

Note that theoretically the pulse shape has an infinite extension in time. In practice it therefore has to be truncated. A pulse length of 3–4 symbols often provide an adequate approximation.

LREC (rectangular pulse shape with length $L \times T$)

$$p_D(t) = \begin{cases} \dfrac{1}{2LT}, & 0 \le t \le LT \\[2ex] 0, & \text{otherwise} \end{cases}$$

5.5.2 COMPUTING SPECTRA FOR SIGNALS WITH LARGE STATE DIAGRAMS

A method especially suited for numerical computer calculations of spectra for modulated signals is described below. The complexity of this method is independent of which modulation scheme is being considered. The method is given without proof, since this is beyond the scope of this book. The power spectral density is

$$
S(f) = 2\mathrm{Re}\Bigg\{ \int_0^{LT} R(\tau)\exp(-j2\pi f\tau)d\tau
$$

$$
+ \frac{\exp(-j2\pi fLT)}{1 - C_\alpha\exp(-j2\pi fT)} \int_0^T R(\tau + LT)\exp(-j2\pi f\tau)d\tau \Bigg\} \tag{5.140}
$$

where the autocorrelation holds for equally probable symbols

$$
R(\tau) = R(\tau' + mT) = \frac{1}{T}\int_0^T \prod_{n=1-L}^{m+1}
$$

$$
\times \Bigg\{ \sum_{k=-(M-1),\,k\ \mathrm{odd}}^{M-1} \exp\{j2\pi hk[q(t + \tau' - (n-m)T) - q(t - nT)]\} \Bigg\}dt
$$

$$
\tag{5.141}
$$

where $0 \le \tau' < 1$, $m = 0,1,\dots$ and

$$
C_\alpha = \frac{1}{M} \sum_{k=-(M-1),\,k\ \mathrm{odd}}^{M-1} \exp(jh\pi k) \tag{5.142}
$$

Exercise 5.1

Show the effect of a phase and frequency error in the receiver oscillator when AM-DSBSC and AM-SSB are demodulated. The phase and frequency difference between the received signal and the receiver oscillator signal are φ and $\Delta\omega$, respectively.

Exercise 5.2

Compute the spectrum for a digitally modulated signal with a rectangular frequency pulse, $p_D(t) = 1\mathrm{REC}$ see below. The information bits are pre-coded by the operation $I_n = x_n + x_{n+1}$ and I_n is used as input signal at the modulation. The symbols x_n are uncorrelated information symbols with the values $+1$ or -1.

$$
p_D(t) = \begin{cases} \dfrac{1}{2T}, & 0 \le t \le T \\[2mm] 0, & \text{otherwise} \end{cases}
$$

Exercise 5.3

The signal $x(t)$ is the input signal to a modulator for AM with carrier. The signal $x(t)$ can be modelled as a stationary Gaussian signal with mean $\bar{x} = 0$ and variance $\overline{x^2} = S_x$. Determine max. S_x for over modulation to happen at most 1% of the time. Suppose $\mu = 1$ (by over modulation is meant that the AM signal cannot be demodulated without distortion by an envelope detector).

Exercise 5.4

Suppose the signal $s(t) = 10\cos\left(90 \times 10^6 t + 200\cos(2000t)\right)$. This signal can represent two possible kinds of analogue modulation; which? What does the information signal, or message, look like for each type of modulation? What is the necessary bandwidth on the transmission channel?

Exercise 5.5

The figure below shows the positions for the signal points in the I–Q plane for 32QAM. Partition the signal set into 8 subgroups. Give the smallest Euclidean distance between two signal points in the partitioned signal set. Give a block diagram showing how a suitable TCM modulator. Each symbol on the modulator input is represented by 3 binary bits.

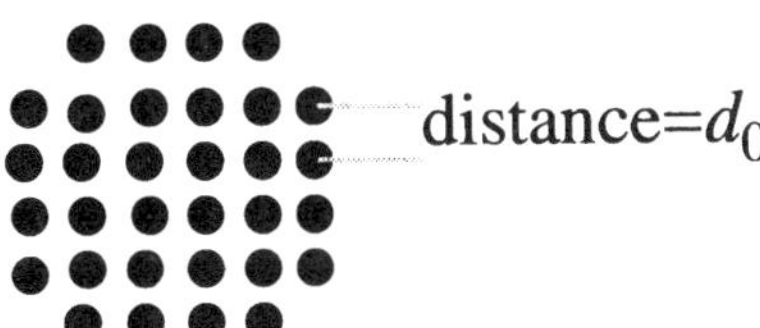

Exercise 5.6

A rectangular pulse at a carrier with the frequency f_c:

$$x(t) = \begin{cases} A\cos(2\pi f_c t), & 0 \le t \le T \\ 0, & \text{otherwise} \end{cases}$$

is the input a signal to a filter with the impulse response $h(t) = x(T - t)$ and f_c is a large integer multiple of $1/T$. Determine the output signal.

Exercise 5.7

The frequency response for a bandpass filter is

$$H(j\omega) = \frac{j\omega}{\left(j\omega + (1/2)\right)^2 + (3/4)}$$

What is the complex lowpass equivalent for the impulse response? The centre frequency, for the system where the filter is included, is f_c.

Exercise 5.8

The equivalent noise bandwidth can be used as a definition of the bandwidth of a signal. What is the noise bandwidth in Hz for BPSK, QPSK and MSK, respectively?

Exercise 5.9

Determine the spectrum for BPSK- and QPSK-modulated signals, respectively. The transmission rate is 10 kbit/s. Rectangular pulse forming is used.

6 DETECTION IN NOISE

In transmission of information through a channel, interference and noise are always added to a larger or smaller extent. For example the mobile telephony channel is the victim of severe interference while an optic fibre often has a moderate level of interference. Demodulation of information which has been transmitted over noisy channels is therefore a fundamental problem which has concerned telecommunication experts for more than a century. Several results from mathematical statistics have thereby been applied to this problem. A relatively large part of this book is therefore devoted to this important problem. In analogue modulation, demodulation can be seen as an estimation problem. Digital demodulation can be handled as a detection problem since the number of possible signals is limited, over a limited period of time, and the task for the receiver is to decide which one of these has been sent.

The noise is considered as a stochastic signal added to the signal that has been sent. An important case is the telecommunication systems that can be regarded as narrowband systems, e.g. radio. At the beginning of this chapter some fundamental properties of narrowband noise are therefore presented. This is followed by a section about noise in analogue systems. In this section the question of how the demodulated signal is influenced by the noise level of the channel is investigated. For digital transmission some different variations of receivers are presented, which are optimised with respect to the lowest possible error probability in the demodulated message. In this section, methods for the calculation of the error probability of transmission using binary symbols are also presented. It is possible to generalise these methods to arbitrary symbols but the calculations soon become complicated. In this chapter, a method is therefore presented which provides a relatively simple calculation of the upper limit of the error probability for an arbitrary symbol alphabet. Included also is a short section about using diversity to reduce the interference caused by fading. To put the detector in its technical context, a short presentation of the radio receiver is given at the end of the chapter.

The aim of this chapter is to provide tools for the calculation of the quality of transmitted information in terms of signal to noise ratio or bit error rate versus the signal to noise ratio of the channel. The purpose is also to give the reader an understanding of the ideas that underlie the design of the block diagram for optimal digital detectors.

6.1 Fundamentals of narrowband noise

Interference coming from external sources, thermal noise, etc. will be added to the wanted signal in the channel. For such a channel the additive white Gaussian noise (AWGN) is a proper channel model. For a signal, $s(t)$, which has passed an additive noise channel, the signal reaching the receiver, $r(t)$, will be the sum of the signal sent and the noise

$$r(t) = s(t) + n(t) \tag{6.1}$$

The signal to noise ratio has been defined in Chapter 2 and is

$$SNR = \frac{P_R}{N_R} \tag{6.2}$$

where P_R is the power of the wanted signal at the detector input and N_R is the power or variance of the noise in the same place. The noise bandwidth has usually been limited by a bandpass filter prior to the detector or demodulator. An important factor, for the analysis of the noise properties of the system is how S/N is influenced by demodulation, which is a non-linear operation. Below, the signal to noise ratio on the demodulator input, $(S/N)_R$, is therefore compared with the signal to noise ratio on its output, $(S/N)_D$. In the bandpass filters preceding the demodulator, the frequency band is limited as much as possible without influencing the spectrum of the wanted signal. The noise level thereby becomes minimised without affecting the information. In practice, a negligible influence has to be accepted since the wanted signal is not strictly band limited. The bandwidth limitations which are done ahead of the demodulator are considered to belong to the channel (Figure 6.1).

If $H_R(f)$ is the frequency response of the bandpass filter, the power density spectrum at the filter output becomes

$$N_0|H_R(f)|^2 = N_0 G(f) \tag{6.3}$$

where N_0 is the single-sided noise density in W/Hz, i.e. the same as would be

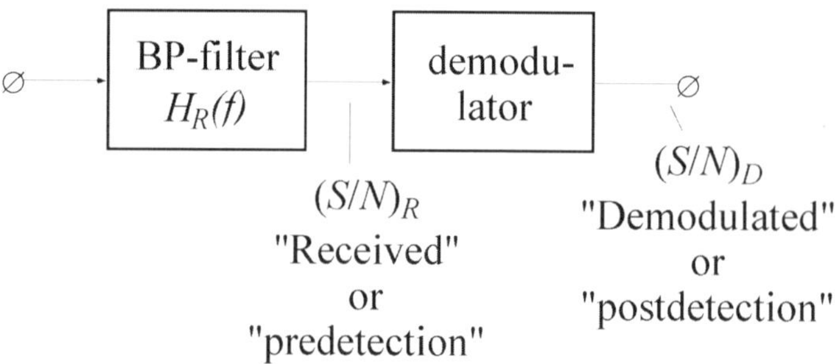

Figure 6.1　The demodulator is preceded by a bandpass filter.

measured by a spectrum analyser. The noise power at the filter output becomes

$$N_R = N_0 \int_0^{\infty} G(f)df = G_{max}N_0B_N \tag{6.4}$$

where B_N is the noise bandwidth of the system. Assume a Gaussian distribution of the noise amplitude. (Note that the amplitude concept is used in two ways in the literature. The amplitude either corresponds to the length of the signal vector and is then always positive, or corresponds to the instantaneous amplitude of a function, such as the voltage of an electric signal. In this case the amplitude can be both positive and negative.) Gaussian distribution is a satisfactory approximation for several types of interference and is given by the central limit theorem when there are several, statistically independent interference sources, none of which dominates. Thermal noise is generated by a large number of independent electron movements and thus has a Gaussian distribution. At the same time this noise model gives simple mathematical solutions. The amplitude probability density function is

$$p_N(n) = \frac{1}{\sqrt{2\pi N_R}}\exp\left(-\frac{n^2}{2N_R}\right), \quad -\infty < n < \infty \tag{6.5}$$

where N is a stochastic variable. Due to the bandpass filtering, the noise can be regarded as narrowband around the centre frequency ω_c (or $2\pi f_c$) and can thus be divided into in phase and quadrature components.

$$N(t) = N_i(t)\cos\omega_c t - N_q(t)\sin\omega_c t = A_n(t)\cos[\omega_c t + \Phi_n(t)] \tag{6.6}$$

The components N_i and N_q are independent and stochastic processes having Gaussian distributions and zero expectation value. For the N_i component, for instance, the following relation holds:

$$p_{N_i}(n) = \frac{1}{\sqrt{2\pi\,N_R}}\exp\left(-\frac{n^2}{2N_R}\right) \tag{6.7}$$

The variance or power of the noise components is $E\{N_i^2\} = E\{N_q^2\} = E\{N^2\} = N_R$. Note that the variance is the same for all parts. This is due to the fact that the in phase and quadrature components of the noise consist of a stochastic part, whose density function is given by eq. (6.7), and a deterministic cosine or sine part, respectively. The expectation value becomes time dependent. For one of the quadrature components, such as the in phase part, the average of the expectation value becomes

$$E\{N_i^2\cos\omega_c t\} = E\{N_i^2\}\int_{-\pi/\omega_c}^{\pi/\omega_c} \frac{\cos^2(\omega_c t)}{T}\,dt = \frac{1}{2}E\{N_i^2\} \tag{6.8}$$

The mean value of the periodic part of the signal is taken over one period, $T = 2\pi/\omega_c$. The envelope of the noise is the vector sum of the noise components

$$A_n = \sqrt{N_i^2 + N_q^2} \tag{6.9}$$

The phase angle, Φ_n, of the resultant noise is a stochastic phase angle having an instantaneous value of

$$\Phi_n = \arctan\frac{n_q}{n_i} \tag{6.10}$$

Since the noise components are independent, the probability density of the joint event becomes the product of the individual density functions (cf. eq. (6.7))

$$p_{N_iN_q}(n_i, n_q) = \frac{1}{2\pi N_R}\exp\left[-\overbrace{(n_i^2 + n_q^2)}^{A_n^2}/2N_R\right] \tag{6.11}$$

To study the envelope of the noise, a transformation from linear to polar coordinates is carried out. This gives

$$P_{N_iN_q}(\Omega) = \frac{1}{2\pi N_R}\iint_\Omega \exp(-(x^2 + y^2)/2N_R)dx\,dy$$

$$= \frac{1}{2\pi N_R}\int_0^{A_n}\int_0^\phi \exp(-r^2/2N_R)r\,dr\,d\theta$$

$$= P_{A_n,\Phi_n}(A_n,\phi) \tag{6.12}$$

where the substitutions $x = r\cos\theta$, $y = r\sin\theta$ and $dxdy = r\,drd\theta$ have been made. The integration area, Ω, shown graphically in Figure 6.2, is given by the values of A_n and ϕ.

Written in mathematical form, the integration area in the first quadrant is (equivalently defined for the other quadrants)

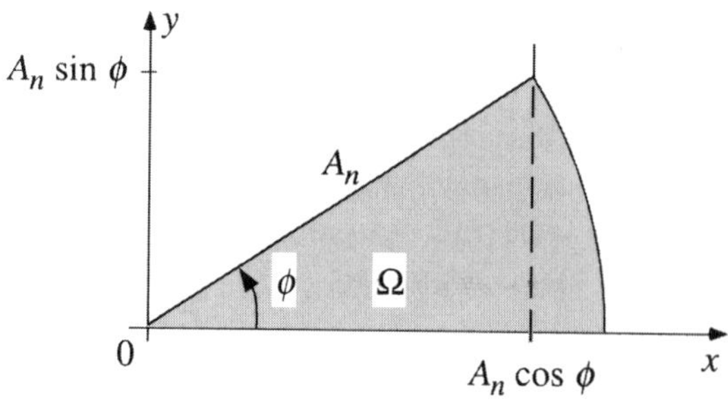

Figure 6.2 Integration area.

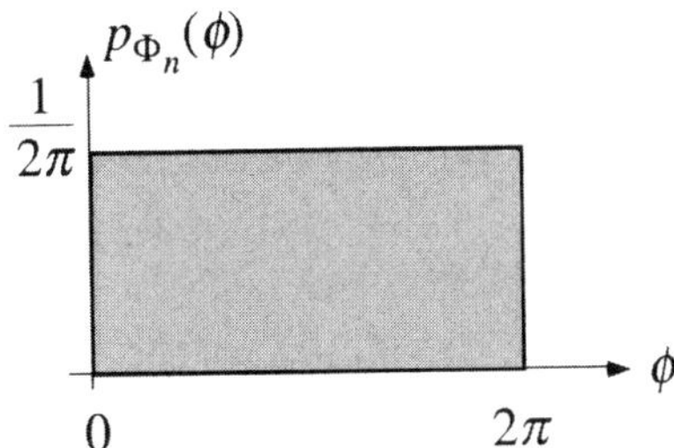

Figure 6.3   The density function of the phase for white Gaussian noise.

$$\Omega = \left\{ (x, y) \left| \begin{array}{l} (0 \le y \le x\sin\phi, \ 0 \le x \le A_n\cos\phi)\vee \\ \vee\left(0 \le y \le \sqrt{A_n^2 - x^2}, \ A_n\cos\phi < x \le A_n\right) \end{array} \right. \right\}$$

To find the density functions of the envelope and phase of the noise, respectively, the marginal distributions are calculated. This gives for the envelope

$$p_{A_n}(A_n) = \int_0^{2\pi} p_{A_n,\Phi_n}(A_n, \phi)\, d\phi = \frac{A_n}{N_R}\exp(-A_n^2/2N_R) \tag{6.13}$$

This means that the envelope has a so called Rayleigh distribution. The marginal distribution of the phase becomes

$$p_{\Phi_n}(\phi) = \int_0^{\infty} p_{A_n,\Phi_n}(A_n, \phi)dA_n = \frac{1}{2\pi} \tag{6.14}$$

for $0 \le \phi < 2\pi$. Thus the noise has a rectangular density between 0 and 2π (Figure 6.3).

In the plane spanned by the in phase and quadrature parts the noise can thus be regarded as a vector whose length is Rayleigh distributed and with all directions of equal probability.

6.2 Analogue systems

6.2.1 INFLUENCE OF NOISE AT AMPLITUDE MODULATION

In demodulation of modulated signals, the influence of the noise on the baseband signal will differ depending on which method is used. The influence of the noise on two methods for amplitude demodulation is investigated below. The first method is synchronous demodulation, or coherent demodulation. To perform coherent demodulation the receiver needs to have access to a carrier reference which is synchronous to the received carrier. Such a reference can, for instance, be generated by using a phase-locked loop, and is further discussed in Chapter 10. For the demodulation of AM with double side bands and suppressed carrier (AM-

DSB-SC) it is necessary to generate such a reference. The noise properties of coherent as well as incoherent demodulation of AM-DSB-SC and AM with carrier are analysed.

Coherent demodulation of AM-DSB-SC

Assume a modulated signal $s(t) = A_c x(t)\cos\omega_c t$. A coherent detector for this type of signal can be designed as illustrated in Figure 6.4. The received signal, $r(t)$, is multiplied by the carrier reference. The high frequency signal components are cancelled in the lowpass filter so that only a base band signal remains.

The transmission is considered narrowbanded. The noise can therefore be divided into quadrature components. The received signal becomes

$$r(t) = s(t) + n(t) = \underbrace{(A_c x(t) + n_i(t))\cos\omega_c t}_{\text{in phase part}} - n_q(t)\sin\omega_c t \qquad (6.15)$$

Multiplication by the carrier frequency gives

$$2r(t)\cos\omega_c t = A_c x(t) + n_i(t) + n_q(t)\underbrace{\sin(\omega_c - \omega_c)t}_{=0} + \text{(HFterms)} \qquad (6.16)$$

The lowpass filter at the output of the demodulator eliminates, as been pointed out before, the high frequency components (HF terms). The quadrature part of the noise is multiplied by a factor of zero and is therefore cancelled. By employing the respective parts in eq. (6.16) the signal to noise ratio, $(S/N)_D$, after the detector becomes

$$\left(\frac{S}{N}\right)_D = \frac{A_c^2 E\{x^2(t)\}}{E\{n_i^2(t)\}} = \frac{2A_c^2 E\{x^2(t)\cos^2\omega_c t\}}{E\{n^2(t)\}} = 2\left(\frac{S}{N}\right)_R \qquad (6.17)$$

To obtain the second equality in eq. (6.17), we have used the relations $E\{x^2(t)\} = 2E\{x^2(t)\cos^2\omega_c t\}$ and $E\{n_i^2(t)\} = E\{n^2(t)\}$ (cf. eq. (6.8)). The conclusion is that the signal to noise ratio at the output of a synchronous detector is twice that of the input.

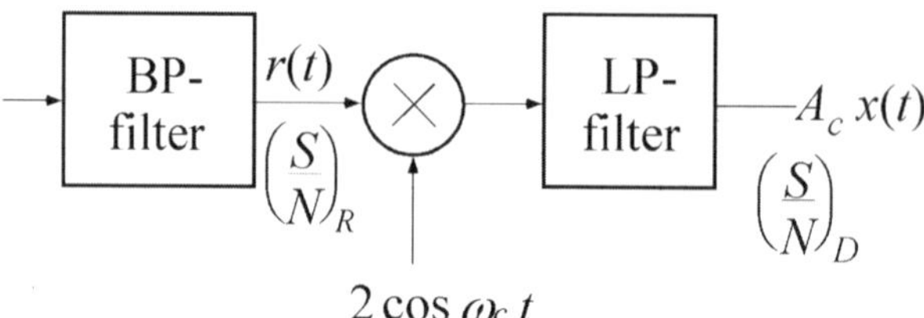

Figure 6.4 Coherent AM demodulator.

Coherent demodulation of AM with carrier

The calculation of the signal to noise ratio of AM with carrier can be achieved in the same way as for AM-DSB-SC. The received signal to noise ratio must, however, be written differently to include the carrier in the calculations. From eq. (5.45) we obtain

$$\left(\frac{S}{N}\right)_R = \frac{A_c^2 E\left\{(1 + \mu x(t))^2 \cos^2(\omega_c t)\right\}}{E\{n^2(t)\}}$$

$$= \frac{A_c^2 \frac{1}{2}\left(1 + 2\mu \overbrace{E\{x(t)\}}^{=0} + \mu^2 E\left\{x^2(t)\right\}\right)}{E\{n^2(t)\}} \tag{6.18}$$

The expectation value, i.e. the DC component, of the information signal, $x(t)$, is assumed to be zero. The unity term inside the bracket in eq. (6.18) represents no information and does not belong to the wanted signal power. In practice this term corresponds to a DC level, which can be eliminated by a highpass filter. The relation between the signal to noise ratio at the demodulator input and output can be obtained by first identifying the ratio between the wanted signal and the noise at the output, followed by a transformation where different components can be identified:

$$\left(\frac{S}{N}\right)_D = \frac{A_c^2 \mu^2 E\left\{x^2(t)\right\}}{E\{n_i^2(t)\}} = \frac{\mu^2 E\left\{x^2(t)\right\}}{1 + \mu^2 E\{x^2(t)\}} \frac{A_c^2\left(1 + \mu^2 E\left\{x^2(t)\right\}\right)}{E\{n_i^2(t)\}}$$

$$= \underbrace{\frac{\mu^2 E\left\{x^2(t)\right\}}{1 + \mu^2 E\left\{x^2(t)\right\}}}_{\eta_{AM}} 2\left(\frac{S}{N}\right)_R \tag{6.19}$$

It can be seen that the same relation for the signal to noise ratio is obtained for AM with carrier as for AM-SSB-SC apart from a factor η_{AM} denoting the modulation efficiency. Since $0 \le \mu^2 E\{x^2(t)\} \le 1$, the value of the modulation efficiency is $0 < \eta_{AM} \le 0.5$. Inserting this in eq. (6.19) we see that the penalty for having a carrier component in the modulation spectrum is a loss of at least 3 dB in the signal to noise ratio of the demodulated signal.

Non-coherent demodulation

Another method for demodulation of AM with carrier is envelope demodulation. In this case a circuit, which can track the envelope of the signal is used. Since no

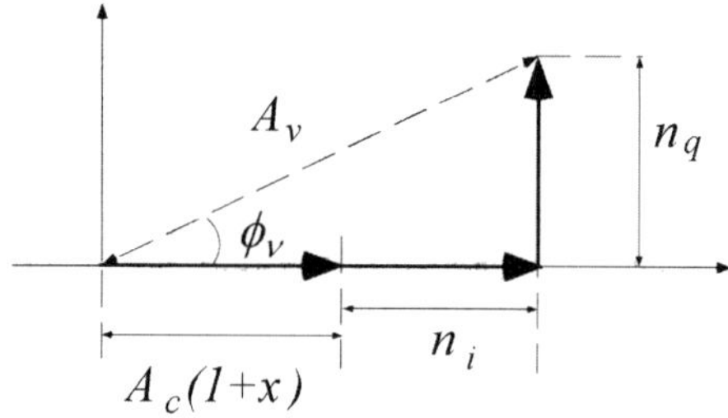

Figure 6.5 AM signal with noise separated into its quadrature components.

phase reference signal is needed in the demodulation this is called a non-coherent method. The method described below can only be used for the demodulation of AM with carrier. Assume an AM signal with modulation index μ, which gives $s(t) = A_c(1 + \mu{\cdot}x(t)\cos\omega_c t)$. Also assume that $|x(t)| \le 1$. The received signal including noise becomes

$$r(t) = A_c(1 + \mu\, x(t))\cos(\omega_c t) + \left[n_i(t)\cos\omega_c t - n_q(t)\sin\omega_c t \right] \tag{6.20}$$

Understanding is easier if a vector model is used for the signals. The signal in eq. (6.20) can then be seen as a vector sum (cf. Figure 6.5).

The envelope of the signal is the length of the resultant

$$A_v(t) = \sqrt{[A_c(1 + \mu x(t)) + n_i(t)]^2 + \left[n_q(t)\right]^2} \tag{6.21}$$

and the phase angle between carrier and resultant becomes

$$\phi_v(t) = \arctan \frac{n_q(t)}{A_c(1 + \mu x(t)) + n_i(t)} \tag{6.22}$$

To study $(S/N)_D$, the signal is separated into two intervals. In the first the signal to noise ratio is assumed to be high, i.e. $A_c^2 \gg E\{n^2\}$. The resultant is then mainly in phase with the signal. The signal can be rewritten

$$A_v(t) = A_c(1 + \mu x(t)) \sqrt{1 + \underbrace{\frac{2n_i(t)}{A_c(1 + \mu x(t))}}_{\text{small}} + \underbrace{\frac{n_i^2(t) + n_q^2(t)}{A_c^2(1 + \mu x(t))^2}}_{\text{very small put}\,=0}}$$

$$\approx A_c(1 + \mu x(t))\left(1 + \frac{n_i(t)}{A_c(1 + \mu x(t))}\right) \approx A_c(1 + \mu x(t)) + n_i(t) \tag{6.23}$$

To obtain eq. (6.23), we have used the series expansion $\sqrt{1 + x} = 1 + (1/2)x - (1/8)x^2 + (3/48)x^3 + \cdots$. The last expression in eq. (6.23) consists of a

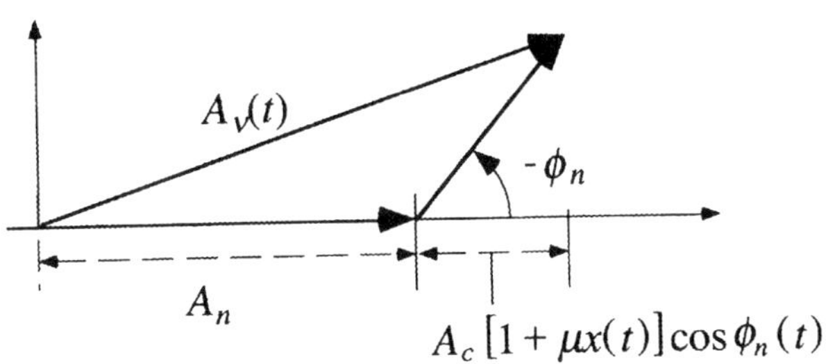

Figure 6.6 Illustration of resultant and noise being mainly in phase.

signal term and a noise term. The expectation value of the square of the two terms gives the power of the signal and noise, respectively. It can be seen that the expressions for the power of the useful signal and noise components are the same as the numerator and denominator in eq. (6.19), respectively. Provided $(S/N)_R$ is large enough, the relation between the signal to noise ratio on the input and output of the demodulator is therefore the same as for coherent detection, i.e.

$$\left(\frac{S}{N}\right)_D \approx 2\eta_{AM}\left(\frac{S}{N}\right)_R \quad \text{when } A_c^2 \gg E\{n^2\} \tag{6.24}$$

In the second interval the signal to noise ratio is assumed to be low, i.e. $A_c^2 \ll E\{n^2\}$. The resultant is then mainly in phase with the noise, as shown in Figure 6.6.

As before, the noise can be written $n(t) = A_n(t)\cos(\omega_c t + \phi_n(t))$. The wanted signal and the noise are then separated by a phase angle $-\phi_n(t)$. This gives the total signal

$$A_v(t) \approx A_n(t) + A_c[1 + \mu x(t)]\cos\phi_n(t) \tag{6.25}$$

The envelope detected signal becomes

$$A_d(t) \approx A_n(t) + A_c[1 + \mu x(t)]\cos\phi_n(t) - \overline{A_v} \tag{6.26}$$

The third term, $\overline{A_v}$, is a DC component equal to the time average of the noise envelope. From the properties of bandpass noise it is given that the phase stochastically oscillates between 0 and 2π. Thus, the average of the second term is zero. In eq. (6.26) there is no term which is proportional only to the message signal $x(t)$. This means that a threshold has been crossed and no information signal can pass. The signal to noise ratio rapidly approaches zero.

A conclusion of this analysis is that when the signal to noise ratio is high, $(S/N)_D$ is proportional to $(S/N)_R$ with a proportionality constant $2\eta_{AM}$, which holds for both coherent and incoherent AM demodulation. For the incoherent envelope demodulator there is a knee below which $(S/N)_D$ decreases rapidly and reception becomes virtually impossible. Such a knee does not exist for the coherent demo-

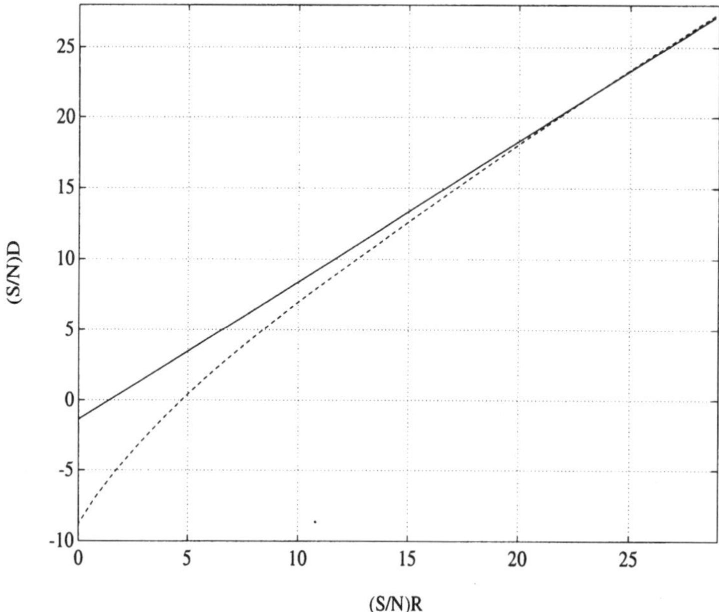

Figure 6.7 Simulated AM demodulation. The solid line corresponds to coherent demodulation and the dashed line to non-coherent AM demodulation. The baseband signal is a sinusoidal signal having unity amplitude. The modulation index is also unity. This gives $\eta_{AM} = 0.33$.

dulator, which can provide a decent output signal for much smaller values of $(S/N)_R$ provided the carrier phase can be exactly reproduced (Figure 6.7).

6.2.2 NOISE IN ANGLE MODULATION

The FM or PM modulated signal is a sinusoidal function with constant envelope but with the phase angle dependent on the transmitted information. The power in the modulated signal is therefore independent of the information signal. The power is the same as for the unmodulated carrier, $A_c^2/2$, and is denoted C. The signal to noise ratio before the demodulator (carrier-to-noise ratio) can be written in standard notation:

$$\frac{C}{N} = \left(\frac{S}{N}\right)_R = \frac{A_c^2}{2}\frac{1}{N_0 B_T} = \frac{S_R}{N_0 B_T} \tag{6.27}$$

where B_T is the noise bandwidth of the parts preceding the demodulator. The received angle modulated signal is

$$r(t) = s(t) + n(t)$$

$$= A_c \cos(\omega_c t + \phi(t)) + A_n(t)\cos(\omega_c t + \phi_n(t)) \tag{6.28}$$

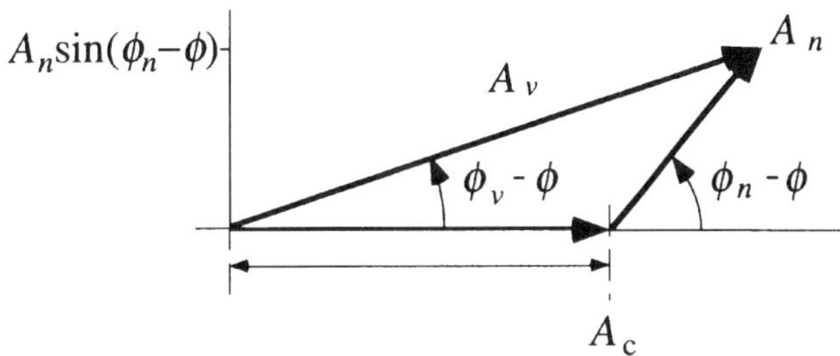

Figure 6.8 The modulated signal, as reference vector, plus a noise vector.

where the angle $\phi(t) = \phi_\Delta x(t)$ for PM and $d\phi/dt = \omega_\Delta x(t)$ for FM. The noisy signal, eq. (6.28), can be illustrated by a vector diagram as shown in Figure 6.8. The phase of the wanted signal, $\omega_c t + \phi(t)$, is used as the reference phase. The deviation of the resultant from the x-axis direction then only depends on the noise. The angle of the noise vector relative to the wanted signal becomes $\omega_c t + \phi_n(t) - \omega_c t - \phi(t) = \phi_n(t) - \phi(t)$.

By using trigonometric formulas the phase of the resultant, A_v, is obtained:

$$\phi_v(t) = \phi(t) + \arctan\left(\frac{A_n(t)\sin[\phi_n(t) - \phi(t)]}{A_c + A_n(t)\cos[\phi_n(t) - \phi(t)]}\right) \tag{6.29}$$

The amplitude is irrelevant since an ideal FM or PM detector is insensitive to the carrier amplitude. The output signal depends only on $\phi_v(t)$. To simplify the calculations of SNR, after detection it is assumed that $(S/N)_R \gg 1$, i.e. $E\{A_c^2\} \gg E\{A_n^2\}$. The phase of the resultant ϕ_v can then be simplified:

$$\phi_v(t) \approx \phi(t) + \frac{A_n(t)\sin[\phi_n(t) - \phi(t)]}{A_c} = \phi(t) + \psi(t) \tag{6.30}$$

The term $\phi(t)$ represents the wanted signal and $\psi(t)$ a phase noise. In order to determine the SNR after the demodulator, $(S/N)_D$, the dependence must be found between $\phi(t)$ and the power of the wanted signal, and between $\psi(t)$ and the noise power. Consider the phase noise

$$\psi(t) = \frac{A_n(t)\sin[\phi_n(t) - \phi(t)]}{A_c} \tag{6.31}$$

The function $\phi_n(t)$ is a stochastic process, with a rectangular distribution between 0 and 2π. Since the phase angle is only calculated modulo 2π, from a statistical point of view nothing will change if ϕ is removed from $\phi_n - \phi$. All angles will still be assumed equally often. The second moment of the stochastic part in the noise term then becomes

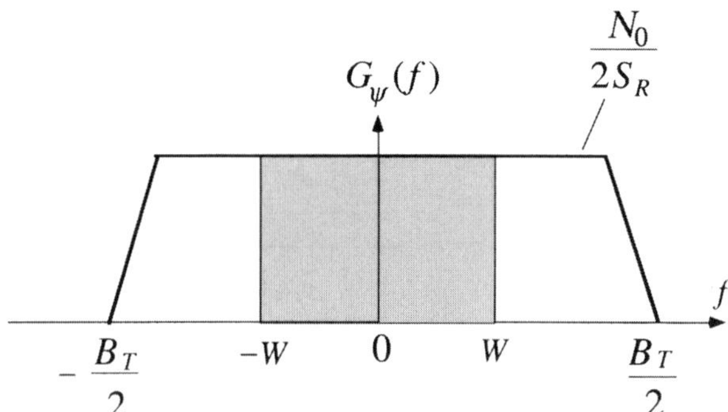

Figure 6.9 Noise spectrum before and after the demodulator. The shaded area corresponds to the noise spectrum after the demodulator.

$$E\left\{A_n^2(t)\sin^2[\phi_n(t) - \phi(t)]\right\} = E\left\{A_n^2(t)\sin^2[\phi_n(t)]\right\} = E\left\{n_q^2(t)\right\} = N_0 B_T \quad (6.32)$$

For a sinusoidal signal the relation between amplitude and power is $A_c^2 = 2S_R$. The variance of the phase noise is then

$$E\left\{\psi^2\right\} = \frac{N_0 B_T}{2S_R} \tag{6.33}$$

The bandwidth of the phase noise is B_T and is due to the filter preceding the detector. The demodulated baseband signal has a narrower bandwidth than the signal at the demodulator input. The filter after the demodulator can therefore have the bandwidth W, where $W < B_T/2$. This is illustrated in Figure 6.9.

The noise power after the PM demodulator thus becomes

$$N_D = 2W\frac{N_0}{2S_R} = \frac{N_0 W}{S_R} \tag{6.34}$$

For the FM detector, on the other hand, the output signal is proportional to the instantaneous frequency in Hz, i.e. $(1/2\pi)d\phi/dt$. The noise then becomes

$$\zeta(t) = \frac{1}{2\pi}\cdot\frac{d\psi(t)}{dt} \tag{6.35}$$

Taking the time derivative of the signal $x(t)$ in the time domain corresponds to the multiplication by $j2\pi fX(f)$ in the frequency domain. For the power spectral density this becomes $(2\pi f)^2 G_x(f)$.

The noise spectrum at the output of the FM demodulator becomes

$$G_\xi(f) = f^2 G_\psi(f) \tag{6.36}$$

which has been plotted in Figure 6.10.The noise power for FM then becomes

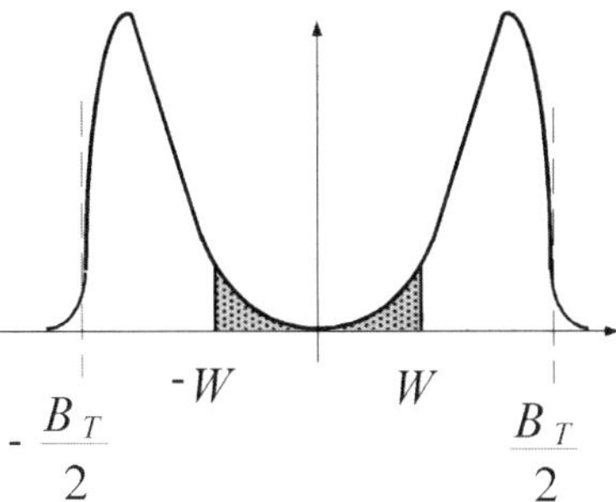

Figure 6.10   Noise spectrum after FM demodulation.

$$N_D = \int_{-W}^{W} f^2 G_\psi(f)df = \frac{N_0}{3S_R}\,W^3 \tag{6.37}$$

Pre-emphasis means that the amplitude is increased at high frequencies in the baseband signal, prior to the modulation. This requires a corresponding decrease after the demodulation, which is called de-emphasis. This technique reduces the peaks of the noise at high frequencies and thereby improves $(S/N)_D$. If the pre-emphasis and de-emphasis are achieved by a first order RC circuit, the frequency modulation at high frequencies will shift towards phase modulation, due to the derivation effect of the circuit at the transmitter side.

For the wanted signal in PM, we can write

$$\phi(t) = \phi_\Delta x(t) \tag{6.38}$$

By using eqs. (6.34) and (6.38) the signal to noise ratio after demodulation becomes

$$\left(\frac{S}{N}\right)_D = \frac{\phi_\Delta^2\, E\{x^2(t)\}}{N_0 W/S_R} = \phi_\Delta^2 S_x\, \frac{S_R}{N_0 W} \tag{6.39}$$

which gives with eq. (6.27) inserted,

$$\left(\frac{S}{N}\right)_D = \phi_\Delta^2 S_x\, \frac{B_T}{W}\left(\frac{C}{N}\right) \tag{6.40}$$

In other words, the signal to noise ratio becomes proportional to the square of the swing, ϕ_Δ^2. Increased swing leads to improved signal to noise ratio after the demodulator as compared to before the demodulator, so-called modulation gain. Since the phase is calculated modulo 2π, it is necessary for $\phi_\Delta \leq \pi$ the possible improvement has therefore an upper limit of around 10 dB.

For FM without pre-emphasis and de-emphasis the signal after the demodulator is

$$f(t) = \frac{1}{2\pi} \frac{d\phi(t)}{dt} = f_\Delta x(t) \tag{6.41}$$

By means of eqs. (6.41) and (6.37), the signal to noise ratio after the FM demodulator becomes

$$\left(\frac{S}{N}\right)_D = \frac{f_\Delta^2 S_x}{N_0 W^3/(3S_R)} = 3\left(\frac{f_\Delta}{W}\right)^2 S_x \frac{S_R}{N_0 W} \tag{6.42}$$

with eq. (6.27) inserted it is found that

$$\left(\frac{S}{N}\right)_D = 3\left(\frac{f_\Delta}{W}\right)^2 S_x \frac{B_T}{W}\left(\frac{C}{N}\right) \tag{6.43}$$

For FM, the signal to noise ratio increases as the square of f_Δ, and can in this case theoretically increase indefinitely. This opens up the possibility of exchanging bandwidth against signal to noise ratio. An approximate modulation gain versus modulation index can be calculated as follows. According to Carson's rule, the required bandwidth of the FM signal is $B_T \approx 2(f_\Delta + W)$. Inserted in eq. (6.43) this gives

$$\left(\frac{S}{N}\right)_D \approx 3\left(\frac{f_\Delta}{W}\right)^2 S_x 2\left(\frac{f_\Delta}{W} + 1\right)\frac{C}{N} \tag{6.44}$$

The concept of modulation index is related to a sinusoidal baseband signal with an amplitude of $A_m = 1$ and a frequency, W (Hz). To let this signal pass through a lowpass filter, a filter bandwidth of at least W (Hz) is required. The power of a sinusoidal signal is $S_x = A_m^2/2$, so in this case we have $S_x = 1/2$. If this is inserted in eq. (6.44) we obtain (β is defined in eq. (5.71)):

$$\left(\frac{S}{N}\right)_D = 3\beta^2(\beta + 1)\frac{C}{N} \approx 3\beta^3 \frac{C}{N} \tag{6.45}$$

The approximation is valid if $f_\Delta \gg W$.

The relations above were derived for large values of carrier-to-noise ratio. As can be seen in Figure 6.11, the signal to noise ratio of the baseband signal will be drastically reduced at a certain value. This knee can be explained by a phenomenon called FM click. For low values of S/N, i.e. $E\{A_n^2\} \gg A_c^2$, small changes in the amplitude or phase of the noise vector, A_n or ϕ_n, respectively, can cause large changes in the phase of the resultant (see Figure 6.12). The wanted signal provides the reference phase in the figure. Assume now that the noise vector A_n is added to the wanted signal at time t_1, as indicated by the black vector in the figure. A relatively small change in phase and amplitude of the noise vector to the position indicated by the shaded vector will, for example, cause the resultant, A_v, to move along the irregular curve to position 2 at time t_2. The resultant has

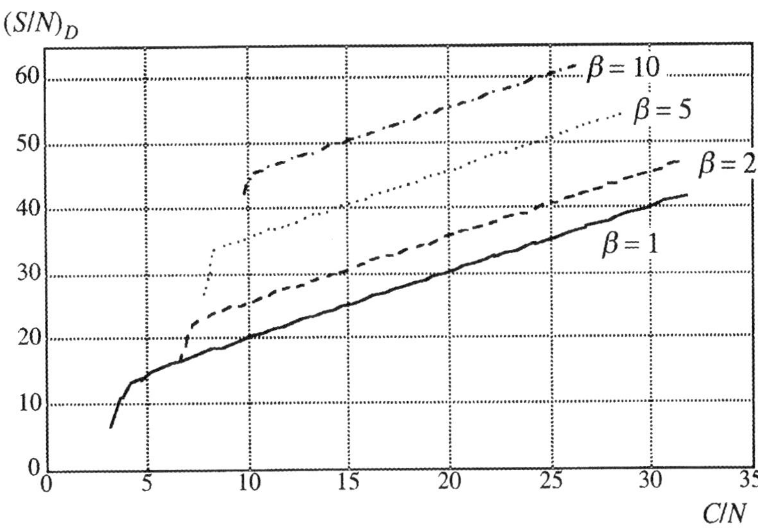

Figure 6.11 Computer simulation of the modulation gain in FM modulation. It can be seen that eq. (6.45) is a good approximation. For $\beta = 2$, for instance, $(S/N)_D = C/N + 16$ in dB, according to eq. (6.45).

thereby changed phase by 180°, and the amplitude has changed drastically compared to position 1. This is illustrated in the phase diagram in Figure 6.13.

The signal at the output of the FM detector will then take the shape of a transient called FM click, which is shown in Figure 6.14.

The probability of such a click occurring increases considerably when the carrier-to-noise ratio decreases, and this leads to a drastic reduction of the base-band signal to noise ratio. The phenomenon above causes a knee below which the modulation gain of angle modulated signals rapidly disappears. The signal to

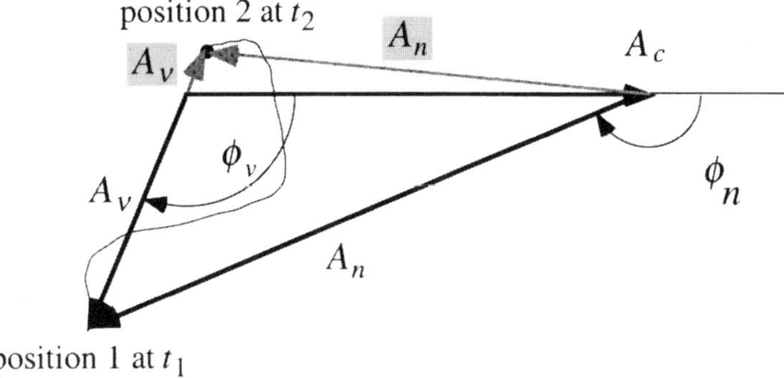

Figure 6.12 Small changes in the noise vector can cause large changes in the phase of the resultant.

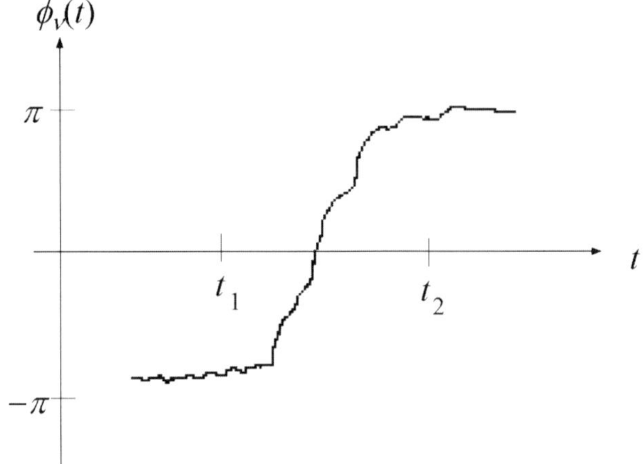

Figure 6.13   Phase versus time for the variation of the noise vector shown in Figure 6.12.

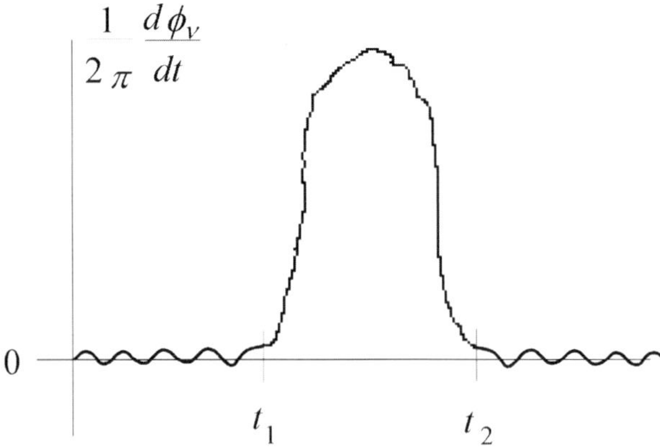

Figure 6.14   FM click.

noise ratio after the demodulator is considerably reduced and the demodulated signal becomes practically useless.

There are various electronic circuits realising the PM or FM demodulator. As the purpose of this book is not to present implementations, no realisation is described here. One way to realise a demodulator is to use a phase locked loop, which is described in Chapter 10. The control voltage to the VCO forms the demodulated signal.

6.3 Digital systems

The remaining part of this chapter is devoted to the influence of noise in digital transmission. The conditions are radically different from those of analogue systems, where the information signal sent is estimated in the receiver demodulator. When digital transmission takes place, a limited number of symbols or symbol sequences, which are known both by transmitter and receiver, are used. The task of the receiver is to decide which of the symbols was sent. The decision made by the receiver is called its hypothesis. From an analytical point of view the problem thus changes from an estimation problem to a hypothesis testing problem. Since a growing fraction of all communication is digital, the presentation in this section is more thorough than for the analogue modulation.

The queries to be answered are: given a certain S/N, how many symbol errors do we get in the transmission? How should the receiver be designed in order to minimise the error probability, i.e. how is the optimal receiver designed? Some tools from hypothesis testing in mathematical statistics are used for the analysis. The mathematical description is similar to the one used in pattern recognition.

The received signal shall be converted to a symbol stream which is as similar as possible to the symbol stream sent, i.e. the receiver shall be optimal. In the section about optimal receivers three types of detectors are presented. These are the basis for receivers in different kinds of communication systems, where the implementation is done in hardware, software or both.

6.3.1 OPTIMAL RECEIVER

A suitable strategy for the optimal receiver, given a certain received signal $y(t)$, is to select the most probable of the possible sent signals $s_i(t)$ as its hypothesis. Since the signal is directly connected to a symbol, x, by the modulation process, the hypothesis corresponds to the most probable symbol. A receiver having this strategy is called a maximum a posteriori (MAP) receiver. Mathematically the condition is written

$$\hat{s} = \arg\max_i \ P(s_i|y) \qquad (6.46)$$

The function $P(s_i|y)$ is the probability of obtaining the signal s_i, given an outcome, or received signal, y. The receiver hypothesis, $\hat{s}$, is the signal s_i, which maximises this function. In digital communication only a limited number of symbols are used. Therefore there are a limited number of sent signal shapes, s_i. The channel models used, for example the AWGN, describe y as a function of the signal s_i, which has been sent over the channel. The function in eq. (6.46) expresses it the other way around. This problem can be solved by rewriting $P(s_i|y)$ using Bayes' theorem, giving

$$P(s_i|y) = \frac{P(y|s_i)P(s_i)}{P(y)} \qquad (6.47)$$

$P(y|s_i)$ is the probability of the signal y being received, given that the signal s_i has been sent. $P(s_i)$ is the probability of the respective symbol and $P(y)$ is the probability of y being received. For an AWGN channel, for instance, y is a continuous stochastic variable, which means that $P(Y = y) = 0$. To avoid this problem a small interval around y is studied. For very small intervals, $P(y) = p_Y(y)dy$ holds with sufficiently high accuracy. The same is valid for the conditional probability, which gives $P(y|s_i) = p_Y(y|s_i)dy$. For continuous channels, eq. (6.47) can therefore be written in the form

$$P(s_i|y) = \frac{p_Y(y|s_i)dy\, P(s_i)}{p_Y(y)dy} = \frac{p_Y(y|s_i)\, P(s_i)}{p_Y(y)} \tag{6.48}$$

The reader should therefore be aware of the fact that the above adjustment must be done for continuous channels. Both $P(y)$ and $p_Y(y)$ are completely independent of the sent signal, s_i, and do not influence the hypothesis testing. Hence, they do not need to be taken into account in the continued calculations, either for discrete channels or for continuous ones.

A stochastic process is not completely characterised by the marginal density function $p(y|s_i)$. Compare with Figure 6.15 where the time dependent functions of two different implementations of a stochastic process have been plotted. The density function is in both cases $p(x) = 1/2a$ for $-a < x \leq a$.

In order to sort this out, the signal can be transformed to a number of stochastic variables, for instance by sampling the signals involved. The samples can be arranged in a column vector, giving for the received signal: $\mathbf{y} = (y_1 \cdots y_N)^T$. The elements y_n are made up from samples taken at the received signal. The total signal, usually over one symbol interval, is then expressed by using the joint density function. The task of the receiver is then to find the signal

$$\max_i p_Y(\mathbf{y}|\mathbf{s}_i)P(\mathbf{s}_i) \tag{6.49}$$

To simplify the analysis as well as the implementation of the receiver, eq. (6.49)

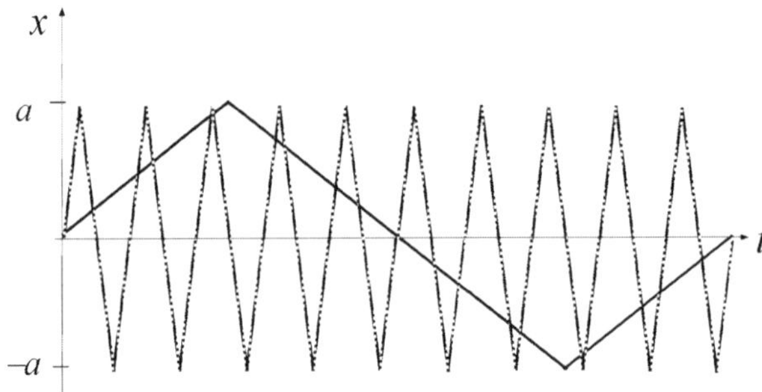

Figure 6.15   Different implementations of the same stochastic process can give signals with totally different time functions.

is transformed by using the transformation $-\ln(\cdot)$. Since ln is a monotone function we get a 1:1 mapping. This means that no new extreme values are introduced which could cause problems for the optimisation. The minus sign makes the maximisation change into minimisation:

$$\min_i \left(-\ln p_Y(\mathbf{y}|\mathbf{s}_i) - \ln P(\mathbf{s}_i) \right) \tag{6.50}$$

This expression will be used in the following discussion. First, however, another way of simplifying the problem deserves mention. When the coding has been performed in an optimised way, all M symbols are equally probable, i.e.

$$P(\mathbf{s}_i) = \frac{1}{M} \tag{6.51}$$

for all $i = 1,\ldots,M$. In this case the term is constant and independent of i, so it does not need to be included in the hypothesis evaluation. When $P(\mathbf{s}_i)$ not is a part of the hypothesis test, the maximisation in eq. (6.49) is done over $p_Y(\mathbf{y}|\mathbf{s}_i)$ only. This detection criterion gives the maximum likelihood, or ML, receiver.

Gaussian distributed interference

Assume that the sampled data in the vector $\mathbf{y}$ are identically Gaussian distributed stochastic variables. This is expressed by a multivariate normal distribution, which for N samples is

$$p_Y(\mathbf{y}|\mathbf{s}_i) = \frac{1}{(2\pi)^{N/2}\sqrt{\det(\Sigma)}} \exp\left(-\frac{1}{2}(\mathbf{y} - \mathbf{s}_i)^H \Sigma^{-1}(\mathbf{y} - \mathbf{s}_i) \right) \tag{6.52}$$

where Σ is the covariance matrix which expresses the correlation between the noise samples. The product $(\mathbf{y} - \mathbf{s}_i)^H \Sigma^{-1}(\mathbf{y} - \mathbf{s}_i)$ is called the quadratic Mahalanobis distance.

When elements outside the diagonal of the covariance matrix are non-zero, the noise samples are statistically dependent. The noise is then said to be coloured since its spectrum is not constant. The noise can then be whitened by the operation $\mathbf{y}_w = \mathbf{A}^H\mathbf{y}$. This operation makes up a linear filtering and is therefore often called a whitening filter. The filter coefficients, or elements in the matrix $\mathbf{A}$, are solutions to the equation system (the Karhunen–Loeve or Hotelling transform) $\mathbf{A}^H\Sigma\mathbf{A} = \sigma_w^2\mathbf{I}$. The equation system is always solvable due to the structure of the autocovariance matrix. For a channel with additive interference the signal becomes $\mathbf{y}_w = \mathbf{A}^H\mathbf{s}_i + \mathbf{n}_w$ after whitening. Therefore, remember that the wanted signal is now $\mathbf{A}^H\mathbf{s}_i$. After filtering of the input, the noise can always be regarded as white. In the following presentation, noise is therefore always assumed to be white.

Assume white noise, i.e. uncorrelated noise samples. The covariance matrix is then a diagonal matrix and eq. (6.52) is simplified to

$$p_Y(\mathbf{y}|\mathbf{s}_i) = \frac{1}{(\pi N_0)^{N/2}} \exp\left(-\frac{(\mathbf{y} - \mathbf{s}_i)^H(\mathbf{y} - \mathbf{s}_i)}{N_0}\right) \tag{6.53}$$

Distance measuring receiver

The argument of the exponential function in eq. (6.53) is $(\mathbf{y} - \mathbf{s}_i)^H(\mathbf{y} - \mathbf{s}_i) = \|\mathbf{y} - \mathbf{s}_i\|^2$, which is a quadratic or Euclidean norm. The Gaussian density inserted in the minimising condition (6.50) gives

$$\min_i \left(\frac{\|\mathbf{y} - \mathbf{s}_i\|^2}{N_0} - \ln(P(\mathbf{s}_i)) + \frac{N}{2}\ln(\pi N_0)\right) \tag{6.54}$$

As previously mentioned $\ln(P(\mathbf{s}_i))$ is usually constant. This term is then independent of i, and does not influence the minimising operation. For the sake of simplicity $\ln(P(\mathbf{s}_i))$ is therefore ignored in the following presentation. The term $(N/2)\ln(\pi N_0)$ is always independent of i and can also be ignored. We are then only left with the norm $\|\mathbf{y} - \mathbf{s}_i\|^2$, which we utilise as a decision variable, i.e. a variable used to establish the hypothesis of the receiver. Since i assumes M different values, there are M decision variables. A distance receiver shall thus, under the conditions given above, find $\min_i\|\mathbf{y} - \mathbf{s}_i\|^2$. This is done by computing the norm for each and every one of all $\mathbf{s}_i$. The index i, giving the lowest value of the norm, indicates the signal which constitutes the receiver hypothesis. The quadratic norm gives the Euclidean distance between the received signal $\mathbf{y}$ and the reference signals $\mathbf{s}_i$. This type of receiver, based on the quadratic norm, is therefore called a distance measuring receiver. For modulation methods without memory the length of the vectors $\mathbf{y}$ and $\mathbf{s}_i$ corresponds to one symbol interval. For each new interval a new collection of samples are taken for the vector $\mathbf{y}$. The vectors $\mathbf{s}_i$ are usually the same from interval to interval.

A continuous time signal can be seen as a signal which has been sampled with an infinitely high sampling rate. Hence the quadratic norm changes into an integral:

$$\|y(t) - s_i(t)\|^2 = \int [y(t) - s_i(t)] \cdot [y(t) - s_i(t)]^* \, dt \tag{6.55}$$

When a modulation without memory has been used the integration interval corresponds to one symbol interval, i.e. T. For modulation methods with memory the integration interval becomes longer (more about that later). A distance measuring receiver based on the calculation of the norm is shown in Figure 6.16.

The block diagram gives the reader a picture of the operations done in the distance receiver. It can also function as a starting point for an analogue realisa-

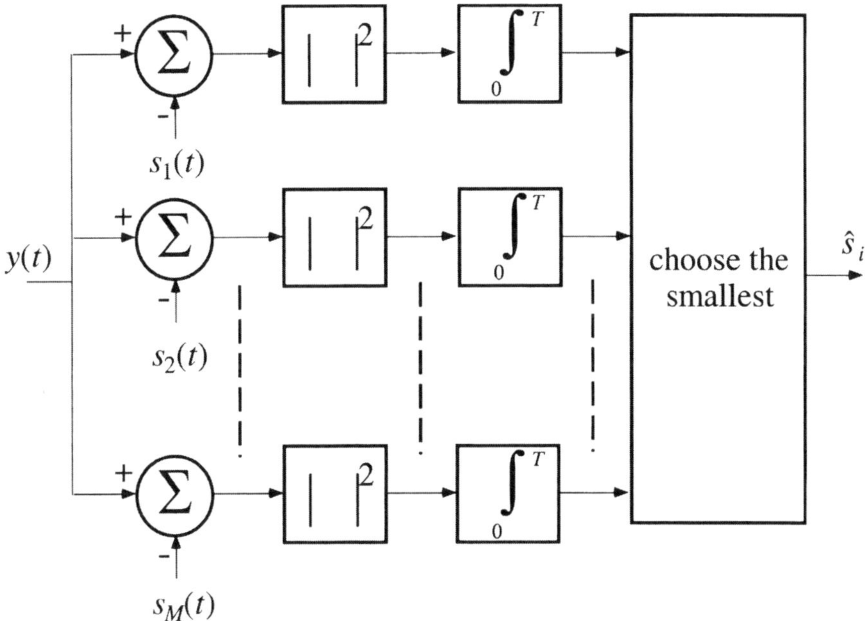

Figure 6.16 Block diagram for the distance measuring receiver.

tion. For a software realisation the mathematical expressions for the distance receiver are used, i.e. eq. (6.54) or its simplified counterpart given in the text.

Correlation receiver

Expanding the norm in eq. (6.54) gives

$$\|\mathbf{y} - \mathbf{s}_i\|^2 = \|\mathbf{y}\|^2 - 2\,\mathrm{Re}\{\langle \mathbf{y}, \mathbf{s}_i\rangle\} + \|\mathbf{s}_i\|^2 \tag{6.56}$$

The term $\|\mathbf{y}\|^2$ represents the norm of the received signal and is independent of i. Hence it does not influence the hypothesis evaluation. This term can therefore be ignored. The term $\langle \mathbf{y}, \mathbf{s}_i\rangle$ is an inner product, i.e. the vector multiplication $\mathbf{y}^H \mathbf{s}_i$. For continuous time signals the inner product is defined as

$$\langle y(t), s_i(t)\rangle = \int y(t)\cdot s_i^*(t)\, dt \tag{6.57}$$

The integration interval is the same as for the distance receiver. The second kind of receiver to be presented is based on this inner product. Since eq. (6.57) corresponds to a correlation, this type of receiver is called correlation receiver (cf. Figure 6.17). In the block diagram shown in Figure 6.17 it has been assumed that the signal energy is $\|\mathbf{s}_1\|^2 = \cdots = \|\mathbf{s}_M\|^2$. If this is not the case an additional term appears, corresponding to a parallel shift of the respective decision variable (see Example

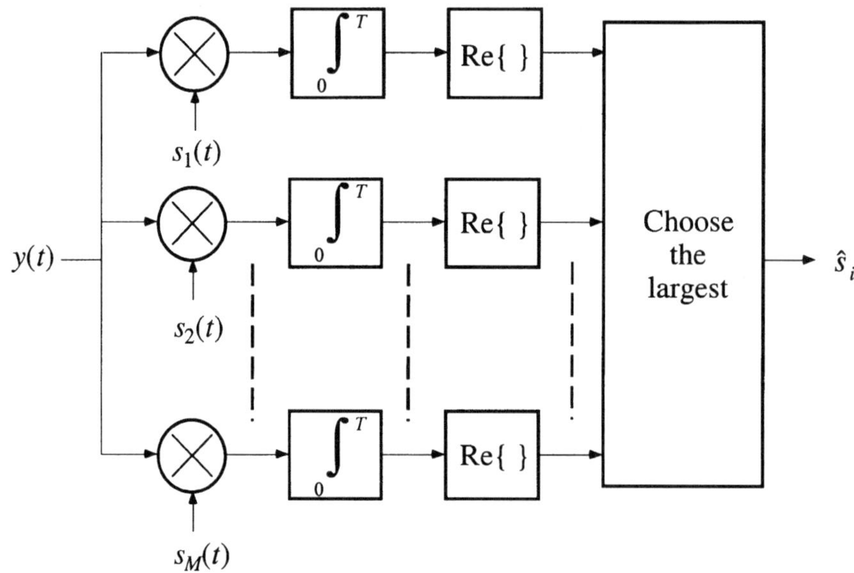

Figure 6.17   Correlation receiver.

6.3). Since the correlation is preceded by a minus sign, the correlation receiver must select the signal which maximises the correlation as its hypothesis.

Matched filter receiver

Another way of implementing the correlation receiver is obtained if the correlation operation is replaced by a filter having the impulse response $h_i(t) = s_i^*(T - t)$. Since the output signal from a linear filter is the convolution between the input signal and the impulse response, the output signal becomes

$$\int_0^t y(\tau)h_i(t - \tau)d\tau = \int_0^t y(\tau)s_i^*(T - t + \tau)d\tau \qquad (6.58)$$

The filter thereby performs the same operation as the correlation receiver. This type of receiver is called a matched filter receiver and is the third basic type of receiver.

It should be noted that according to eq. (6.58), all previous signals in the filter must have decayed; the initial state must be zero. The input signal is connected to the filter exactly at time $t = 0$ and is read at $t = T$, so the need for synchronisation does not vanish when a matched filter is used to realise a receiver.

In an analogue realisation it can be difficult to implement a filter with an impulse response corresponding to the reversed time function of the signal. The possibilities are limited to relatively simple signal shapes or an approximation of more complex signals. However, digital realisation of the matched filter

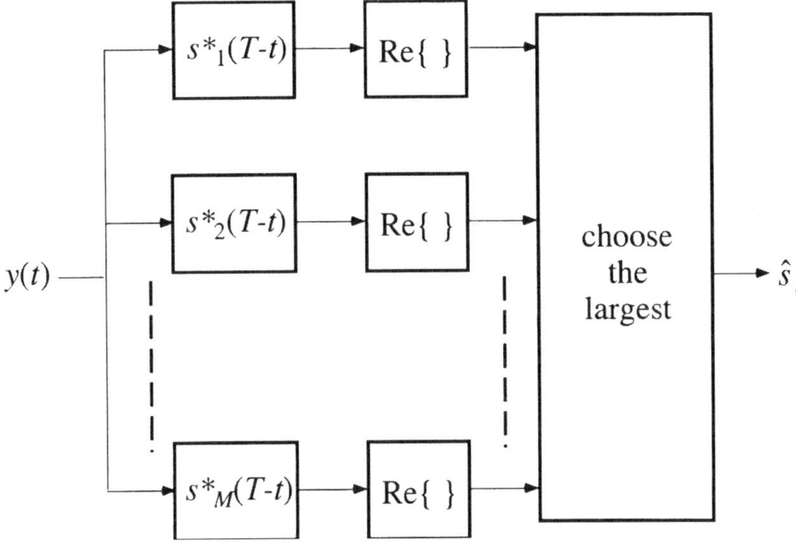

Figure 6.18 Matched filter receiver.

gives great possibilities to realise appropriate impulse response for complicated signal shapes.

The matched filter receiver instantaneously gives the optimal signal to noise ratio. Using Parseval's relation in the frequency domain, the power of the wanted signal after filtering by $H(\omega)$ is given by the nominator of eq. (6.59). The noise power after filtering by $H(\omega)$ is given by the denominator. By use of the Schwarz inequality the following can be obtained:

$$\left(\frac{S}{N}\right)_{t=T} = \frac{\left|\int_{-\infty}^{\infty} S(\omega)\, H(\omega)\, \exp(j\omega T)\, d\omega\right|^2}{\pi N_0 \int_{-\infty}^{\infty} |H(\omega)|^2 d\omega}$$

$$\leq \frac{\int_{-\infty}^{\infty} |S(\omega)|^2 d\omega \, \int_{-\infty}^{\infty} |H(\omega)|^2 d\omega}{\pi N_0 \int_{-\infty}^{\infty} |H(\omega)|^2 d\omega} \tag{6.59}$$

where the equality only holds if $H(\omega) = S^*(\omega)\exp(-j\omega T)$ which gives $s^*(T - t)$ after inverse Fourier transformation. After reducing and using Parseval's relation the maximum signal to noise ratio becomes

$$\frac{\int_{-\infty}^{\infty} |S(\omega)|^2 d\omega}{\pi N_0} = \frac{\int_{-\infty}^{\infty} |s(t)|^2 dt}{N_0/2} = \frac{E}{N_0/2} = \frac{2E}{N_0} \tag{6.60}$$

Compare this result with eq. (2.29).

Sequential decoding

For modulation methods without memory the length of the vectors $\mathbf{y}$ and $\mathbf{s}_i$, corresponds to one symbol interval, as has been mentioned previously. In the continuous time case the signal is similarly integrated over one symbol. It is therefore sufficient to consider one symbol at a time, in the detection of the symbol sequence. The reason for this is that the successive modulating symbols are statistically independent. It would not, therefore, give more information to consider a longer interval. For methods with memory the successive signals are not independent and a gain can be done by considering several symbol intervals when a decision should be taken. The dependence is described by a trellis (see e.g. Figure 5.48, valid for MSK). A number of possible paths, consisting of different state sequences, run through the trellis. The state transitions at each symbol interval correspond to a given signal shape, $s_i(t)$. There are S_f different states in the trellis; for the definition cf. eq. (5.98). Let k be a time index, the states are denoted by the integers σ_k, in the interval $1 \leq \sigma_k \leq S_f$. A path that is travelled during the time 0 to k is described by the vector $\Theta_k = \{\sigma_0, \sigma_1, \ldots, \sigma_k\}$. For every sequence of length k there are $S_f M^k$ different signal paths. The set of these signal paths is called B. For a maximum likelihood estimation of dependent symbols, eq. (6.54) is computed over an infinite signal interval. From eqs. (6.54) and (6.55) it is given that for continuous time signals with equal probability the estimated state sequence, $\{\hat{\sigma}_k\}_{k=-\infty}^{\infty}$, is obtained from

$$\{\hat{\sigma}_k\}_{k=-\infty}^{\infty} = \underset{\Theta \in B}{\arg \min} \int_{-\infty}^{\infty} |y(t) - s_\Theta(t)|^2 \, dt \tag{6.61}$$

for every possible path through the trellis, B. The computation of eq. (6.61) can be done sequentially for each symbol interval and be given a recursive formulation. The estimation of the state sequence is simplified by means of the Viterbi algorithm. By cancelling impossible paths in the way described in Chapter 4, one single path through the trellis is finally found. The method is identical to the one used for convolutional codes, except that, instead of using the Hamming distance as a metric the Euclidean distance metric is used. This gives

$$L_{k+1} = \int_{-\infty}^{(k+1)T} \left| y(t) - s_{\Theta_{k+1}}(t) \right|^2 dt$$

$$= \underbrace{\int_{-\infty}^{kT} \left| y(t) - s_{\Theta_k}(t) \right|^2 dt}_{L_k} + \underbrace{\int_{kT}^{(k+1)T} |y(t) - s_i(t)|^2 dt}_{d_{k+1}^2}$$

$$= L_k + d_{k+1}^2 \tag{6.62}$$

When several paths are joined at one state, at time k, the paths having a larger metric, L_k, are cancelled. To every state, only one definite path runs $\Theta_k = \{\sigma_0, \sigma_1, \ldots, \sigma_k\}$. Instead of making a decision after each symbol, the value of the metric, L_k, is saved together with the path, Θ_k. When $k \rightarrow \infty$ the least metric value corresponds to the ML estimate in eq. (6.61). In practice, this would mean an unacceptable delay in the system. Therefore, a decision is made already when a path has diverted from the rest by having a radically smaller metric. Then, with a large probability this path is the correct ML decision. Since a particular signal sequence, or chain of state transitions, corresponds to a certain sequence of symbols, a maximum likelihood estimate of the symbols is obtained. This is called maximum likelihood sequence estimation (MLSE). For the correlation receiver a similar expansion can be done:

$$
\begin{aligned}
J_{k+1} &= \mathrm{Re}\left\{ \int_{-\infty}^{(k+1)T} y(t) s^*_{\Theta_{k+1}}(t)\, dt \right\} \\[2mm]
&= \underbrace{\mathrm{Re}\left\{ \int_{-\infty}^{kT} y(t) s^*_{\Theta_k}(t)\, dt \right\}}_{J_k} + \underbrace{\mathrm{Re}\left\{ \int_{kT}^{(k+1)T} y(t) s^*_i(t)\, dt \right\}}_{\rho_{k+1}} \\[2mm]
&= J_k + \rho_{k+1}
\end{aligned}
\tag{6.63}
$$

The decision block, to the right in Figures 6.16–6.18 contain, for MLSE, in addition a Viterbi processor which keeps track of the estimated signal path through the state trellis.

One application, except detection of modulation with memory, is detection of signals exposed to intersymbol interference. When the signal arrives from several paths with different delay the symbols from different time instants add together, giving intersymbol interference. In the same way as for convolutional codes and CPM signals the signal in this case can be described by a trellis. A detector based on MLSE gives the optimal estimation of signals with intersymbol interference, given that the multi-path has been accurately estimated.

6.3.2 SIGNAL SPACE

Sampling is one way of transferring a signal in continuous time, to a number of stochastic variables, as has been described before. Another way is to represent the signal by orthonormalised basis functions. This also provides a good model for the understanding of error probability calculations.

A signal can be written as the sum of a limited number of basis functions, $\varphi_k(t)$:

$$
s_i(t) = \alpha_{i1}\varphi_1(t) + \cdots + \alpha_{iK}\varphi_K(t) = \sum_{k=1}^{K} \alpha_{ik}\varphi_k(t)
\tag{6.64}
$$

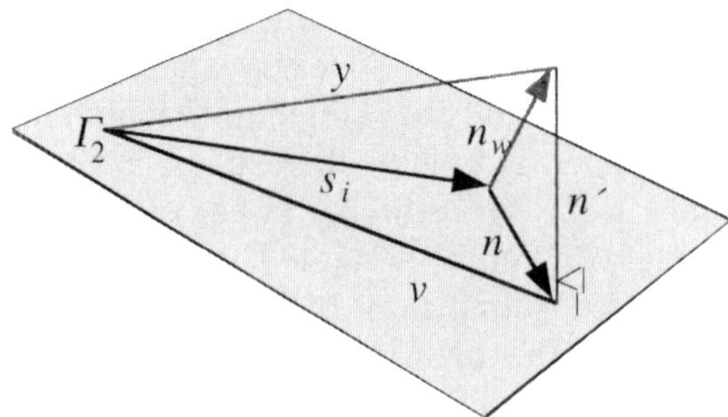

Figure 6.19 Example of a signal space in two dimensions. The space is then a plane and part of the noise cannot be expressed in this plane. Vectors in the plane are black and vectors outside the plane are shaded.

The objective is to achieve an exact representation of the signal using a limited value of K. The set of basis functions is then said to be complete. The coefficients are obtained through

$$\alpha_{ik} = \int_{-\infty}^{\infty} s_i(t)\varphi_k^*(t)dt = \underbrace{\langle s_i, \varphi_k \rangle}_{\text{inner product}} \tag{6.65}$$

The value of the coefficient α_{ik} is therefore determined by the projection of the signal s_i on the basis function φ_k. The basis functions can, for instance, be obtained by Gram–Schmidt orthogonalisation. The signal s_i can then be represented in a space, Γ_K, which is spanned by the basis functions. This space is called the signal subspace. The values of the coefficients α_{ik} give the signal location in that space. Then, the signal can be expressed by a vector notation, $\mathbf{s}_i = [\alpha_{i1}\,\alpha_{i2}\,\ldots\,\alpha_{iK}]$. The input signal at the receiver is $y = s_i + n_w$. The noise n_w has an arbitrary pattern and can therefore usually not be represented only by the basis vectors of the signal space. The noise is, in other words, not always inside the signal space (Figure 6.19).

The noise vector is then separated into two components; one part, n, within the signal subspace and one part, n', which cannot be expressed inside the signal subspace, and is thus orthogonal to it. The component n' is in other words a vector in the orthogonal subspace. The corresponding series expansion of the noise gives the coefficients

$$\beta_k = \langle n, \varphi_k \rangle = \langle n_w - n', \varphi_k \rangle = \langle n_w, \varphi_k \rangle \tag{6.66}$$

The last equality holds as n' is orthogonal to all φ_k. The series becomes

$$n_w = n + n' = \sum_{k=1}^{K} \beta_k \varphi_k + n' \tag{6.67}$$

The series contains different terms where n is the projection of n_w on the signal subspace. This is the part of the noise which influences the detection, since the orthogonal component cannot be seen in the signal subspace. The component of the input signal, y, which is inside and can therefore be observed in the subspace, is denoted v. It is written

$$v = s_i + n \tag{6.68}$$

Provided the average value of the noise is zero, the expectation value of the coefficients β_k becomes

$$E\{\beta_k\} = 0 \tag{6.69}$$

The variance becomes

$$E\{\beta_k\beta_j\} = E\left\{\int_{-\infty}^{\infty} n_w(t)\varphi_k(t)dt \int_{-\infty}^{\infty} n_w(\lambda)\varphi_j(\lambda)d\lambda\right\}$$

$$= \int_{-\infty}^{\infty}\int_{-\infty}^{\infty} \underbrace{E\{n_w(t)n_w(\lambda)\}}_{\text{white noise} \Rightarrow \frac{N_0}{2}\delta(t-\lambda)} \varphi_k(t)\varphi_j(\lambda)dtd\lambda$$

$$= \frac{N_0}{2}\int_{-\infty}^{\infty}\int_{-\infty}^{\infty} \delta(t-\lambda)\varphi_k(t)\varphi_j(\lambda)dtd\lambda \tag{6.70}$$

Given that the noise is white and because φ_k and φ_j are orthogonal for $k \neq j$, after integration we are left with

$$E\{\beta_k\beta_j\} = \begin{cases} N_0/2 & j = k \\ 0 & j \neq k \end{cases} \tag{6.71}$$

From eq. (6.71) it can be seen that the variance is the same for all j. Since the operations above are linear, all β_k will be Gaussian distributed stochastic variables if n_w is a Gaussian process. This gives for the probability density of the coefficient values

$$p_{\beta_k}(\beta) = \frac{1}{\sqrt{\pi N_0}}\exp\left(-\frac{\beta^2}{N_0}\right) \tag{6.72}$$

As the coefficients β_k have been obtained through inner products by orthogonal functions, they will be mutually independent stochastic variables. The joint probability density of the noise is then the product of the individual probabilities

$$p_n(n) = p_n(\beta_1, \ldots, \beta_k) = \frac{1}{(\pi N_0)^{K/2}}\exp\left(-\frac{\sum \beta_k^2}{N_0}\right) = \frac{1}{(\pi N_0)^{K/2}}\exp\left(-\frac{||n||^2}{N_0}\right)$$

$$\tag{6.73}$$

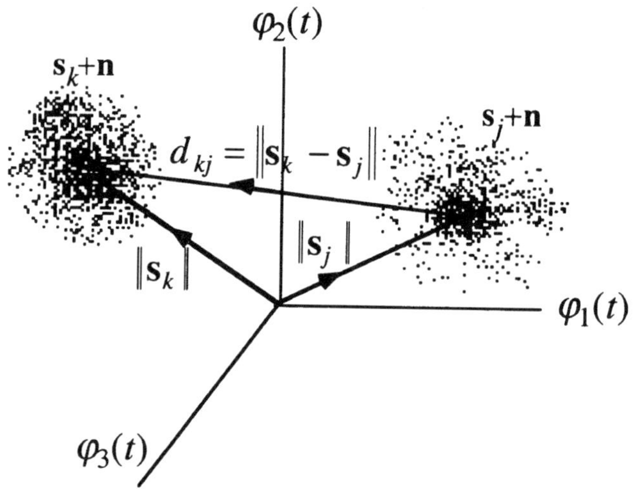

Figure 6.20   Example of a signal subspace with noise.

A constant value of the density $p_n(n)$ then gives a hypersphere in the space spanned by the basis functions. This leads to, for instance, noise with rotational symmetry. Given the signal s_i the conditional probability density becomes

$$p(v|s_i) = \frac{1}{(\pi N_0)^{K/2}} \exp\left(-\frac{\|v - s_i\|^2}{N_0}\right)$$

(6.74)

This is depicted in Figure 6.20.

Compare the expressions for $p(y|s_i)$ in eq. (6.53) and $p(v|s_i)$ in eq. (6.74). The part of the noise not found in v is orthogonal to s_i and does not influence the detection process which means that these two expressions are equivalent models for the same matter.

Receiver based on orthogonal basis functions

The separation of the received signal into orthogonal basis functions can also be the basis of a receiver. In the same way as before, the ML receiver should minimise $\|v - s_i\|^2$ and a receiver structure as in Figure 6.21 is obtained.

The parameter K, i.e. the number of required basis functions, is smaller than the number of possible characters M, so this receiver may provide advantages for the implementation, but is nevertheless not frequently used. In the same way as before, the distance measure can be expanded and this gives $\|v - s_i\|^2 = \|v\|^2 - 2\text{Re}\{\langle v, s_i \rangle\} + \|s_i\|^2$. The first term on the right hand side is independent of s_i. The middle term corresponds to correlation with the input signal

$$\langle v, s_i \rangle = \langle y - n', s_i \rangle = \langle y, s_i \rangle$$

(6.75)

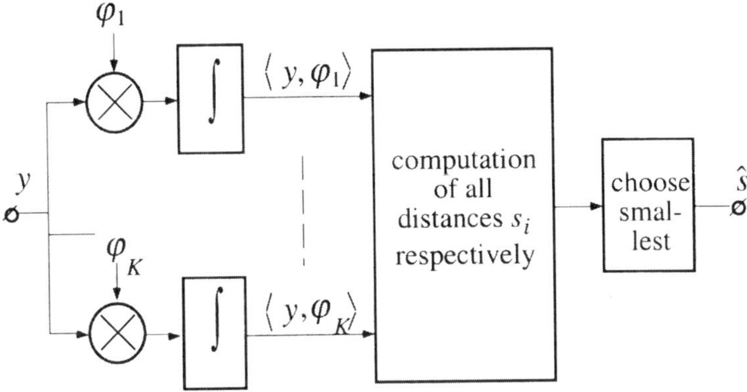

Figure 6.21 Receiver based on orthogonal basis functions.

since $\langle n', s_i \rangle = 0$. This results in the same correlation type of receiver as was derived by another method in eq. (6.56). The hypothesis of the receiver is the reference signal, s_i, which has the shortest distance to the received signal. We now define $\gamma_k = \langle y, \varphi_k \rangle$, which together with eq. (6.64) and the linearity of the series expansion give that the difference between the received signal and s_i can be written

$$\sum_{k=1}^{K} \gamma_k \varphi_k - \sum_{k=1}^{K} \alpha_{ik} \varphi_k = \sum_{k=1}^{K} (\gamma_k - \alpha_{ik}) \varphi_k \tag{6.76}$$

Since the received signal and the reference signals are expressed in the same coordinate system, the distance can be determined by considering the coordinates in the common system. Since the basis functions are orthonormalised we get

$$d_{y,s_i} = \sum_{k=1}^{K} |\gamma_k - \alpha_{ik}|^2 \tag{6.77}$$

which make up the decision variables that are the basis for the receiver decision.

Signal space representation of modulated signals

Signal space representation is, for instance, useful for illustrating different types of modulation. Assume the basis functions

$$\varphi_1(t) = \sqrt{\frac{2}{D}} \cos \omega_c t \tag{6.78}$$

and

$$\varphi_2(t) = \sqrt{\frac{2}{D}} \sin \omega_c t \tag{6.79}$$

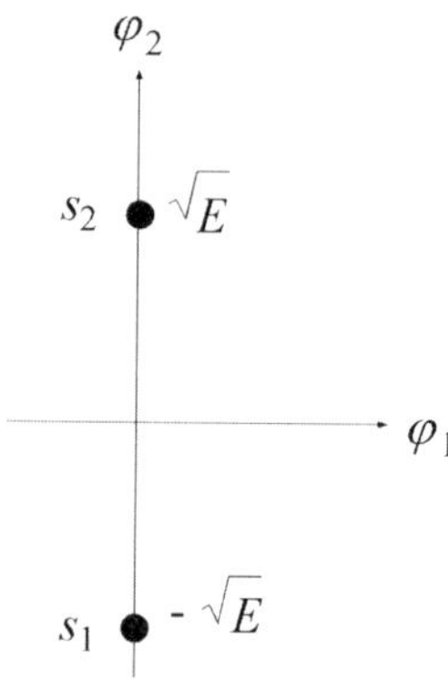

Figure 6.22　Signal space representation of antipodal BPSK.

The basis functions happen to coincide with the signals on the I and Q channels. These basis functions have been chosen because they are suitable for the signals studied below. Of course other sets of orthogonal functions could also have been used. For BPSK, with the carrier in phase against the sinusoidal signal, the following relation holds:

$$s_{1,2} = \pm\sqrt{\frac{2E}{D}}\sin\omega_c t \tag{6.80}$$

The representation in the signal space is shown in Figure 6.22.

The parameter E is the signal energy over one symbol interval. Only one basis vector is really needed here, i.e. $K = 1$. The modulation can then be regarded as one-dimensional. A signal space representation of binary orthogonal signalling is shown in Figure 6.23.

It should be pointed out that the signal locations shown above represent different types of BPSK and hold for the particular choice of basis functions made here. For other choices of basis functions the signal points can be positioned in the same way as above, although other types of modulation are represented. The

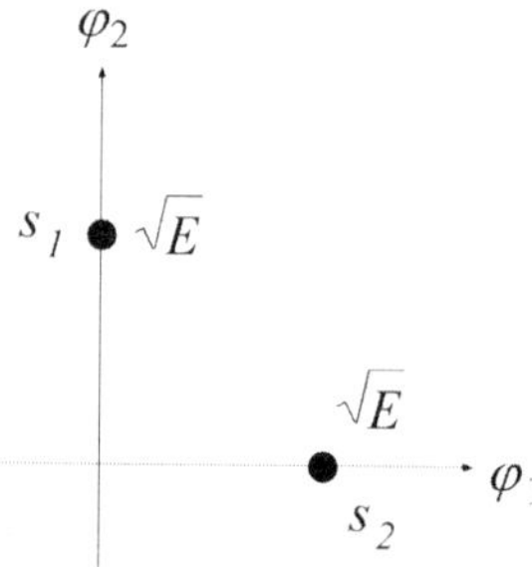

Figure 6.23　Orthogonal BPSK in signal space.

advantage with this is that the methods for the calculation of error probabilities, discussed in the following section, are generally useful for different modulation types. Another example of basis functions, which are suitable for other situations, are the ones often used for orthogonal FSK: $\varphi_k(t) = \sqrt{2/T}\cos([\omega_c + k\cdot\pi/(2T)]t)$. In this case the signal set makes up the basis vectors.

Example 6.1

Use the Gram–Schmidt method to find an orthogonal base for the signals

$$s_1(t) = 1 \quad s_2(t) = t \quad s_3(t) = t^2 \tag{6.81}$$

defined over $-1 \leq t \leq 1$. Then write the signals using the basis vectors.

Solution Gram–Schmidt gives a set of orthogonal basis functions. For a more detailed description of the method, refer to suitable literature in mathematics. The expansion is not unambiguous since it depends on the choice of basis functions. One of the signals can be used as the first basis function. Use, for instance, s_1 as the basis function φ_1, which gives

$$\varphi_1 = \frac{s_1}{\|s_1\|} = \frac{1}{\sqrt{2}} \tag{6.82}$$

The norm is defined as

$$\|s_n\| = \sqrt{\int_a^b |s_n(t)|^2 dt} \tag{6.83}$$

where $[a, b]$ is the signal definition interval or one period. In this case we then have

$$\|s_1\| = \sqrt{\int_{-1}^1 1\, dt} = \sqrt{2} \tag{6.84}$$

Select the next signal

$$\varphi_2 = \frac{s_2 - \alpha_{21}\varphi_1}{\|s_2 - \alpha_{21}\varphi_1\|} = \sqrt{\frac{3}{2}}\cdot t \tag{6.85}$$

where the inner product:

$$\alpha_{21} = \langle s_2, \varphi_1 \rangle = \int_a^b s_2\varphi_1^* dt = \int_{-1}^1 t\cdot\frac{1}{\sqrt{2}} = 0 \tag{6.86}$$

$$\|s_2\| = \sqrt{\frac{2}{3}} \tag{6.87}$$

The remaining signal gives the basis function

$$\varphi_3 = \frac{s_3 - \alpha_{31}\varphi_1 - \alpha_{32}\varphi_2}{\|s_3 - \alpha_{31}\varphi_1 - \alpha_{32}\varphi_2\|} = \frac{t^2 - \dfrac{\sqrt{2}}{3}\dfrac{1}{\sqrt{2}}}{\sqrt{\dfrac{8}{45}}} = \sqrt{\frac{45}{8}}\left(t^2 - \frac{1}{3}\right) \tag{6.88}$$

where

$$\alpha_{31} = \frac{\sqrt{2}}{3}, \quad \alpha_{32} = 0 \quad \Rightarrow$$
$$\|s_3 - \alpha_{31}\varphi_1 - \alpha_{32}\varphi_2\|^2 = \int_{-1}^{1}\left(t^2 - \frac{1}{3}\right)^2 dt = \frac{8}{45} \tag{6.89}$$

which gives the basis functions

$$\varphi_1 = \frac{1}{\sqrt{2}}, \quad \varphi_2 = \sqrt{\frac{3}{2}}\,t \quad \text{and} \quad \varphi_3 = \sqrt{\frac{45}{8}}\left(t^2 - \frac{1}{3}\right) \tag{6.90}$$

The signals expressed as a series by using the basis functions become

$$s_n(x) = \sum_{\text{all } i} \alpha_{ni}\varphi_i(x) \tag{6.91}$$

and the coefficients

$$\alpha_{ni} = \langle s_n(x), \varphi_i(x)\rangle \tag{6.92}$$

If we start from the end, the signals above then become for s_3

$$s_3 = \alpha_{31}\varphi_1 + \alpha_{32}\varphi_2 + \alpha_{33}\varphi_3 \tag{6.93}$$

$$\alpha_{31} = \int_{-1}^{1} t^2 \frac{1}{\sqrt{2}}\, dt = \frac{\sqrt{2}}{3} \tag{6.94}$$

$$\alpha_{32} = 0 \tag{6.95}$$

$$\alpha_{33} = \sqrt{\frac{45}{8}}\int_{-1}^{1} t^2\left(t^2 - \frac{1}{3}\right) dt = \sqrt{\frac{8}{45}} \tag{6.96}$$

thus

$$s_3 = \frac{\sqrt{2}}{3}\varphi_1 + \sqrt{\frac{8}{45}}\varphi_3 \tag{6.97}$$

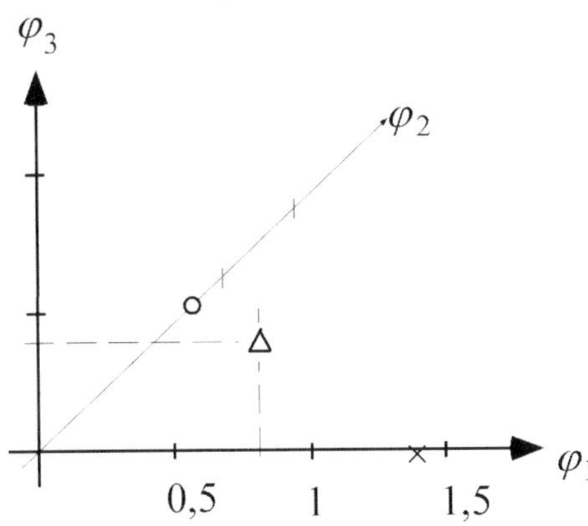

Figure 6.24   Positions of the signals in the signal space, $s_1 = \times$, $s_2 = \bigcirc$ and $s_3 = \triangle$.

and

$$s_2 = \sqrt{\frac{2}{3}}\,\varphi_2 \tag{6.98}$$

$$s_1 = \sqrt{2}\,\varphi_1 \tag{6.99}$$

The signal space drawn in a diagram is shown in Figure 6.24.

6.3.3 CALCULATION OF BIT ERROR RATE

One important measure of quality for a digital communication system is the risk of interpreting the received signal as other than the one that has been sent. Then, the detected symbol does not coincide with the modulating symbol and an error has occurred in the transmission. This happens when the decision variable is smallest for a signal other than the one that really has been sent, which can occur as a result of the noise in the channel. According to eq. (6.50) the symbol i is chosen if the received signal $\mathbf{y}$ lies within the area in the signal subspace where

$$\left(-\ln p_Y(\mathbf{y}|\mathbf{s}_i) - \ln P(\mathbf{s}_i)\right) < \left(-\ln p_Y(\mathbf{y}|\mathbf{s}_j) - \ln P(\mathbf{s}_j)\right) \tag{6.100}$$

for all $j \neq i$. To compute the probability for the receiver to choose the wrong symbol, the probability density function of the received signal is integrated over the area where an erroneous signal alternative is chosen according to the decision rule of the detector. By using the fact that the integral over the entire sample space is one and given that the signal $\mathbf{s}_i$ has been sent, the error probability is

$$P(e|\mathbf{s}_i) = 1 - \int_{\Omega_i} p_Y(\boldsymbol{\psi}|\mathbf{s}_i)d\boldsymbol{\psi} \qquad (6.101)$$

Ω_i is the area, in the signal space, where the correct signal is chosen. Given that $\mathbf{s}_i$ is sent the correct signal alternative will be chosen whenever the statistic in eq. (6.50) is smaller for $\mathbf{s}_i$ than for any other alternative $\mathbf{s}_j$, $i \neq j$. The area of integration therefore becomes

$$\Omega_i = \left\{ \boldsymbol{\psi} | (-\ln p_Y(\boldsymbol{\psi}|\mathbf{s}_i) - \ln P(\mathbf{s}_i)) < (-\ln p_Y(\boldsymbol{\psi}|\mathbf{s}_j) - \ln P(\mathbf{s}_j)), \ \forall j \neq i \right\} \quad (6.102)$$

The boundary of the area where the signal $\mathbf{s}_i$ is chosen in favour of some other one is the limit of the integration area Ω_i. How the error probability is determined for two basic cases is described below.

Error probability of one-dimensional signalling

In this section calculations of the bit error rate in Gaussian noise are discussed given an optimum receiver. For a one-dimensional modulation the probability density of the received baseband signal or decision variable will consist of Gaussian-shaped bells centred around s_1–s_M. The numerical values correspond to the square root of the energies of the respective signal alternative. An example is antipodal BPSK (cf. Figure 6.25).

For a one-dimensional signal with Gaussian interference the decision rule is, according to eq. (6.50): choose s_1 if

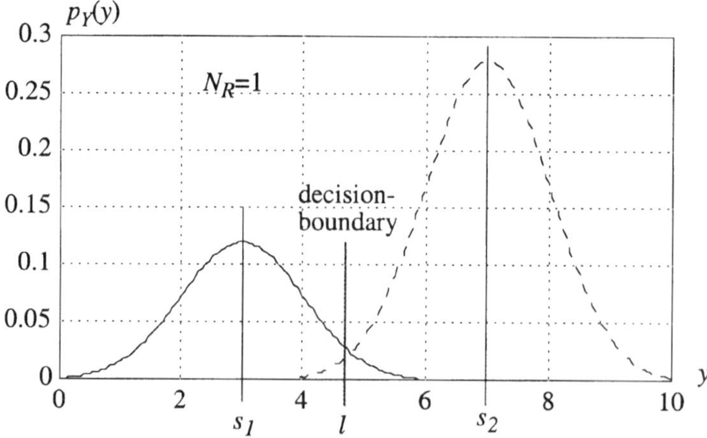

Figure 6.25   The received signal probability density given two alternative signals, s_1, s_2 and Gaussian interference. Due to different probabilities of the signals, the decision boundary is shifted a bit to the left of the point where the lines intersect.

$$\frac{(y - s_1)^2}{N_0} - \ln(P(s_1)) < \frac{(y - s_2)^2}{N_0} - \ln(P(s_2)) \tag{6.103}$$

otherwise chose s_2. The edge of the area is the decision boundary l. The edge is the solution to the equation

$$\frac{(l - s_1)^2}{N_0} - \ln(P(s_1)) = \frac{(l - s_2)^2}{N_0} - \ln(P(s_2)) \tag{6.104}$$

which is

$$l = \frac{N_0 \ln\left(\frac{P(s_1)}{P(s_2)}\right)}{2(s_2 - s_1)} + \frac{s_1 + s_2}{2} \tag{6.105}$$

The probability of making an error, i.e. of interpreting the received signal as s_m when s_n has been sent out, is denoted $P(s_m|s_n)$, where $m \neq n$. Given the sent signal s_1, the error probability is obtained by integrating the tail of eq. (6.74), between the decision boundary l to plus infinity. Since the signal is one-dimensional $K = 1$ holds. For a two signal case the probability of making an error is

$$P_e = P(s_1)P(s_2|s_1) + P(s_2)P(s_1|s_2)$$

$$= \frac{P(s_1)}{\sqrt{\pi N_0}} \int_l^{\infty} \exp\left(-\frac{(v - s_1)^2}{N_0}\right) dv + P(s_2) \underbrace{\frac{1}{\sqrt{\pi N_0}} \int_{-\infty}^{l} \exp\left(-\frac{(v - s_2)^2}{N_0}\right) dv}_{1 - \int_l^{\infty} \ldots dv}$$

$$= P(s_2) + \frac{1}{\sqrt{\pi N_0}} \int_l^{\infty} \left(P(s_1)\exp\left(-\frac{(v - s_1)^2}{N_0}\right) - P(s_2)\exp\left(-\frac{(v - s_2)^2}{N_0}\right)\right) dv \tag{6.106}$$

When both signals have equal probability, i.e. when $P(s_1) = P(s_2) = 1/2$, the decision boundary is obtained from eq. (6.105):

$$l = \frac{s_1 + s_2}{2} \tag{6.107}$$

This means that the decision boundary should be where the curves of the respective PDF intercept. This can easily be realised if it is remembered that the decision rule in this case is that s_1 is chosen if $(y - s_1)^2 < (y - s_2)^2$. The boundary between the areas is the given by eq. (6.107).

The error probability for equally probable symbols becomes

$$P_e = \frac{1}{2} + \frac{1}{2\sqrt{\pi N_0}}\left[\int_{(s_1 + s_2)/2}^{\infty} \exp\left(\frac{-(v - s_1)^2}{N_0}\right)dv\right.$$

$$\left. - \int_{(s_1 + s_2)/2}^{\infty} \exp\left(\frac{-(v - s_2)^2}{N_0}\right)dv\right]$$

$$= \frac{1}{2} + \frac{1}{2\sqrt{2\pi}}\left[\int_{(s_2 - s_1)/\sqrt{2N_0}}^{\infty} \exp\left(\frac{-\lambda^2}{2}\right)d\lambda - \int_{(s_1 - s_2)/\sqrt{2N_0}}^{\infty} \exp\left(\frac{-\lambda^2}{2}\right)d\lambda\right]$$

$$= \frac{1}{2} + \frac{1}{2}\left[Q\left(\frac{s_2 - s_1}{\sqrt{2N_0}}\right) - Q\left(\frac{s_1 - s_2}{\sqrt{2N_0}}\right)\right] \tag{6.108}$$

where the variable substitution

$$\lambda = (v - s_x)\sqrt{\frac{2}{N_0}}, \quad d\lambda = \sqrt{\frac{2}{N_0}}\,dv \tag{6.109}$$

has been utilised in the second line. Since $Q(x) = 1 - Q(-x)$, eq. (6.108) can be further simplified:

$$\frac{1}{2} + \frac{1}{2}\left[Q\left(\frac{s_2 - s_1}{\sqrt{2N_0}}\right) - Q\left(\frac{s_1 - s_2}{\sqrt{2N_0}}\right)\right] = Q\left(\frac{s_2 - s_1}{\sqrt{2N_0}}\right)$$

$$= Q\left(\sqrt{\frac{2E_b}{N_0}}\right) = Q\left(\frac{d}{\sqrt{2N_0}}\right) \tag{6.110}$$

In the former expression, the Euclidean distance, d, of the signal points in BPSK can be identified and inserted in the last expression. From Figure 6.22 it is found that the distance is $d = 2\sqrt{E_b}$.

The probability of having a symbol error in BPSK is then

$$P_e = Q\left(\sqrt{\frac{2E_b}{N_0}}\right) \tag{6.111}$$

which in this case is the same thing as bit error rate, since in BPSK the modulation symbols are binary. (The total error probability, eq. (6.106), can also be regarded as a cost function with zero cost for a correct decision and one for an incorrect decision. A probability ratio can then be introduced and Bayes' hypothesis used. This leads to the same result as the discussion above.)

Error probability of two-dimensional signalling

For two-dimensional signals of the QAM, MPSK types, etc., the density func-

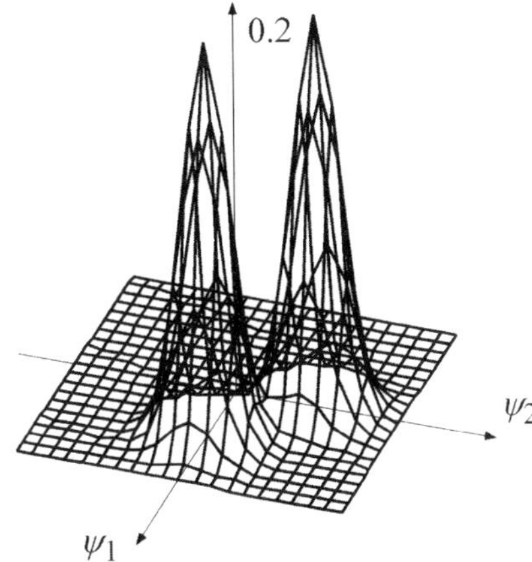

Figure 6.26 Orthogonal signals with additive Gaussian noise.

tions span a surface. The calculations must therefore be performed in two dimensions. As we shall see below, this problem can be simplified. Suppose that the locations for the signals are $\mathbf{s}_1 = [\,0 \quad s_1\,]^T$ and $\mathbf{s}_2 = [\,s_2 \quad 0\,]^T$. A density function with two orthogonal signals in two dimensions is shown in Figure 6.26. Since the axes are orthogonal, the PDF becomes the product of the individual probability densities.

For Gaussian noise we obtain from eq. (6.74), with $K = 2$,

$$p(\psi_1, \psi_2 | \mathbf{s}_1) = \frac{1}{\pi N_0} \exp\left(\frac{-\psi_1^2}{N_0}\right) \exp\left(\frac{-(\psi_2 - s_1)^2}{N_0}\right)$$

$$= \frac{1}{\pi N_0} \exp\left(-\frac{\psi_1^2 + (\psi_2 - s_1)^2}{N_0}\right) \tag{6.112}$$

A corresponding expression is valid for $p(\psi_1, \psi_2 | \mathbf{s}_2)$. The decision rule in eq.

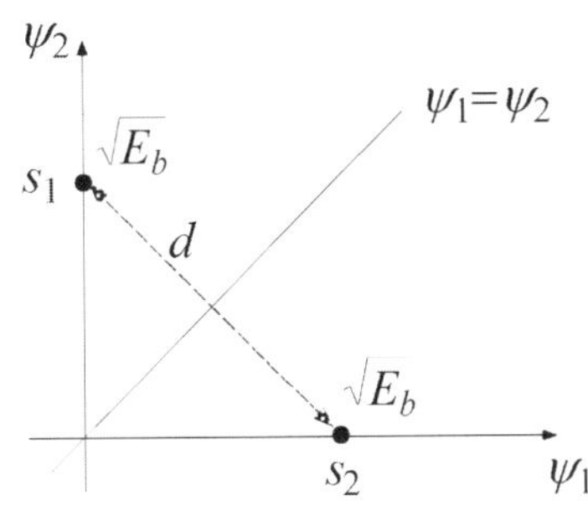

Figure 6.27 Signal configuration and decision boundary of orthogonal signals. In this case $s_1 = s_2 = \sqrt{E_b}$.

(6.50) says: choose $\mathbf{s}_1$ as the receiver hypothesis if

$$\frac{\|\mathbf{y} - \mathbf{s}_1\|^2}{N_0} - \ln(P(\mathbf{s}_1)) < \frac{\|\mathbf{y} - \mathbf{s}_2\|^2}{N_0} - \ln(P(\mathbf{s}_2)) \tag{6.113}$$

otherwise chose $\mathbf{s}_2$. This means that the boundary between the decision areas corresponds to the ψ_1 and ψ_2 values fulfilling (the quadratic norms have been expanded)

$$N_0 \ln \frac{P(\mathbf{s}_2)}{P(\mathbf{s}_1)} = -\left(\psi_1^2 + (\psi_2 - s_1)^2 \right) + (\psi_1 - s_2)^2 + \psi_2^2 \tag{6.114}$$

which gives the boundary as the straight line

$$\frac{s_1}{s_2} \psi_2 + \left(-\frac{N_0}{2s_2} \ln \frac{P(\mathbf{s}_2)}{P(\mathbf{s}_1)} - \frac{s_1^2 - s_2^2}{2s_2} \right) = \psi_1 \tag{6.115}$$

For signal alternatives having equal probabilities $P(s_1) = P(s_2) = 1/2$, we get after simplification

$$\psi_1 s_2 - \psi_2 s_1 + \frac{s_1^2 - s_2^2}{2} = 0 \tag{6.116}$$

With $s_1 = s_2 = \sqrt{E_b}$ inserted in eq. (6.116) we obtain the integration boundary

$$\psi_2 - \psi_1 = 0 \quad \Leftrightarrow \quad \psi_1 = \psi_2 \tag{6.117}$$

(Figure 6.27). P_e then becomes as before (cf. eq. (6.108))

$$P_e = \frac{1}{2} + \frac{1}{2\pi N_0} \left[\int_{-\infty}^{\infty} \exp\left(-\frac{\psi_1^2}{N_0} \right) \int_{-\infty}^{\psi_1} \exp\left(\frac{-(\psi_2 - s_1)^2}{N_0} \right) d\psi_2 d\psi_1 \right.$$
$$\left. - \int_{-\infty}^{\infty} \exp\left(-\frac{(\psi_1 - s_2)^2}{N_0} \right) \int_{-\infty}^{\psi_1} \exp\left(-\frac{\psi_2^2}{N_0} \right) d\psi_2 d\psi_1 \right] \tag{6.118}$$

The integration above is difficult to perform, and some simplification must be done. This can be achieved by the following transformation. The starting point is the usual signal space shown in Figure 6.28. This is rotated $\pi/4$ rad. as shown in Figure 6.29. After a parallel shift we obtain the result in Figure 6.30. A new coordinate system is obtained having the basis functions

$$\alpha = \psi_1 \cos\theta - \psi_2 \sin\theta + \sqrt{\frac{E_b}{2}}$$

$$\beta = \psi_1 \sin\theta + \psi_2 \cos\theta - \sqrt{\frac{E_b}{2}} \tag{6.119}$$

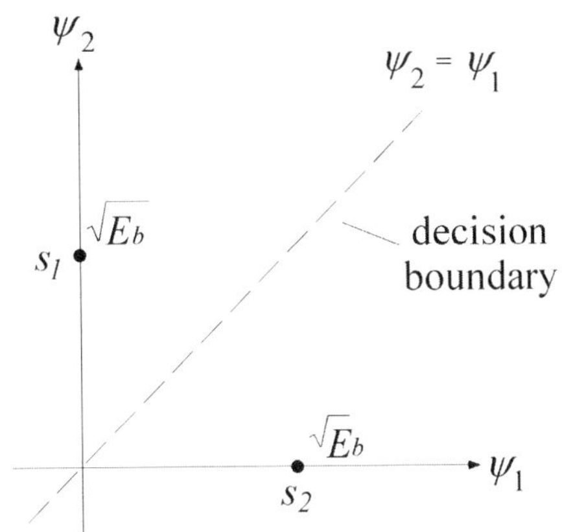

Figure 6.28   Original signal configuration.

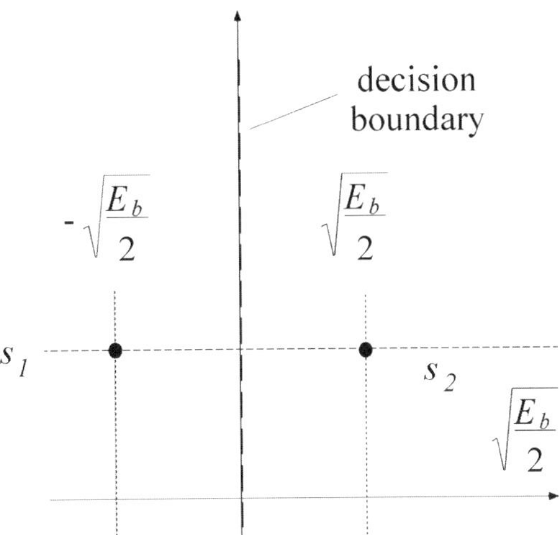

Figure 6.29   Signal configuration after rotation by $\pi/4$ rad.

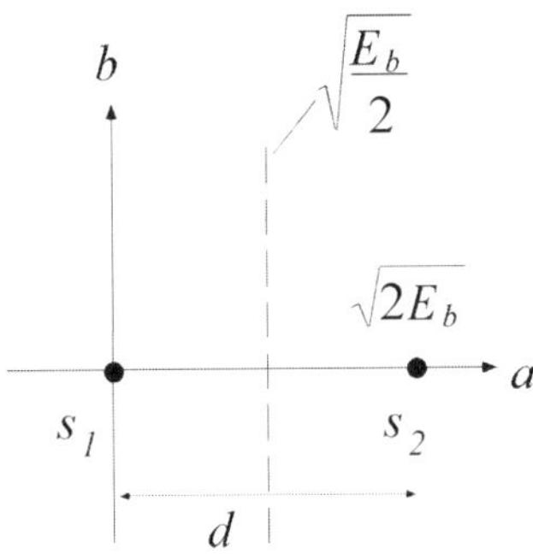

Figure 6.30   Final signal configuration after rotation and parallel shift.

which means that s_1, s_2 and the decision areas have been transferred to

$$\alpha_{s_2} = \sqrt{E_b}\frac{1}{\sqrt{2}} + \sqrt{\frac{E_b}{2}} = \sqrt{2E_b}$$

$$\alpha_{s_1} = \sqrt{E_b}\frac{-1}{\sqrt{2}} + \sqrt{\frac{E_b}{2}} = 0 \qquad (6.120)$$

$$\alpha_{dec.} = \sqrt{\frac{E_b}{2}}, \quad -\infty \leq \beta_{dec.} \leq \infty$$

and that $\beta_{s_1} = \beta_{s_2} = 0$. We only consider linear operations and noise with circular symmetry. This means that the probability of the received signal ending up to the right of the decision boundary in Figure 6.30 is

$$P\left(\alpha > \sqrt{\frac{E_b}{2}}; -\infty < \beta < \infty \,\middle|\, \mathbf{s}_1\right) \cdot P(\mathbf{s}_1) = P(\mathbf{s}_2|\mathbf{s}_1) \cdot P(\mathbf{s}_1) \qquad (6.121)$$

With $P(\mathbf{s}_1) = 1/2$ inserted in eq. (6.121) we obtain

$$\frac{1/2}{2\pi} \int_{-\infty}^{\infty} \int_{\sqrt{E_b/N_0}}^{\infty} \exp\left(-\frac{\alpha^2}{2}\right)\exp\left(-\frac{\beta^2}{2}\right) d\alpha\, d\beta = \frac{1}{2}\cdot 1\cdot Q\left(\sqrt{\frac{E_b}{N_0}}\right)$$

$$= \frac{1}{2}Q\left(\frac{d}{\sqrt{2N_0}}\right) \qquad (6.122)$$

Here d is the Euclidean distance of orthogonal BPSK, $d = \sqrt{2E_b}$. Due to the symmetry, $P(e|\mathbf{s}_1) = P(e|\mathbf{s}_2)$ and the total error probability of orthogonal BPSK becomes

$$P_e = P(\mathbf{s}_2|\mathbf{s}_1)P(\mathbf{s}_1) + P(\mathbf{s}_1|\mathbf{s}_2)P(\mathbf{s}_2) \Rightarrow P_e = Q\left(\sqrt{\frac{E_b}{N_0}}\right) = Q\left(\frac{d}{\sqrt{2N_0}}\right) \quad (6.123)$$

Note that expressed in d this result is the same as for the one-dimensional signalling that is given in eq. (6.110). Since the transformations above can be done for all two-dimensional signals, the conclusion can be drawn for the computation of error probability, that there is no need for a multi-dimensional error probability function. The only necessary parameters are the signals relative to Euclidean distances, d, and the noise density N_0, which are inserted in eq. (6.123). This result is generally valid also for signals expressed in spaces with more dimensions. The Q-function (alternatively erf. or the Φ-function), the Euclidean distances of the signals and the noise density, are what is needed in an arbitrary case to compute the error probability in Gaussian interference.

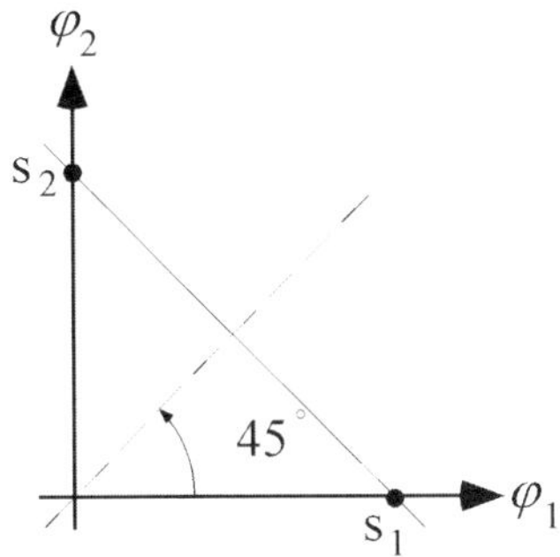

Figure 6.31 Signal configuration and decision boundary, dashed line.

Example 6.2

Consider the following set of binary signals defined by the orthogonal basis functions φ_1 and φ_2. Determine decision areas and give the decision statistics of each signal. Simplify and draw the optimum correlation receiver.

Do this for two different situations. The first is case (a), the signals are $s_1 = \alpha\,\varphi_1$ and $s_2 = \alpha\,\varphi_2$. Both signals have equal probability and the same energy. The second is case (b), where the signals are instead $s_1 = \alpha\varphi_1$ and $s_2 = -\alpha\varphi_1$. Also in this case, both signals have equal probability and the same energy.

Solution (a) Shown in signal space the signals are located as in Figure 6.31. The decision boundary becomes the straight line $\varphi_1 = \varphi_2$, i.e. the line perpendicular to the axis through the signals s_1 and s_2. The decision variables are

$$z_1 = \mathrm{Re}\{\langle y, s_1\rangle\} \quad \text{and} \quad z_2 = \mathrm{Re}\{\langle y, s_2\rangle\} \tag{6.124}$$

respectively, where y is the received signal. The receiver is schematically outlined in Figure 6.32.

The decision rules give the following hypotheses.

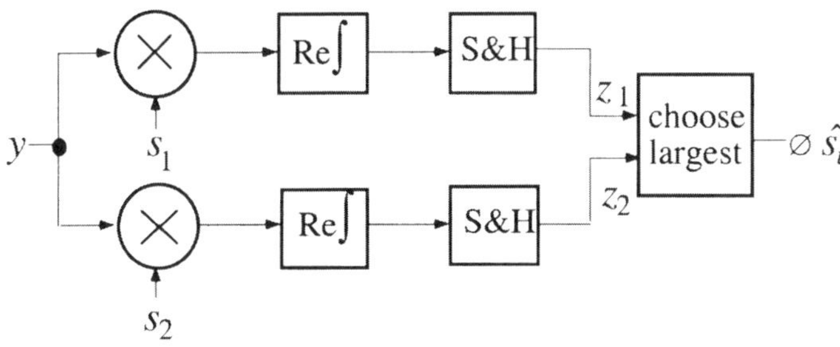

Figure 6.32 Receiver for binary orthogonal signals.

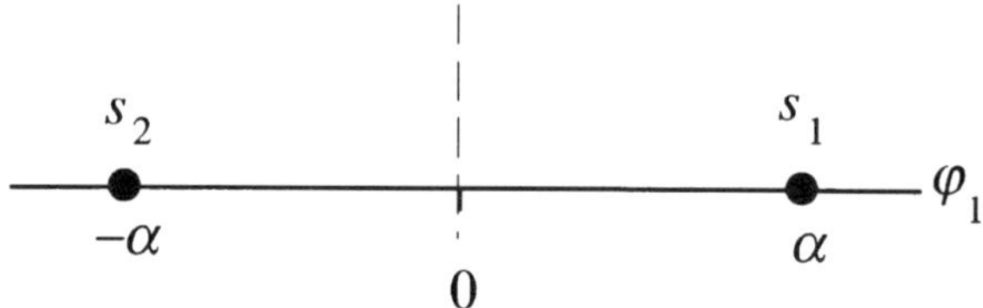

Figure 6.33 Signal configuration for antipodal signals.

$$\text{if } z_1 > z_2 \Rightarrow \hat{s}_1 \tag{6.125}$$

$$\text{if } z_1 < z_2 \Rightarrow \hat{s}_2 \tag{6.126}$$

(b) For the antipodal signals, see Figure 6.33. The decision boundary is the zero line halfway between and perpendicular to the line connecting the signals s_1 and s_2. The decision variable becomes $z = \text{Re}\{\langle y, s_1 \rangle\}$. The decision rules give the following hypotheses:

$$z > 0 \Rightarrow \text{choose } s_1 \tag{6.127}$$

$$z < 0 \Rightarrow \text{choose } s_2 \tag{6.128}$$

which means that the detector makes the decision according to $\hat{s}_i = \text{sgn}(z)$, where

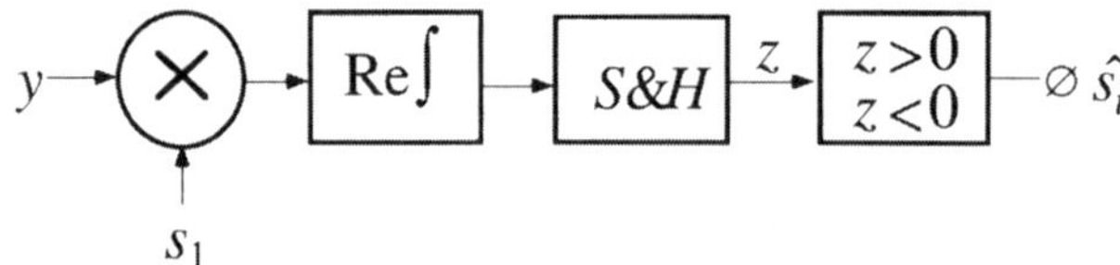

Figure 6.34 Receiver for binary antipodal signals.

sgn is the signum function, which can be realised as a limiter. The receiver is outlined in Figure 6.34.

Example 6.3

Consider the following two sets of quadruple signals ($M = 4$) defined by the orthogonal basis functions φ_1 and φ_2:

$$\text{in case (a)}: \quad s_1 = a\varphi_1 \quad s_2 = a\varphi_2 \quad s_3 = -a\varphi_1 \quad s_4 = -a\varphi_2$$

$$\text{in case (b)}: \quad s_1 = a\varphi_1 \quad s_2 = a\varphi_2 \quad s_3 = a(\varphi_1 + \varphi_2) \quad s_4 = 0$$

Determine decision variables for both sets and boundaries for a maximum likelihood decision. Define the optimal correlation receivers.

Solution (a) The signal configuration and decision boundaries in the signal space are

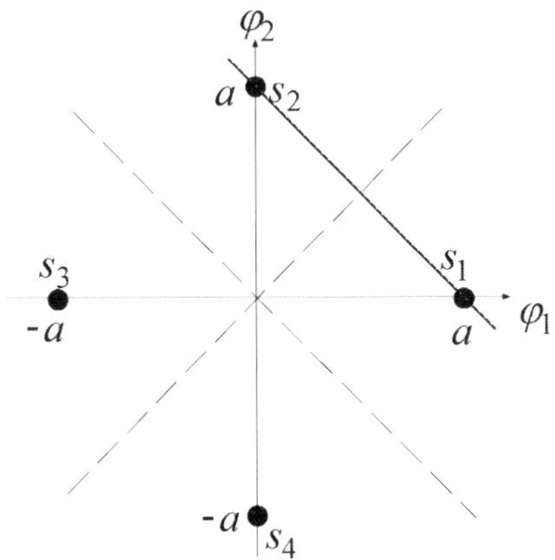

Figure 6.35 Signal configuration QAM. Decision boundaries are shown by dashed lines and the axis between s_1 and s_2 by the shaded line.

shown in Figure 6.35. In the signal space the signal set can be said to constitute a QAM or QPSK signal.

All signals are assumed to have equal probability and energy. This gives us the decision boundary between two signals as the perpendicular line halfway on the axis between the signals located closest to each other. The decision variables are:

$$z_1 = \text{Re}\{\langle y_1, s_1 \rangle\} \tag{6.129}$$

$$z_2 = \text{Re}\{\langle y_1, s_2 \rangle\} \tag{6.130}$$

$$z_3 = \text{Re}\{\langle y_1, s_3 \rangle\} = -\text{Re}\{\langle y_1, s_1 \rangle\} \tag{6.131}$$

$$z_4 = \text{Re}\{\langle y_1, s_4 \rangle\} = -\text{Re}\{\langle y_1, s_2 \rangle\} \tag{6.132}$$

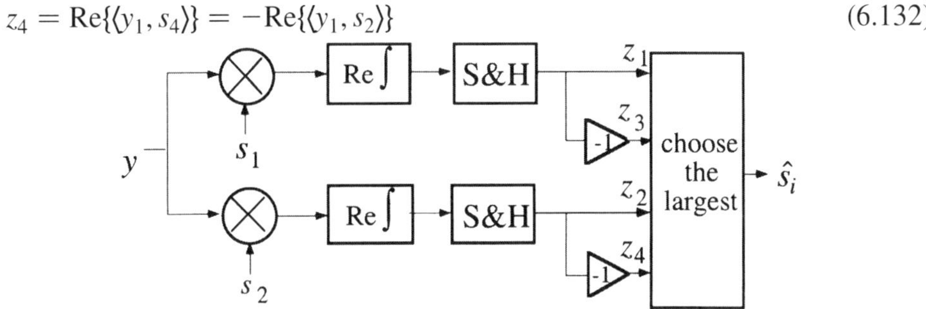

Figure 6.36 Optimal detector QAM.

The receiver is schematically outlined in Figure 6.36.

(b) The signal configuration in the signal space is shown in Figure 6.37. The correlation receiver is shown in Figure 6.38. Note that as the signals have differ-

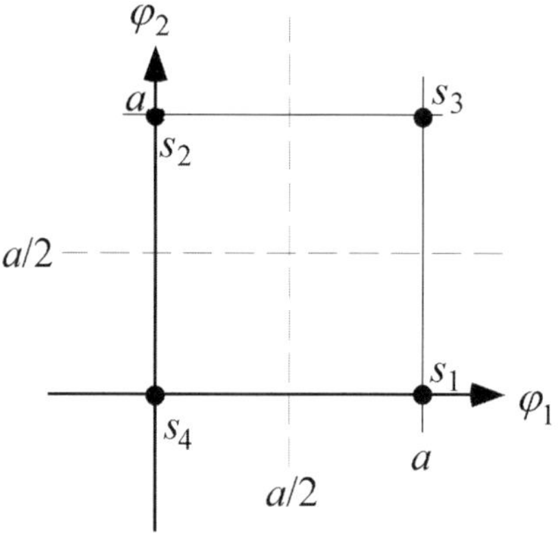

Figure 6.37 Signal configuration for offset QAM. Decision boundaries are shown by dashed lines.

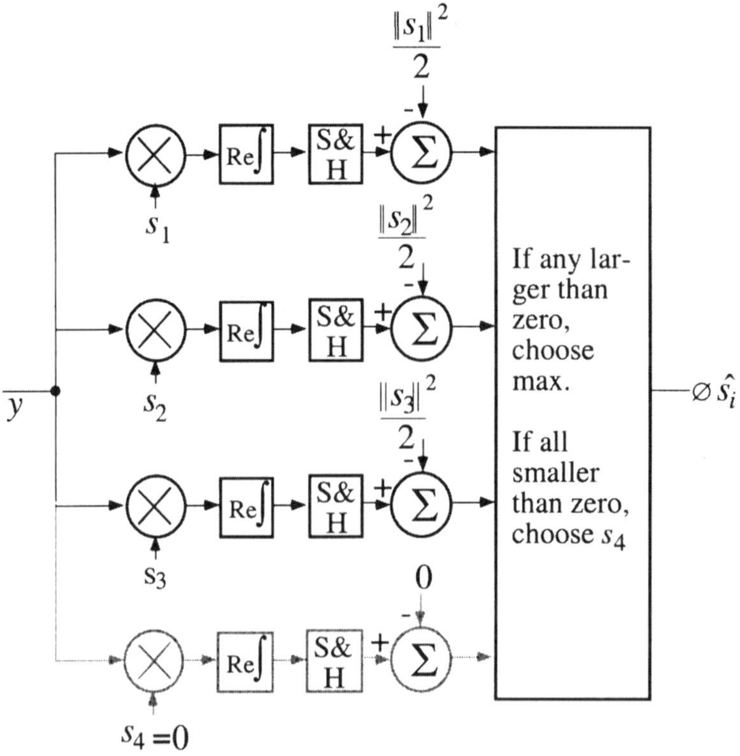

Figure 6.38 The receiver outline for offset QAM. The shaded section corresponds to multiplication by zero and can thus be ignored.

ent energies, thus a constant must be subtracted when the decision variable is formed (cf. eq. (6.56)).

The decision variables of a receiver based on the orthogonal basis functions instead, become

$$z_1 = \mathrm{Re}\{\langle y_1, s_1 \rangle\} - \frac{\|s_1\|^2}{2} = \mathrm{Re}\{\langle y_1, a\varphi_1 \rangle\} - \frac{a^2}{2} \tag{6.133}$$

$$z_2 = \mathrm{Re}\{\langle y_1, s_2 \rangle\} - \frac{\|s_2\|^2}{2} = \mathrm{Re}\{\langle y_1, a\varphi_2 \rangle\} - \frac{a^2}{2} \tag{6.134}$$

$$z_3 = \mathrm{Re}\{\langle y_1, a\varphi_1 \rangle\} + \mathrm{Re}\{\langle y_1, a\varphi_2 \rangle\} - a^2 \tag{6.135}$$

$$z_4 = 0 \tag{6.136}$$

The solution based on these basis functions is shown in Figure 6.39.

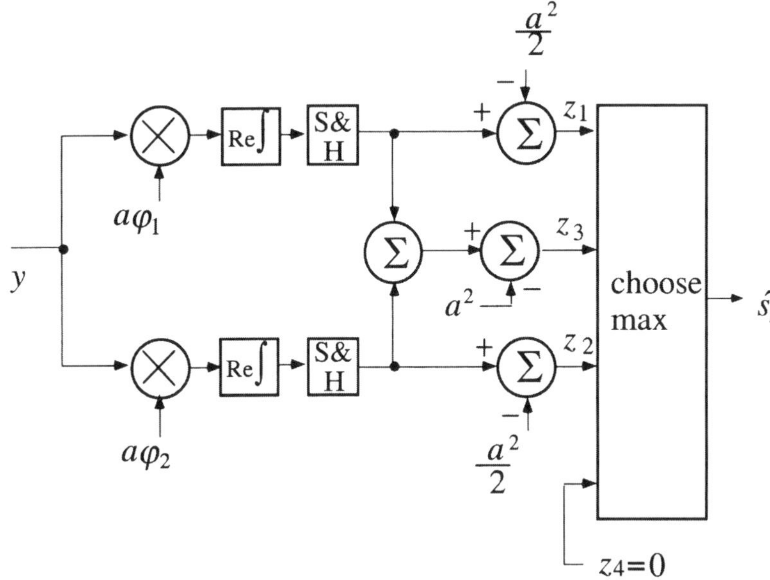

Figure 6.39   Receiver for offset QAM based on orthogonal basis functions.

Example 6.4

Which frequency deviation gives the lowest possible error probability for binary FSK over an AWGN? Derive the expression for P_e as a function of E_b/N_0.

Solution   For a binary FSK signal there are two signal alternatives:

$$s_1(t) = A_c\cos(2\pi(f_c + f_d)t), \quad s_2(t) = A_c\cos(2\pi(f_c - f_d)t) \tag{6.137}$$

To obtain the smallest possible error probability, P_e, the Euclidean distance, d, between the signals should be as large as possible. The distance is

$$d^2 = \|s_1(t) - s_2(t)\|^2 = A_c^2 \int_0^T \cos^2(2\pi(f_c + f_d)t)$$

$$- \underbrace{2\cos(2\pi(f_c + f_d)t)\cos(2\pi(f_c - f_d)t)}_{\cos(4\pi f_d t) + \cos(4\pi f_c t)} + \cos^2(2\pi(f_c - f_d)t)dt$$

$$= A_c^2\left[\frac{T}{2} + \frac{\sin[4\pi(f_c + f_d)T]}{8\pi(f_c + f_d)} - \frac{\sin(4\pi f_d T)}{4\pi f_d} - \frac{\sin(4\pi f_c T)}{4\pi f_c}\right.$$

$$\left. + \frac{T}{2} + \frac{\sin[4\pi(f_c - f_d)T]}{8\pi(f_c - f_d)} \right] \tag{6.138}$$

Assume for simplicity that $f_c \gg f_d$. The dominating terms then become

$$d^2 \approx A_c^2 T\left(1 - \frac{\sin(4\pi f_d T)}{4\pi f_d T}\right) = 2E_b\left(1 - \frac{\sin 4\pi f_d T}{4\pi f_d T}\right) \tag{6.139}$$

Assume $x = 4\pi f_d T$. The distance d then has its maximum when the function $\sin x/x$ has a minimum. The minimum value should occur when x is close to $3\pi/2$. Assume a value of $3\pi/2 + \Delta x$, where Δx is small. The following equation can be written:

$$\frac{\sin x}{x} = \frac{\sin\left(\dfrac{3}{2}\pi + \Delta x\right)}{\dfrac{3}{2}\pi + \Delta x} = \frac{2}{3\pi}\frac{-\cos(\Delta x)}{1 + \dfrac{2}{3\pi}\Delta x}$$

$$\approx \frac{-2}{3\pi}\left(1 - \frac{2}{3\pi}\Delta x\right)\left(1 - \frac{\Delta x^2}{2}\right) \tag{6.140}$$

The approximations

$$\cos(\Delta x) \approx 1 - \frac{\Delta x^2}{2} \quad \text{and} \quad \frac{1}{1 + \dfrac{2}{3\pi}\Delta x} \approx 1 - \frac{2}{3\pi}\Delta x$$

have been used. To find the optimum, we take the derivative of the last line in eq. (6.140), which gives

$$\frac{d}{d\Delta x}\left(1 - \frac{\Delta x^2}{2} - \frac{2}{3\pi}\Delta x + \frac{1}{3\pi}\Delta x^3\right) = 0$$

$$\Rightarrow \ \Delta x \approx -0.2 \ \Rightarrow \ x = \frac{3\pi}{2} - 0.2 \approx 4.51 \tag{6.141}$$

We solve for f_d, which gives

$$4\pi f_d T = 4.51 \quad \Rightarrow \quad f_d = \frac{4.51}{4\pi} r_b \approx 0.36 r_b \tag{6.142}$$

where $r_b = 1/T$. Inserted in eq. (6.108) the error probability becomes

$$P_e = Q\left(\frac{d}{\sqrt{2N_0}}\right) = Q\left(\sqrt{1.22\frac{E_b}{N_0}}\right) \tag{6.143}$$

Note that orthogonal signalling gives a larger value of P_e since the error probability of orthogonal signalling is $Q(\sqrt{E_b/N_0})$. In that case the smallest deviation is $f_d = 0.25 r_b$.

Example 6.5

Determine the bit error probability of QPSK.

Solution The decision areas are transformed according to Figure 6.40. Given that the signal s_1 has been sent, the probability of the receiver making an incorrect decision is

$$P(e|s_1) = \underbrace{\frac{1}{2\pi}\int_{\sqrt{E/N_0}}^{\infty}\int_{-\infty}^{-\sqrt{E/N_0}}\exp\left(\frac{-\alpha^2}{2}\right)\exp\left(\frac{-\beta^2}{2}\right)d\alpha\,d\beta}_{\text{area A}}$$

$$+ \underbrace{Q\left(-\sqrt{\frac{E}{N_0}}\right)Q\left(\sqrt{\frac{E}{N_0}}\right)}_{\text{area B}} + \underbrace{\left(1 - Q\left(\sqrt{\frac{E}{N_0}}\right)\right)Q\left(\sqrt{\frac{E}{N_0}}\right)}_{\text{area C}}$$

$$= 2Q\left(\sqrt{\frac{E}{N_0}}\right) - Q^2\left(\sqrt{\frac{E}{N_0}}\right) \tag{6.144}$$

Here the substitutions $\alpha = x\sqrt{2/N_0}$ and $\beta = y\sqrt{2/N_0}$ have been made. To prevent the equation from becoming too long, the integral is written using the Q-function after the first row. It should also be noted that $Q(-x) = 1 - Q(x)$. The total error probability becomes

$$P_e = P(e|s_1)P(s_1) + P(e|s_2)P(s_2) + P(e|s_3)P(s_3) + P(e|s_4)P(s_4) \tag{6.145}$$

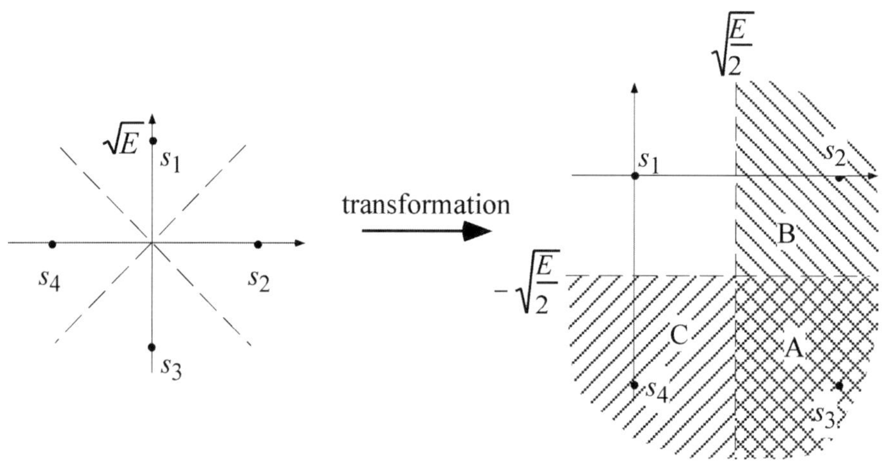

Figure 6.40   Transformation for the calculation of error probability of QPSK.

Since all symbols are assumed to have equal probability $P(s_i) = 1/4$ for $i = 1, 2, 3, 4$. For symmetry reasons all $P(e|s_i)$ are equal and the error probability becomes

$$P_e = 2Q\left(\sqrt{\frac{E}{N_0}}\right) - Q^2\left(\sqrt{\frac{E}{N_0}}\right) \tag{6.146}$$

This is the probability of making an error for a quadruple symbol with $M = 4$. The task is to calculate the bit error probability. A bit is a binary symbol and, consequently, a 4-ary symbol can be seen as two binary symbols. Assume that the 4-ary modulation symbols are binary coded by using Gray coding as shown in Figure 6.41. The most probable error is that a symbol is mistaken for one of its closest neighbours. For this symbol error only one bit error occurs when Gray coding is used.

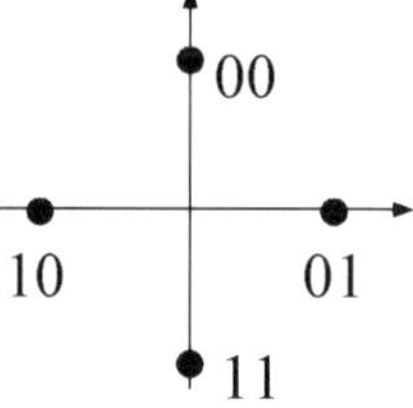

Figure 6.41   Gray-coded QPSK.

The signal s_1 is interpreted as 00 has been sent. In the areas B and C one bit error per group of two is made. This means the probability of a certain bit in the group being incorrect is 1/2. In area A both bits are incorrect and the probability of a bit being incorrect is equal to one. The total bit error probability of QPSK is therefore

$$P_{be} = \overbrace{1 \cdot Q^2\left(\sqrt{\tfrac{E}{N_0}}\right)}^{\text{area A}} + \overbrace{\tfrac{1}{2}\left[Q\left(\sqrt{\tfrac{E}{N_0}}\right) - Q^2\left(\sqrt{\tfrac{E}{N_0}}\right)\right]}^{\text{area B}}$$

$$+ \overbrace{\tfrac{1}{2}\left[Q\left(\sqrt{\tfrac{E}{N_0}}\right) - Q^2\left(\sqrt{\tfrac{E}{N_0}}\right)\right]}^{\text{area C}} = Q\left(\sqrt{\frac{E}{N_0}}\right) \tag{6.147}$$

Since the symbol energy $E = E_b \log_2(M)$, $E = 2E_b$ and the bit error probability of QPSK becomes

$$P_{be} = Q\left(\sqrt{\frac{2E_b}{N_0}}\right) \tag{6.148}$$

which is the same as for antipodal BPSK. The similarity can be interpreted in many ways. But given a constant rate, in bits/s, for the information transmission, the symbol length can be doubled since each modulation symbol contains two bits of information. The energy, E, in the symbols are thereby larger and in this way balance the effect of the increased number of signal points in the signal space as compared to BPSK.

6.3.4 CALCULATION OF ERROR RATE FOR A COMPLICATED DECISION SPACE

Union bound

In many cases an exact determination of the error probability is impossible or too complicated to perform. In this case, estimating an upper limit can be a solution. A boundary called the union bound is presented below. The name emanates from the fact that the union of all error events is used for the estimation. This is illustrated in Figure 6.42. Assume a value of $M = 3$. The decision boundaries between pairs of signals are made up of the straight lines which are halfway between these points and perpendicular to the axis joining them.

Given the sent signal s_1, the error probability is written

$$P(e|s_1) = P(B|s_1) + P(C|s_1) - P(B \cap C|s_1) \tag{6.149}$$

The probabilities $P(B|s_1)$ and $P(C|s_1)$ represent the probabilities of the received signal being inside the areas B and C, respectively. The area $B \cap C$ will therefore be counted twice. To correct for this, the probability $P(B \cap C|s_1)$ has to be subtracted in eq. (6.149). The point of using the union bound is to disregard

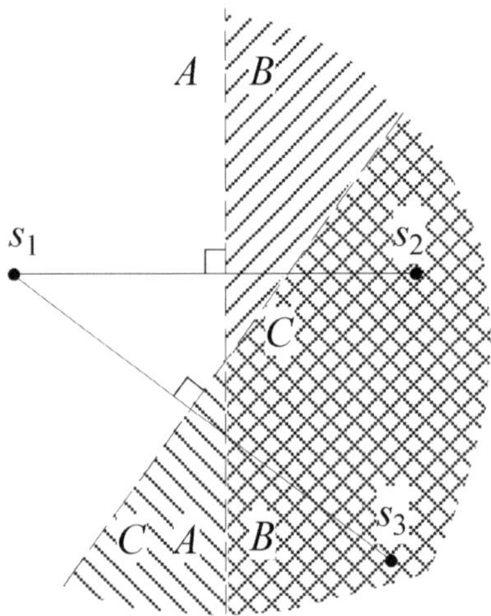

Figure 6.42   Decision areas for three signals. $B \cap C$ is marked by the square pattern.

the fact that some areas are counted several times and to let the total error probability be the sum of only the pairwise error probabilities. In the example above the union bound becomes

$$P(e|s_1) \leq P(B|s_1) + P(C|s_1) = Q\left(\frac{d_{12}}{\sqrt{2N_0}}\right) + Q\left(\frac{d_{13}}{\sqrt{2N_0}}\right) \tag{6.150}$$

In the general case we obtain

$$P(e|s_j) \leq \sum_{i=1,\ i\neq j}^{M} P(\Omega_i|s_j) = \sum_{i=1,\ i\neq j}^{M} Q\left(\frac{d_{ij}}{\sqrt{2N_0}}\right) \tag{6.151}$$

In the example above we have $A = \Omega_1$ and $B = \Omega_2$. An even simpler but looser bound is obtained if the shortest Euclidean distance, d_{min}, between two signals is used. This gives the maximum error probability as

$$P(e|s_j) \leq (M-1)Q\left(\frac{d_{min}}{\sqrt{2N_0}}\right) \tag{6.152}$$

The total error probability is as usual the sum of all possible error events

$$P_e = \sum_{j=1}^{M} P(e|s_j)P\left(s_j\right) \tag{6.153}$$

When all signals s_j have equal probability we get $P(s_j) = 1/M$. Inserting eqs. (6.151) and (6.152) into eq. (6.153) gives

$$P_e = \frac{1}{M}\sum_{j=1}^{M} P(e|s_j) \le \frac{1}{M}\sum_{j=1}^{M}\sum_{i=1, i\ne j}^{M} Q\left(\frac{d_{ij}}{\sqrt{2N_0}}\right) \le (M-1)Q\left(\frac{d_{min}}{\sqrt{2N_0}}\right)$$

$$(6.154)$$

The union bound, in one appearance or another, is very useful and makes it fairly easy to obtain reasonable expressions for the error probability even for complicated modulation schemes.

Union bound of MPSK

The technique of using union bounds for evaluation can be applied when calculating the bit error probability of MPSK. The shortest Euclidean distance between signal points can be obtained as in Figure 6.43. The distances between the signal points can then be obtained by using the geometrical consideration shown in Figure 6.44. Inserted in eq. (6.154) the union bound for the symbol error prob-

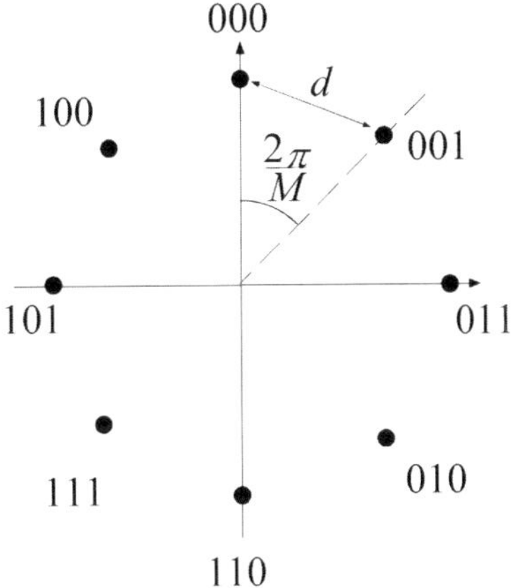

Figure 6.43   Gray-coded 8PSK.

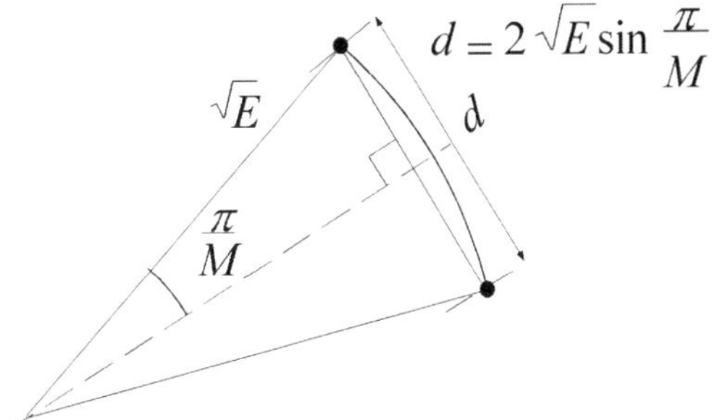

Figure 6.44   Geometry of MPSK signal.

ability is obtained

$$P_e \leq (M - 1)Q\left(\sqrt{\frac{2E\sin^2(\pi/M)}{N_0}}\right) \tag{6.155}$$

This expression can be compared with a generally established approximation, obtained by using the density function for the variation of the phase angle of a sinusoidal signal in noise (can be obtained from eq. (6.169))

$$P_e \approx 2Q\left(\sqrt{\frac{2E\sin^2(\pi/M)}{N_0}}\right) \tag{6.156}$$

for $M > 2$. The above shows that the error probability is dominated by the fact that the detector selects one of the two signal points closest to the correct one, which is to be expected. The union bound always overestimates the error probability but provides a good approximation for small values of M, although for higher values of M the estimation becomes less tight. Then the union bound based eq. (6.151) would have been better. The value P_e is the probability of having a symbol error. To calculate the bit error probability, P_{be}, at M-ary transmission, the number of bit errors generated at each symbol error must be taken into account. The most probable error is that a symbol is mistaken by one of its two nearest neighbours. As the signal is Gray coded, only one bit error occurs for each symbol error. Every symbol is coded into K bits, where $K = \log_2 M$. The bit error probability then becomes

$$P_{be} \approx \frac{1}{K}P_e \tag{6.157}$$

Furthermore, the energy should be normalised

$$E_b = \frac{E}{K} \tag{6.158}$$

This gives an approximate bit error probability for MPSK

$$P_{be} \approx \frac{1}{K}2Q\left(\sqrt{\frac{2KE_b\sin^2(\pi/M)}{N_0}}\right) \leq \frac{M - 1}{K}Q\left(\sqrt{\frac{2KE_b\sin^2(\pi/M)}{N_0}}\right) \tag{6.159}$$

6.3.5 NON-COHERENT RECEIVERS

In the previous sections concerning digital receivers it is assumed that the phase is known so a coherent detection can be made. The phase of the received carrier depends among other things on the distance to the transmitter, which usually is unknown. Assume that the received signal is noiseless and that the phase differs $-\theta$ radians from the receiver reference signal. With a complex representation the received signal can be expressed as $y(t) = \sqrt{E_b/T}\, s(t) \exp(-j\theta)$. Consider the decision variable for example in the matched filter receiver (Figure 6.18). The matched filter has been normalised such that the energy of the matched filter impulse response $E_h = 1$, when $s(t) = \pm 1$. At the sampling instant $t = T$ the decision variable becomes

$$z = \mathrm{Re}\left\{ \int_0^T \sqrt{\frac{E_b}{T}}\, s(\tau) \exp(-j\theta)\cdot\sqrt{\frac{1}{T}}\, s^*(\tau)d\tau \right\} = \mathrm{Re}\left\{ \exp(-j\theta)\sqrt{E_b} \right\} = \sqrt{E_b}\cos\theta$$

$$(6.160)$$

Apparently, the decision variable depends on the phase difference, θ, between the received signal and the receiver reference. The value even becomes zero for some angles. This is not acceptable and therefore requires special arrangements to find the carrier phase; cf. Chapter 10 which deals with phase-locked loops. In certain applications it is desirable to avoid such arrangements. Another possibility, at the detection, is then to utilise the magnitude of the matched filter output signal instead:

$$z = \left| \int_0^T \sqrt{\frac{E_b}{T}}\, s(\tau) \exp(-j\theta)\cdot\sqrt{\frac{1}{T}}\, s^*(\tau)d\tau \right| = \left| \exp(-j\theta)\sqrt{E_b} \right| = \sqrt{E_b} \qquad (6.161)$$

As is clear from eq. (6.161) the decision variable now becomes independent of the phase difference. This is called a non-coherent receiver. A detector where the carrier phase is of no concern offers a simplification at the implementation of the receiver, but has its price, as is evident when the error probability is studied. The modulation schemes requiring the phase to be known, such as different types of phase modulation, cannot be detected by a non-coherent receiver without special arrangements, like for example differential coding of the symbols. The block diagram for a typical non-coherent receiver with a matched filter is shown in Figure 6.45. The matched filter receiver is a common realisation of a non-coherent receiver. As has been shown earlier the impulse response of the matched filter is $s^*(T - t)$.

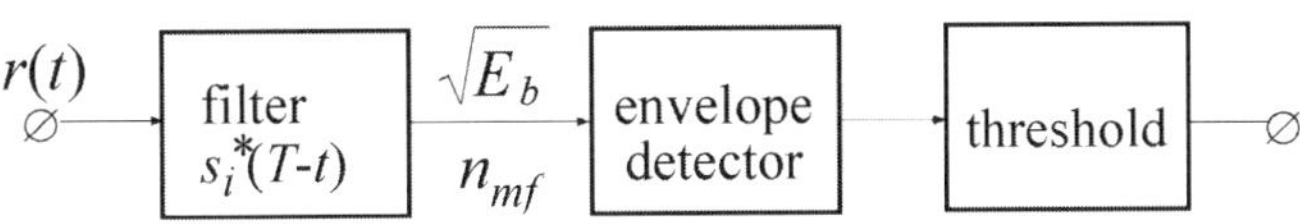

Figure 6.45   Receiver of the amplitude detector type.

The magnitude operation in eq. (6.161) is done by an envelope detector. This means that only the amplitude of the output signal from the filter is used for the subsequent decision, made by a threshold detector. An envelope detector can be realised as in Figure 5.17.

Non-coherent receiver and noise

The input signal, $r(t)$, to the receiver consists of a wanted signal plus bandpass Gaussian noise, $n(t)$. For the complex representation, the noise is divided into its quadrature parts:

$$r(t) = \sqrt{\frac{E_b}{T}}\, s(t) + n_i(t) + j\, n_q(t) \tag{6.162}$$

The signal passes through the matched filter. After that, the magnitude of the filter output signal is taken. Since $|A{\cdot}B| = |A|{\cdot}|B|$ the phase term $\exp(-j\theta)$ in eq. (6.161) can be broken out in the mathematical analysis. The magnitude is then formed from the square of the remaining real and imaginary parts:

$$z = \sqrt{\left(\sqrt{E_b} + \tilde{n}_{mfi}\right)^2 + \left(\tilde{n}_{mfq}\right)^2} \tag{6.163}$$

Each part is the result of a linear filtering. As is clear from eq. (6.161) the contribution from the desired signal is $\sqrt{E_b}$. The noise power is obtained by examining the variance of the real and imaginary parts in the filter output signal. For the real part

$$n_{mfi}^2 = \mathrm{Var}\left\{\tilde{n}_{mfi}\right\} = \mathrm{Var}\left\{\int_0^T n_i(t)e(-j\theta){\cdot}\sqrt{\frac{1}{T}}s^*(t)dt\right\}$$

$$= E\left\{\int_0^T \int_0^T n_i(t)\exp(-j\theta)n_i^*(u)\exp(+j\theta){\cdot}\sqrt{\frac{1}{T}}s(t)\sqrt{\frac{1}{T}}s^*(u)dtdu\right\}$$

$$= \frac{1}{T}\int_0^T \int_0^T E\{n_i(t)n_i^*(u)\}s(t)\,s^*(u)dtdu \tag{6.164}$$

As usual white noise is assumed, which gives

$$E\{n_i(t)n_i^*(u)\} = \frac{N_0}{2}\delta(t - u)$$

Since $\delta(t - u) \neq 0$ only for $t = u$ and the energy of the impulse response $E_h = 1$ for the matched filter, the noise power is

$$n_{mfi}^2 = \frac{1}{T}\int_0^T \int_0^T \frac{N_0}{2}\delta(t - u)s(t)\,s^*(u)dtdu = \frac{N_0}{2}\underbrace{\frac{1}{T}\int_0^T s(t)\,s^*(t)dt}_{E_h} = \frac{N_0}{2}$$

$$\tag{6.165}$$

The noise variance for the quadrature part is computed the same way and the result is identical.

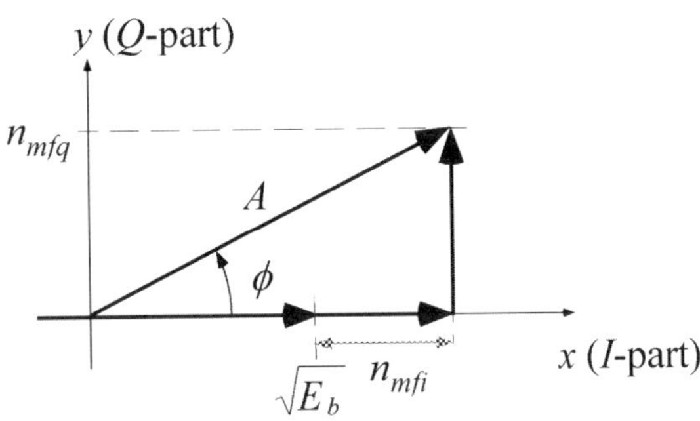

Figure 6.46 Signal and noise in a coordinate system where the phase of the wanted signal is the reference.

Expressed by vectors the signal can be represented as in Figure 6.46. The amplitude and phase of the received signal are, respectively

$$A = \sqrt{\left(\sqrt{E_b} + n_{mfi}\right)^2 + n_{mfq}^2} \tag{6.166}$$

and

$$\phi = \arctan\left(\frac{n_{mfq}}{\sqrt{E_b} + n_{mfi}}\right) \tag{6.167}$$

The signal space is one-dimensional and the decision variable consists of the amplitude A. To determine the amplitude probability density, the probability density in the point (x,y) is first computed. Since the stochastic variables are independent along the x and y axes, respectively (see Section 6.1 on narrowband noise), the joint density function becomes

$$p_{IQ}(x, y) = p_I(x)p_Q(y) = \frac{1}{\pi N_0}\exp\left(-\frac{(x - \sqrt{E_b})^2}{N_0}\right)\exp\left(-\frac{y^2}{N_0}\right) \tag{6.168}$$

Above, the density function is expressed as a function of x and y. To express the density function versus more suitable units, the signal amplitude A and phase angle ϕ are used. The Jacobian becomes $dxdy = A\, dA\, d\phi$, and inserted in the expression for the density function we obtain

$$p_{A,\phi}(A, \phi) = \frac{1}{\pi N_0}A\,\exp\left(-\frac{A^2 - 2A\sqrt{E_b}\cos\phi + E_b}{N_0}\right) \tag{6.169}$$

Since the phase angle ϕ is useless for an amplitude detector, only the marginal density for the amplitude is of interest:

$$p_A(A) = \frac{1}{\pi N_0} \int_{-\pi}^{\pi} A \, \exp\left(- \frac{A^2 - 2A\sqrt{E_b}\cos\phi + E_b}{N_0}\right) d\phi$$

$$= \frac{A}{\pi N_0} \exp\left(- \frac{A^2 + E_b}{N_0}\right) \int_{-\pi}^{\pi} \exp\left(\frac{2A\sqrt{E_b}\cos\phi}{N_0}\right) d\phi$$

$$= \frac{2A}{N_0} \exp\left(- \frac{A^2 + E_b}{N_0}\right) I_0\left(\frac{2A\sqrt{E_b}}{N_0}\right), \quad A \geq 0 \tag{6.170}$$

where $I_0(\cdot)$ is the modified Bessel function of the first kind and order zero. The density function described by eq. (6.170) is called Rice distribution. It is a classical density function in detection theory and gives the amplitude density for the envelope of the signal plus noise, when the signal has the amplitude $\sqrt{E_b}$. A special case is when $\sqrt{E_b} = 0$; the distribution then becomes a Rayleigh distribution, which has been shown previously.

Error probability in OOK

The error probability of a binary signal where one symbol is indicated by the presence of a sinusoidal signal and the other by the absence of a signal is a fundamental case. This is called on-off keying (OOK), and the error probability is calculated in the following way. For OOK the signal is

$$s(t) = \begin{cases} \sqrt{2E_b/T}\cos(\omega_c t + \theta) & \text{when a 1 is transmitted} \\ 0 & \text{when a 0 is transmitted} \end{cases} \tag{6.171}$$

The density functions for the signals from the envelope detector output are shown in Figure 6.47, for the different signal alternatives.

The optimal threshold is calculated as before. Due to the amplitude density function the calculation will be algebraically more difficult and has to be approximated. The result is

$$V_{opt} \approx \frac{\sqrt{E_b}}{2}\sqrt{1 + \frac{4N_0}{E_b}} \approx \frac{\sqrt{E_b}}{2}, \quad \frac{E_b}{N_0} \gg 1 \tag{6.172}$$

The total error probability is

$$P_e = \frac{1}{2}\left(P(1|0) + P(0|1)\right) \tag{6.173}$$

The error probability when a zero is sent is obtained by using eq. (6.170) and putting $\sqrt{E_b} = 0$, which gives

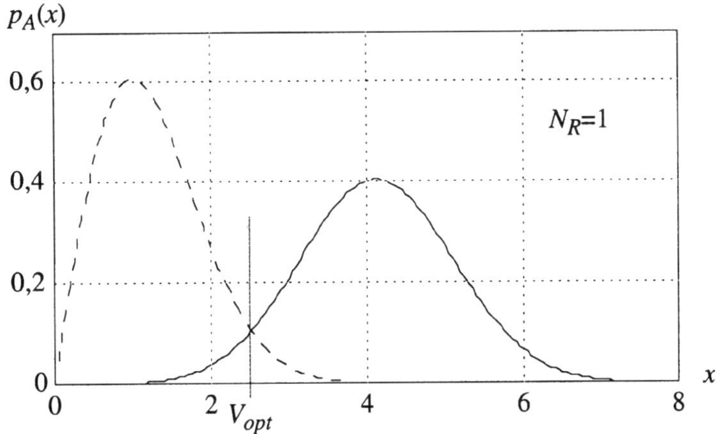

Figure 6.47 Amplitude density function for one and zero, respectively.

$$P(1|0) = \int_{\sqrt{E_b}/2}^{\infty} p_A(y)\,dy = \exp(-E_b/4N_0) \tag{6.174}$$

For a 'one' sent, we have $\sqrt{E_b} \neq 0$ and by means of eq. (6.170) we obtain

$$P(0|1) = \int_{0}^{\sqrt{E_b}/2} p_A(y)\,dy = 1 - Q\left(\sqrt{\frac{2E_b}{N_0}}, \sqrt{\frac{E_b}{2N_0}}\right) \tag{6.175}$$

In eq. (6.175), $Q(\alpha, \beta)$ is the generalised Q-function, or Marcum's Q-function. This function cannot be expressed in a closed form. For numerical values the reader is referred to standard tables or to numerical integration. To anyhow obtain an analytical expression, $I_0(x)$ can be approximated by

$$I_0(x) \approx \begin{cases} \exp\dfrac{x^2}{4} & \text{for small } \dfrac{S}{N} \\[2ex] \dfrac{\exp(x)}{\sqrt{2\pi x}} & \text{for large } \dfrac{S}{N} \end{cases} \tag{6.176}$$

For large S/N the density function then becomes

$$p_A(A) \approx \sqrt{\frac{A}{\pi \sqrt{E_b}N_0}}\, \exp\left(-\frac{(A - \sqrt{E_b})^2}{N_0}\right) \tag{6.177}$$

This density function can be used for the calculation of error probabilities. When integrating eq. (6.177), the main contribution to the result comes from the peak of the function. The factor $\sqrt{(A/\pi\sqrt{E_b}N_0)}$ varies only slowly, relative to the clearly peaked exponential factor. Hence a further simplification can be made by setting $A = \sqrt{E_b}$ under the square root. The integral can then be expressed by using the previously used Q-function, so that

$$P(0|1) \approx 1 - Q\left(\sqrt{\frac{2}{N_0}}\left[\frac{\sqrt{E_b}}{2} - \sqrt{E_b}\right]\right) = Q\left(\sqrt{\frac{E_b}{2N_0}}\right) \approx \sqrt{\frac{N_0}{\pi E_b}}\exp\left(-\frac{E_b}{4N_0}\right)$$

(6.178)

Equations (6.174) and (6.178) inserted in eq. (6.173) give the total error probability for non-coherent detected OOK

$$P_e \approx \frac{\exp(-E_b/4N_0)}{2}\left[1 + \sqrt{\frac{N_0}{\pi E_b}}\right] \approx \frac{1}{2}\exp\left(-\frac{E_b}{4N_0}\right)$$

(6.179)

The approximation is useful if the ratio $\sqrt{E_b/N_0}$ is large enough. Note that the last approximation is only the probability to interpret a '1' as a '0'. This can be done because the density function when a zero has been sent decreases quickly for increasing E_b/N_0. If more exact values are required it is necessary to go backwards in the approximation chain or utilise numerical integration.

Non-coherent detection of FSK

FSK is a type of modulation where the possibility of simplifying the receiver and eliminating the carrier recovery is sometimes used. In this section non-coherent detection of FSK is discussed. The design of the receiver for this case is shown in Figure 6.48.

The decision variable, i.e. the signal on the input of the threshold detector, is $y_1 - y_0$. In the case of FSK the energy of a zero and a one is the same. This gives the energy for a rectangular pulse shape

$$E_b = E_1 = E_0$$

(6.180)

Inserted in eq. (6.170) the distribution function of y_1 given a one is

$$P_{y_1}(y_1|1) = \frac{2y_1}{N_0}\exp\left(-\frac{y_1^2 + E_b}{N_0}\right)I_0\left(\frac{2y_1\sqrt{E_b}}{N_0}\right)$$

(6.181)

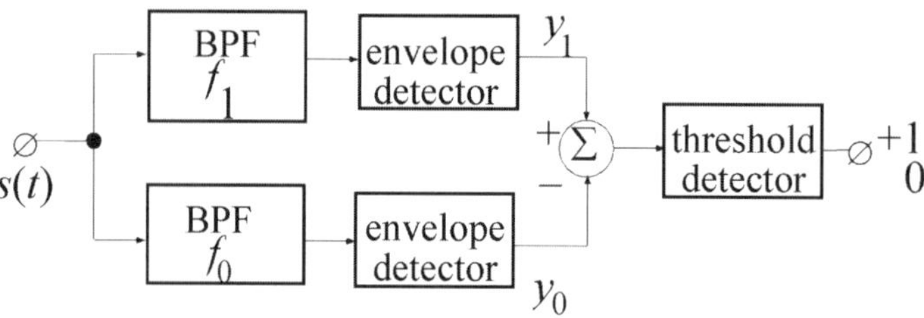

Figure 6.48   Receiver for non-coherent detection of FSK.

For y_0 given a one we obtain

$$P_{y_0}(y_0|1) = \frac{2y_0}{N_0}\exp\left(-\frac{y_0^2}{N_0}\right)$$
(6.182)

The same distribution function, only reversed, holds for a zero. Because of this symmetry the decision boundary becomes zero, i.e.

$$y_1 - y_0 \underset{0}{\overset{1}{\gtrless}} 0$$
(6.183)

The error probability therefore becomes $P(0|1)$, and due to the symmetry we get $P(1|0) = P(0|1)$. The total error probability is obtained in the usual way as

$$P_e = \frac{1}{2}P(1|0) + \frac{1}{2}P(0|1) \quad \Rightarrow \quad P_e = P(0|1)$$
(6.184)

Inserting this in eqs. (6.181) and (6.182) gives the error probability

$$P(0|1) = P(y_0 > y_1|1) = \int_0^\infty P_{y_1}(y_1|1)\underbrace{\int_{y_1}^\infty P_{y_0}(y_0|1)\,dy_0 dy_1}_{\exp(-\,y_1^2/N_0)}$$

$$= \int_0^\infty \frac{2y_1}{N_0}\exp\left(-\frac{(y_1^2 + E_b)}{N_0}\right)I_0\left(\frac{2y_1\sqrt{E_b}}{N_0}\right)\exp\left(-\frac{y_1^2}{N_0}\right)dy_1$$

$$\left[\text{substitute } z = 2y_1\right]$$

$$= \frac{1}{2}\exp\left(-\frac{E_b}{2N_0}\right)\underbrace{\int_0^\infty \frac{z}{N_0}\exp\left(-\frac{\left(z^2 + E_b\right)}{2N_0}\right)I_0\left(\frac{z\sqrt{E_b}}{N_0}\right)dz}_{=1}$$

$$= \frac{1}{2}\exp\left(-\frac{E_b}{2N_0}\right)$$
(6.185)

Equation (6.185) inserted in eq. (6.184) gives the error probability of non-coherent detected FSK. This should be compared with the previously derived expression for coherent detection of orthogonal FSK signalling (eq. (6.123))

$$P_e = Q\left(\sqrt{\frac{E_b}{N_0}}\right)$$
(6.186)

Differentially coded BPSK

Since the information about the current symbol is contained in the phase of the signal, detection of a transmitted symbol in phase-modulated signals is not possible without a phase reference. One way of creating a phase reference is to let the previous symbol constitute this reference. The difference in the signal phase for two consecutive symbol intervals is observed, and the method is therefore called differential phase shift keying (DPSK). It is useful for MPSK with arbitrary M. The explanation below, however, is based on binary phase shifts, BPSK.

Consider a BPSK-modulated signal $s(t) = \sqrt{2E_b/T_b}\cos(\omega_c t + \phi_k)$, where the phase angle is either $\phi_k = 0$ or $\phi_k = \pi$. The message must always start with a given symbol, say 1. The signal is then coded at the transmitter so that $a_k = a_{k-1}$ if $x_k = 1$ and $a_k \neq a_{k-1}$ if $x_k = 0$. An example is

- message, x_k 1   0   1   1   0   1   0   0
- coded message, a_k 1   1   0   0   0   1   1   0   1
- signal phase, ϕ_k π π 0   0   0 π π 0 π

This means that if the phase of the signal changes between two consecutive symbol intervals, the receiver detects the symbol zero otherwise one. The basic structure of the receiver is shown in Figure 6.49.

The received signal is $r(t) = \sqrt{2E_b/T_b}\cos(\omega_c t + \phi_k + \theta) + n(t)$. It is phase shifted by an unknown angle θ, which changes slowly enough to be regarded as constant over two symbol intervals. Without noise the signal $y(t) = E_b/2 \times \cos(\phi_k - \phi_{k-1})$. To analyse the effect of noise in the signal, the noise can be separated into its quadrature components; see eq. (6.162). Given that the carrier frequency f_c $(= \omega_c/2\pi)$ is much larger than the symbol rate $1/T_b$, the integral of the desired signal will be $\pm\sqrt{E_b}$. The signal, which in the receiver has been delayed by the length of a symbol interval, is denoted $\sqrt{E'_b}$ and the delayed

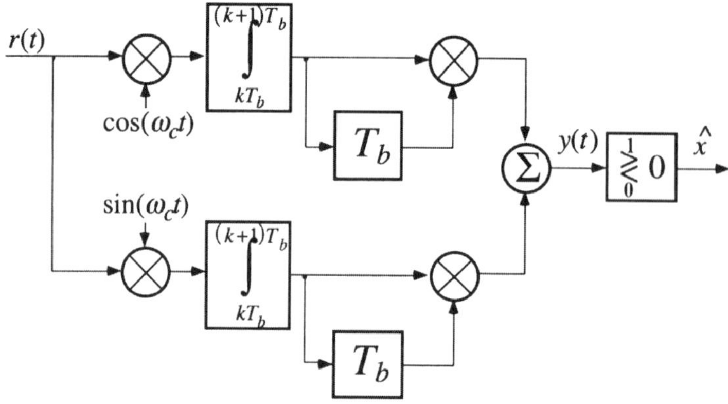

Figure 6.49   Receiver for binary DPSK.

noise is n_i' and n_q', respectively. The signal $y(t)$ then becomes

$$y(T_b) = \frac{1}{2}\left[\sqrt{E_b E_b'} + \sqrt{E_b}n_i' + \sqrt{E_b'}n_i + n_i'\cdot n_i - n_q'\cdot n_q\right] \tag{6.187}$$

All noise components are mutually uncorrelated. Because of the symmetry in this problem it is sufficient to study the case when $a_k = a_{k-1}$. In this case $\sqrt{E_b'} = \sqrt{E_b}$ and eq. (6.187) becomes

$$2y(T_b) = \underbrace{E_b + \sqrt{E_b}(n_i' + n_i) + n_i'\cdot n_i}_{\text{I}} - \underbrace{n_q'\cdot n_q}_{\text{II}} \tag{6.188}$$

An incorrect decision is made here if $y(T_b) < 0$. The error probability can be directly calculated from eq. (6.188), but there are mathematical difficulties involved in the integration of the density function for the product of two variables with Gaussian distribution. A solution of the problem is to complete the squares in eq. (6.188). The equation is then written in a form which make the results from previous chapters useful. This provides a simple solution to the problem. Completion of the squares in part I gives

$$E_b + \sqrt{E_b}(n_i' + n_i) + n_i'\cdot n_i = \left(\sqrt{E_b} + \frac{1}{2}(n_i' + n_i)\right)^2 - \left(\frac{1}{2}(n_i' - n_i)\right)^2 \tag{6.189}$$

Completion of the squares in part II of eq. (6.188) gives

$$n_q'\cdot n_q = \left(\frac{1}{2}(n_q' + n_q)\right)^2 - \left(\frac{1}{2}(n_q' - n_q)\right)^2 \tag{6.190}$$

The expressions with completed squares are inserted into eq. (6.188). Rearranging the terms and using that $y(T_b) < 0$ we can see that an incorrect decision is made when

$$\left(\sqrt{E_b} + \frac{1}{2}(n_i' + n_i)\right)^2 + \left(\frac{1}{2}(n_q' - n_q)\right)^2 < \left(\frac{1}{2}(n_q' + n_q)\right)^2 + \left(\frac{1}{2}(n_i' - n_i)\right)^2 \tag{6.191}$$

Each side of the inequality sign corresponds to an envelope-detected signal (cf. eq. (6.163)). The left-hand side corresponds to a signal, with bit energy E_b, and the noise separated into its in phase and quadrature components. The right hand side corresponds to noise only. The decision problem is then exactly the same as for non-coherent detection of FSK, which has been discussed earlier.

The noise components in eq. (6.191), e.g. $1/2(n_q' + n_q)$, are a half times the sum of two Gaussian distributed noise components, each having the variance σ^2. The result is a Gaussian distribution with a variance of $2\sigma^2/2^2 = \sigma^2/2$. This means that the noise power density in each noise component corresponds to a density which is half of the density of the input signal, N_0. The error probability is

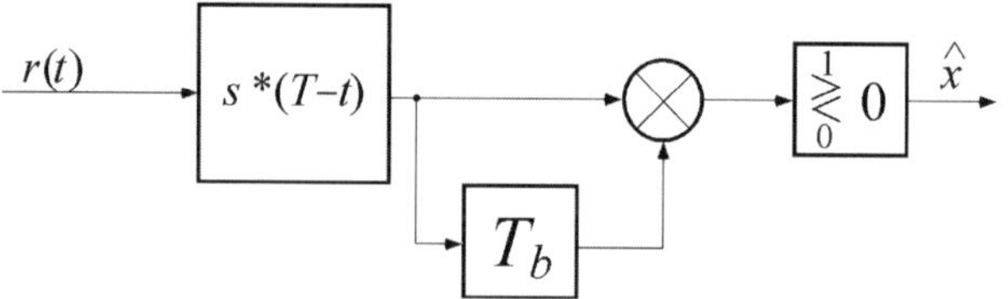

Figure 6.50 Alternative block diagram for binary DPSK.

given by the same expression as in eq. (6.185), except that the noise power density is reduced by half. The error probability of binary DPSK therefore becomes

$$P_e = \frac{1}{2}\exp\left(-\frac{E_b}{N_0}\right) \tag{6.192}$$

An alternative block diagram for the detector, constructed around a matched filter, is shown in Figure 6.50. This may be more suitable for implementation than the block diagram which has been used as the model for the calculations.

6.3.6 COMPARISON OF SOME TYPES OF MODULATION

Some of the expressions for bit error rate that have been derived in this chapter are plotted in Figure 6.51.

As can be seen, the antipodal BPSK modulation gives the lowest bit error rate. A small price, in the form of increased bit error probability, has to be paid in order to spare the carrier recovery, which is possible when DPSK is used. If we then go to the orthogonal modulation methods it is found that the noise properties are relatively bad, both for coherent and non-coherent detection. This is valid for binary modulation. Orthogonal modulation with a large M is considerably better than M-PSK and M-QAM, which is clear from Figure 6.52. The reason for the improvement is that the number of dimensions in the signal space is increased. The improvement has its price in the form of increased bandwidth of the signal which means a low bandwidth efficiency. If 8PSK is considered it is evident that the error probability is definitely higher than for BPSK. But on the other hand, 8PSK only needs to occupy only a third of the bandwidth compared to BPSK at the same transmission rate in bits per second.

Noise and bandwidth efficiency

When designing a communication system the bandwidth and noise level of the transmission channel are often limiting factors. It is therefore important to choose the type of modulation which is best suited for the properties of the channel and

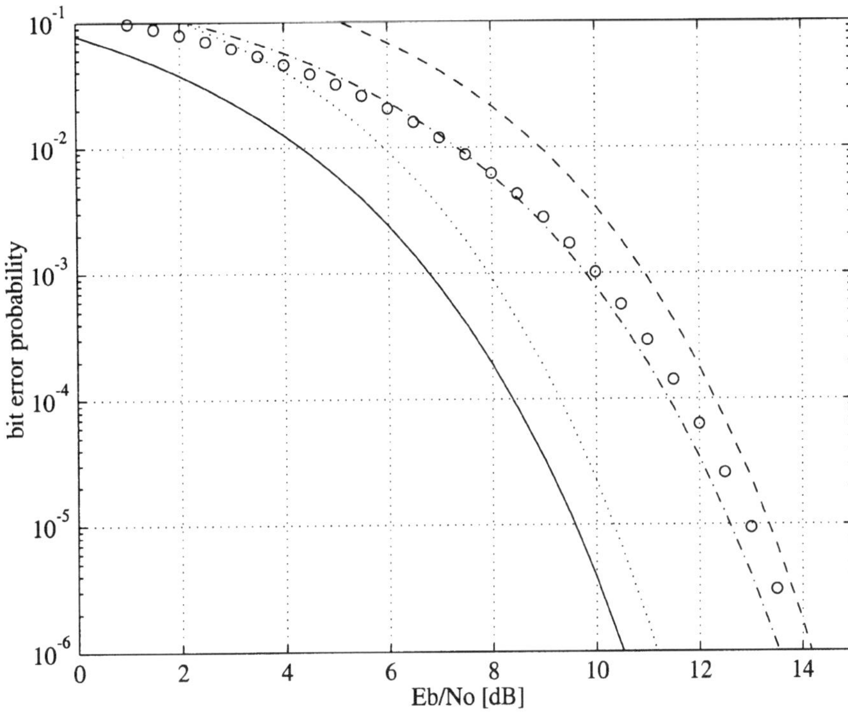

Figure 6.51 Plot of the bit error probability as a function of the signal to noise ratio. Solid, BPSK, QPSK, MSK; dash- dotted, orthogonal BPSK; dotted, DBPSK; dashed, non-coherent FSK; rings, 8PSK.

the requirements of the communication system. In some channels noise is the limiting factor. For example in satellite communication one has a large bandwidth available but a low power due to the limited space in the satellite. In this case a modulation type, such as coherent demodulated orthogonal 16FSK, giving low bit-error probability has to be chosen, possibly in combination with channel coding. In other channels the bandwidth is the limiting factor, for example in a telephone wire. A bandwidth efficient modulation, such as 64QAM, then has to be chosen. A third kind of channel can be both noise and bandwidth limited, in which case TCM could be a suitable alternative. A way of illustrating the different properties of the modulation schemes is to indicate their bandwidth efficiency and required signal to noise ratio for a certain bit error probability in a diagram, as in Figure 6.52.

Consider the single-sided bandwidth of a modulated signal which has been frequency shifted so that the centrum frequency f_c is at 0 Hz. For BPSK, for instance, the signal bandwidth B can be defined as the distance between the null frequency and the position of the first spectrum null. In the case of BPSK, the first null is at $f = 1/T$, which gives $B = 1/T$. The time T is the time interval between the

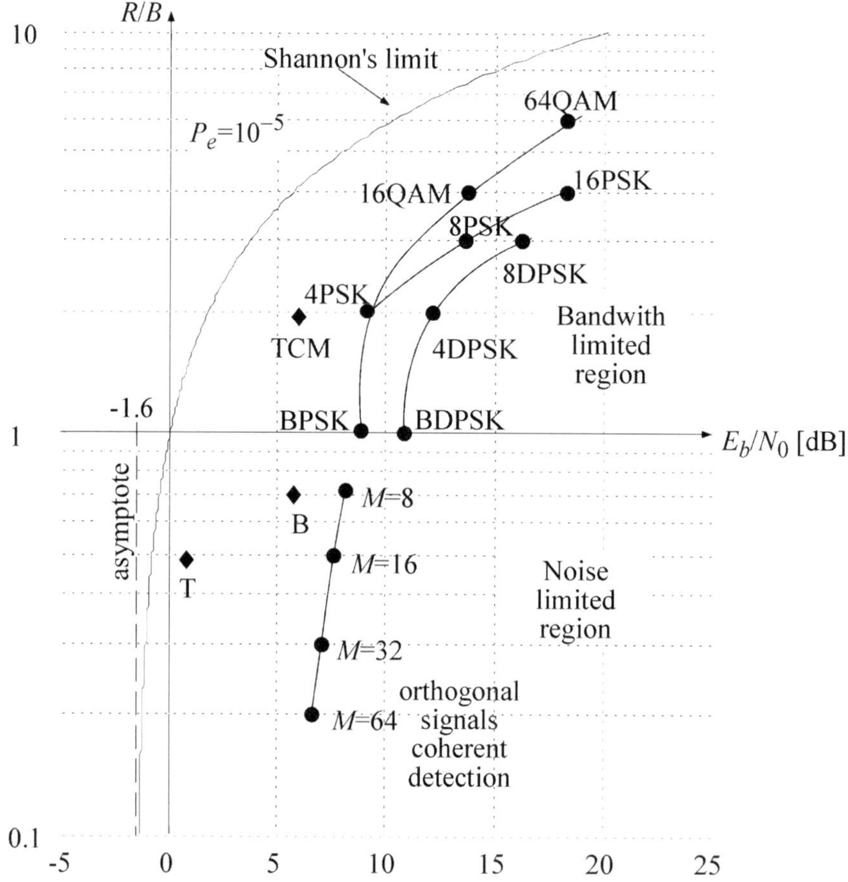

Figure 6.52   Comparison of different modulation schemes at a symbol error probability of 10^{-5}. For T, TCM and B see the text.

symbols. The information transmission per symbol is at most $\log_2 M$, which for the case of BPSK gives 1 bit/symbol. The information rate is $R = \log_2 M/T$ (bits/s). The modulated efficiency is defined as the number of transmitted bits per second and unit bandwidth (bits/s per Hz). For BPSK this ratio then becomes $R/B = 1$, which has been indicated at the value of E_b/N_0 corresponding to a bit error rate of 10^{-5}. For $M = 4$ the information transmission per symbol is doubled, which means the transmission rate, and thus the bandwidth, can be halved for an unchanged transmission capacity in bits/s. Hence, the value of QPSK in the diagram is at $R/B = 2$. In general the modulated signal bandwidth is reduced by a factor of $1/\log_2 M$.

Point B indicates what can be done by a (127,92) BCH code and its location is obtained from Figure 4.5a. The point T indicates what can be achieved by a rate 1/2 turbo code with the generator polynomials $\mathbf{g}_1 = (37)_8$ and $\mathbf{g}_2 = (21)_8$. TCM shows an example of TCM coded signal where an 8PSK signal and a 64 state Ungerboeck code have been used. The coding gain compared to QPSK is 3.6 dB. Compared to the uncoded modulations, all three cases are closer to the Shannon limit.

A limit for what is possible to achieve with regard to the transmission rate is set by Shannon's formula, eq. (3.47), which after rewriting becomes

$$\frac{C}{B} = \log_2\left(1 + \frac{P_R}{N_R}\right) = \log_2\left(1 + \frac{C\,E_b}{B\,N_0}\right) \tag{6.193}$$

The signal power P_R is equal to the energy per bit, E_b, multiplied by the number of bits per second, C, possible to transmit, i.e. $P_R = C \cdot E_b$. For the noise power $N_R = B \cdot N_0$. Rewriting eq. (6.193) makes the connection between E_b/N_0 and C/B clearer.

$$\frac{E_b}{N_0} = \frac{2^{C/B} - 1}{C/B} \tag{6.194}$$

This defines a boundary in the diagram which it is not possible to cross over to the left. The aim should be to come as close to this boundary as possible. For a signal with infinite bandwidth, i.e. $B \rightarrow \infty$, E_b/N_0 approaches ln 2 in the limit, which in dB gives an asymptote at -1.6 dB. Thus, this is the smallest signal to noise ratio needed to transmit one bit of information without errors.

Example 6.6

Binary data are transmitted at a rate of $r_b = 500$ kbps on a radio channel with a 400 kHz bandwidth.

(a) Find the modulation method requiring the smallest signal energy and calculate the value of E_b/N_0 in dB required to achieve $P_{be} \leq 10^{-5}$.

(b) What is the answer if non-coherent detection is used?

Solution   The ratio between required transmission capacity and available bandwidth is

$$\frac{r_b}{B_T} = \frac{500}{400} = 1.25 \tag{6.195}$$

(a) Cf. Figure 6.52. Choose a modulation type having $r_b/B_T = 2$, such as QAM or QPSK. DPSK could also have been used but requires a larger value of E_b/N_0.

The bit error probability is

$$P_{be} = Q\left(\sqrt{\frac{2E_b}{N_0}}\right) \leq 10^{-5} \quad \Rightarrow \quad \frac{E_b}{N_0} \geq \frac{1}{2}4.26^2 = 9.09 \rightarrow 9.59 \text{ (dB)}$$

$$\tag{6.196}$$

(b) Choose a self-synchronising method, for instance DPSK with $M = 4$. An expression for the bit error rate is given in Section 6.6.1. This is

$$P_e \approx \frac{2}{K} Q\left(\sqrt{\frac{4KE}{N_0} \sin^2 \frac{\pi}{2M}}\right) \qquad (6.197)$$

where $K = \log_2 M = 2$. With numerical values inserted we obtain the required signal to noise ratio,

$$P_e \approx \frac{2}{2} Q\left(\sqrt{\frac{4 \cdot 2 E_b}{N_0} \sin^2 \frac{\pi}{8}}\right) \leq 10^{-5} \quad \Rightarrow \quad \frac{E_b}{N_0} \geq \frac{1}{8}\left(\frac{4.26}{0.383}\right)^2 = 15.5$$

$$(6.198)$$

which corresponds to 11.9 dB.

Example 6.7

(a) Which modulation type and which E_b/N_0 are required if the transmission requirement instead is $r_b = 1$ Mbps. As in Example 6.6, the radio channel has a bandwidth of 400 kHz and a highest error probability of $P_{be} \leq 10^{-5}$.
(b) What is the answer if non-coherent detection is used?

Solution   In this case the ratio becomes

$$\frac{r_b}{B_T} = \frac{10^6}{400 \times 10^3} = 2.5 \qquad (6.199)$$

(a) Cf. Figure 6.52. A modulation type with $r_b/B_T = 3$ is required, which gives 8PSK i.e. $M = 8$. The bit error probability is

$$P_e \approx \frac{2}{K} Q\left(\sqrt{2K \frac{E_b}{N_0} \sin^2 \frac{\pi}{M}}\right) \qquad (6.200)$$

where $K = \log_2 M = 3$. The required signal to noise ratio becomes

$$P_e \approx \frac{2}{K} Q\left(\sqrt{2 \times 3 \frac{E_b}{N_0} \sin^2 \frac{\pi}{8}}\right) \leq 10^{-5} \quad \Rightarrow$$

$$(6.201)$$

$$\frac{E_b}{N_0} \geq \frac{1}{6}\left(\frac{4.26}{0.383}\right)^2 = 20.7 \rightarrow 13.2 \text{ dB}$$

(b) Self-synchronising, non-coherent DPSK. The bit error probability is

$$P_e \approx \frac{2}{K} Q\left(\sqrt{\frac{4KE_b}{N_0} \sin^2 \frac{\pi}{2M}} \right) \tag{6.202}$$

which gives the required signal to noise ratio

$$P_e \approx \frac{2}{3} Q\left(\sqrt{4 \cdot 3 \frac{E_b}{N_0} \sin^2 \frac{\pi}{16}} \right) \leq 10^{-5} \quad \Rightarrow \tag{6.203}$$

$$\frac{E_b}{N_0} \geq \frac{1}{12} \left(\frac{4.26}{0.195} \right)^2 = 39.8 \rightarrow 16.0 \text{ dB}$$

Example 6.8

A binary transmission system with an output power of $S_T = 200$ mW, losses of $L = 90$ dB and a single-sided noise density of $N_0 = 10^{-15}$ W/Hz should have an error probability of $P_e \leq 10^{-4}$. What is the highest transmission rate if: (a) non-coherent FSK, (b) DPSK or (c) coherent BPSK, is used?

Solution The loss expressed in a linear value becomes

$$L = 10^{90/10} = 10^9 \tag{6.204}$$

The received power becomes

$$\frac{S_T}{L} = 0.2 \times 10^{-9} = E_b r_b \quad \text{(W)} \tag{6.205}$$

and SNR

$$\gamma_b = \frac{E_b}{N_0} = \frac{0.2 \times 10^{-9}}{r_b 10^{-15}} = \frac{0.2}{r_b} 10^6 \tag{6.206}$$

(a) For non-coherent FSK (see table in Section 6.6.1.)

$$P_e = \frac{1}{2} \exp\left(-\frac{\gamma_b}{2} \right) \quad \Rightarrow \quad -2\ln(2P_e) = \gamma_b = 17 \tag{6.207}$$

The maximum allowed bit rate for $P_e \leq 10^{-4}$ becomes

$$17 = \frac{0.2 \times 10^6}{r_b} \quad \Rightarrow r_b = \frac{0.2 \times 10^6}{17} = 11.7 \text{ kbaud} \tag{6.208}$$

The baud unit means the number of information symbols per second. Since each symbol in this case corresponds to one bit, the unit here is kbit/s = kbaud.
(b) For DPSK, eq. (6.192) gives

$$\gamma_b = -\ln(2P_e) = 8.5 \quad \Rightarrow \quad r_b = \frac{0.2 \times 10^6}{8.5} = 23.5 \text{ kbaud} \tag{6.209}$$

(c) For BPSK

$$10^{-4} = Q\left(\sqrt{2\gamma_b}\right) \quad \Rightarrow \quad \gamma_b \approx 6.9 \quad \Rightarrow \quad r_b = \frac{0.2 \times 10^6}{6.9} = 28.9 \text{ kbaud}$$

$$\tag{6.210}$$

It is interesting to compare with the prediction from Shannon's formula. An E_b/N_0 of 6.9 corresponds to 8.4 dB. If the Shannon limit for this value is read in Figure 6.52, it is clear that the ratio R/B can be approximately 5.2. For BPSK the required channel bandwidth is $B_T \approx r_b$, which means that $R/B = 1$. The upper limit for transmission of information is therefore located at 5.2 times the transmission rate at BPSK, i.e. $28.9 \times 5.2 = 150$ kbaud. Thus, there is a lot to gain with a more efficient coding and modulation. A switch to QPSK would be one way to approach the limit. For QPSK the ratio is $R/B = 2$.

Example 6.9

Determine the smallest Euclidean distance, d_{min}, for MSK. Binary signalling, rectangular frequency pulse and modulation index $h = 1/2$ are assumed.

Solution Since the modulated waveform depends on earlier symbols, a longer section of the signal has to be studied. Usually a transfer function, $T(W,I)$, is made up from the trellis, the same way as in convolutional codes. In this case, it is apparent already from a visual inspection of the trellis. The paths which have the smallest mutual distance are those that have been depicted by bold lines in Figure 6.53 and correspond to symbols $+1-1$ and $-1+1$, respectively. From the table given in

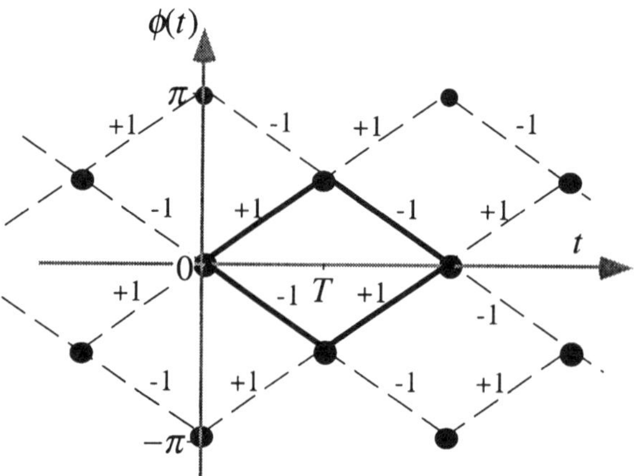

Figure 6.53 The paths giving d_{min} have been depicted by fat lines.

Example 5.10, the signals for corresponding state transitions are obtained. Other combinations than the ones chosen here can also give d_{min}, but none of these have a smaller distance.

Suppose that the number of symbols between the paths split up and until they join again is N, which gives for the total energy (the signal is defined in Example 5.10)

$$\int_0^{NT} |u(t)|^2 dt = \int_0^{NT} |A_c \exp(j\phi(t))|^2 dt = A_c^2 NT = NE \tag{6.211}$$

For a CPM signal the Euclidean distance between two signals $u_I(t)$ and $u_{II}(t)$ then becomes

$$d^2 = \int_0^{NT} |u_I(t) - u_{II}(t)|^2 dt = 2NE\left(1 - \frac{1}{NE}\int_0^{NT} \cos[\phi_I(t) - \phi_{II}(t)] dt\right) \tag{6.212}$$

For the case studied here, the number of symbols from the split to the merger is 2, which gives the upper integration limit $2T$. The smallest Euclidean distance for MSK therefore becomes

$$d_{min}^2 = \int_0^{2T} |u_{+1,-1}(t) - u_{-1,+1}(t)|^2 dt$$

$$= 4E\left(1 - \frac{1}{2E}\left[\int_0^T \underbrace{\cos\left(\frac{\pi t}{2T} + \frac{\pi t}{2T}\right)}_{\cos(\pi t/T)} dt + \int_0^T \underbrace{\cos\left(\frac{\pi}{2} - \frac{\pi t}{2T} + \frac{\pi}{2} - \frac{\pi t}{2T}\right)}_{-\cos(\pi t/T)} dt\right]\right)$$

$$= 4E_b \tag{6.213}$$

Note that the signals as they are given in the table, Example 5.10, are defined on the interval $0 \leq t \leq T$. The integration limit has been adjusted to that. The result of the computation gives $d_{min} = 2\sqrt{E_b}$, which is the same as for BPSK. But, as is clear from Chapter 5 the modulated signal has a much narrower bandwidth. These good properties are gained at a price of higher complexity in the transmitter and receiver. The latter has to be implemented by using for example the Viterbi algorithm.

Example 6.10

In many cases one single noise source, generating impulsive interference, is dominating. Examples of objects generating such interference are: switching circuits in telephony systems, motors etc. Impulse noise is often modelled as filtered white noise with a power of σ^2 and a Laplace distributed density function, PDF (Figure 6.54).

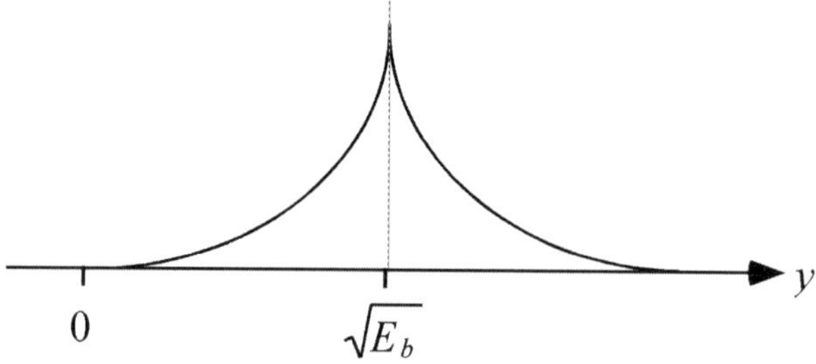

Figure 6.54 Density function for impulse noise around $\sqrt{E_b}$.

$$p_N(n) = \frac{1}{\sqrt{2\sigma^2}}\exp(-\sqrt{2}|n|/\sigma) \tag{6.214}$$

Suppose a system where binary orthogonal modulation is used. Compute an expression for the error rate P_e if a receiver which has been optimised for Laplace noise is used.

Solution Both signal alternatives, s_1 and s_2, are supposed to have the same probability. Let the signals make up basis vectors in the signal space such that $s_1 = \psi_1$ and $s_2 = \psi_2$. By using eq. (6.50) it is given that the receiver will make the following decision:

$$|\psi_1| + |\psi_2 - s_2| \underset{s_2}{\overset{s_1}{\gtrless}} |\psi_1 - s_1| + |\psi_2| \tag{6.215}$$

To compute the bit error rate the decision boundaries have to be determined. The decision boundaries are obtained by finding the ψ_1 and ψ_2 fulfilling $|\psi_1| + |\psi_2 - s_2| = |\psi_1 - s_1| + |\psi_2|$. The signal space must be divided in different regions. In Figure 6.55 a plot of the signal space is shown, with the decision boundaries depicted.

Region	Definition region	Equation	Decision boundary
I	$\psi_1 > s_1 \lor \psi_2 > s_2$	$\psi_1 + \psi_2 - s_2 = \psi_1 + \psi_2 - s_1$	No boundary
II	$s_1 > \psi_1 > 0 \lor \psi_2 > 0$	$\psi_1 + \psi_2 - s_2 = -\psi_1 + \psi_2 + s_1$	$\psi_1 = (s_1 + s_2)/2$
III	$\psi_1 > 0 \lor s_2 > \psi_2 > 0$	$\psi_1 - \psi_2 + s_2 = -\psi_1 + \psi_2 + s_1$	$\psi_1 - \psi_2 = (s_1 - s_2)/2$
IV	$\psi_1 > s_1 \lor s_2 > \psi_2 > 0$	$\psi_1 - \psi_2 + s_2 = \psi_1 + \psi_2 - s_1$	$\psi_2 = (s_1 + s_2)/2$
V	$\psi_1 < 0 \lor \psi_2 < 0$	$-\psi_1 - \psi_2 + s_2 = -\psi_1 - \psi_2 + s_1$	No boundary
VI	$s_1 > \psi_1 > 0 \lor \psi_2 < 0$	$\psi_1 - \psi_2 + s_2 = \psi_1 - \psi_2 + s_1$	$\psi_1 = (s_1 - s_2)/2$
VII	$\psi_1 < 0 \lor \psi_2 > s_2$	$\psi_1 - \psi_2 + s_2 = \psi_1 - \psi_2 - s_1$	No boundary
VIII	$\psi_1 < 0 \lor s_2 > \psi_2 > 0$	$-\psi_1 - \psi_2 + s_2 = -\psi_1 + \psi_2 + s_1$	$\psi_2 = (s_1 - s_2)/2$
IX	$\psi_1 < 0 \lor \psi_2 > 0$	$-\psi_1 + \psi_2 - s_2 = -\psi_1 + \psi_2 + s_1$	No boundary

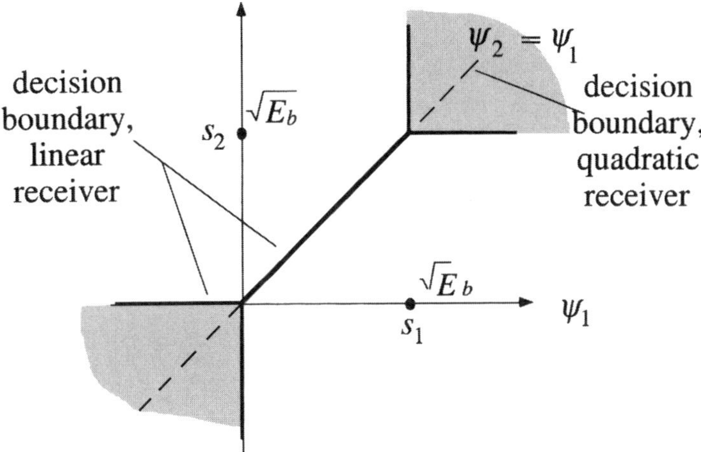

Figure 6.55 Signal space. Thick lines indicate decision boundaries for a receiver optimised for Laplace noise. Dashed line corresponds to receiver optimised for Gaussian noise. In the shaded area the probability is equal for both signal alternatives.

Note that both signal alternatives are equally probable in the shaded area. This means that in these areas one can just as well toss a coin about which of the signals was sent. Another strategy is to steadily chose one of the alternatives. In both cases one will have 50% errors. The probability to do an error, given that the signal s_1 has been sent, is derived in the following way. Each region, I–IX, gives one function to integrate. For region I is obtained

$$P_e(\mathrm{I}|s_1) = \frac{1}{2\sigma^2} \int_{s_2}^{\infty} \int_{s_1}^{\infty} \exp(-\sqrt{2}(\psi_1 + \psi_2 - s_1)/\sigma)\,d\psi_1 d\psi_2 = \frac{1}{4}\exp(-\sqrt{2}s_2/\sigma)$$

(6.216)

The error probability for s_1 is obtained by computing the error probability in every region lying on the other side of the boundary compared to the signal s_1. In the regions where both symbols have equal likelihood some of the strategies giving 50% error is chosen, which means that the integral for those regions shall be multiplied by 1/2. The error probabilities in all regions are thereafter summed. The probability of doing an error when s_2 is sent is equally large as for s_1 due to the symmetry of the problem. Since $s_1 = s_2 = \sqrt{E_b}$, the total error probability is

$$P_e = \frac{\exp(-(\sqrt{2E_b})/\sigma)}{2}\left(\frac{1}{\sigma}\sqrt{\frac{E_b}{2}} + 1\right)$$

(6.217)

Example 6.11

For a downlink, i.e. from a satellite to a receiver on the earth, we have the following parameters:

Satellite transmission power:	35 dBW
Carrier frequency:	7.7 GHz
Satellite effective antenna aperture:	1 m^2
Receiver noise temperature, incl. antenna:	1000 K
Receiver antenna gain:	6 dB
Signal bandwidth:	1 MHz
Distance transmitter-receiver:	41000 km
Miscellaneous system losses:	5 dB

(a) What is the received signal power in dBW?

(b) What is the receiver noise level in dBW?

(c) Calculate the signal to noise ratio in dB.

(d) What is the error probability of: binary BPSK, binary DPSK, binary non-coherent FSK and QPSK with the SNR calculated in (c).

Solution   (a) The received signal power is

$$10\log_{10}P_R = 20\log_{10}\frac{\lambda}{4\pi d} + 10\log_{10}P_T - 10\log_{10}L$$

$$+ 10\log_{10}G_T + 10\log_{10}G_R$$

$$= -202 + 35 - 5 + 39 + 6 = -127 \text{ dBW} \qquad (6.218)$$

where

$$G_T = 4\pi\frac{A_T}{\lambda^2} = 4\pi\frac{1}{1.5\times 10^3} = 8.28\times 10^3 \quad \text{and} \quad \lambda = \frac{3\times 10^8}{7.7\times 10^9} \approx 0.04$$

(b) The noise power is

$$10\log_{10}N_R = 10\log_{10}\left[kT_0\frac{T_e}{T_0}B\right] = \overbrace{10\log_{10}(kT_0)}^{\text{standard term: }-204} + 10\log_{10}\left(\frac{T_e}{T_0}\right) + 10\log_{10}B$$

$$= -204 + 5.38 + 60 = -139 \text{ dBW}$$

$$(6.219)$$

(c) SNR $= -127 + 139 = 12$ dB, which corresponds to 16 times. For the calculation of the bit error rate also refer to eq. (2.29) stating that $E_b/N_0 = P_R/(2N_R\log_2 M)$.

(d) The bit error rate becomes (see table in Appendix 6.6.1)

$$\text{for BPSK}: \quad Q\left(\sqrt{\frac{2E_b}{N_0}}\right) = Q\left(\sqrt{16}\right) = 3.17 \times 10^{-5} \tag{6.220}$$

$$\text{for DPSK}: \quad \frac{1}{2}\exp\left(-\frac{E_b}{N_0}\right) = 1.68 \times 10^{-4} \tag{6.221}$$

$$\text{for FSK}: \quad \frac{1}{2}\exp\left(-\frac{E_b}{2N_0}\right) = 9.16 \times 10^{-3} \tag{6.222}$$

$$\text{for QPSK}: \quad Q\left(\sqrt{\frac{2E_b}{N_0}}\right) = 3.17 \times 10^{-5} \tag{6.223}$$

6.4 Diversity

When the signal propagates along many different paths the signal level will vary depending on the receiver location, as has been shown in Chapter 2. The phenomenon is called fading and means that the bit error rate can be low in one place and high in another place not far away. Even if the receiver is stationary the transmission quality can change through changes in the environment, for example by objects being moved. There are statistical models for how the signal changes can be formulated. One such model is the Rice density:

$$p_\alpha(\alpha) = \frac{\alpha}{\sigma^2}\exp[-(\alpha^2 + A^2)/2\sigma^2]I_0\left(\frac{\alpha A}{\sigma^2}\right), \quad \alpha \geq 0 \tag{6.224}$$

which has been described earlier in eq. (6.170). Here it describes the probability density for the signal amplitude α in a specific point given a direct signal with the amplitude A and many reflections. The variance σ is adjusted to the variations in signal level of the channel to be modelled and depends on wavelength, environment such as city, rural, etc. The ratio $K = A^2/(2\sigma^2)$ is usually denoted the Ricean factor and is often independent of the signal strength. If no direct wave exists but only reflected signals reaches the receiver, then $A = 0$ and the Rice density changes to a Rayleigh density. In that case, we name it a Rayleigh fading channel (another model for fading is the Nakagami-m density).

To mitigate fading, a system can be equipped with diversity, i.e. the communication is done over a number of transmission channels whose fading patterns are statistically independent. The probability that at least some of the channels will have low bit error rate increases with the number of independent channels. Three important diversity methods are:

- *Time diversity:* when the channel is time variant it can be beneficial to transmit the same message at many different occasions. Channel coding and interleaving helps to keep the bit error rate down.

- *Frequency diversity:* since the pattern with destructive and constructive interference looks different at various carrier frequencies it can be beneficial to transmit the same message at different frequencies.
- *Antenna diversity:* since the fading pattern is geographically induced it can be beneficial to have several antennas which are geographically separated. This can be used at both transmitter and receiver. Most common, however, is at the receiver and is in this case sometimes called receiver diversity. The separation necessary between the antennas depends on the spread of reflection points. In city environment it can be enough with half a wavelength separation to achieve statistically independent channels. Another possibility to achieve antenna diversity is to have the antennas at the same place but using different direction of polarisation for the received electromagnetic wave. This is called polarisation diversity.

6.4.1 METHODS OF COMBINATION

Assume L independent diversity channels and that the wanted signal in channel l has been attenuated by a factor α_l and phase shifted by an angle θ_l. The signal to noise ratio on the channel output therefore is $\alpha_l^2 E_b/N_0$. There are different ways to combine the signals from the different channels. The optimal one is maximal ratio combining, where the wanted signals in the different diversity channels are summed after being co-phased by multiplication with the weight $w_l = g_l\exp(-j\theta_l)$. The total signal level becomes

$$\sqrt{E_{b,L}} = \sum_l |w_l|\, \alpha_l\sqrt{E_b} = \sum_l g_l\, \alpha_l\sqrt{E_b}$$

Since the noise is assumed independent and equally large on the various channels, the total noise power is

$$N_L = \sum_l g_l^2 N_0$$

The total signal to noise ratio then becomes

$$\gamma_b = \frac{E_{b,L}}{N_L} = \frac{\left(\sum_{l=1}^{L} g_l\, \alpha_l\sqrt{\dfrac{E_b}{N_0}}\right)^2}{\sum_{l=1}^{L} g_l^2} \leq \frac{\left(\sum_{l=1}^{L} g_l^2\right)\sum_{l=1}^{L}\left(\alpha_l\sqrt{\dfrac{E_b}{N_0}}\right)^2}{\sum_{l=1}^{L} g_l^2} \tag{6.225}$$

where Schwarz inequality has been used. The upper limit is reached if

$g_l = \alpha_l \sqrt{E_b/N_0}$. In short, the amplitude of the signal from diversity channel l is weighted by g_l. Thereafter the signals are co-phased and summed. The best possible signal to noise ratio is thus obtained in the summed signal.

The maximal ratio combining described above gives the best possible signal to noise ratio given that the signal to noise ratio can be determined separately at each diversity channel, or branch. There are more ways to choose the combination of the signals which are easier to implement. Among these equal gain should be mentioned; this combines all branches with equal weight, or selection diversity, which chooses the strongest of the signals.

6.4.2 BIT ERROR RATE

The mean value of the bit error rate given L diversity branches is

$$\bar{P}_b = \int_0^\infty \cdots \int_0^\infty P_b\left(\frac{E_b}{N_0}, \{\alpha_l\}_{l=1}^L\right) p_{\alpha_1}(\alpha_1) \cdots p_{\alpha_L}(\alpha_L) d\alpha_1 \cdots d\alpha_L \tag{6.226}$$

Assume binary signalling, i.e. $x_k = \pm 1$. The bit error rate for maximal ratio combining then becomes

$$P_b\left(\frac{E_b}{N_0}, \{\alpha_l\}_{l=1}^L\right) = Q\left(\sqrt{\frac{2E_b}{N_0} \sum_{l=1}^L \alpha_l^2}\right) = \frac{1}{\pi} \int_0^{\pi/2} \exp\left(-\frac{\frac{E_b}{N_0} \sum_{l=1}^L \alpha_l^2}{\sin^2(\theta)}\right) d\theta \tag{6.227}$$

An alternative way to write the Q function (see Section 6.6.2), which simplifies the integration is used. If eq. (6.227) is inserted in eq. (6.226) this gives

$$\bar{P}_b = \int_0^\infty \cdots \int_0^\infty \frac{1}{\pi} \int_0^{\pi/2} \exp\left(-\frac{\frac{E_b}{N_0} \sum_{l=1}^L \alpha_l^2}{\sin^2(\theta)}\right) p_{\alpha_1}(\alpha_1) \cdots p_{\alpha_L}(\alpha_L) d\theta \, d\alpha_1 \cdots d\alpha_L \tag{6.228}$$

The strength with this formulation of the Q function is that the integrals can be separated and one integral for each α_l is obtained. Assume equal fading statistics for all diversity branches which, after changing the order of integration, gives

$$\bar{P}_b = \frac{1}{\pi} \int_0^{\pi/2} \left[\int_0^\infty \exp\left(-\frac{E_b \alpha^2}{N_0 \sin^2(\theta)}\right) p_\alpha(\alpha) d\alpha\right]^L d\theta \tag{6.229}$$

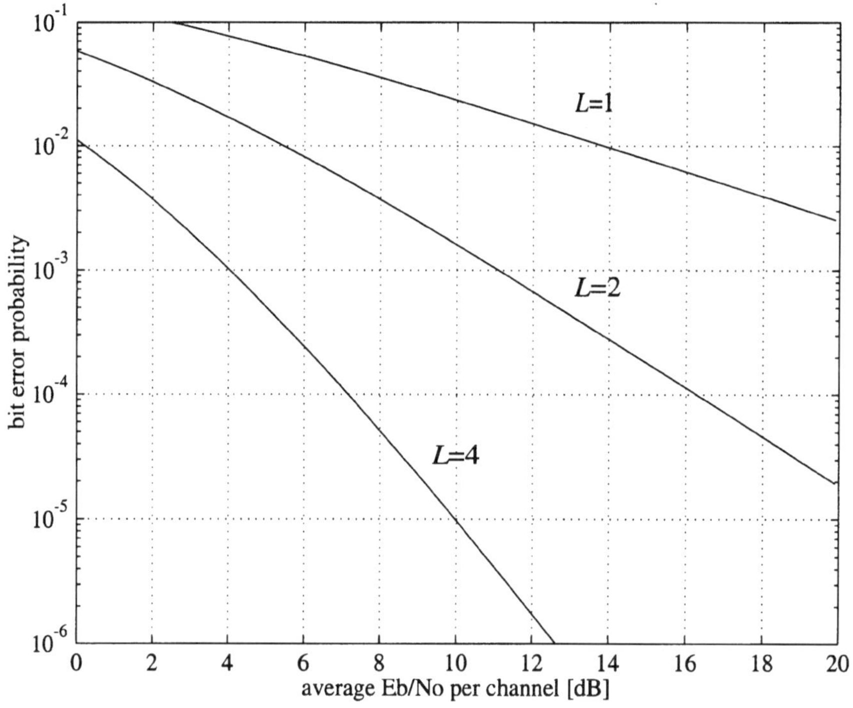

Figure 6.56 Bit error rate for BPSK as a function of the number of diversity channels in Rayleigh fading and with maximal ratio combining.

If eq. (6.224) is used, at Rayleigh fading the inner integral is

$$\int_0^\infty \exp\left(-\frac{E_b\alpha^2}{N_0\sin^2(\theta)}\right)p_\alpha(\alpha)d\alpha = \frac{1}{\left(1 + \dfrac{2\sigma^2 E_b}{N_0\sin^2\theta}\right)} \qquad (6.230)$$

With eq. (6.230) inserted in eq. (6.229), an expression for computation of the bit error rate at binary signalling in Rayleigh fading and maximal ratio combining, with L diversity channels, is obtained. In eq. (6.230) the value $2\sigma^2 E_b/N_0$ corresponds to the average of the signal to noise ratio in each channel (Figure 6.56).

An interesting observation is that a fading channel is turned into an AWGN channel by increasing the number of diversity branches.

6.5 The radio receiver

The task of a radio receiver is to select the desired signal among a large number of radio signals in space and to amplify and transform the signal to suit the user of the message. Some important properties are:

- *Adjacent channel attenuation*, i.e. the filter limiting the frequency band that is received, must have a stop band attenuation that is high enough, so that a weak wanted signal can be received without interference from a strong signal on the adjacent frequency. The capability to separate signals on different frequencies is called selectivity.
- *Spurious*. In amplifiers, mixers, etc. non-linearities, other than the wanted ones, occur (cf. Example 5.5). This is the origin to non-desired frequencies being transformed into the receiver pass band. For example, suppose that the receiver local oscillator signal is not a perfect sinusoidal and therefore has frequency components at $n_0 f_{LO}$, where n_0 is an integer larger than one. If n_1 is an integer and f_{MF} is the frequency that will pass the receiver filter, radio signals with frequencies fulfilling $n_1 f_{RF} = n_0 f_{LO} \pm f_{MF}$ will pass into the receiver and interfere with the wanted signal. Also the amplifier non-linearities can cause spurious signals by the mixing of two strong signals which gives a signal with a frequency passing the receiver filter.

A receiver which in most cases is capable of having high selectivity and low spurious is the super-heterodyne receiver, illustrated in Figure 6.57. Most types of receivers discussed in this book are to some extent based on this receiver. A brief description is therefore justified.

The electromagnetic waves in the surrounding space are captured by the antenna. There are several signals and to prevent the first amplifier from saturating and thereby distorting the signals, the number of signals must be limited. This is achieved by a bandpass filter at the input which picks out the desired frequency band. The first low-noise amplifier amplifies the signal to a level suitable for the mixer. The second input signal of the mixer is a sinusoidal signal generated by the local oscillator. In the mixer the input signal is down-converted to a lower frequency than the original one (this operation is described in Section 5.3.1.). The next stage in the chain is the intermediate frequency (IF) filter, which is a

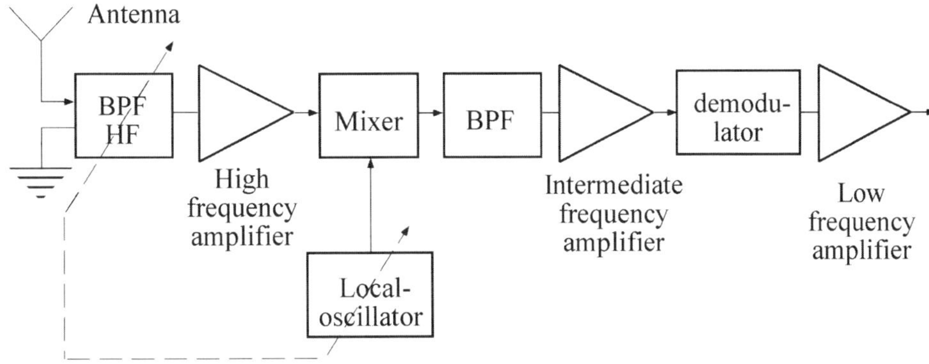

Figure 6.57 Super heterodyne receiver.

bandpass filter. It has a very high stopband attenuation and removes frequencies other than the desired ones. This filter corresponds to one of the filters indicated BPF n in Figure 1.7. Due to the large requirements on the attenuation, the frequency limits of the filter are fixed. To be able to change the receiving frequency, the frequency of the local oscillator and the HF filter are instead changed. This is indicated by the dashed arrows in Figure 6.57. After the IF filter follows the IF amplifier, with a very high gain. The signal then has a level suitable for the demodulator or the detector. The last stage adapts the signal to the succeeding usage. In the case of analogue communication this stage is usually a low-frequency amplifier.

The progress is moving towards an increased use of digital signal processing. This is utilised for the detector, the equaliser, if any, and successive signal processing. In such a receiver, the signal is down-converted to baseband. To take care of all phases in the received narrowband signal, the signal is divided into I and Q components. Such a receiver can be realised as a homodyne, i.e. the signal is converted directly from high frequency to baseband, or heterodyne where the signal takes the leap down to baseband via an intermediate frequency, as above. A digital receiver is shown in Figure 6.58. For the homodyne case, the diagram in the figure is connected directly after the high frequency amplifier (cf. Figure 6.57), and replaces the parts following this one. The receiver selectivity now depends on the stopband attenuation of the lowpass filters. In the heterodyne case the diagram from Figure 6.58 is connected before the demodulator and replaces this one.

More receiver blocks will be made digitally in the future. The A/D conversion will be made at higher and higher frequencies. This offers great versatility in the receiver which will be able to work with several different communication standards only by a change of software.

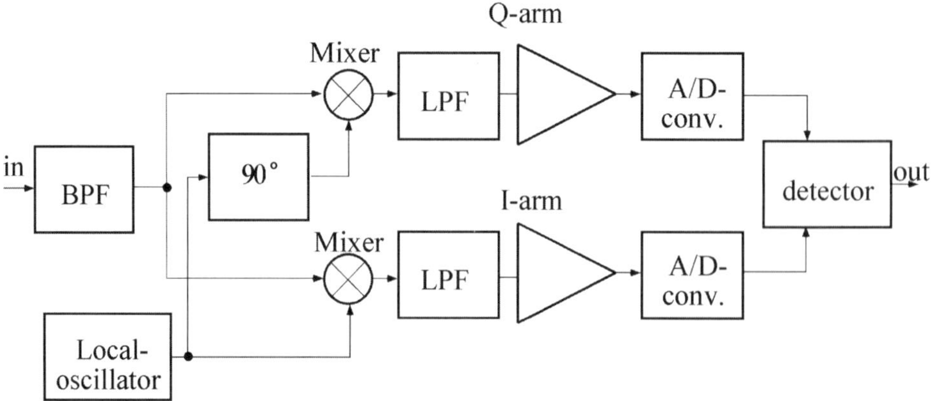

Figure 6.58 Receiver for digital communication. The detector block may consist of a distance, correlator or matched filter receiver, described in this chapter.

6.6 Appendix

6.6.1 ERROR PROBABILITIES FOR SOME CASES OF MODULATION

Modulation type:	Bit error probability:
Coherently detected: BPSK, QPSK, MSK	$P_b = Q\left(\sqrt{\dfrac{2E_b}{N_0}}\right)$
Coherent detection: generally for binary signalling with signal distance d.	$P_b = Q\left(\dfrac{d}{\sqrt{2N_0}}\right)$
Coherently detected: binary orthogonal FSK	$P_b = Q\left(\sqrt{\dfrac{E_b}{N_0}}\right)$
Non-coherently detected: DPSK (differential PSK)	$P_b = \dfrac{1}{2}\exp\left(-\dfrac{E_b}{N_0}\right)$
Non-coherently detected: binary FSK	$P_b = \dfrac{1}{2}\exp\left(-\dfrac{E_b}{2N_0}\right)$
MPSK, $M \geq 4$ $K = \log_2 M$	$P_b \approx \dfrac{2}{K}Q\left(\sqrt{\dfrac{2KE}{N_0}}\sin^2\dfrac{\pi}{M}\right)$
MDPSK $K = \log_2 M$	$P_b \approx \dfrac{2}{K}Q\left(\sqrt{\dfrac{4KE}{N_0}}\sin^2\dfrac{\pi}{2M}\right)$

6.6.2 THE Q-FUNCTION

$$Q(x) = \frac{1}{\sqrt{2\pi}}\int_x^\infty \exp\left(-\frac{\lambda^2}{2}\right)d\lambda$$

$$\mathrm{erf}(x) = \frac{2}{\sqrt{\pi}}\int_0^x \exp(-\lambda^2)d\lambda = 1 - 2Q(\sqrt{2}\,x)$$

$$\text{erfc}(x) = \frac{2}{\sqrt{\pi}} \int_x^{\infty} \exp(-\lambda^2)d\lambda = 1 - \text{erf}(x) = 2Q(\sqrt{2}\,x)$$

$$\Phi(x) = \frac{1}{\sqrt{2\pi}} \int_0^x \exp\left(-\frac{\lambda^2}{2}\right)d\lambda = \frac{1}{2} - Q(x)$$

$$Q(x) = \frac{1}{2}\text{erfc}\left(\frac{x}{\sqrt{2}}\right)$$

$$Q(x) = \frac{1}{\pi} \int_0^{\pi/2} \exp\left(\frac{-x^2}{2\sin^2(\theta)}\right) d\theta, \quad x \geq 0$$

The Q-function is bounded by

$$\frac{x}{\sqrt{2\pi}(1 + x^2)}\exp(-x^2/2) \leq Q(x) \leq \frac{1}{\sqrt{2\pi}x}\exp(-x^2/2)$$

For large values of x, the approximation below can be used:

$$Q(x) \approx \frac{1}{\sqrt{2\pi}x}\exp\left(-\frac{x^2}{2}\right), \quad x > 3$$

Numerical calculation of Q can be carried out by using a suitable length of the

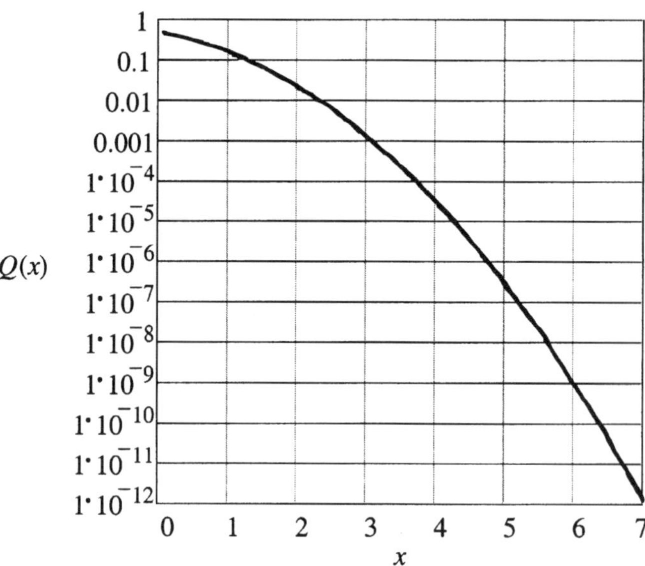

Figure 6.59 Q-function plot.

series:

$$Q(x) = \frac{1}{2} - \frac{2}{\pi} \sum_{n=1,\ n\ \text{odd}}^{\infty} \frac{\exp\left(-4n^2\pi^2/2T^2\right)\sin(2\pi nx/T)}{n}$$

T is a parameter influencing the accuracy. For $T = 28$ and with 17 terms in the series, the relative error is smaller than 6.3×10^{-7} for $x < 6$ and 5×10^{-10} for $x < 4.85$.

6.6.3 SOME IMPORTANT DENSITY FUNCTIONS IN COMMUNICATION

Name:	Density function	Example of application:
Normal (Gaussian)	$\dfrac{1}{\sqrt{2\pi\sigma^2}}\exp\left(-\dfrac{x^2}{2\sigma^2}\right)$	Interference model
Rayleigh	$\dfrac{2x}{E\{X^2\}}\exp\left(-\dfrac{2x^2}{E\{X^2\}}\right)$	Amplitude of normal distributed interference, fading, non-coherent detection
Rice	$x\exp\left(-\dfrac{1}{2}\left(x^2+s^2\right)\right)I_0(sx)$	Fading with component of direct signal, non-coherent detection
Log-normal	$\dfrac{1}{\sqrt{2\pi}ax}\exp\left(-\dfrac{(1/2)\ln^2(x/x_m)}{a^2}\right)$	Variation of SNR owing to slow amplitude variations at mobile communication
ε-contaminated	$(1-\varepsilon)f_1 + \varepsilon f_2$	A signal with a main density of f_1 but with an additional fraction ε of another density f_2

6.6.4 NORM

In this book the names distance and norm occur. Unless otherwise stated, Euclidean distance or quadratic norm is intended (however not in Chapter 4 where Hamming distance is commonly used). For two vectors $\mathbf{x}$ and $\mathbf{y}$ the concept above is defined:

$$d(\mathbf{x}, \mathbf{y}) = \|\mathbf{x} - \mathbf{y}\| = \sqrt{(\mathbf{x}-\mathbf{y})(\mathbf{x}-\mathbf{y})^H} = \sqrt{\sum_{i=1}^{n}|x_i - y_i|^2}$$

For continuous time signals or functions it is given that

$$d(f,g) = \|f(t) - g(t)\| = \sqrt{\int_a^b |f(t) - g(t)|^2 dt}$$

The inner product, or scalar product, of two vectors is defined

$$\langle \mathbf{x}, \mathbf{y} \rangle = \mathbf{x} \cdot \mathbf{y}^H = \sum_{i=1}^{n} x_i y_i^*$$

The corresponding for continuous functions is

$$\langle f, g \rangle = \int_a^b f(t) g^*(t) dt$$

In both cases $\langle \mathbf{x}, \mathbf{y} \rangle = \|\mathbf{x}\| \, \|\mathbf{y}\| \cos(\theta)$. When $\langle \mathbf{x}, \mathbf{y} \rangle = 0$ or $\langle f, g \rangle = 0$ the respective vectors or functions are orthogonal.

Exercise 6.1

Compute a union bound for the error probability when the signal set shown in the figure is used.

- without using d_{min}
- by using d_{min}

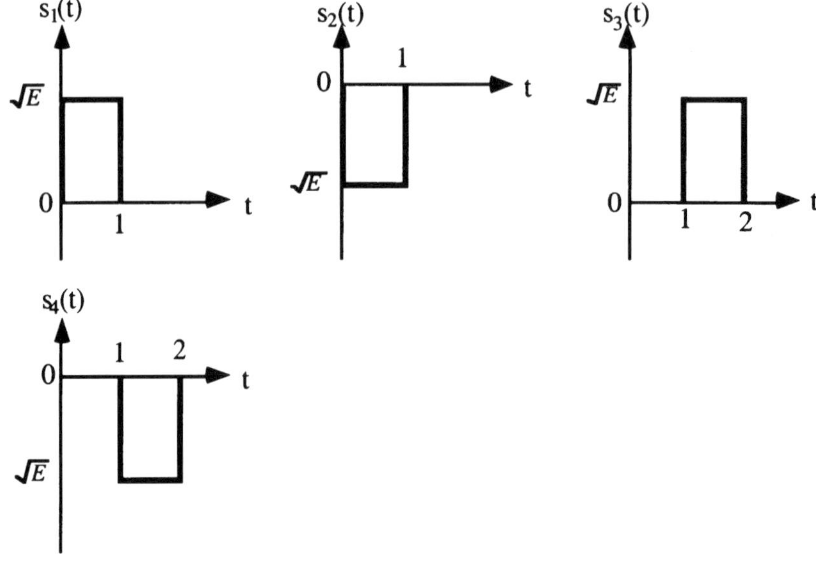

Exercise 6.2

Compute the improvement in error probability relatively to an uncoded signal if a double error correcting (24,12) block code is used. Assume BPSK modulation, coherent detection and a received $E_b/N_0 = 10$ dB.

Exercise 6.3

The distance between transmitter and receiver in a radio connection is 50 km. Free space between transmitter and receiver can be assumed. The antenna gain at the transmitter is 3 dB and at the receiver 3 dB. The receiver noise factor is 3.5 dB. The carrier frequency is 500 MHz and the bandwidth 250 kHz. The transmitter power is 4 W. What is the bit error rate in the transmitted message if BPSK is used? Maximal signalling rate is assumed.

Exercise 6.4

A data transmissions system uses the four signal alternatives $s_i(t)$, $i = 1,2,3,4$.

$$s_1(t) = \begin{cases} A\cos(2\pi f_c t) & 0 \le t \le D \\ 0 & \text{otherwise} \end{cases}$$

$$s_2(t) = \begin{cases} A\cos(2\pi f_c t + \dfrac{\pi}{2}) & 0 \le t \le D \\ 0 & \text{otherwise} \end{cases}$$

$$s_3(t) = \begin{cases} A\cos(2\pi f_c t + \pi) & 0 \le t \le D \\ 0 & \text{otherwise} \end{cases}$$

$$s_4(t) = \begin{cases} A\cos(2\pi f_c t + \dfrac{3\pi}{2}) & 0 \le t \le D \\ 0 & \text{otherwise} \end{cases}$$

The bit rate is $R_b = 19.2$ kbit/s. The interference is additive white noise with the spectral density $N_0/2$.
(a) What is this modulation scheme called?
(b) Determine the optimal decision regions of the receiver.
(c) Draw a detailed block diagram for the optimal receiver.
(d) Which magnitude should A assume for the transmitted energy per bit, E_b, to be 10^{-3}?

Exercise 6.5

In FM broadcasting the frequency deviation $f_\Delta = 75$ kHz. Maximal modulation
frequency is 15 kHz (in the mono channel). What is the required bandwidth of the
transmission channel, with regard to the parameters above? What is the modulation gain in dB?

Exercise 6.6

Suppose a transmission rate of 9600 bits/s with bit error rate of 10^{-5} is required.
A channel with a bandwidth of 4000 Hz is available. Find the modulation scheme
which fulfils the above and requires the smallest S/N ratio. State the smallest S/N
ratio. Also compare the transmission rate with the theoretically feasible rate.

Exercise 6.7

Suppose a transmission rate of 9600 bits/s with a bit error rate of 10^{-5} is required.
A channel with a bandwidth of 45,000 Hz is available. Find the modulation
scheme which fulfils the above and requires the smallest S/N ratio. State the
smallest S/N ratio. Also compare the transmission rate with the theoretically
feasible rate.

Exercise 6.8

We would like to establish a radio connection between a base station and a
mobile phone inside a building. Measurements have shown that due to absorption
in walls, etc. the propagation attenuation to a receiver on the bottom floor is 50
dB larger than the propagation in free space. The output power of the base station
is 20 W, its antenna gain is 5 dB and the distance to the receiver is 8 km. The
carrier frequency is 900 MHz and the bandwidth of the sent signal is 200 kHz.
The receiver has a noise factor of 5 dB and an antenna gain of 2 dB. What is the
bit error rate if BPSK is used?

Exercise 6.9

Consider the signal set:

$$
s_i(t) = \begin{cases} \sqrt{\dfrac{2E}{T}} \cos\left(2\pi f_c t + i\dfrac{\pi}{4}\right), & 0 \le t \le T \\ 0, & \text{otherwise} \end{cases}
$$

where $i = 1, 2, 3, 4$ and $f_c = n/T$ for some integer n.
  (a) What is the dimension of the signal space, spanned by the signals?
  (b) Find orthonormal basis functions to represent the signal set.

    (c) Determine the coefficients when the signal is expressed by the basis
       functions.
    (d) Plot the signal locations in the signal space.

Exercise 6.10

In one application you have the requirement that the bit error rate should not be larger than 10^{-9} when the signal to noise ratio is $E_b/N_0 = 9$ dB. Find a method to solve the transmission problem.

Exercise 6.11

Compute the S/N in dB for the demodulated signal in an AM system. Speech and music are transmitted with an upper frequency limit of 8 kHz. The modulation index is 0.3 and the message signal power is 0.7 W. The power of the signal that is received from the transmitter is -27 dBm. The noise has been measured to 60×10^{-9} W by means of a spectrum analyser with a filter bandwidth of 10 kHz.

Exercise 6.12

In a communication system on 450 MHz, binary FSK is used as modulation. The detection is made non-coherently. You want to transmit 1.2 kbits/s. A rate 1/2 convolutional code with the constraint length 5 is used. What is the bit error rate if the signal to noise ratio P_R/N_R is 15 dB? One possible error path, giving one decoded error, is assumed.

Exercise 6.13

A receiver in the GSM system can have a maximal bit error rate of 2% at a received signal power of -129 dBW. The bandwidth of one channel is 200 kHz. What is the maximum noise figure of the receiver to fulfil the requirement above? The GSM system uses a GMSK modulation which has an error probability curve close to BPSK.

7 ADAPTIVE CHANNEL EQUALISERS

As we have seen in Section 5.1.2, there is a risk that the detection process will be perturbed by intersymbol interference when data are being transmitted at a high rate over a channel with time dispersion. In this context, the channel can be seen as a linear filter with an impulse response whose length is in the order of, or longer than, the symbol time. Furthermore, in many situations the transmission channel changes with time. For telephone channels, for instance, parameters may change after each new connection between different terminals. The properties of radio channels change when the communication terminals are not stationary or when the environment changes during the communication. The combination of signals propagated through different paths will then change and cause the signal to fluctuate by alternating constructive or destructive interference. The channel can then be modelled as a time-varying FIR filter, whose coefficients vary randomly with time. The problem of time varying channels has to be solved by using adaptive systems.

Adaptive equalisers are discussed below, but the problems are similar for echo cancellers, Rake receivers, source coding (as discussed in Section 3.3) and adaptive antennas (as discussed in Section 8.2). Three different equalisers are presented. The first is based on the zero forcing criterion. The next two are based on the minimum mean square error (MMSE) criterion, which is also described. Among the MMSE equalisers, one without feedback and one with feedback of the detected signal are presented.

As has been pointed out in Chapter 6 detectors based on MLSE can be used in intersymbol interference conditions. When the length of the intersymbol interference is large, typically 10 symbols, such a detector becomes too computationally demanding. The detectors presented in this chapter require less computational capacity.

The objective of this chapter is to give a short introduction to some adaptive equalisers. The aim is also to give a background to some fundamental and common algorithms for general adaptive systems, not only for the reader to understand the differences in the design of the algorithms, but also to be able to adapt the algorithms to the different applications in question.

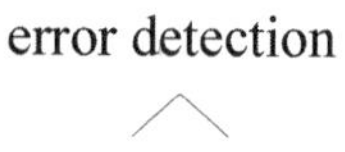

error detection

Figure 7.1 Model for the transmission of a signal on a channel with linear distortion.

7.1 Channel equaliser

The modulated signal is sent between transmitter and receiver via various media such as wires with both inductance and capacitance. Air is another example whereby the signal can reach the receiver along different paths, all with different times of flight, because of reflections from hills, buildings, vehicles, etc. This results in a filtering effect, which alters the amplitude spectrum of the signal and also introduces time dispersion (Figure 7.1).

The signal propagating through the channel will therefore be filtered through some transfer function,

$$C(\omega) = \frac{a_n(j\omega)^n + \cdots + a_1(j\omega) + a_0}{b_m(j\omega)^m + \cdots + b_1(j\omega) + b_0}, \quad m \geq n, \ \omega = 2\pi f \tag{7.1}$$

which leads to intersymbol interference. This can be eliminated by using a channel equaliser. For a linear channel it is equivalent to positioning the equaliser either on the transmitting or on the receiving side of the channel, as illustrated in Figure 7.2. But for channels adding noise, it is often an advantage to place the equaliser before the channel since this eliminates the risk of amplifying or colouring the noise. When the transmission channel is unknown and has to be estimated, however, the equaliser is located at the receiving side.

To reduce the intersymbol interference, the estimated transfer function, $\hat{C}(f)$, is inverted. When the estimation is exact, i.e. when $C(f) = \hat{C}(f)$, a complete inversion of the transfer function of the channel can be made, provided that $\hat{C}^{-1}(f)$ is a stable filter. The equaliser is often an FIR filter of the order of $2N + 1$ (Figure 7.3). The required filter length, i.e. the order, depends on the length of the impulse response.

Most equalisers are implemented as discrete-time systems and since the

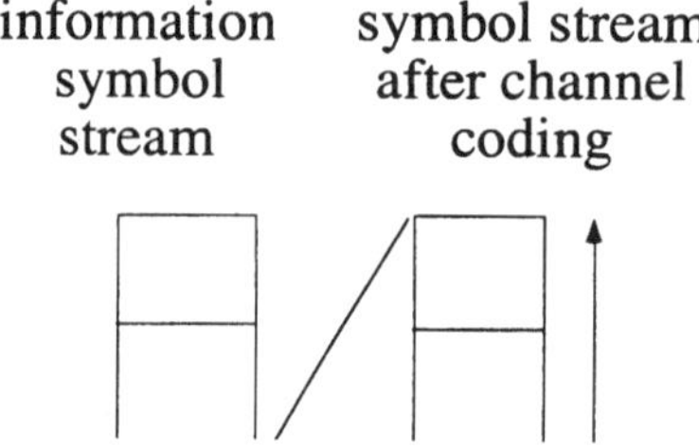

Figure 7.2 The equaliser can be located before or after the channel.

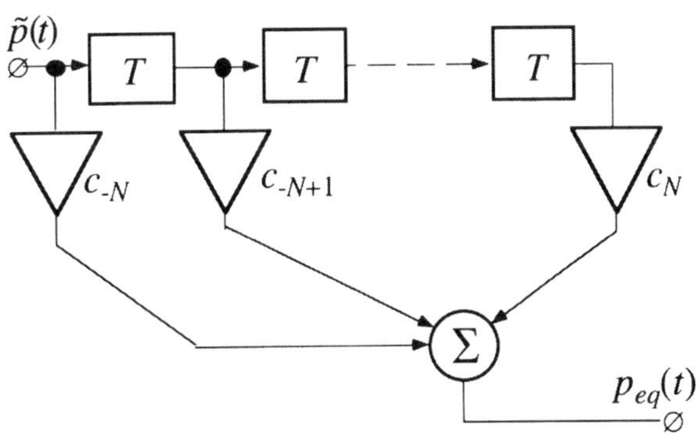

Figure 7.3 An equaliser of FIR filter type.

presentation becomes simpler, discrete time signals are used here. The signal is sampled at times t_k, where $k = 0,1,2,...$, etc. The time between the sample instants is T, which can be equal to the symbol time or a fraction of it. A fractionally spaced equaliser gets better performance but requires larger computational capacity, than when T corresponds to an entire symbol. A spacing of half the symbol time is used in many applications. The output signals at times t_k are given by the convolution between the input signal $\tilde{p}(t)$ and the impulse response of the equaliser,

$$p_{eq}(t_k) = \sum_{n=-N}^{N} c_n^* \tilde{p}(t_{k-n}) \tag{7.2}$$

Using vector notation the output signal in eq. (7.2) can be written in a simpler way, $p_{eq}(t_k) = \mathbf{c}_k^H \tilde{\mathbf{p}}_k$. Hermitian transpose has been used since the vectors are complex and both I and Q channels are being processed. The filter coefficients are $\mathbf{c}_k^T = \{c_{-N,\,k}...c_{N,\,k}\}$ and the received signal is $\tilde{\mathbf{p}}_k^T = \{\tilde{p}(t_{k+N})...\tilde{p}(t_{k-N})\}$

7.1.1 ZERO FORCING EQUALISER

Given that the input signal of the channel is an impulse, the output signal of the equaliser must be

$$p_{eq}(t_k) = \begin{cases} 1 & k = 0 \\ 0 & k = \pm 1, \pm 2, ..., \pm N \end{cases} \tag{7.3}$$

in order to eliminate the intersymbol interference completely in the interval $\pm N$. Any intersymbol interference that may occur outside that interval will remain. The impulse response of the combination of channel and equaliser must therefore be forced towards zero for $k \neq 0$. An equation system can be established, where the input sequence $\tilde{p}(t_k)$ is organised in a $(2N + 1) \times (2N + 1)$ matrix and the filter coefficients c_n in a $(2N + 1) \times 1$ vector:

$$
\begin{bmatrix}
\tilde{p}_0 & & \tilde{p}_{-2N} \\
\vdots & \ddots & \vdots \\
\vdots & \ddots & \vdots \\
\tilde{p}_{2N} & & \tilde{p}_0
\end{bmatrix}
\begin{bmatrix}
c_{-N} \\
\vdots \\
c_0 \\
\vdots \\
c_N
\end{bmatrix}
=
\begin{bmatrix}
0 \\
\vdots \\
1 \\
\vdots \\
0
\end{bmatrix}
\tag{7.4}
$$

Solving with respect to c_n gives a so-called zero-forcing equaliser. For a channel with large attenuation at certain frequencies, this equaliser gives high gain at corresponding frequencies. For a lowpass channel it would give an equaliser with highpass character which would cause an unintentional increase in the noise. The zero forcing equaliser is therefore not used at the receiver, when the channel is noisy.

7.1.2 EQUALISERS BASED ON MINIMUM MEAN SQUARE ERROR

Linear equaliser

The equaliser can also be formulated as a least squares problem. The filter coefficients of the equaliser are selected so that the least square deviation is obtained between the signal sent and the signal at the output of the equaliser. This means the signal from the equaliser should coincide as much as possible with the signal entering the channel. There is now no risk of a large enhancement in noise since this would increase the deviation. A block diagram illustrating this is shown in Figure 7.4.

In Figure 7.4 the sent signal is the reference for the algorithm optimising the filter parameters of the equaliser. We have, of course, no access to the emitted signal at the receiving end, so in practice another reference signal must be chosen. A practical solution where the reference signal is obtained in another way is presented below.

In order to estimate the parameters of the channel and to achieve the initialisation of the receiver during start up, a signal sequence known to the receiver is sent. The received signal is compared to the corresponding reference sequence in the receiver. The known sequence is usually called the training sequence. Such known sequences usually appear in digital communication and are used for coarse tuning of the equaliser or for identification and synchronisation purposes. After the initialisation phase the detected message can constitute the reference, provided that the error rate is small enough. This implies a reasonably, but not

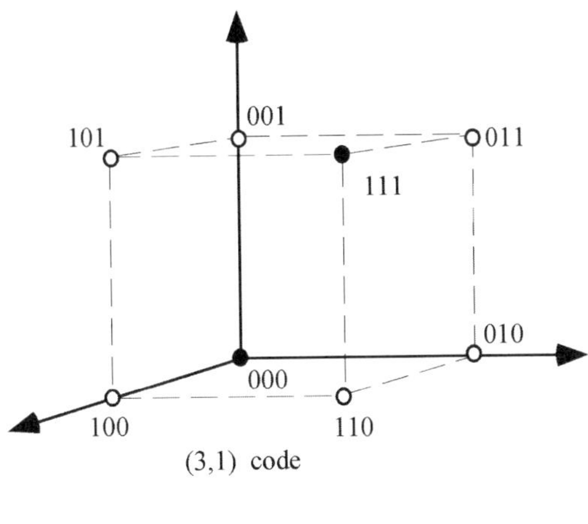

Figure 7.4 Origin of an adaptive equaliser.

necessarily totally, correctly adjusted equaliser so the detection can be done without symbol errors. Provided the transmission channel does not change too rapidly or the noise increase too much, the system will track the variations of the channel without having to receive a new training sequence. Such an equaliser is decision directed and is shown in Figure 7.5. The filter in the equaliser is the same kind as that in Figure 7.3. In comparison with the zero-forcing equaliser only the filter coefficients are calculated differently, which will be shown below.

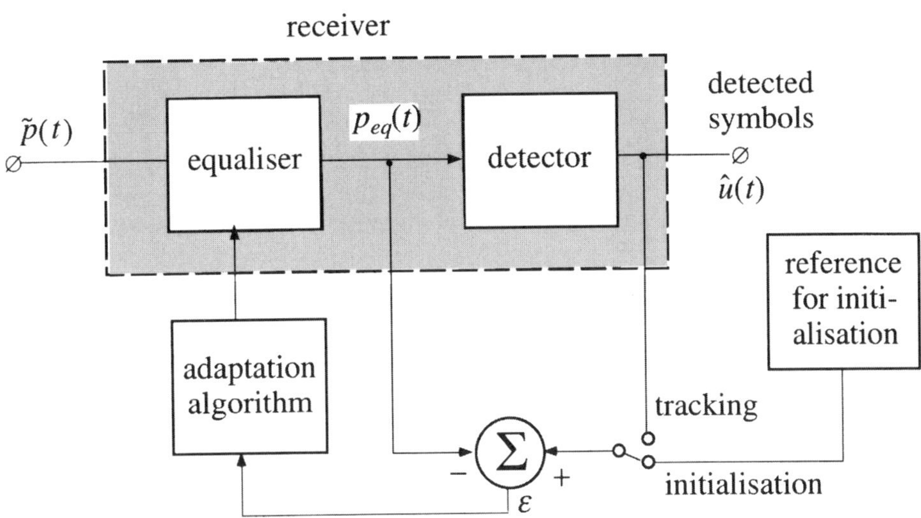

Figure 7.5 Block diagram of an adaptive linear equaliser.

There is nothing in principle prohibiting continuous time signals. But as usual we use samples taken at time instants t_k because implementations are done in discrete-time. By using the decisions made, an error signal can be formed:

$$\varepsilon_k = \hat{u}(t_k) - p_{eq}(t_k) = \hat{u}(t_k) - \mathbf{c}_k^H \tilde{\mathbf{p}}_k \tag{7.5}$$

Assume the mean square error, MSE, is a relevant measure of the error, which can be expressed as a value of goodness, J_k,

$$J_k = E\{|\varepsilon_k|^2\} = E\{\varepsilon_k \varepsilon_k^*\} \tag{7.6}$$

The methods based on finding the minimum of J_k, that is finding the minimum mean square error, are called MMSE methods. Minimising the error signal means that the noise and residual intersymbol interference will be equally weighted since the noise is included as a component of the error signal. With this criterion the noise will therefore not be severely increased.

Decision feedback equaliser

The equaliser described above only has access to zeros for cancelling the intersymbol interference in the received signal. This can be a drawback in equalising, e.g. a radio channel which usually in itself only contains zeros. A much better cancellation of the intersymbol interference can be achieved by means of the decision feedback equaliser, DFE. In this equaliser the detected symbols are looped back through a filter.

The decision feedback equaliser contains two filters, both a feed forward filter, in the same way as the linear equaliser, and also a feedback filter (see Figure 7.6).

The input signal to the detector is now

$$p_{eq}(t_k) = \sum_{n=0}^{N} c_n^{FF} \tilde{p}(t_{k-n}) + \sum_{n=1}^{M} c_n^{FB} \hat{u}(t_{k-n}) = \left(\mathbf{c}_k^{FF}\right)^H \tilde{\mathbf{p}}_k + \left(\mathbf{c}_k^{FB}\right)^H \tilde{\mathbf{p}}_k \hat{\mathbf{u}}_k \tag{7.7}$$

The input signal to the feed forward filter is $\tilde{\mathbf{p}}_k$ but will change designation to $\tilde{\mathbf{p}}_k^{FF}$. In the same way as in previous equaliser, an error signal is formed from the difference between the detector input signal and the, hopefully correct, detected signal

$$\varepsilon_k = \hat{u}(t_k) - p_{eq}(t_k) = \hat{u}(t_k) - \left(\mathbf{c}_k^{FF}\right)^H \tilde{\mathbf{p}}_k^{FF} - \left(\mathbf{c}_k^{FB}\right)^H \hat{\mathbf{u}}_k = \hat{u}(t_k) - \mathbf{c}_k^H \tilde{\mathbf{p}}_k \tag{7.8}$$

where

$$\mathbf{c}_k = \begin{bmatrix} \mathbf{c}_k^{FF} \\ \mathbf{c}_k^{FB} \end{bmatrix} \tag{7.9}$$

and

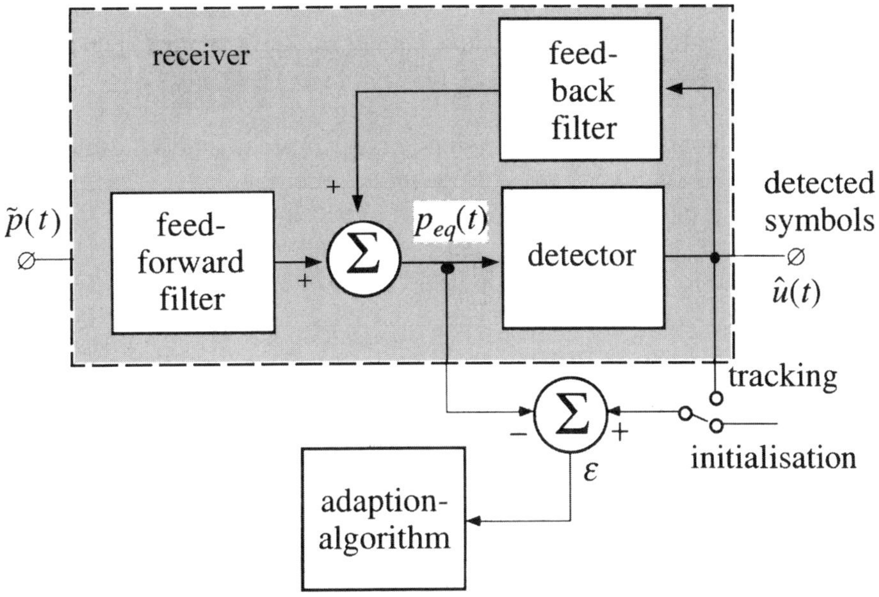

Figure 7.6 Decision feedback equaliser. Block for initialisation and branches between the adaptation block and the two filters have not been drawn.

$$\tilde{\mathbf{p}}_k = \begin{bmatrix} \tilde{\mathbf{p}}_k^{FF} \\ \hat{\mathbf{u}}_k \end{bmatrix}$$

Consequently, in this way the error term can be expressed equally for this equaliser as for the previous one. The mean square error, $J_k = E\{|\varepsilon_k|^2\}$, is used as before to compute the optimal filter coefficients.

The increased efficiency of the feedback equaliser is clarified by the following analysis. The minimum of the error signal occurs when the error signal is zero. The second equality in eq. (7.8) can then be rewritten

$$\left(\mathbf{c}_k^{FF}\right)^H \tilde{\mathbf{p}}_k^{FF} = \hat{u}(t_k) - \left(\mathbf{c}_k^{FB}\right)^H \hat{\mathbf{u}}_k \tag{7.10}$$

and since the equaliser output is (cf. eq. (7.7))

$$\begin{aligned} p_{eq}(t_k) &= \left(\mathbf{c}_k^{FF}\right)^H \tilde{\mathbf{p}}_k^{FF} + \left(\mathbf{c}_k^{FB}\right)^H \hat{\mathbf{u}}_k \\ &= \hat{u}(t_k) - \left(\mathbf{c}_k^{FB}\right)^H \hat{\mathbf{u}}_k + \left(\mathbf{c}_k^{FB}\right)^H \hat{\mathbf{u}}_k \\ &= \hat{u}(t_k) \tag{7.11} \end{aligned}$$

Assuming the detection has been correct $\hat{u}(t_k) = u(t_k)$. Consequently, the intersymbol interference has been completely cancelled and the detector input signal consists of the same signal as the one being fed into the channel (cf. Figure 7.4). If the noise has been taken into our derivations it should have been shown that a term originating from noise filtered through the feed forward filter remains. The feedback path does however not contribute any noise. Note that for the non-feedback equaliser, the corresponding relation cannot be computed since $\hat{u}(t_k)$ is not present in the expression for the equaliser filter. The decision feedback equaliser therefore has a greater potential to eliminate intersymbol interference. A condition is that, in both cases, the feedback decisions are correct.

7.2 Algorithms for adaptation

The adaptation algorithm is supposed to compute the coefficients of the equalising filter and to monitor changes in the channel in order to upgrade the coefficients. The algorithm will compute the filter coefficients which fulfil the minimum mean square error, MMSE, i.e.

$$\mathbf{c}_k^{opt} = \underset{\mathbf{c}_k}{\operatorname{argmin}}\,(J_k) = \underset{\mathbf{c}_k}{\operatorname{argmin}}\left(E\{|\varepsilon_k|^2\}\right) \tag{7.12}$$

Strictly mathematically the error signal in eqs. (7.5) and (7.8) should be expressed by using the sent signal $u(t)$ as reference. In the case described here, however, the decisions $\hat{u}(t)$ made by the detector are used as a reference signal. These two cases are equivalent as long as the detector makes correct decisions.

Two algorithms for solving general minimum mean square error problems are discussed below. Similar algorithms are obtained by solving different least squares problems in general. Therefore they are not just limited to equalisers but are also useful for speech coding (cf. eq. (3.69)) and adaptive antennas (cf. eq. (8.24)). Index k in the equations is used to indicate values of the parameters which hold at time t_k. The most common reason for introducing adaptive systems into communication applications is that these systems operate in an environment with non-stationary signals. It is therefore necessary to indicate the actual time by using a time index for expectation values, vectors, etc. We thus write $\tilde{p}(t_k) = \tilde{p}_k$ for the input signal and $c_n(t_k) = c_{n,\,k}$ for the filter coefficients.

The filter coefficients of the solution are allowed to be complex. The reason is that the equalisation is most often made after down conversion to the baseband. The equaliser therefore deals with the in-phase and quadrature coefficients (I and Q). The equaliser filter will then consist of four FIR filters (see Figure 5.14). The real and imaginary parts of the impulse response will each correspond to the coefficients of the respective filters.

Find the value of c minimising J_k

To find the filter coefficients which minimise J_k, we insert eq. (7.5) in eq. (7.6); this gives

$$J_k = E\left\{\left|\hat{u}(t_k) - p_{eq}(t_k)\right|^2\right\} = E\left\{\left(\hat{u}_k - \mathbf{c}_k^H \tilde{\mathbf{p}}_k\right)\left(\hat{u}_k - \mathbf{c}_k^H \tilde{\mathbf{p}}_k\right)^*\right\}$$

$$= E\left\{\left(\hat{u}_k - \mathbf{c}_k^H \tilde{\mathbf{p}}_k\right)\left(\hat{u}_k^* - \tilde{\mathbf{p}}_k^H \mathbf{c}_k\right)\right\}$$

$$= E\{\hat{u}_k \hat{u}_k^*\} - \mathbf{c}_k^H \overbrace{E\{\tilde{\mathbf{p}}_k \hat{u}_k^*\}}^{\tilde{\mathbf{h}}_k} - \overbrace{E\{\hat{u}_k \tilde{\mathbf{p}}_k^H\}}^{\tilde{\mathbf{h}}_k^H} \mathbf{c}_k + \mathbf{c}_k^H E\{\tilde{\mathbf{p}}_k \tilde{\mathbf{p}}_k^H\}\mathbf{c}_k \tag{7.13}$$

The expectation $E\{\tilde{\mathbf{p}}_k \tilde{\mathbf{p}}_k^H\} = \mathbf{R}_k$ forms the autocorrelation matrix of the input signal. Note that $\tilde{\mathbf{p}}_k$ is a stochastic vector which is not the case for $\mathbf{c}_k$. To further simplify the notation, the vector $\tilde{\mathbf{h}}_k$ is defined in eq. (7.13) and will be used later.

As has been said, the measure of goodness J_k should be as small as possible. Since J_k is a quadratic function of $\mathbf{c}_k$, there is only one single minimum, which is therefore global. An unambiguous vector $\mathbf{c}_k$, i.e. filter coefficients, which minimises eq. (7.13) can then be found by setting the gradient equal to zero.

$$\mathbf{0} = \frac{dJ_k}{d\mathbf{c}}$$

$$= \underbrace{\frac{d}{d\mathbf{c}} E\{\hat{u}_k \hat{u}_k^*\}}_{=\mathbf{0}\text{ as not depending on } \mathbf{c}} - \frac{d}{d\mathbf{c}} \mathbf{c}_k^H E\{\tilde{\mathbf{p}}_k \hat{u}_k^*\}$$

$$- \frac{d}{d\mathbf{c}} E\{\hat{u}_k \tilde{\mathbf{p}}_k^H\}\mathbf{c}_k + \frac{d}{d\mathbf{c}} \mathbf{c}_k^H \mathbf{R}_k \mathbf{c}_k \tag{7.14}$$

The condition for the first term in the second line being independent of $\mathbf{c}$ is obviously that the correct decisions are made. The second term in the second line can be studied more closely by making the expansion

$$\frac{d}{d\mathbf{c}} \mathbf{c}_k^H E\{\tilde{\mathbf{p}}_k \hat{u}_k^*\} = \frac{d}{d\mathbf{c}} \sum_{n=-N}^{N} c_{n,k}^* E\{\tilde{p}_{k-n} \hat{u}_k^*\}$$

$$= \frac{d}{d\mathbf{c}} \underbrace{\sum_{n=-N}^{N} (a_{n,k} - jb_{n,k}) E\{\tilde{p}_{k-n} \hat{u}_k^*\}}_{\text{is denoted } g_k}$$

$$= \frac{dg_k}{d\mathbf{c}} \tag{7.15}$$

For the complex variable $c_{n,k}$, the real and imaginary parts have been written separately, $c_{n,k} = a_{n,k} + jb_{n,k}$. The gradient $dg_k/d\mathbf{c}$ becomes

$$\frac{dg_k}{d\mathbf{c}} = \begin{bmatrix} \dfrac{\partial g_k}{\partial a_1} + j\dfrac{\partial g_k}{\partial b_1} \\ \vdots \\ \vdots \\ \dfrac{\partial g_k}{\partial a_M} + j\dfrac{\partial g_k}{\partial b_M} \end{bmatrix} \tag{7.16}$$

Derivation of the second line in eq. (7.15) gives

$$\left. \begin{aligned} \frac{\partial g_k}{\partial a_n} &= E\{\tilde{p}_{k-n}\,\hat{u}_k^*\} \\ \frac{\partial g_k}{\partial b_n} &= -jE\{\tilde{p}_{k-n}\,\hat{u}_k^*\} \end{aligned} \right\} \quad \Rightarrow \quad \frac{\partial g_k}{\partial c_n} = 2E\{\tilde{p}_{k-n}\,\hat{u}_k^*\} \tag{7.17}$$

which inserted in eq. (7.16) gives the gradient of g_k as

$$\frac{d}{d\mathbf{c}}\,\mathbf{c}_k^H E\{\tilde{\mathbf{p}}_k\hat{u}_k^*\} = 2E\{\tilde{\mathbf{p}}_k\hat{u}_k^*\} \tag{7.18}$$

Using the same argument as above, we find that the third term in the second line of eq. (7.14) is equal to the $\mathbf{0}$-vector.

$$\frac{d}{d\mathbf{c}}\,E\{\hat{u}_k\tilde{\mathbf{p}}_k^H\}\mathbf{c}_k = \mathbf{0} \tag{7.19}$$

The last term in eq. (7.14) contains a scalar which can be denoted f.

$$f = \mathbf{c}_k^H \mathbf{R}_k \mathbf{c}_k \tag{7.20}$$

To calculate the gradient, $d(\mathbf{c}_k^H \mathbf{R}_k \mathbf{c}_k)/d\mathbf{c}$, we rewrite eq. (7.20) and then calculate the partial derivatives. Define the vector

$$\mathbf{q}_1^H = \mathbf{c}_k^H \mathbf{R}_k \tag{7.21}$$

which inserted in eq. (7.20) gives $f = \mathbf{q}_1^H \mathbf{c}_k$. Since $\mathbf{q}_1$ is a $(2N + 1) \times 1$ vector, which is also the case for $E\{\hat{u}_k\tilde{\mathbf{p}}_k^H\}$, f is in this case an equation of the same type as eq. (7.19). The vector $\mathbf{q}_1$ can be treated as a constant with respect to $\mathbf{c}$. Taking the derivative as in eq. (7.19) gives

$$\frac{\partial}{\partial \mathbf{c}}\,\mathbf{q}_1^H \mathbf{c}_k = \mathbf{0} \tag{7.22}$$

Since $\mathbf{q}_1$ obviously depends on $\mathbf{c}$, this cannot be regarded as a complete derivative but only a partial derivative. To find the next component of the derivative we define a new vector

$$\mathbf{q}_2 = \mathbf{R}_k \mathbf{c}_k \tag{7.23}$$

which inserted in eq. (7.20) gives $f = \mathbf{c}_k^H \mathbf{q}_2$. This is an expression of the same kind as eq. (7.18). In the same way as before, $\mathbf{q}_2$ is regarded as a constant and the derivative with regard to $\mathbf{c}$ becomes

$$\frac{\partial}{\partial \mathbf{c}} \mathbf{c}_k^H \mathbf{q}_2 = 2\mathbf{q}_2 = 2\mathbf{R}_k \mathbf{c}_k \tag{7.24}$$

The complete derivative becomes the sum of the derivatives in eq. (7.24) and (7.22), which gives

$$\frac{d}{d\mathbf{c}} \mathbf{c}_k^H \mathbf{R}_k \mathbf{c}_k = \frac{\partial}{\partial \mathbf{c}} \mathbf{q}_1^H \mathbf{c}_k + \frac{\partial}{\partial \mathbf{c}} \mathbf{c}_k^H \mathbf{q}_2 = 2\mathbf{R}_k \mathbf{c}_k \tag{7.25}$$

The vector (7.25) inserted in eq. (7.14) gives the final expression for (7.14):

$$\mathbf{0} = \frac{dJ_k}{d\mathbf{c}} = -2\tilde{\mathbf{h}}_k + 2\mathbf{R}_k \mathbf{c}_k \tag{7.26}$$

The filter coefficients $\mathbf{c}_k^{opt}$, which can be extracted from eq. (7.26), are optimal from a least squares point of view. We therefore solve $\mathbf{c}_k$ from eq. (7.26) which gives

$$\mathbf{c}_k^{opt} = \mathbf{R}_k^{-1} \tilde{\mathbf{h}}_k \tag{7.27}$$

The system of equations in eq. (7.27) gives the solution to the normal equations. The system has a solution if the matrix $\mathbf{R}_k$ is non-singular, i.e. if $\det[\mathbf{R}_k] \neq 0$. (The normal equations can be derived in a simpler way by completion of squares in eq. (7.13). Then the derivative does not need to be calculated. Since the derivative is required for the derivation of the LMS algorithm the more complicated method has been chosen here.)

7.2.1 THE LMS ALGORITHM

Recursive solution of the filter coefficients c

The normal equations can be solved directly by matrix inversion or by some gradient method, such as the steepest descent. Equation (7.13), which is to be minimised, has a quadratic dependence of $\mathbf{c}_k$. For a quadratic curve the minimum is the only position where the derivative is zero, and to find this point one has to move in the negative direction if the derivative is positive and in the positive direction if the derivative is negative. This can be understood by studying a plot of J for a system of one only adaptive parameter, or filter coefficient (cf. Figure 7.7).

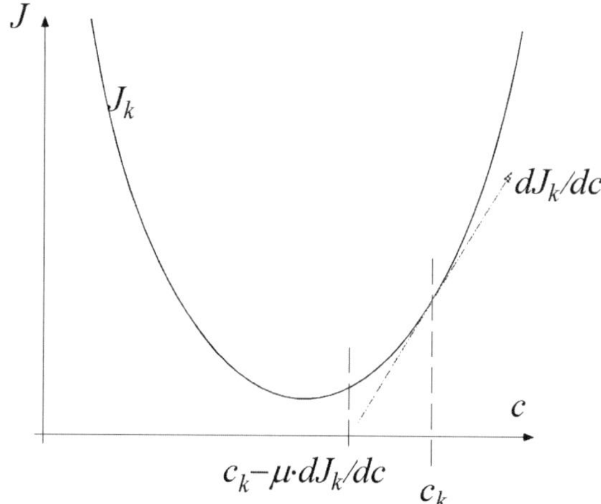

Figure 7.7   The solution of the optimisation problem is found by moving in the negative direction of the derivative. In the figure the position at iteration k is shown, to the left of it is the position at $k + 1$.

According to the steepest descent algorithm, $\mathbf{c}_k$ is therefore recursively updated in the following way:

$$\mathbf{c}_{k+1} = \mathbf{c}_k - \mu \underbrace{\text{grad}\ (J_k)}_{dJ_k/d\mathbf{c}} \tag{7.28}$$

where μ is the step size or the gain factor. To guarantee that the minimum is found in a discrete-time system, μ must in principle approach zero. This gives, however, slow convergence and a compromise must be made. One way is to use large values of μ far away from the minimum and then reduce μ as the measure of goodness gets closer to its minimum.

The gradient has been calculated previously in eq. (7.26) and can be inserted in eq. (7.28), to give

$$\mathbf{c}_{k+1} = \mathbf{c}_k + 2\mu(\tilde{\mathbf{h}}_k - \mathbf{R}_k\mathbf{c}_k) \tag{7.29}$$

Recursive updating of the filter coefficients is an advantage when non-stationary signals are being detected. A mobile terminal, which during communication moves in an environment with multipath propagation, is such an example.

Estimation of included expectation values

The autocorrelation matrix $\mathbf{R}_k$ and the vector $\tilde{\mathbf{h}}_k$ in eq. (7.29) are expectation values. How to calculate the expectation from a time series of samples, instead of using different realisations of the stochastic process, is not completely obvious.

Given an ergodic signal, or process, it is, however, equivalent to calculating the expectation values over an infinite time series. In the different cases considered in this book, it cannot be assumed that the signal is ergodic since, for example, the transmission channel varies over time. Provided this variation is slow compared to the sample rate, the expectation values can be approximated by using finite time series. The easiest way is to use the instantaneous values, i.e. a time series consisting of one single time sample. Thus put

$$\hat{\mathbf{R}}_k = \tilde{\mathbf{p}}_k \tilde{\mathbf{p}}_k^H \quad \text{and} \quad \tilde{\mathbf{h}}_k = \tilde{\mathbf{p}}_k \hat{u}_k^* \tag{7.30}$$

This way of estimating the autocorrelation matrix characterises the adaptation algorithm which is formulated below.

Formulation of the LMS algorithm

The equations in (7.30) inserted in eq. (7.29) give

$$\hat{\mathbf{c}}_{k+1} = \hat{\mathbf{c}}_k + 2\mu \tilde{\mathbf{p}}_k \underbrace{\left[\hat{u}_k^* - \tilde{\mathbf{p}}_k^H \hat{\mathbf{c}}_k\right]}_{e_k^*} \tag{7.31}$$

which is the well-known least mean squares, or LMS algorithm (other names are Widrow–Hopf and stochastic gradient). The expression in (7.31) is usually partitioned into two, one expression for updating of the coefficients,

$$\hat{\mathbf{c}}_{k+1} = \hat{\mathbf{c}}_k + 2\mu \, \tilde{\mathbf{p}}_k \, e_k^* \tag{7.32}$$

and one for the error term,

$$e_k^* = \hat{u}_k^* - \tilde{\mathbf{p}}_k^H \hat{\mathbf{c}}_k \tag{7.33}$$

Consider a single tap in the filter delay chain (tap n). Equation (7.32) then becomes

$$\hat{c}_{n,k+1} = \hat{c}_{n,k} + 2\mu \, \tilde{p}(t_{k+N-n}) e_k^* \tag{7.34}$$

This means that each filter coefficient is updated by the section shown in Figure 7.8.

There are many versions of the LMS algorithm. One common version is to maintain only the sign of the error instead of the error itself, i.e. $\text{sgn}(e_k^*)$ is used instead of e_k^* in eq. (7.34). In other versions $\text{sgn}[\tilde{p}(t_{k+N-n})]$ is used, or the sign function is used in both positions, i.e. both for the error term and for the regressor. These simplifications obviously lead to slower convergence and to larger residual errors. The algorithms are, however, very simple and possess an internal robustness due to the normalisation of the adaptation step.

A diagram of an adaptive linear equaliser based on the LMS algorithm is

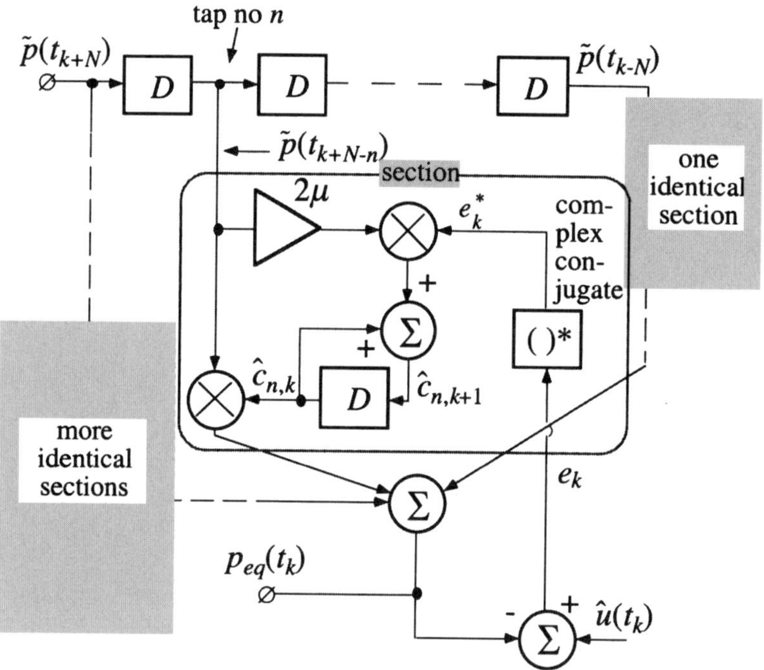

Figure 7.8 One tap of an adaptive FIR filter equaliser updated by using the LMS algorithm.

shown in Figure 7.9. Note that the decision of the detector is fed back and constitutes the reference; a so-called decision directed equaliser.

Convergence properties of the LMS algorithm

The LMS algorithm contains a feedback loop and thus the algorithm can become unstable. It is therefore of interest to study how the properties of the loop are influenced by the parameter μ, which can be said to be a gain factor in the loop. When it comes to properties of filters in general it is common to study the impulse response, i.e. the response at the output to an impulse at the input. To investigate the convergence properties of the LMS algorithm we can study the response of the filter coefficients to a change in the input signal. The impulse response of the coefficients is obtained by finding the homogeneous solution to the difference equation defined by eq. (7.31). After minor rewriting, we obtain

$$\hat{\mathbf{c}}_{k+1} = \left(\mathbf{I} - 2\mu\tilde{\mathbf{p}}\tilde{\mathbf{p}}^H\right)\hat{\mathbf{c}}_k + 2\mu\tilde{\mathbf{p}}\hat{u}_k^* \tag{7.35}$$

An exact value of the derivative is not used in the LMS algorithm, but instead a value is used which can shift up and down depending on the noise in the signal.

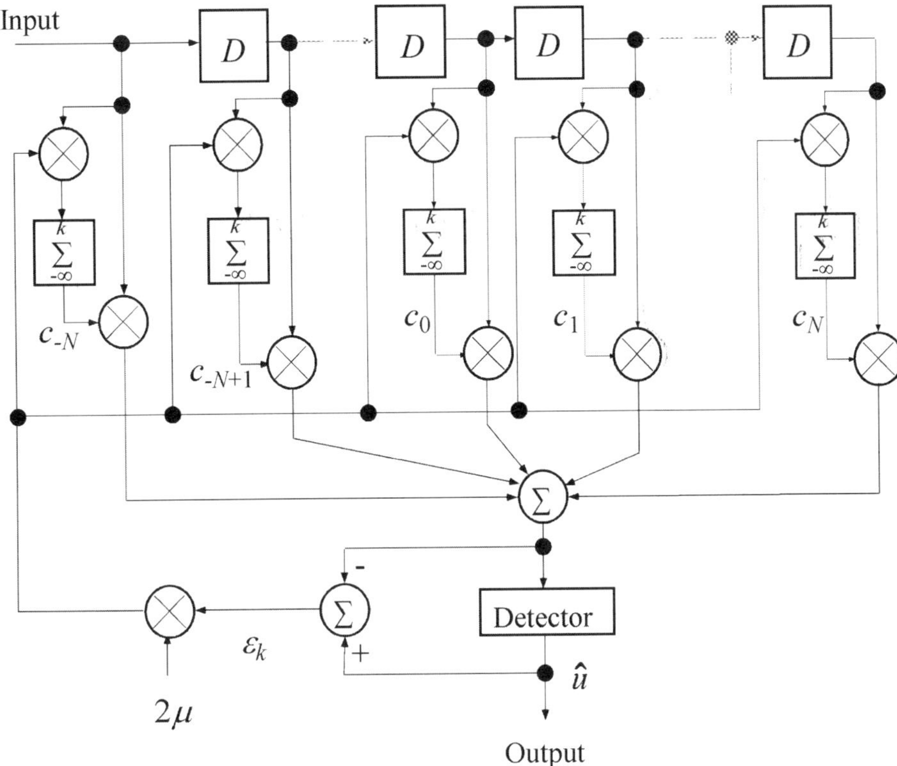

Figure 7.9   Adaptive equaliser, without feedback, based on the LMS algorithm for real signals.

The analysis must therefore be carried out by studying the convergence in average, i.e.

$$E\{\hat{\mathbf{c}}_{k+1}\} = \left(\mathbf{I} - 2\mu E\{\tilde{\mathbf{p}}\tilde{\mathbf{p}}^H\}\right)E\{\hat{\mathbf{c}}_k\} + 2\mu E\{\tilde{\mathbf{p}}\hat{u}_k^*\}$$

$$= (\mathbf{I} - 2\mu\mathbf{R})E\{\hat{\mathbf{c}}_k\} + 2\mu\tilde{\mathbf{h}}_k \tag{7.36}$$

As can be seen, it has been assumed that $\hat{\mathbf{c}}_k$ and $\tilde{\mathbf{p}}$ are uncorrelated. This can be explained by considering eq. (7.32). The vector $\hat{\mathbf{c}}_k$ can be calculated by, for instance, multiplying the vector $\tilde{\mathbf{p}}$ by the error signal e_k. Since the error signal consists of a sequence with independent samples, nothing can be deduced about $\hat{\mathbf{c}}_k$ from the value of $\tilde{\mathbf{p}}$. The vectors $\tilde{\mathbf{p}}$ and $\hat{\mathbf{c}}_k$ are thus statistically independent.

Depending on the autocorrelation matrix $\mathbf{R}$, eq. (7.35) can constitute a coupled difference equation, i.e. different c's depend on each other. The difference equation could therefore be difficult to solve. A way of avoiding this problem is to diagonalise $\mathbf{R}$ through $\mathbf{R} = \mathbf{U}\Lambda\mathbf{U}^H$, where Λ is a diagonal matrix consisting of the eigenvalues of $\mathbf{R}$. The eigenvalues for the diagonalised matrix Λ are the same as for the matrix $\mathbf{R}$. The vector $\mathbf{U}$ contains the corresponding eigenvectors. Furthermore, the matrix $\mathbf{U}$ is in this case unitary, i.e. $\mathbf{I} = \mathbf{U}\mathbf{U}^H$ and consequently $\mathbf{U}^{-1} = \mathbf{U}^H$, which gives

$$E\{\hat{\mathbf{c}}_{k+1}\} = \left(\mathbf{U}\mathbf{U}^H - 2\mu\mathbf{U}\Lambda\mathbf{U}^H\right)E\{\hat{\mathbf{c}}_k\} + 2\mu\tilde{\mathbf{h}}_k$$

$$\Updownarrow$$

$$\underbrace{E\left\{\mathbf{U}^H\hat{\mathbf{c}}_{k+1}\right\}}_{E\{\hat{\mathbf{c}}^U_{k+1}\}} = (\mathbf{I} - 2\mu\Lambda)\underbrace{E\left\{\mathbf{U}^H\hat{\mathbf{c}}_k\right\}}_{E\{\hat{\mathbf{c}}^U_k\}} + 2\mu\underbrace{\mathbf{U}^H\tilde{\mathbf{h}}_k}_{\tilde{\mathbf{h}}^U_k} \tag{7.37}$$

$$\Updownarrow$$

$$E\left\{\hat{\mathbf{c}}^U_{k+1}\right\} = (\mathbf{I} - 2\mu\Lambda)E\left\{\hat{\mathbf{c}}^U_k\right\} + 2\mu\tilde{\mathbf{h}}^U_k$$

The vector element $\hat{c}^U_{n,k}$ is a scaling of $\hat{c}_{n,k}$ and consequently does not affect the convergence properties. The diagonalisation gives a number of uncoupled difference equations. The nth row is

$$E\left\{\hat{c}^U_{n,k+1}\right\} = (1 - 2\mu\lambda_n)E\left\{\hat{c}^U_{n,k}\right\} + 2\mu E\left\{\tilde{p}^U_{n,k}\hat{u}^*_k\right\} \tag{7.38}$$

which is a difference equation of the type $y_{k+1} = C{\cdot}y_k + x_k$. The impulse response of a system described by such a difference equation is given by the homogeneous solution of this equation, i.e. $y^{hom}_k = C^k x_0$. The impulse response of the filter coefficients then becomes $E\{\hat{c}^U_{n,k+1}\} = (1 - 2\mu\lambda_n)^k E\{\hat{c}^U_{n,0}\}$. Note that the time constant is mainly determined by the smallest eigenvalue. The stability criteria are that each such equation must converge as $k \rightarrow \infty$, which means that $|1 - 2\mu\lambda_n| < 1$ for each row. A condition for convergence of the algorithm is thus

$$0 < \mu < \frac{1}{\lambda_{max}} \tag{7.39}$$

where λ_{max} is the largest eigenvalue of the autocorrelation matrix $\mathbf{R}$. The autocorrelation matrix has been assumed to be stationary in the discussion above. As previously mentioned this is only true for short time intervals of the signal. Equation (7.39) is therefore a necessary but not sufficient condition for stability. A more detailed analysis indicates that the eigenvalues of the autocorrelation matrix will influence not only the convergence rate, but also the stationary error of the algorithm.

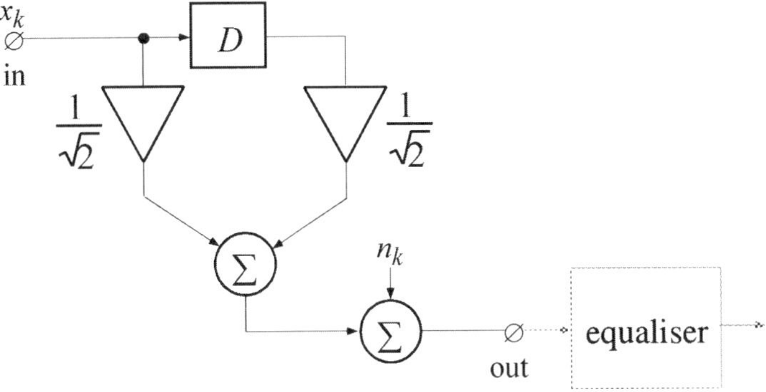

Figure 7.10 Channel model for the example described in the text.

Example 7.1

The received message has propagated through a transmission channel with multi-path. A model of the used channel is shown in Figure 7.10. The model is a so-called two-ray model with one direct signal and one reflex. An adaptive equaliser with three taps is used in the receiver. Binary, independent symbols, $x_i = \pm 1$, are used for the transmission, where both signs have equal probability. The noise is white with variance N_R. The statistical parameters of the input signal are assumed to be stationary.

(a) Find the optimum coefficients, $\mathbf{c}^{opt}$, of an equaliser optimised according to the MSE criterion.

(b) Determine the maximum value of μ.

Solution (a) The output signal from the channel can be written as

$$\mathbf{y}_k^T = \begin{pmatrix} y_{k+1} \\ y_k \\ y_{k-1} \end{pmatrix} = \begin{pmatrix} \dfrac{1}{\sqrt{2}}x_{k+1} + \dfrac{1}{\sqrt{2}}x_k + n_{k+1} \\ \dfrac{1}{\sqrt{2}}x_k + \dfrac{1}{\sqrt{2}}x_{k-1} + n_k \\ \dfrac{1}{\sqrt{2}}x_{k-1} + \dfrac{1}{\sqrt{2}}x_{k-2} + n_{k-1} \end{pmatrix} \tag{7.40}$$

Since the equaliser has three taps, three consecutive samples from the channel output are analysed. With the conditions valid for the input signal, the autocorrelation matrix $\mathbf{R} = E\{\mathbf{y}_k^T \mathbf{y}_k\}$ becomes

$$\mathbf{R} = \begin{pmatrix} 1 + N_R & \dfrac{1}{2} & 0 \\[2mm] \dfrac{1}{2} & 1 + N_R & \dfrac{1}{2} \\[2mm] 0 & \dfrac{1}{2} & 1 + N_R \end{pmatrix} \tag{7.41}$$

and $\tilde{\mathbf{h}} = E\{\mathbf{y}_k^{\mathrm{T}}\mathbf{x}_k\}$ becomes

$$\tilde{\mathbf{h}} = \begin{pmatrix} \dfrac{1}{\sqrt{2}} \\[2mm] \dfrac{1}{\sqrt{2}} \\[2mm] 0 \end{pmatrix} \tag{7.42}$$

To obtain $\mathbf{c}^{opt}$, the autocorrelation matrix must be inverted. Instead of the methods described earlier, we now use Cramer's rule, which is better suited when using pen and paper. According to Cramer's rule the inverse of $\mathbf{R}$ can be found by replacing the respective column by the known $\tilde{\mathbf{h}}$ vector for each element in $\mathbf{c}^{opt}$, taking the determinant and then dividing by det[$\mathbf{R}$]. The determinant of $\mathbf{R}$ is in this particular case

$$\det[\mathbf{R}] = \begin{vmatrix} 1 + N_R & \dfrac{1}{2} & 0 \\[2mm] \dfrac{1}{2} & 1 + N_R & \dfrac{1}{2} \\[2mm] 0 & \dfrac{1}{2} & 1 + N_R \end{vmatrix} = (1 + N_R)^3 - \dfrac{1}{2}(1 + N_R) \tag{7.43}$$

We then have for the respective element,

$$c_{-1}^{opt} = \dfrac{\begin{vmatrix} \dfrac{1}{\sqrt{2}} & \dfrac{1}{2} & 0 \\[2mm] \dfrac{1}{\sqrt{2}} & 1 + N_R & \dfrac{1}{2} \\[2mm] 0 & \dfrac{1}{2} & 1 + N_R \end{vmatrix}}{\det[\mathbf{R}]} = \dfrac{\dfrac{1}{\sqrt{2}}\left(N_R^2 + \dfrac{3}{2}N_R + \dfrac{1}{4}\right)}{(1 + N_R)^3 - \dfrac{1}{2}(1 + N_R)} \tag{7.44}$$

$$c_0^{opt} = \frac{\begin{vmatrix} 1+N_R & \dfrac{1}{\sqrt{2}} & 0 \\[2mm] \dfrac{1}{2} & \dfrac{1}{\sqrt{2}} & \dfrac{1}{2} \\[2mm] 0 & 0 & 1+N_R \end{vmatrix}}{\det[\mathbf{R}]} = \frac{\dfrac{1}{\sqrt{2}}\left(N_R^2 + \dfrac{3}{2}N_R + \dfrac{1}{2}\right)}{(1+N_R)^3 - \dfrac{1}{2}(1+N_R)} \tag{7.45}$$

$$c_1^{opt} = \frac{\begin{vmatrix} 1+N_R & \dfrac{1}{2} & \dfrac{1}{\sqrt{2}} \\[2mm] \dfrac{1}{2} & 1+N_R & \dfrac{1}{\sqrt{2}} \\[2mm] 0 & \dfrac{1}{2} & 0 \end{vmatrix}}{\det[\mathbf{R}]} = \frac{\dfrac{1}{\sqrt{2}}\left(-\dfrac{N_R}{2} - \dfrac{1}{4}\right)}{(1+N_R)^3 - \dfrac{1}{2}(1+N_R)} \tag{7.46}$$

Combining these into a vector gives

$$(\mathbf{c}^{opt})^T = \frac{1}{\sqrt{2}\left(N_R^3 + 3N_R^2 + \dfrac{5}{2}N_R + \dfrac{1}{2}\right)}\begin{pmatrix} N_R^2 + \dfrac{3}{2}N_R + \dfrac{1}{4} \\[2mm] N_R^2 + \dfrac{3}{2}N_R + \dfrac{1}{2} \\[2mm] -\dfrac{1}{2}N_R - \dfrac{1}{4} \end{pmatrix} \tag{7.47}$$

(b) In order to find the maximum value of μ, we must first determine the eigenvalues of the correlation matrix, $\mathbf{R}$, according to eq. (7.39).

The eigenvalues are given by the solution to the secular equation $\det[\mathbf{R} - \lambda\mathbf{I}] = 0$, which gives

$$(1 + N_R - \lambda)^3 - \frac{1}{2}(1 + N_R - \lambda) = 0$$
$$\Updownarrow \tag{7.48}$$
$$(1 + N_R - \lambda)\left((1 + N_R - \lambda)^2 - \frac{1}{2}\right) = 0$$

Equation (7.48) has the following solutions:

$$\lambda_1 = 1 + N_R, \quad \lambda_2 = 1 + N_R + \frac{1}{\sqrt{2}}, \quad \lambda_3 = 1 + N_R - \frac{1}{\sqrt{2}} \tag{7.49}$$

The maximum value of μ during adaptation becomes

$$\mu < \frac{1}{1 + N_R + \dfrac{1}{\sqrt{2}}} \tag{7.50}$$

because $\lambda_{\max} = \lambda_2$

7.2.2 THE RECURSIVE LEAST SQUARES ALGORITHM

In this section the recursive least squares (RLS) algorithm is described. Just as LMS, this is an algorithm for a recursive solution of MMSE problems. The difference is that the RLS algorithm can provide an optimum solution from the least squares point of view provided that the parameter λ is given the value 1, which is not always the case for various reasons. The algorithm is based on the optimum MMSE solution in eq. (7.27). The autocorrelation is estimated by using time averages (sample averages) after which a direct inversion of the autocorrelation matrix is carried out. As can be deduced from the name, the inversion is performed recursively. Compared to LMS, the algorithm will adapt the filter coefficients to the optimum solution faster and with less asymptotic variation. The price for the good properties is a higher degree of complexity.

Estimation of the expectation values

Compared to the LMS algorithm, a better estimation of the expectation values included in the autocorrelation matrix is obtained by taking the sample average of consecutive estimations (cf. eq. (7.30)):

$$\hat{\mathbf{R}}_k = \sum_{l=0}^{k} \lambda^{k-l} \tilde{\mathbf{p}}_l \tilde{\mathbf{p}}_l^H \tag{7.51}$$

Forming the average really requires a normalising factor of $1/(k + 1)$ which for simplicity has been excluded since this factor disappears later. A forgetting factor, $\lambda \leq 1$, has been introduced partly because the input signal is not always stationary and partly to reduce the influence of an incorrect starting point at the optimisation. When $\lambda = 1$ a time average is obtained. When $\lambda < 1$ the oldest samples will successively be given less weight and the estimation then depends more strongly on the later signal points. This is an advantage when data are non-stationary because of, for instance, a varying transmission channel. By writing the terms where $l = k$ separately, we obtain

$$\hat{\mathbf{R}}_k = \lambda \left[\sum_{l=0}^{k-1} \lambda^{k-1-l} \tilde{\mathbf{p}}_l \tilde{\mathbf{p}}_l^H \right] + \tilde{\mathbf{p}}_k \tilde{\mathbf{p}}_k^H = \lambda \hat{\mathbf{R}}_{k-1} + \tilde{\mathbf{p}}_k \tilde{\mathbf{p}}_k^H \tag{7.52}$$

In the same way we get a better estimation of $\tilde{\mathbf{h}}$,

$$\hat{\tilde{\mathbf{h}}}_k = \lambda \left[\sum_{l=0}^{k-1} \lambda^{k-1-l} \tilde{\mathbf{p}}_l \hat{u}_l^* \right] + \tilde{\mathbf{p}}_k \hat{u}_k^* = \lambda \hat{\tilde{\mathbf{h}}}_{k-1} + \tilde{\mathbf{p}}_k \hat{u}_k^* \tag{7.53}$$

Efficient inversion of autocorrelation matrix

The inversion of the autocorrelation matrix in eq. (7.27), which is done in real time, is demanding with respect to computer power. The demands can be reduced by utilising a relation called the matrix inversion lemma, which is a result from the matrix algebra. Let $\mathbf{A}$ and $\mathbf{B}$ be two positive definite $M \times M$ matrices, where

$$\mathbf{A} = \mathbf{B}^{-1} + \mathbf{C}\mathbf{D}^{-1}\mathbf{C}^H \tag{7.54}$$

$\mathbf{D}$ is a positive definite $N \times N$ matrix and $\mathbf{C}$ is an $M \times N$ matrix. According to the matrix inversion lemma, we have

$$\mathbf{A}^{-1} = \mathbf{B} - \mathbf{B}\mathbf{C}\left(\mathbf{D}^H + \mathbf{C}^H\mathbf{B}\mathbf{C}\right)^{-1}\mathbf{C}^H\mathbf{B} \tag{7.55}$$

This may seem a way of making the problem more complicated, but if we identify the terms in eq. (7.54) by the corresponding terms in eq. (7.52) we obtain

$$\mathbf{A} = \hat{\mathbf{R}}_k, \quad \mathbf{B}^{-1} = \lambda \hat{\mathbf{R}}_{k-1}, \quad \mathbf{C} = \tilde{\mathbf{p}}_k, \quad \mathbf{D} = \mathbf{I} \tag{7.56}$$

Inserting this in eq. (7.55), we obtain an algorithm for recursive updating of the inverse autocorrelation matrix,

$$\hat{\mathbf{R}}_k^{-1} = \frac{1}{\lambda}\left(\hat{\mathbf{R}}_{k-1}^{-1} - \mathbf{k}_k \tilde{\mathbf{p}}_k^H \hat{\mathbf{R}}_{k-1}^{-1}\right) \tag{7.57}$$

where

$$\mathbf{k}_k = \frac{\hat{\mathbf{R}}_{k-1}^{-1}\tilde{\mathbf{p}}_k}{\lambda + \tilde{\mathbf{p}}_k^H \hat{\mathbf{R}}_{k-1}^{-1}\tilde{\mathbf{p}}_k} \tag{7.58}$$

The only matrix inversion included now is the one which is carried out in the initialisation stage, and then the algorithm delivers an inverted autocorrelation matrix. Note that the denominator in eq. (7.58) is a scalar, so this division is relatively simple. A recursive upgrading of the filter coefficients can now be obtained from eqs. (7.27) and (7.53).

$$\mathbf{c}_k = \hat{\mathbf{R}}_k^{-1}\lambda\hat{\tilde{\mathbf{h}}}_{k-1} + \hat{\mathbf{R}}_k^{-1}\tilde{\mathbf{p}}_k\hat{u}_k^* \tag{7.59}$$

By inserting eq. (7.57) into the first term of eq. (7.59), we obtain

$$\mathbf{c}_k = \hat{\mathbf{R}}_{k-1}^{-1}\hat{\tilde{\mathbf{h}}}_{k-1} - \mathbf{k}_k\tilde{\mathbf{p}}_k^H\hat{\mathbf{R}}_{k-1}^{-1}\hat{\tilde{\mathbf{h}}}_{k-1} + \hat{\mathbf{R}}_k^{-1}\tilde{\mathbf{p}}_k\hat{u}_k^* = \mathbf{c}_{k-1} + \hat{\mathbf{R}}_k^{-1}\tilde{\mathbf{p}}_k\hat{u}_k^* - \mathbf{k}_k\tilde{\mathbf{p}}_k^H\mathbf{c}_{k-1}$$

$$(7.60)$$

Furthermore, through multiplication by the denominator and reorganising the terms, eq. (7.58) can be rewritten as

$$\mathbf{k}_k = \underbrace{\left[\frac{1}{\lambda}\left(\hat{\mathbf{R}}_{k-1}^{-1} - \mathbf{k}_k\tilde{\mathbf{p}}_k^H\hat{\mathbf{R}}_{k-1}^{-1}\right)\right]}_{\hat{\mathbf{R}}_k^{-1}}\tilde{\mathbf{p}}_k = \hat{\mathbf{R}}_k^{-1}\tilde{\mathbf{p}}_k \qquad (7.61)$$

Finally, by using eqs. (7.61) and (7.60) we have

$$\mathbf{c}_k = \mathbf{c}_{k-1} + \mathbf{k}_k\hat{u}_k^* - \mathbf{k}_k\tilde{\mathbf{p}}_k^H\mathbf{c}_{k-1} = \mathbf{c}_{k-1} + \mathbf{k}_k\left(\hat{u}_k^* - \tilde{\mathbf{p}}_k^H\mathbf{c}_{k-1}\right) \qquad (7.62)$$

Formulation of the RLS algorithm

The last equality of eq. (7.62) constitutes an algorithm for a recursive updating of the filter coefficients. The term within parentheses is an error term called the a priori estimation error, which can be compared to the a posteriori error in eq. (7.33). The a priori error is then

$$\varepsilon_k = \hat{u}_k^* - \tilde{\mathbf{p}}_k^H\mathbf{c}_{k-1} \qquad (7.63)$$

Note that for the RLS algorithm the filter output signal $p_{eq}(t_k) = \tilde{\mathbf{p}}_k^H\mathbf{c}_{k-1}$ at time k. The final formulation of the algorithm then becomes

$$\mathbf{c}_k = \mathbf{c}_{k-1} + \mathbf{k}_k\varepsilon_k \qquad (7.64)$$

Choice of parameters and initiation

The choice of the parameter λ (the forgetting factor) is important for the convergence properties of the algorithm. A low value means that the old samples are quickly forgotten and that the new ones are given more weight. This is important for the ability of the algorithm to follow quick changes in the signal. The larger the value the slower the response. The value 'one' gives an exact MMSE solution for ergodic signals. The parameter λ can be transformed to a time constant T_0, where

$$T_0 = \frac{1}{1-\lambda} \qquad (7.65)$$

The samples from a time T_0 back in time will only contribute by 36% compared to the latest sample. The value of λ is often chosen as 0.98–0.995.

When starting up the algorithm there must be values of $\hat{\mathbf{R}}_0^{-1}$ and $\mathbf{c}_0$. A good starting point for the autocorrelation matrix is obtained if an initial estimation of the autocorrelation is made. If this is not possible $\hat{\mathbf{R}}_0^{-1} = \rho\mathbf{I}$ can be chosen. The parameter ρ is a small positive constant, which, for instance, can be chosen equal to the noise variance $N_0/2$. For the filter coefficients we can either choose a probable value of $\mathbf{c}$ or put $\mathbf{c}_0 = \mathbf{0}$.

Implementation

The LMS algorithm is characterised by its structural simplicity, which makes it easy to implement in real systems. The computational load becomes reasonable and increases approximately linearly with $2N$. The constant N gives the order $2N + 1$ of the equaliser filter. The RLS algorithm, on the other hand, demands more computing capacity and the complexity increases as $4N^2 + 4N + 2$. Note, however, that for this level of computing demand, the matrix multiplications

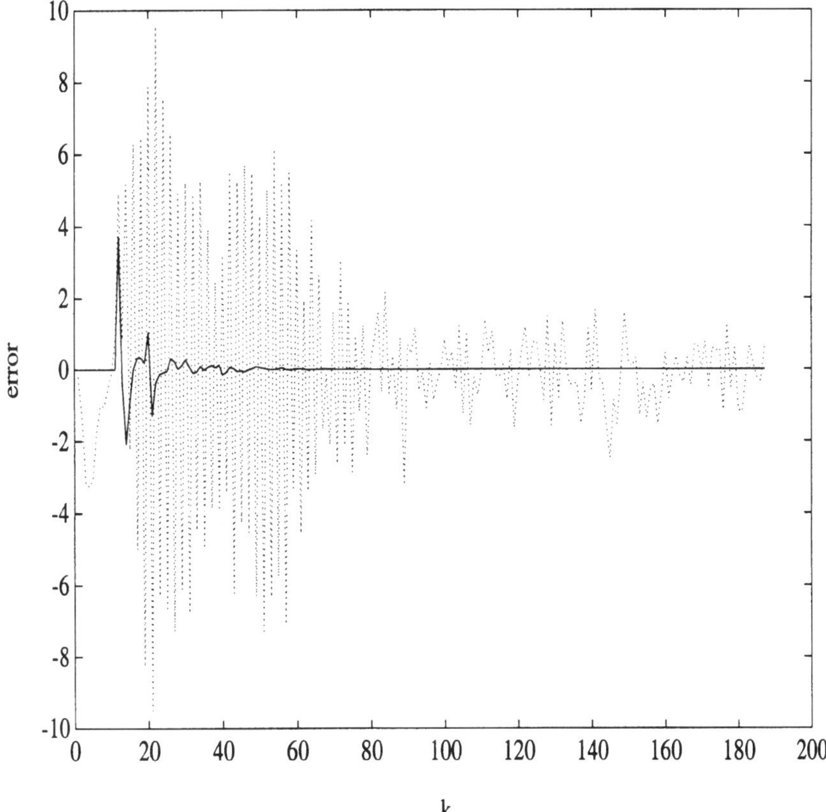

Figure 7.11 Response to a sudden change in the transfer function of the channel occurring at sample no. 10 in the diagram. The dashed curve corresponds to LMS and the solid line to RLS.

must be performed in the order given by eqs. (7.63) and (7.64). If we insert eq. (7.63) in (7.64), multiply by $\mathbf{k}_k$ and perform the multiplications from left to right, the number of multiplications can be proportional to N^3. The RLS algorithm usually converges both faster and with less variance in the error than the LMS algorithm (Figure 7.11). The distribution of the eigenvalues of the autocorrelation matrix is conclusive for the convergence properties of the LMS algorithm. To improve the convergence rate the signal can be transformed to equalise the eigenvalues. This increases the number of mathematical operations and thereby the complexity. The implementation is, as always, a compromise between good properties and complexity. Because of its simplicity, it is possible for the LMS algorithm to complete a larger number of recursions than RLS in the same time in a given computer system. LMS can therefore produce better results than a step by step comparison would suggest. When the algorithms are converged, both are very capable of tracking the changes in the channel provided that μ and λ are carefully chosen.

7.3 Appendix

7.3.1 SYSTOLIC ARRAY

Earlier in this chapter it has been described how matrix inversion can be done recursively in connection with solution of MMSE problems, eq. (7.27). Of course this inversion can also be carried out in blocks. In a sequential computer, the calculation load becomes rather heavy and since in several telecommunication applications real time function is required, block-wise inversion may become impossible to realise with general sequential processors. Certain algorithms can be divided in such a way that many operations can be carried out in parallel. The algorithm can then be implemented using special hardware built according to the principle of systolic array. A systolic array consists of calculation units located in a network. The word systolic is used in medicine. When the heart is compressed the blood is pushed out into the circulatory system. This movement is called the pump phase or the systole. In a systolic array the role of the heart is replaced by a clock signal which controls all units in the array. The units communicate locally, i.e. with their nearest neighbours. Data are fed in and out of the net through the outer units of the net. A global clock controls the communication between the units and data are shifted at a given rate. A systolic array is therefore suitable to realise in an integrated circuit. Below, a systolic array for solving equation systems is described.

The algorithm used is called Gaussian elimination with backward substitution. The method works in such a way that when the equation system $\mathbf{Ax} = \mathbf{y}$ is to be solved the matrix $\tilde{\mathbf{A}}^{(1)} = [\mathbf{A}|\mathbf{y}]$ is first formed by combining the matrix $\mathbf{A}$ and the vector $\mathbf{y}$. The object is then to make the matrix $\mathbf{A}$ upper triangular. The bottom row in $\mathbf{A}$ will then only contain one element $\neq 0$ and represents an equation with only one unknown, which is easily solved. The row next to the bottom one has two unknowns. If the solution from the bottom row equation is inserted the remaining unknown is easily found. The process continues upwards along the rows in $\tilde{\mathbf{A}}^{(1)}$ until all unknowns have been found.

Assume the system of equations $\mathbf{A}\mathbf{x} = \mathbf{y}$, where

$$\mathbf{A} = \begin{bmatrix} 5 & -3 & -2 \\ 7 & 11 & -6 \\ 3 & -1 & 4 \end{bmatrix}, \quad \mathbf{y} = \begin{bmatrix} 11 \\ -31 \\ 31 \end{bmatrix} \tag{7.66}$$

In the first step, $k = 1$, $\tilde{\mathbf{A}}^{(1)} = [\mathbf{A}|\mathbf{y}]$ is formed. For this example, we then obtain

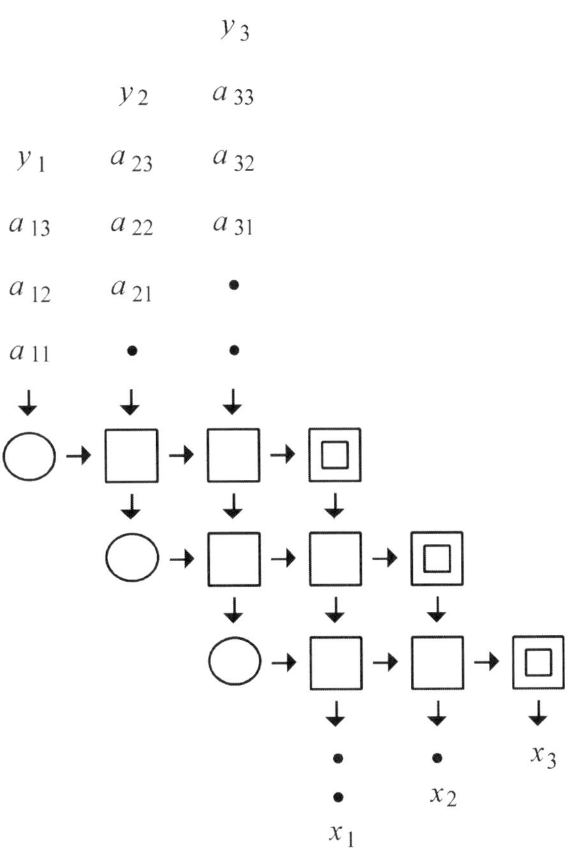

Figure 7.12 Systolic array for solving an equation system with three unknowns.

$$\tilde{\mathbf{A}}^{(1)} = \begin{bmatrix} 5 & -3 & -2 & 11 \\ 7 & 11 & -6 & -31 \\ 3 & -1 & 4 & 31 \end{bmatrix} \tag{7.67}$$

For $k > 1$ a new matrix $\tilde{\mathbf{A}}^{(k)}$ is formed through linear operations on rows and columns. After two further steps, $k = 3$ and we get

$$\tilde{\mathbf{A}}^{(3)} = \begin{bmatrix} 5 & -3 & -2 & 11 \\ 0 & \dfrac{76}{5} & -\dfrac{16}{5} & -\dfrac{232}{5} \\ 0 & 0 & \dfrac{102}{19} & \dfrac{510}{19} \end{bmatrix} \tag{7.68}$$

The last row represents the equation $(102/19)x_3 = 510/19$, which gives $x_3 = 5$. Then $x_2 = -2$ and $x_1 = 3$ are solved successively. The calculation method described above can be separated into a matrix of calculation elements which constitutes the systolic array (see Figure 7.12).

The calculation elements are represented by circles and squares in the figure. In a real case these can be specially designed elements placed on an integrated circuit and in the same pattern as in the figure. To solve the equation system according to the method described above the elements must perform the operation shown in Figure 7.13.

The first time a unit receives data it carries out Init operations and afterwards it carries out Else operations. The solution requires $4n - 1$ clock cycles where n is the number of unknowns, which in this case means 11 clock cycles. After 11 clock pulses through the system, x_1, x_2 and x_3 have become available on the outputs of the systolic array.

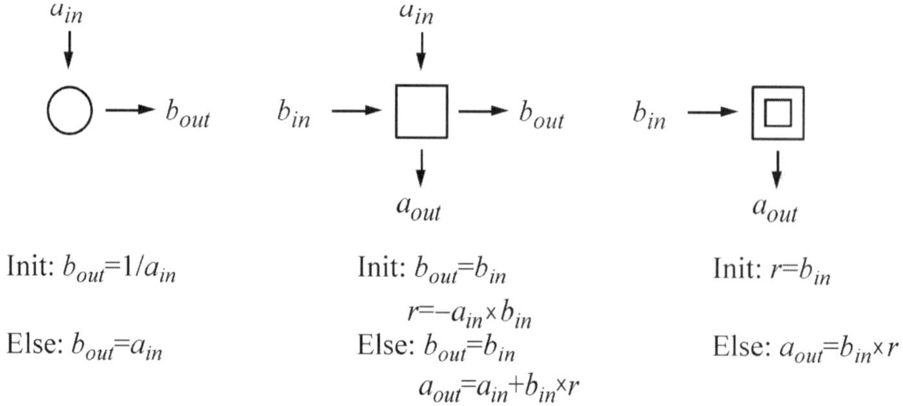

Figure 7.13 Computation units in the systolic array.

8   ADAPTIVE ANTENNAS

Communication systems using radio waves are sometimes jammed by other systems emitting electromagnetic radiation, for instance other radio transmitters using the same frequency band. In civilian applications such interference is unintentional but sometimes unavoidable. In military systems the interference is often intentional with the aim of making communication impossible. In both these cases the interference scenario often varies with time, in the sense that interference sources crop up or disappear, or change their positions. The desired signal is also often sent from a mobile source. A way of reducing interference is to use antennas which have their largest sensitivity in the direction of the desired signal source and close to zero sensitivity in the directions where interference sources are located. Since the scenario changes with time the antenna must possess the ability to adapt to changing conditions. Adaptive antennas are therefore used to suppress interference in communication systems. The theory of adaptive antennas is extensive and this chapter therefore only contains an introduction to the subject. An adaptive antenna consists mainly of a signal processing part in combination with an array antenna, which in turn consists of several individual antenna elements as shown in Figure 8.1.

The fundamental theory of array antennas is given below. This is followed by a presentation of two different adaptation algorithms based on different a priori information about the desired signal. The presentation is based on what was said about adaptive systems in Chapter 7. Two methods for estimating the angle of arrival for signals are also given. Most effective for detection of signals with multipath and discrete noise sources spread in the space are methods that join equalisation in time and space, so called spatio-temporal algorithms. Two methods for spatio-temporal equalisers are presented. The aim is to give the reader a brief introduction to the utilisation of adaptive antenna systems in telecommunication.

8.1 Array antennas

An array antenna consists of several antennas, or antenna elements, from which the output signals are combined by summation. The antenna radiation pattern, or radiation diagram, describes the sensitivity of the antenna in different directions.

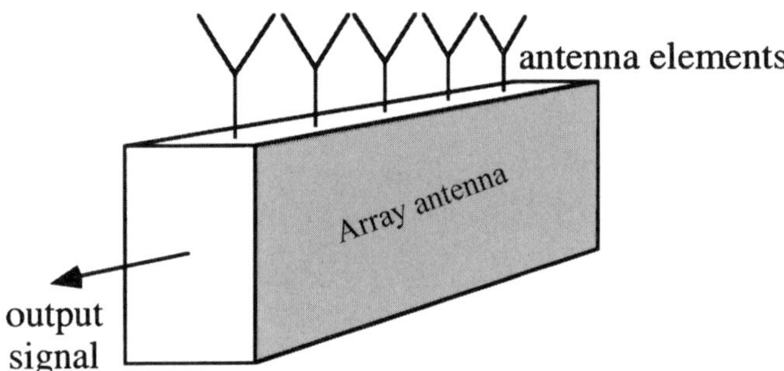

Figure 8.1   Array antenna.

The antenna radiation pattern can be either completely adaptive, or only have a steerable main lobe which can be directed towards the desired signal source.

The total antenna diagram, $U(\theta)$, of an array antenna can be written as the individual antenna diagrams, $G(\theta)$, of the antenna elements multiplied by the so-called array factor, $AF(\theta)$:

$$U(\theta) = G(\theta)\cdot AF(\theta) \tag{8.1}$$

For a spherically radiating antenna element $G(\theta) = 1$ for all directions. In the rest of the chapter all antenna elements are assumed to be spherically radiating.

For the analysis in Section 8.1 a plane wave front is assumed to be incident to the array antenna. In other words, it is necessary that the wave front is coherent, which is in contrast to the requirements at diversity. Diversity improves SNR by obtaining several statistically independent signal paths, using for instance several antennas. This requires other evaluation methods. However, some of the methods presented below give optimal solutions even in diversity.

8.1.1 ANTENNAS WITH STEERABLE MAIN BEAM

Antenna radiation pattern

To begin with the presentation is limited to an array antenna with a steerable main lobe. Assume N antenna elements are arranged as in Figure 8.2. The impinging electromagnetic wave is individually captured by each antenna element.

To be able to steer the main lobe in the desired direction, a phase shift of the captured signal is introduced, which increases by b (rad) for each element. The output signals from the antenna elements are then added.

The wave front is a plane wave if the signal source is at a distance much greater than the length of the array antenna, $(N - 1)d$, and that possible multi-path propagation is limited to a narrow sector around the signal source. It is assumed that no electromagnetic coupling between the antenna elements exists.

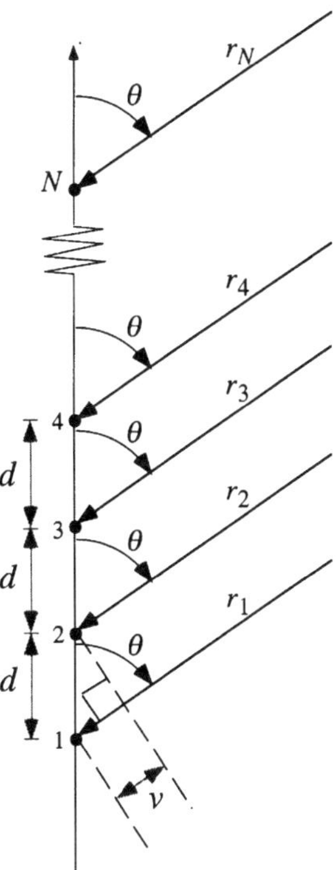

Figure 8.2 Impinging signals captured by the antenna elements in the positions 1,...,N.

The incident signal is captured by all antenna elements. There is a difference, v, in path length for the incident signal between adjacent elements. From the geometry in Figure 8.2 it can be seen that this path difference is

$$v = -d\sin\left(\theta - \frac{\pi}{2}\right) = d\cos\theta \qquad (8.2)$$

The phase difference per metre travelled distance, i.e. the wave number, is $k = 2\pi f/c = 2\pi/\lambda$ (rad/m) where λ is the wavelength (m) and c is the speed of light $= 2.997925 \times 10^8$ ms^{-1}.

The total phase difference, between the output signals from adjacent antenna elements, becomes the sum of the difference in path length and the extra phase difference introduced for the steering of the main lobe, i.e. $kd\cos\theta + b$ rad.

$AF(\theta)$ is the vector sum of the contributions from each element. The output signal from each antenna element can in complex notation be written $r_n = A_n\exp(j\Phi_n)$ where A_n is the amplitude and Φ_n a phase angle. Assume the amplitude is the same for all output signals. This is a reasonable assumption when the source is assumed to be far away and when all incident power is tapped from the respective antenna element, i.e. all $A_n = 1$. Using eq. (8.2) the array factor is obtained:

$$AF = 1 + \exp[j(kd\cos\theta + b)] + \exp[j2(kd\cos\theta + b)]$$
$$+ \exp[j3(kd\cos\theta + b)] + \ldots + \exp[j(N-1)(kd\cos\theta + b)] \qquad (8.3)$$

According to the formula for the sum of a geometrical series, eq. (8.3) gives

$$AF = \frac{1 - \exp(jN\psi)}{1 - \exp(j\psi)} = \exp[j((N-1)/2)]\psi\frac{\sin\dfrac{N}{2}\psi}{\sin\dfrac{\psi}{2}} \qquad (8.4)$$

where

$$\psi = kd\cos\theta + b \qquad (8.5)$$

If the reference point is the geometrical centre of the antenna and if the maximum value is normalised to one through multiplication by $1/N$, a normalised array factor is obtained,

$$AF_n = \frac{1}{N}\frac{\sin(N/2)\psi}{\sin(\psi/2)} \qquad (8.6)$$

which for small angles ψ can be approximated by

$$AF_n \approx \frac{1}{N}\frac{\sin(N/2)}{(\psi/2)} \qquad (8.7)$$

For examples of radiation patterns, see Figures 8.3 and 8.4.

Direction and width of the main lobe

Equations (8.6) and (8.7) always have maximas for $\psi = 0$, since the incident waves then interfere constructively. If this is defined as the direction of the main lobe, we obtain the direction θ_m ($m =$ maxima) by solving eq. (8.5) for $\psi = 0$:

$$\theta_m = \pm\arccos\left(\frac{\lambda b}{2\pi d}\right) \qquad (8.8)$$

We see that if $b = 0$, the main lobes will occur for $90°$ and $270°$. An array antenna where the elements have been arranged on a line can not discriminate between front and back. Therefore a backward beam pattern which is a mirror image of the pattern towards the front is obtained, giving two main beams. To discriminate between front and back the antenna elements have to be arranged in a two-dimensional pattern. An antenna radiating at a $90°$ angle relative to the axis through the antenna elements is called a broadside array (see Figures 8.3 and 8.4).

By changing the value of the phase shift, b, between the antennas, the main

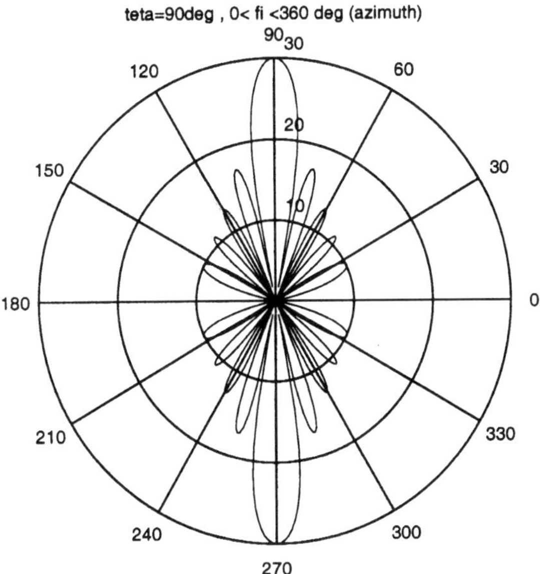

Figure 8.3 Example of antenna radiation pattern of a linear 10 element array antenna radiating at 90°.

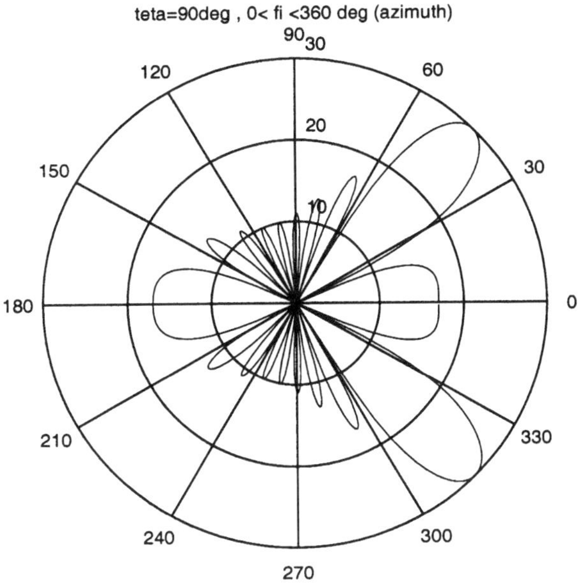

Figure 8.4 Example of antenna radiation pattern of a linear 10 element array antenna radiating at 45°.

lobe can be steered in different directions. The result is an antenna for which the angle of the main lobe can be scanned with the direction given by eq. (8.8).

If b is replaced by $2\pi f \tau$, the lobe angle is obtained when the output signal from the antenna element is delayed by a time, τ. The advantage of using time delays instead of phase shifts is that the lobe angle becomes independent of the signal frequency.

Since eq. (8.6) has zeros for $\psi = 2\pi n = kd\cos\theta + b$, where $n = 0, 1, 2,\ldots$, the antenna radiation pattern will have maxima for integer values of n. To avoid multiple main lobes, so-called grating lobes, the distance between the antenna elements must fulfil $d < (2\pi - b)/k$. The lobe width can be defined by the point where eq. (8.6) is reduced by 3 dB on both sides of the main lobe maximum. This is obtained by solving eq. (8.6) which gives the 3 dB points at ($h =$ half power)

$$\theta_h = \cos^{-1}\left[\frac{\lambda}{2\pi d}\left(-b \pm \frac{2.782}{N}\right)\right] \tag{8.9}$$

The lobe width then becomes

$$\theta_{lb} = 2|\theta_m - \theta_h| \tag{8.10}$$

8.1.2 VECTOR DESCRIPTION OF THE OUTPUT SIGNAL OF ARRAY ANTENNAS

Only the direction of the main lobe is steerable in the discussion above. An even better interference suppression can be obtained if the complete antenna radiation pattern can be shaped. The maximum sensitivity of the antenna can then be directed towards the desired signal source and the antenna radiation pattern can be designed to have nulls in the directions where there are interference sources. To simplify the mathematical analysis we now change to a vector and matrix representation. Assume the terms in the array factor, eq. (8.3), are taken as the elements of a vector, $\mathbf{a}$. The parameter b is removed as it can be included in a vector to be defined later. At each antenna element an incident wave receives a phase difference which depends on the spatial extension of the antenna. Given a linear array with constant separation between the antenna elements the phase shifts relative to the first element are

$$\mathbf{a}(\theta, k) = \begin{bmatrix} 1 & \exp[j(kd\cos\theta)] & \exp[j2(kd\cos\theta)] & \cdots & \exp[j(N-1)(kd\cos\theta)] \end{bmatrix}^H \tag{8.11}$$

This vector is called the array response vector and is a way to take account in the computations of the element positions and the direction of the impinging wave. In eq. (8.11) the constituent elements are assumed to have spherically symmetrical antenna radiation patterns. If this is not the case, $\mathbf{a}$ must be supplemented by a factor for each antenna element describing the amplitude sensitivity in different directions, $f_n(\theta)$, of this element. This gives

$$\mathbf{a}(\theta, k) = \begin{bmatrix} f_1(\theta) & f_2(\theta)\exp[j(kd\cos\theta)] & \cdots & f_N(\theta)\exp[j(N-1)(kd\cos\theta)] \end{bmatrix}^H$$

$$(8.12)$$

In the previous section only the phase difference between each antenna element was varied by changing the parameter b, which provides a way of changing the direction of the main lobe. If the phase can be changed arbitrarily and if the amplitude of the output signal from each antenna can also be controlled, the shape of the antenna lobe can be varied more freely. The shape of the main lobe, the side lobe level, etc., can be controlled within the limits which are theoretically possible. The size of the antenna, for instance, sets a limit for how narrow the main lobe can be. Phase and amplitude control is achieved by multiplying the response from each antenna by complex weight factors, which then constitute the weight vector $\mathbf{W}$,

$$\mathbf{W}^H = \begin{bmatrix} w_1^* & w_2^* & \cdots w_N^* \end{bmatrix} \tag{8.13}$$

Using the vector representation for AF in eq. (8.3), the elements of the weight vector would be $w_n^* = \exp[j(n-1)b]$ and $|w_n|^2 = 1$. The total response from the array antenna to an input signal from a direction given by θ and having the wave number k, is denoted the antenna radiation pattern and it becomes (cf. eq. (8.1))

$$U(\theta, k) = \mathbf{W}^H \mathbf{a}(\theta, k) \tag{8.14}$$

Usually the arguments θ and k are not written out explicitly. Assume a bandpass signal source, the signal of which can be described in the usual way as $s(t) = A_d\exp[j(\omega_d t + \phi_d)]$ where A_d is a slowly varying envelope and ϕ_d is a slowly varying phase. The direction to this signal source is θ_d. The output signal from the antenna becomes

$$y(t) = \mathbf{W}^H \mathbf{a}(\theta_d, k) A_d\exp[j(\omega_d t + \phi_d)] = \mathbf{W}^H \mathbf{X} \tag{8.15}$$

The vector $\mathbf{X}$ not only describes the input signal and the angle of arrival, but also contains parameters for the array antenna, such as distance between the antenna elements (Figure 8.5).

When there are several signals impinging on the array antenna, i.e. $s_m(t)$ where $m = 1,\ldots,M$, and when noise interferes with the signal from each antenna element, $\mathbf{X}$ can be represented by an $N \times 1$ matrix where all input signals are included. For M incident signals we obtain

$$\mathbf{X} = \mathbf{A}(\theta, \mathbf{k})\mathbf{s}(t) + \mathbf{n}(t) \tag{8.16}$$

where $\mathbf{A}(\theta, \mathbf{k}) = \begin{bmatrix} \mathbf{a}(\theta_1, k_1) & \cdots & \mathbf{a}(\theta_M, k_M) \end{bmatrix}$ which is called the array manifold, $\mathbf{s}(t) = \begin{bmatrix} s_1(t) & \cdots & s_M(t) \end{bmatrix}^H$ and $\mathbf{n}(t)$ is an $N \times 1$ matrix with statistically independent

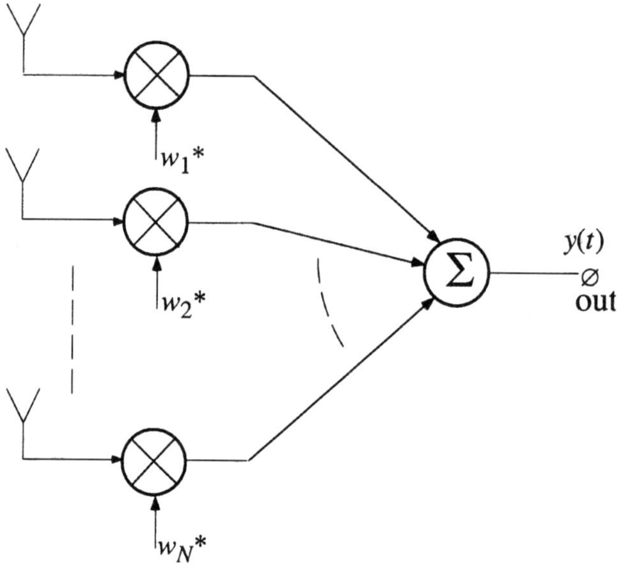

Figure 8.5   Array antenna with adjustable weights **W**.

elements which model the noise, e.g. the receiver noise, at each antenna input in the system. Equation (8.16) inserted in eq. (8.15) gives the output signal from an antenna given the input signals $\mathbf{s}(t)$. This representation is used below to find the weights, i.e. the phase and amplitude controls, which are the solutions to certain optimum criteria of which some will be discussed later.

8.1.3 MULTILOBE ANTENNAS

A multilobe antenna consists of a number of antenna elements for which the output signals are transformed, either by using digital signal processing or by using an analogue network. The transformation means that a number of outputs, each with a separate antenna lobe, are formed (cf. Figure 8.6).

Depending on which transformation is chosen, different lobe patterns can be created. A linear transform, or mapping, is achieved by

$$\mathbf{B} = \mathbf{T}\mathbf{X} \tag{8.17}$$

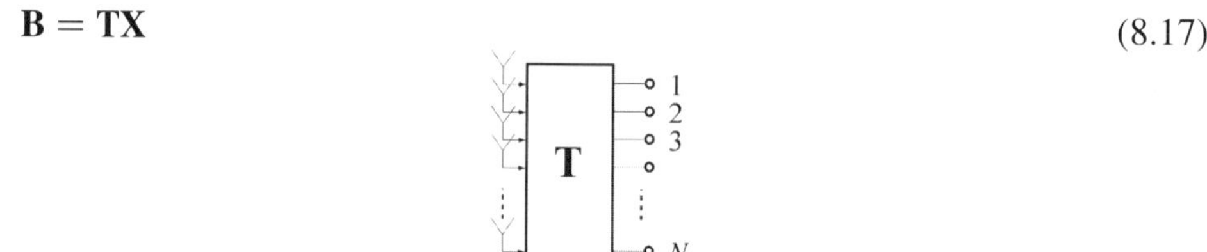

Figure 8.6   Multilobe antenna with transforming network and N outputs.

where $\mathbf{T}$ is a transformation matrix and the elements of $\mathbf{B}$ are the coordinates of $\mathbf{X}$ in the new coordinate system. In this context $\mathbf{X}$ is the input signal of the antenna elements and has been defined in eq. (8.16). To find out how the transformation influences the antenna lobes the array response vector can be used. Calculate $\mathbf{B}_d = \mathbf{T}\mathbf{a}(\theta)$ and vary θ, which corresponds to the incidence angle of the signal. Each element in the vector $\mathbf{B}_d$ corresponds to one output from the network performing the transformation.

A very interesting transform is the discrete Fourier transform (DFT) which can be written in matrix form as

$$\mathbf{T}_F = \begin{bmatrix} 1 & 1 & 1 & \cdots & 1 \\ 1 & W_N^1 & W_N^2 & & W_N^{N-1} \\ 1 & W_N^2 & W_N^4 & & W_N^{2(N-1)} \\ \vdots & & & \ddots & \vdots \\ 1 & W_N^{N-1} & W_N^{2(N-1)} & \cdots & W_N^{(N-1)(N-1)} \end{bmatrix} \tag{8.18}$$

where W_N is a so-called twiddle factor

$$W_N = \exp(-j2\pi/N) \tag{8.19}$$

The element in column n and row l is raised to the power of $(n-1)\times(l-1)$. After the transformation, the lth output corresponds to the lth vector element in $\mathbf{B}_d$, which is

$$b_d(\theta, l) = \sum_{n=0}^{N-1} \exp[j(k\,n\,d\cos\theta)]W_N^{n(l-1)} = \sum_{n=0}^{N-1} \exp\left[j2\pi n\left(\frac{d}{\lambda}\cos\theta - \frac{l-1}{N}\right)\right]$$

$$= \exp\left[j\pi(N-1)\left(\frac{d}{\lambda}\cos\theta - \frac{l-1}{N}\right)\right] \frac{\sin\left(\pi N\left[\frac{d}{\lambda}\cos\theta - \frac{l-1}{N}\right]\right)}{\sin\left(\pi\left[\frac{d}{\lambda}\cos\theta - \frac{l-1}{N}\right]\right)} \tag{8.20}$$

where $l = 1,2,\dots N$. The constants k, d and λ have been defined earlier. The lth output gets a directional sensitivity which can be written

$$|b_d(\theta, l)| = \left| \frac{\sin\left(\pi N\left[\frac{d}{\lambda}\cos\theta - \frac{l-1}{N}\right]\right)}{\sin\left(\pi\left[\frac{d}{\lambda}\cos\theta - \frac{l-1}{N}\right]\right)} \right| \tag{8.21}$$

Compare eq. (8.4). By the Fourier transformation a directional maximum has been created for the lth output at the angle θ_l. By using eq. (8.21) we obtain

$$\pi\left[\frac{d}{\lambda}\cos\theta_l - \frac{l-1}{N}\right] = \pi m \quad \Rightarrow \quad \theta_l = \arccos\left(\frac{\lambda}{d}\left[m + \frac{(l-1)}{N}\right]\right) \quad (8.22)$$

where m is an integer defined so that the argument of the arccos function is in the interval -1 to 1. The sensitivity decreases on both sides of the maximum. After the transformation, the outputs form lobes in different directions as illustrated in Figure 8.7. By noticing from which of the outputs a certain signal comes, its angle of arrival is determined.

Note the duality between the determination of the frequency components of a time-sampled signal using Fourier transformation and the determination of the incidence direction of a space-sampled signal. The variable f has the same function at a Fourier transformation in time as $\cos(\theta)/\lambda$ at a Fourier transformation in space.

The similarities to signal processing in time do not end with this. A common implementation of DFT is the fast Fourier transform (FFT). In Figure 8.8 a decimation in frequency variant of FFT is shown.

The signals can be continuous in time without any problem; it is after all in

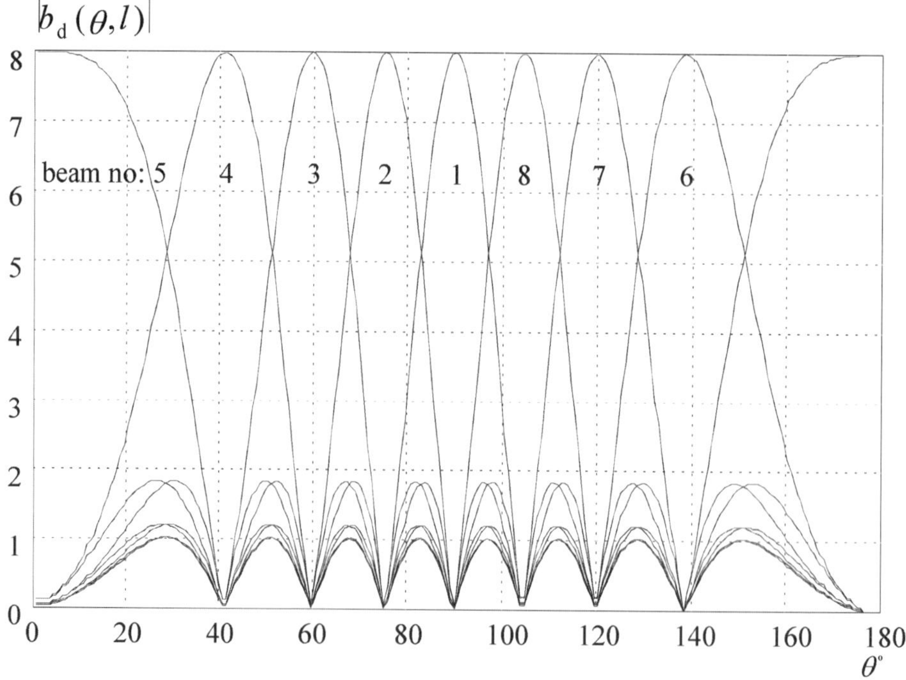

Figure 8.7 Beam patterns at each output for a transformed 8 element array antenna. Note that all curves intercept at the same level. The positions of the peaks are given by eq. (8.22).

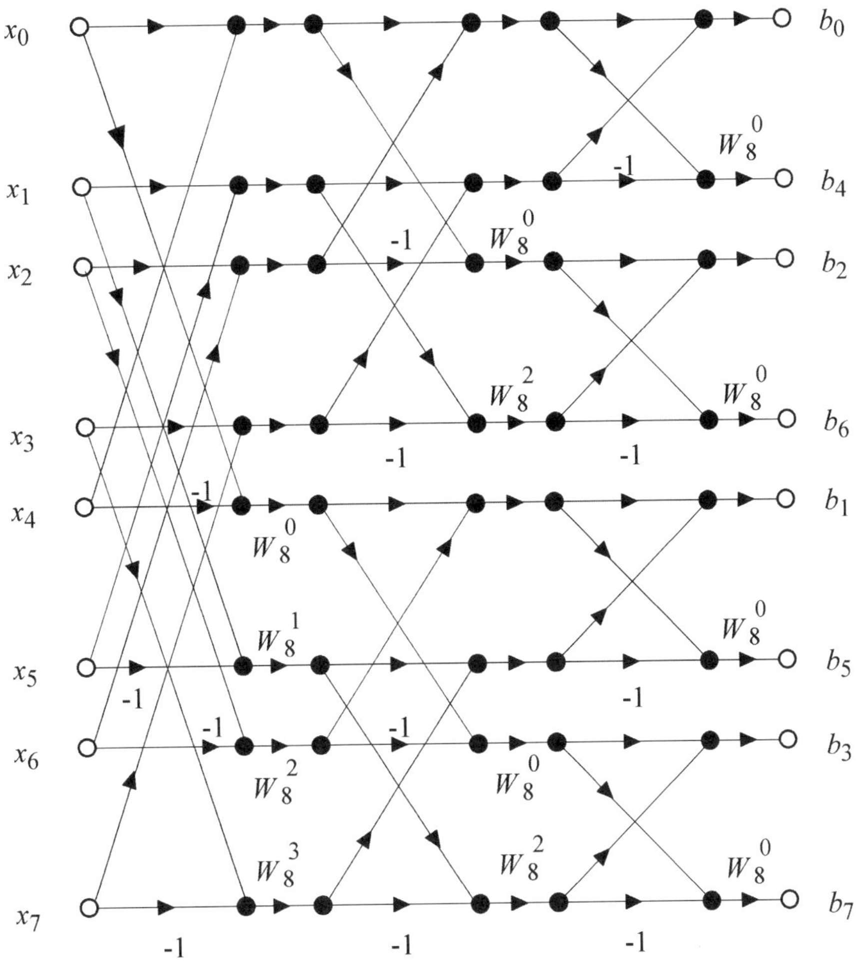

Figure 8.8 The Butler matrix and a decimation in the frequency FFT algorithm.

space that the sampling takes place. The factors W_N consist of complex exponential functions, which for a narrowband high frequency signal corresponds to a phase shift equal to the argument of the complex value. A phase shift of a high frequency signal can easily be arranged by hardware. Branch and summation points can easily be realised by couplers or power dividers. It is, in other words, very easy to realise a flow chart as in Figure 8.8 using hardware for high frequency signals. The hardware realisation gives a network called the Butler matrix. (The FFT algorithm and the Butler matrix were independently invented during the second half of the 1950s; officially the similarity was not pointed out until around 1965.) When spherically radiating antenna elements are connected to the inputs $x_1 - x_N$, in Figure 8.8, directed lobes in different angles are obtained at the outputs $b_1 - b_N$ of the network. Note that the beams are not in numerical order at the outputs of the network.

8.2 Signal processing with array antennas

By controlling both phase and amplitude of the signal from each of the antenna elements in an array antenna, the antenna radiation pattern can be controlled. The control is adaptive in a way that the output signal from the antenna is optimised according to some criterion. To make adaptation to a desired signal possible, the antenna must have the necessary information to distinguish the desired signal from interference. This can be achieved in different ways, such as using a reference signal similar to the desired one, providing information about the direction to the desired signal source or about its statistical parameters. There are a number of such criteria. The ones presented below optimise the signal to noise ratio of the output signal and are therefore called max. SNR.

8.2.1 THE REFERENCE SIGNAL METHOD

In the reference signal method it is assumed that in the wanted signal there is a known pattern which can function as a reference for the receiver. The goal is that the output signal of the adaptive antenna should match the reference as closely as possible. The desired signal is, however, not alone at the receiver input but is received together with interference and other distortions which can occur in the channel. In many applications certain properties of the desired signal are known well enough to make it possible to create a reference signal which can be the objective function of the optimisation algorithm (Figure 8.9).

The reference signal can be denoted $s_d(t)$ and an error term can be obtained as

$$\varepsilon(t) = y(t) - s_d(t) \tag{8.23}$$

This can be compared to the error term in eq. (7.5). By performing the same matrix algebra as in Section 7.2, the result of the optimum weights becomes (cf. eq. (7.27))

$$\mathbf{W}^{opt} = \mathbf{R}_X^{-1}\mathbf{r}_{Xd} \tag{8.24}$$

where $\mathbf{r}_{Xd}$ is called the steering vector and is the cross correlation between the reference signal and the input signal $\mathbf{X}$:

$$\mathbf{r}_{Xd} = E\{\mathbf{X}\, s_d^*(t)\} \tag{8.25}$$

As can be seen in eq. (8.24), inversion of the autocorrelation matrix is required to find the optimal weights. This can be achieved by, for instance, the LMS or RLS algorithms described in Chapter 7.

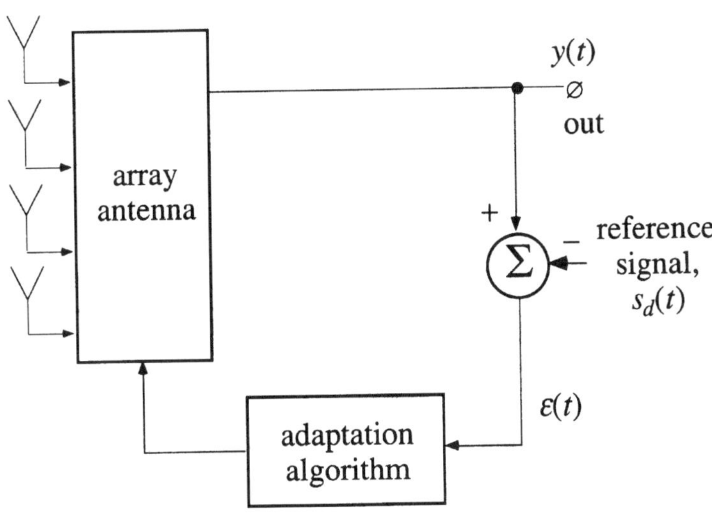

Figure 8.9 Adaptive antenna system based on the reference signal method.

To compute the signal to noise and interference ratio (SNIR) the powers of the different signal components on the antenna output are computed. As can be seen from eq. (8.15) the output signal from the array antenna is the product between the input signal and the antenna weights, the power then is

$$E\{y(t)y^*(t)\} = \mathbf{W}^H \mathbf{X} \mathbf{X}^H \mathbf{W} = \mathbf{W}^H \mathbf{R}_X^H \mathbf{W} \qquad (8.26)$$

where $\mathbf{R}_X$ is the autocorrelation matrix of the input signal. For the wanted signal, the autocorrelation matrix is $\mathbf{R}_d$ and for the interference, $\mathbf{R}_i$. Suppose that the noise is white, with the spectral density $N_0/2$ and uncorrelated with the other signals. The signal to noise and interference ratio in the output signal then becomes (cf. eq. (8.28))

$$SNIR = \frac{\mathbf{W}^H \mathbf{R}_d \mathbf{W}}{\mathbf{W}^H \mathbf{R}_i \mathbf{W} + (N_0/2)\mathbf{W}^H \mathbf{W}} \qquad (8.27)$$

8.2.2 THE LINEAR CONSTRAINT MINIMUM VARIANCE METHOD

To find the optimal antenna weights in this method (other names are Capon or minimum variance distortionless response (MVDR)), it is assumed that the direction, θ_d, to the desired signal source is known, but the directions to the interfering signal sources are assumed unknown. The antenna is forced to receive input signals from the direction of the desired source, by locking the main lobe in the direction where the desired signal is located. The antenna will also receive unwanted signals, or interference, which contribute to the output power of the antenna signal. One optimisation criterion is therefore to reduce the power, or variance, of the antenna signal as much as possible, given that the main lobe is locked. For statistically independent signal and interference, the output power is

$$P_y = E\{|y|^2\} = E\{|\mathbf{W}^T(\mathbf{X}_d + \mathbf{X}_i + \mathbf{X}_n)|^2\}$$

$$= \mathbf{W}^H\Big(E(\mathbf{X}_d\mathbf{X}_d^H) + E(\mathbf{X}_i\mathbf{X}_i^H) + E(\mathbf{X}_n\mathbf{X}_n^H)\Big)\mathbf{W}$$

$$= \mathbf{W}^H(\mathbf{R}_d + \mathbf{R}_i + \mathbf{R}_n)\mathbf{W} = \mathbf{W}^H\mathbf{R}_X\mathbf{W} \qquad (8.28)$$

where $\mathbf{X}_d$ is the desired signal, $\mathbf{X}_i$ is the impinging interference, for instance from other radio transmitters, and $\mathbf{X}_n$ is the noise on the antenna inputs. Since the antenna is locked in the direction of the desired signal source, adjust the weights to get the lowest possible power in the antenna output signal. This is the power of the desired signal, given that there is no noise. The value of 1 (one) is given to the antenna radiation pattern in the direction of the desired signal, which gives a constraint. The optimisation criterion therefore becomes

$$\min_{\mathbf{W}} \mathbf{W}^H\mathbf{R}_X\mathbf{W} \text{ subject to the constraint } \mathbf{a}^H(\theta_d, k)\mathbf{W} = 1 \qquad (8.29)$$

This method is therefore usually called linear constraint minimum variance (LCMV). A usual way of handling the optimisation of a function under certain constraints is by using Lagrange's multiplier method. This is based on the fact that if f is an optimisation criterion, g_n is a constraint and x_0 is a solution of this optimisation problem, then the derivative of f is a linear combination of the derivative of the constraints:

$$f'(x_0) = \sum_n \lambda_n \, g_n'(x_0) \qquad (8.30)$$

where λ_n is called Lagrange's multipliers. For a multidimensional optimisation problem the derivatives are replaced by gradients. The gradient of the matrix $\mathbf{A}$ is written grad$\mathbf{A}$. In the case discussed here there is only one constraint and the following equation system can be obtained:

$$\text{grad}\Big[\mathbf{W}^H\mathbf{R}_X\mathbf{W}\Big] = \lambda_1 \text{grad}\Big[\mathbf{a}^H(\theta_d, k)\mathbf{W}\Big] \qquad (8.31)$$

This gives (cf. Section 7.2.)

$$2\mathbf{R}_X\mathbf{W} = \lambda_1\mathbf{a}(\theta_d, k) \qquad (8.32)$$

This means that the optimal weights must fulfil

$$\mathbf{W}^{opt} = \frac{\lambda_1}{2}\mathbf{R}_X^{-1}\mathbf{a}(\theta_d, k) \qquad (8.33)$$

According to the constraint the optimal weights must also fulfil

$$1 = \mathbf{a}^H(\theta_d, k)\mathbf{W}^{opt} = \frac{\lambda_1}{2}\mathbf{a}^H(\theta_d, k)\mathbf{R}_X^{-1}\mathbf{a}(\theta_d, k) \tag{8.34}$$

which gives

$$\lambda_1 = \frac{2}{\mathbf{a}^H(\theta_d, k)\mathbf{R}_X^{-1}\mathbf{a}(\theta_d, k)} \tag{8.35}$$

Equation (8.35) inserted into eq. (8.33) then gives the solution for the optimal weights:

$$\mathbf{W}^{opt} = \frac{\mathbf{R}_X^{-1}\mathbf{a}(\theta_d, k)}{\mathbf{a}^H(\theta_d, k)\mathbf{R}_X^{-1}\mathbf{a}(\theta_d, k)} \tag{8.36}$$

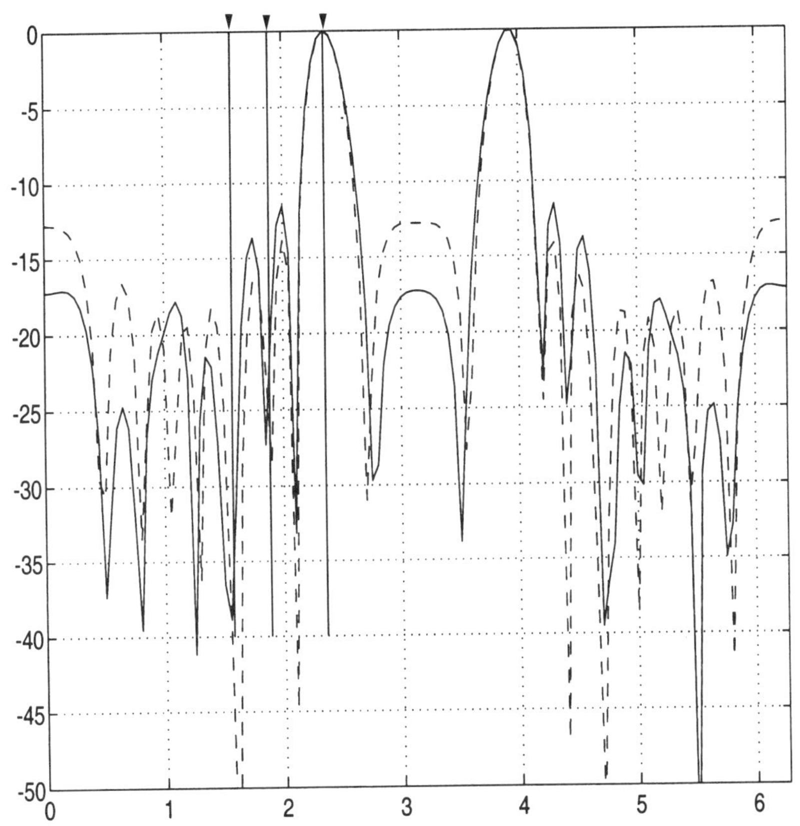

Figure 8.10 Antenna radiation pattern of a 10 element array antenna. The reference signal method is given by the dashed line, the LCMV method by the solid line. Desired signal at $3\pi/4 \approx 2.4$ interfering signals at $\pi/2 \approx 1.6$ and $3\pi/5 \approx 1.9$ rad. These are indicated by vertical lines in the pattern diagram.

An example of an antenna radiation pattern with the reference signal method as well as the LCMV method is shown in Figure 8.10. The antenna radiation pattern is plotted in dB as a function of the signal incident angle, in rad.

The figure shows the situation when all initial transients have faded out. Note that the main lobe has been directed towards the desired signal, which in one case has been identified by some known signal sequence and in the other case by its angle of incidence. In both cases nulls have been created in the directions of the interference sources. The differences between the two patterns are in practice negligible.

In the same way and with the same assumptions as in Section 8.2.1, a signal to interference and noise ratio can be computed. For the LCMV method this is

$$SNIR = \frac{\|s_d(t)\|_2^2}{\mathbf{W}^H \mathbf{R}_i \mathbf{W} + (N_0/2)\mathbf{W}^H \mathbf{W}} \tag{8.37}$$

As soon as the angle of arrival for a signal deviates somewhat from θ_d the algorithm will attempt to place nulls in the direction of this signal. This can cause problems if the estimation of the arrival angle is poor. This sensitivity can be reduced by setting constraints on the antenna pattern derivative, which is not treated here.

8.2.3 MUSIC

Multiple signal classification (MUSIC) is a method to estimate, with great accuracy, the angle of arrival of some impinging uncorrelated signals. For example, this can be used to compute the values, θ_d, for the angle of wanted signal arrival, needed in the LCMV method. MUSIC is based on the fact that the signal components of the array output signal lie in a low-rank subspace, provided that the number of antenna elements is strictly greater than the number of signal sources.

A narrowband signal received by antenna element n is $s_i(t, n) = A_i\cos(\omega_c t + nkd\cos(\theta_i))$. Expressed by means of its complex lowpass equivalent, the signal is $A_i\exp(jnkd\cos(\theta_i))$. Suppose K signals are impinging on the array antenna. The total received signal by antenna element n then is

$$x_n(t) = \sum_{i=1}^{K} A_i\exp[jn(kd\cos\theta_i)] + z(t) \tag{8.38}$$

where $z(t)$ is a complex white noise. Take a snapshot of the output signals from the antenna elements by sampling the signal at some instant. The sampled signals are arranged in a vector $\mathbf{X}$. The received signal on each element can be written, by using the array response vector and eq. (8.16)

$$\mathbf{X} = \sum_{i=1}^{K} A_i \mathbf{a}(\theta_i, k) + \mathbf{z} \tag{8.39}$$

The autocorrelation for the received signal becomes, where $\mathbf{a}(\theta_i, k) = \mathbf{a}_i$,

$$\mathbf{R} = E\left\{\mathbf{X}\,\mathbf{X}^H\right\} = \sum_{i=1}^{K} |A_i|^2 \mathbf{a}_i \mathbf{a}_i^H + \sigma^2 \mathbf{I} = \mathbf{R}_s + \sigma^2 \mathbf{I} \tag{8.40}$$

Note that every element in the matrix $\mathbf{R}$ consists of the sum of the various signals plus noise. The noise is white and will therefore only contribute to the diagonal elements in the matrix. Since the rank of $\mathbf{R}_s$, defined in eq. (8.40), is equal to the number of impinging uncorrelated signals, $\mathbf{R}_s$ has K eigenvalues different from zero. The other eigenvalues, for $\mathbf{R}$, come solely from the noise part $\sigma^2 \mathbf{I}$. The eigenvalues, λ_i, can therefore be divided into two groups: one coming only from noise and the other from signal plus noise, and they are consequently larger than σ^2. The autocorrelation matrix is Hermitian and its eigenvectors are then orthogonal. One can therefore create an orthonormal base by computing the eigenvectors for $\mathbf{R}$ and order them such that the eigenvalues are decreasing for increasing index:

$$\mathbf{E} = [\mathbf{E}_s, \ \mathbf{E}_n] = \begin{bmatrix} \mathbf{e}_1 & \mathbf{e}_2 & \cdots & \mathbf{e}_K & \mathbf{e}_{K+1} & \cdots & \mathbf{e}_N \end{bmatrix} \tag{8.41}$$

The first K eigenvectors, $\mathbf{E}_s$, spans the signal plus noise subspace and $\mathbf{E}_n$ the noise subspace. The signals therefore lie in the space spanned by the K first eigenvectors. In accordance with the spectral theorem of the matrix calculus, the matrix $\mathbf{R}$ can be expressed by its eigenvectors:

$$\mathbf{R} = \sum_{i=1}^{K} |A_i|^2 \mathbf{a}_i \mathbf{a}_i^H + \sigma^2 \mathbf{I} = \sum_{i=1}^{K} \lambda_i \mathbf{e}_i \mathbf{e}_i^H + \sum_{i=K+1}^{N} \sigma^2 \mathbf{e}_i \mathbf{e}_i^H \tag{8.42}$$

Since the noise subspace is orthogonal to the signal subspace and the signals are completely inside the signal subspace, the projection of the signal on the noise subspace will be $\mathbf{e}_i^H \mathbf{a}(\theta_i, k) = 0$ for all $K + 1 \leq i \leq N$. The constant A_i has not been included. If one performs the multiplication $\mathbf{e}_i^H \mathbf{a}(\theta, k)$ and let θ vary from 0 to 2π, the product is 0 for the values of θ corresponding to the angles of arrival, θ_i, of the signals. The product $\mathbf{e}_i^H \mathbf{a}(\theta, k) = B_i(\theta)$ is called the spatial spectrum. Compute a mean of this spectrum, taken over all eigenvectors in the noise subspace. Thereafter, invert the received signal spatial spectrum,

$$S(\theta) = \cfrac{1}{\cfrac{1}{K-N} \displaystyle\sum_{i=K+1}^{N} |B_i(\theta)|^2} = \cfrac{1}{\cfrac{1}{K-N} \mathbf{a}^H(\theta, k) \left[\displaystyle\sum_{i=K+1}^{N} \mathbf{e}_i \mathbf{e}_i^H \right] \mathbf{a}(\theta, k)} \tag{8.43}$$

The maximum points of the new spectrum, $S(\theta)$, then constitute the angles of arrival of the wanted signals.

Usually we do not have access to the precise autocorrelation matrix, therefore it has to be estimated. This is done from a number of snapshots, $\mathbf{x}_m$, of the antenna signals:

$$\hat{\mathbf{R}} = \frac{1}{M} \sum_{m=0}^{M-1} \mathbf{x}_m \cdot \mathbf{x}_m^H \tag{8.44}$$

When the autocorrelation matrix has been estimated, the eigenvalues are computed by solving $\det\left[\hat{\mathbf{R}} - \lambda\mathbf{I}\right] = 0$, and the eigenvectors, by solving the eigenvalue equation $\left(\hat{\mathbf{R}} - \lambda_i\mathbf{I}\right)\mathbf{e}_i = \mathbf{0}$. The eigenvectors belonging to the noise subspace are chosen. The spatial spectrum is computed by eq. (8.43). The peaks in the spectrum correspond to the angles of arrival of the signals (cf. Figure 8.11). As is clear from the figure, the capability to resolve close signals is much better when using the MUSIC method than using the Fourier transform. The same

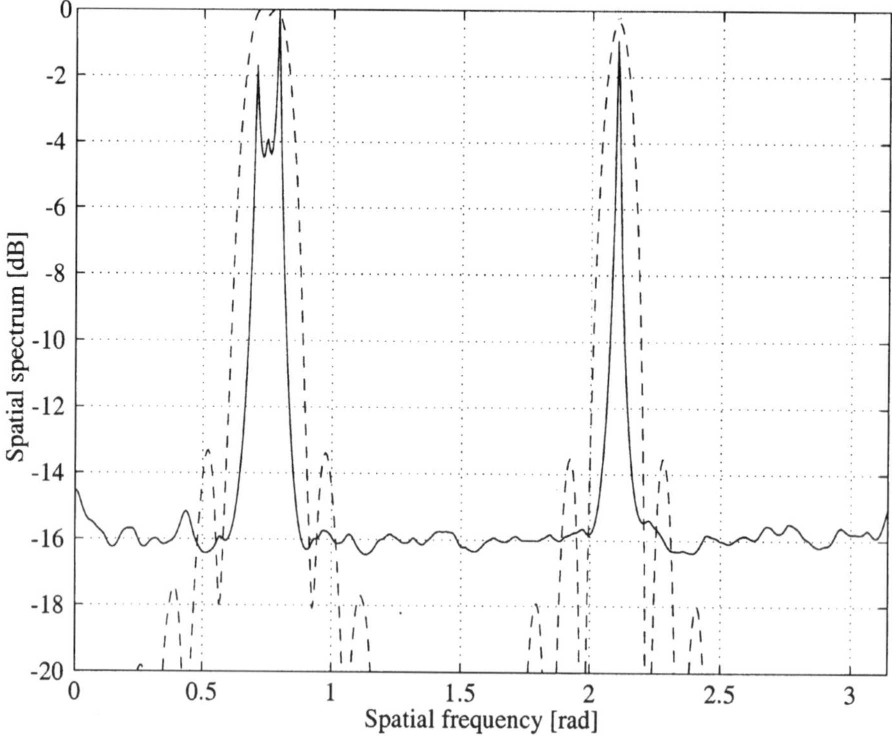

Figure 8.11   Spatial spectrum of three received signals with the normalised spatial frequency $\Omega_1 = 0.7$, $\Omega_2 = 0.79$ and $\Omega_3 = 2.1$. The solid line is MUSIC- and the dashed line is the Fourier spectrum. The spatial frequency is $\Omega = kd\cos(\theta)$.

signal propagating along several paths and arriving from different directions cannot be resolved as separate signals, since the linear dependence in the signals does not increase the rank of $\mathbf{R}_s$. If transmitter or receiver is in motion, the different signal paths will vary in different ways, which will make the received signals statistically independent anyway. Then the chance will still be good to resolve the signals.

Example 8.1

Determine the optimal weights with the reference signal method when a single signal impinges on a two element adaptive antenna. The system is assumed to be noise free.

Solution   The impinging signal is by means of eq. (8.16):

$$\mathbf{X} = \begin{bmatrix} 1 \\ \exp[-j(kd\cos\theta)] \end{bmatrix} s(t) \tag{8.45}$$

For computation of the optimal weights $\mathbf{R}_X^{-1}$ and $\mathbf{r}_{Xd}$ are needed. The autocorrelation matrix becomes

$$\mathbf{R}_X = E\{\mathbf{X}\,\mathbf{X}^H\} = \begin{bmatrix} 1 & \exp[j(kd\cos\theta)] \\ \exp[-j(kd\cos\theta)] & 1 \end{bmatrix} P_s \tag{8.46}$$

where $P_s = E\{s(t)s^*(t)\}$. Note that the matrix is singular since $\det[\mathbf{R}_X] = 0$. The steering vector becomes

$$\mathbf{r}_{Xd} = E\{\mathbf{X}\, s^*(t)\} = \begin{bmatrix} 1 \\ \exp[-j(kd\cos\theta)] \end{bmatrix} P_s \tag{8.47}$$

If the equation for the optimal weights is still solved, in spite of the singular autocorrelation matrix, by means of eq. (8.24) it is found that

$$\mathbf{R}_X \mathbf{W}^{opt} = P_s \begin{bmatrix} 1 & \exp[j(kd\cos\theta)] \\ \exp[-j(kd\cos\theta)] & 1 \end{bmatrix} \begin{bmatrix} w_1 \\ w_2 \end{bmatrix}$$

$$= \begin{bmatrix} 1 \\ \exp[-j(kd\cos\theta)] \end{bmatrix} P_s = \mathbf{r}_{Xd} \tag{8.48}$$

The second equality in eq. (8.48) constitutes an equation system that can be solved. The solution becomes

$$\mathbf{W}^{opt} = \begin{bmatrix} w_1 \\ w_2 \end{bmatrix} = \begin{bmatrix} 1 \\ 0 \end{bmatrix} + w_2 \begin{bmatrix} -\exp[j(kd\cos\theta)] \\ 1 \end{bmatrix} \qquad (8.49)$$

This means that the optimal weights are positioned along a line in the w_1–w_2 plane. As long as the weights are chosen from this line an optimal solution is obtained but some unambiguous solution does not exist. Return to eq. (8.46) and analyse what is required to obtain $\det[\mathbf{R}_X] \neq 0$ and thereby a definite point as an optimal solution. If the signal for example contained noise, the variance σ^2 would be added to the diagonal elements and the determinant should be different from zero. An additional signal, independent of the first one, would also give a determinant which is different from zero. Generally, the shape of the bowl formed by $J_k = E\{|\varepsilon_k|^2\}$ is dependent of the ratio between the least and the largest eigenvalues of the autocorrelation matrix. When the ratio $\lambda_{\max}/\lambda_{\min} = \infty$ is the area stretched so that the optimal point becomes a line. In the example above the two eigenvalues are $\lambda_1 = 2$ and $\lambda_2 = 0$, which gives an infinite ratio and thereby the kind of solution that was obtained.

8.3 Spatio-temporal equalisers

We have shown earlier how it is possible by using several antennas to form an antenna beam in such a way that the required signal is strengthened and interfering signals are suppressed. We have also looked at how the time dispersion of the transfer channel can be equalised. If these two functions are joined, a very efficient receiver is obtained, which simultaneously equalises the signal both in the space and time domains, a so-called spatio-temporal equaliser. We start by studying a transmission channel with multipath, in more detail than we did in Chapter 2. We then describe two detectors that equalise the signal in time and space: ST-MLSE, based on a vectorised Viterbi algorithm, and ST-MMSE, an interference cancelling least square detector.

8.3.1 THE TRANSMISSION CHANNEL

Suppose that a transmitter with a single antenna element transmits a signal to a receiver with an array antenna. Since the transmitter has only one antenna element it can be seen as if the channel has one input, and since each antenna element in the array antenna can be seen as an individual output, the channel has multiple outputs. Such a channel is called single input multiple output (SIMO). If several transmitters are using the same channel, i.e. multiple user (MU), there is a risk that they may interfere with each other. In Figure 8.12 there are for instance Tx_1 and Tx_2. In the transmission channel there is not only additive white Gaussian noise, but also physical objects that obstruct the signal propagation. The signal is

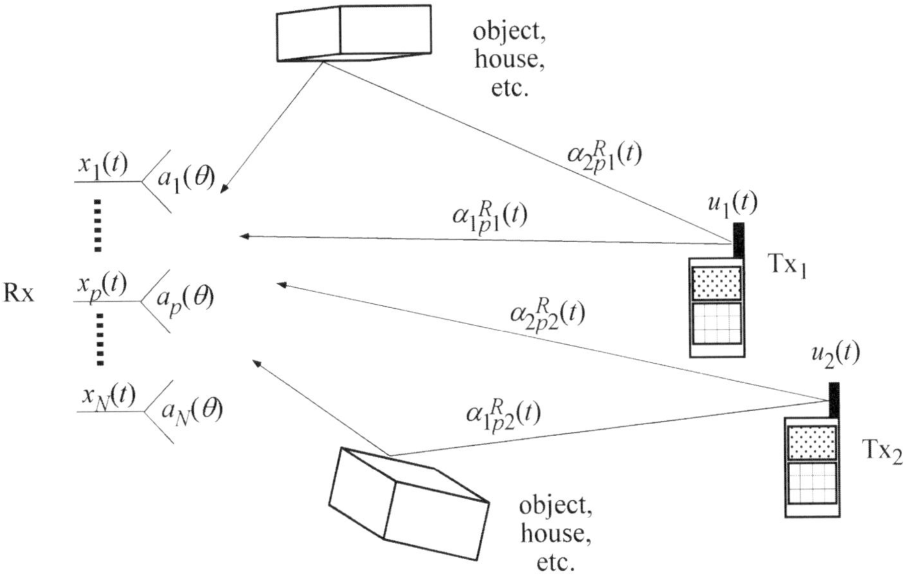

Figure 8.12 Example of the MU-SIMO environment where signals from the transmitters, Tx, are reflected by various objects on their way to the receiver.

reflected in the objects and reaches the receiver along several paths. See for instance the signal paths between Tx$_1$ and Rx. The algorithms presented below represent detection methods, designed with a communication system in mind operating in an environment which can be described by a MU-SIMO channel model.

Suppose there are M active transmitters, Tx$_m$ where $m \in \{1, 2, ..., M\}$. The signals are received by N antenna elements. The output signal from each antenna element is $x_p(t)$ where $p \in \{1, 2, ..., N\}$. Along the path l, between transmitter m and antenna element p, the signal has been attenuated and phase shifted. This is represented by the complex parameters $\alpha_{lpm}^{R}(t)$. The letter R denotes reverse, i.e. the direction back to the base station. If the signal from a transmitter, such as Tx$_1$, received by antenna element p is studied, a signal which has arrived along L different signal paths is observed. The signal can thus be written as the sum

$$x_p(t) = \sum_{l=1}^{L} \alpha_{lp1}^{R}(t) a_p(\theta_{l1}) u_1(t - \tau_l) + n_p(t) \tag{8.50}$$

where $a_p(\theta_{l1})$ is a complex number, for which the phase depends on the position of the antenna and the incident angle of the signal, and the magnitude depends on the antenna gain in the direction from which signal l arrives (cf. eq. (8.11)). The noise term $n_p(t)$ is complex, Gaussian and spatially white, i.e. it corresponds to a signal arriving from all directions. This noise mainly represents the receiver noise, but can also represent diffuse interference from the surrounding space. The variable $u_m(t)$ is the complex representation of the output signal from transmitter Tx_m. Note that if $\tau_l > T$, then $u_m(t - \tau_l)$ will be the preceding symbol and for each time τ_l becomes larger than a multiple of T, symbols from further and further back in time are obtained. Since the signal processing in connection with equalisers can usually only be realised by digital methods, we now change from a continuous time variable t to a discrete time variable with index k where $t = kT_s$. Note that this formalism also includes fractional sampling, as the sampling period T_s may be a fraction of the symbol time T. The discrete time representation of the received signal at the antenna element p becomes

$$x_p(k) = \sum_{l=1}^{L} \alpha_{lpm}^{R}(k) a_p(\theta_{lm})\, u_m(k - l + 1) + n_p(k) \tag{8.51}$$

The parameter θ_{lm} can be seen as a short term average of the incident angle of signal m at the delay l and does not at all have to represent the real angle to the transmitter. The received signal may have been reflected by an object from another direction than the transmitter, or it may consist of the sum of two or more signals having the same delay, but coming from different directions. The incident angle of the total sum of all signals can then be totally different than the incident angles of the individual signals. If then the signals from all M transmitters in the surrounding are added, each having L_m propagation paths, we obtain another sum

$$x_p(k) = \sum_{m=1}^{M} \sum_{l=1}^{L_m} \alpha_{lpm}^{R}(k) a_p(\theta_{lm}) u_m(k - l + 1) + n_p(k) \tag{8.52}$$

An $N \times N$ matrix labelled $\alpha_{lm}^{R}(k)$ can be formed. The channel coefficients $\alpha_{lpm}^{R}(k)$ are arranged along the diagonal, with $\alpha_{l1m}^{R}(k)$ as top left. Then, a signal model, where the output signals from the antenna elements $1,\ldots,N$ are arranged in a column vector, can be written

$$\mathbf{x}(k) = \begin{bmatrix} x_1(k) \\ x_2(k) \\ \vdots \\ x_N(k) \end{bmatrix} = \sum_{m=1}^{M} \sum_{l=1}^{L_m} \alpha_{lm}^{R}(k) \mathbf{a}(\theta_{lm}) u_m(k - l + 1) + \mathbf{n}(k) \tag{8.53}$$

$\mathbf{a}(\theta_{lm})$ is the array response vector for the incident angle of the signal lm. The elements of the $N \times 1$ vector $\mathbf{x}(k)$ give the output signal from the respective antenna. The notation $\mathbf{n}(k)$ represents a complex $N \times 1$ vector, which gives the

Gaussian noise. One way of writing the signal model is to let the inner sum in eq. (8.53) be represented by a matrix multiplication. For simplicity the channel is from now on time invariant, so that index k can be ignored. Form the vector $\mathbf{h}_{lm} = \alpha_{lm}^{R}(k)\mathbf{a}(\theta_{lm})$ and the matrix $\mathbf{H}_{m}$ through the operation

$$\mathbf{H}_{m} = \begin{bmatrix} \mathbf{h}_{1m} & \mathbf{h}_{2m} & \cdots & \mathbf{h}_{L_{m}m} \end{bmatrix} \tag{8.54}$$

If the signal source and all reflections are far away from the receiver, the signal paths to all antenna elements are mainly the same, which means that we can put $\alpha_{l1m}^{R} = \alpha_{l2m}^{R} = \cdots = \alpha_{lL_{m}m}^{R}$. The incident wave front is then a plane wave. If, on the other hand, the reflections are close in relation to the size of the receiving antenna, we have considerably different signal paths to each antenna element, i.e. $\alpha_{l1m}^{R} \neq \alpha_{l2m}^{R} \neq \cdots \neq \alpha_{lL_{m}m}^{R}$. We then have a situation with antenna diversity, i.e. a number of independent propagation paths. Suppose, furthermore, that the desired signal is the Mth in the sequence and label it with index s instead. It can then be taken outside the summation and will constitute a term of its own. We then have an expression which is easier to implement in the rest of the procedure:

$$\mathbf{x}(k) = \mathbf{H}_{s}\mathbf{u}_{s}(k) + \underbrace{\sum_{m=1}^{M-1} \mathbf{H}_{m}\mathbf{u}_{m}(k) + \mathbf{n}(k)}_{\mathbf{n}_{M}(k)} \tag{8.55}$$

The output signal from transmitter m is written

$$\mathbf{u}_{m}(k) = [u_{m}(k) \cdots u_{m}(k - K + 1)]^{T}$$

where $u_{m}(\cdot)$ is the complex representation of the signal, as previously mentioned. The columns in the matrix $\mathbf{H}_{m}$ represent the impulse response and spatial signature of the channel for the different signal paths.

We have now obtained an efficient way of expressing the received signal, in spite of the signal conditions being rather complicated.

8.3.2 ST-MLSE

The acronym ST-MLSE stands for space time maximum likelihood sequence estimator. When this detector is defined, it is assumed there are several equally strong interference sources, so that the central limit theorem can be applied and the amplitude density of the signal sum becomes Gaussian. The signal model can therefore be written

$$\mathbf{x}(k) = \mathbf{H}_{s}\mathbf{u}_{s}(k) + \mathbf{n}_{M}(k) \tag{8.56}$$

The noise vector $\mathbf{n}_{M}(k)$ now consists of Gaussian distributed samples with a variance, σ_{M}^{2}, which corresponds to the sum of received noise and other unwanted signals (cf. eq. (8.55)). Even if there are several interference sources, the interference sometimes arrives from certain distinct directions. The interference from other sources therefore introduces a signal which is not white, especially in its spatial dimension. The autocorrelation matrix, $\mathbf{R}_n$, of the interference is therefore not an identity matrix. In the same way as described in Section 6.3.1, the interference can easily be whitened, after which the signal processing is applied to the whitened signal. The probability density of the whitened signals, $\mathbf{x}'(k)$, can then be written

$$p(\mathbf{x}'(k)|\mathbf{H}'_s\mathbf{u}_s(k)) = \frac{1}{\left(2\pi\sigma_M^2\right)^{N/2}} \exp\left[\frac{\|\mathbf{x}'(k) - \mathbf{H}'_s\mathbf{u}_s(k)\|_2^2}{2\sigma_M^2}\right] \tag{8.57}$$

The maximum likelihood (ML) receiver is obtained by maximising eq. (8.57) with regard to the different possible signals $\mathbf{u}_s(k)$. In the same way as in Section 6.3.1 the maximum is obtained by minimising the norm in the exponent, $\|\mathbf{x}'(k) - \mathbf{H}'_s\mathbf{u}_s(k)\|_2^2$. When the norm is expanded we obtain

$$\|\mathbf{x}'(k) - \mathbf{H}'_s\mathbf{u}_s(k)\|_2^2 = \sum_{p=1}^{N}\left[x'_p(k) - \sum_{l=1}^{L}\alpha'^R_{lps}u_s(k - l + 1)\right]^2 \tag{8.58}$$

Eq. (8.58) implies that the distance between the received signal at all antenna elements and a weighted sequence of possible signals is minimised. The weights are determined by the parameters, α^R_{lps}, of the transmission channel. In eq. (8.58) only the received signal at time instant k is used. A more efficient detector is obtained if a longer sequence of the received signal is used (cf. sequential decoding in Section 6.3.1).

As in Section 6.3.1, we now assume that we have a sequence of K successive time samples at each antenna output. The matrix $\mathbf{X}'$ is then formed by combining the time samples so that $\mathbf{X}' = [\mathbf{x}'(k)\ \mathbf{x}'(k - 1)\ ...\ \mathbf{x}'(k - K)]$. Let us also combine the reference signals so that $\mathbf{U}_s = [\mathbf{u}_s(k), \mathbf{u}_s(k - 1), ..., \mathbf{u}_s(k - K)]$. The received signal then consists of a time sequence of data along the rows of the matrix and a spatial data sequence arranged along the columns of the matrix. The received signal can be written

$$\mathbf{X}' = \mathbf{H}'_s\mathbf{U}_s + \mathbf{N}'_M \tag{8.59}$$

where $\mathbf{N}'_M = [\mathbf{n}'_M(k), \mathbf{n}'_M(k - 1), ..., \mathbf{n}'_M(k - K)]$. The way of finding the ML receiver is the same as in Section 6.3.1, with the difference that the argument $\mathbf{X}'$ is a matrix instead of a vector. The most probable signal is then found through the search

$$\hat{\mathbf{U}} = \underset{s}{\arg\min} \|\mathbf{X}' - \mathbf{H}'_s\mathbf{U}_s\|_F^2 \tag{8.60}$$

where F means the Frobenius norm. This is the sum of the squares of all matrix elements, i.e. for the $I \times J$ matrix $\mathbf{A}$, we have $\|\mathbf{A}\|_F^2 = \sum_{i=1}^{I} \sum_{j=1}^{J} |a_{ij}|^2$. To find the minimum of the transmitted signal, eq. (8.60), all possible signal alternatives, $\mathbf{U}_s$, must be tested. If the delay is large enough to extend over several successive symbols, every possible combination of symbol sequences must be tested. The number of different $\mathbf{U}_s$, for which the search must be carried out, increases exponentially with the length of the symbol sequence. A way of reducing the computational load is to cancel symbol sequences that cannot possibly be the most probable ones as early as possible. This is achieved by using the Viterbi algorithm, which has been described earlier in Chapters 4 and 6. The complexity is then exponentially increased with the time dispersion of the channel as measured by the number of symbols, i.e. $\max\{L_m\}$. Take for instance the logarithm of the probability, eq. (8.57), apply the condition (8.60) and form the distance receiver. If the signal sequence contains several symbols, the computation can still be made on matrix blocks corresponding to the length of one symbol as in eq. (6.62). A metric to be used in the Viterbi algorithm can then be formed for the whitened signal:

$$\mu_{LL}(N_T) = \sum_{k=1}^{N_T} \left[\mathbf{x}'(k) - \mathbf{H}'_s \mathbf{u}_s(k)\right]^H \left[\mathbf{x}'(k) - \mathbf{H}'_s \mathbf{u}_s(k)\right] \tag{8.61}$$

The algorithm is called vector Viterbi since the metric is based on vectors. The difference compared to Section 6.3.1 is that the reference signal becomes more complicated, since the signal we wish to identify has passed through a dispersive transmission channel and since there is a received signal at each antenna element. The ML receiver then selects the symbol sequence giving the smallest metric value (cf. Figure 8.13). In practice it turns out that the Gaussian requirement, mentioned in the beginning of this section, is not very strict since the ST-MLSE detector works well anyway.

In the same way as was shown earlier, the receiver can be implemented as a correlation receiver, which means that the metric is calculated through a correlation operation. If there are a large number of antennas it may be advantageous to use this type of metric instead of the distance metric as shown above.

If the receiver above is compared to an adaptive equaliser it is understood that, instead of inverting the signal, the distance to a reference signal is minimised. The reference signal has passed through a filter, corresponding to the channel over which the received signal has passed. The advantage is obvious, since we do not have to invert a channel containing zeros. To perform ideal inverse filtering, the inverse filter would need to contain poles which could cause an unstable filter. Although a DFE (see Section 7.1.2) does not need to invert the channel, the symbol decisions are made symbol by symbol, which increases the risk of making incorrect decisions. The feedback can then cause a long sequence of incorrect decisions. The decisions of a sequential detector are instead based on the observation of several symbols which

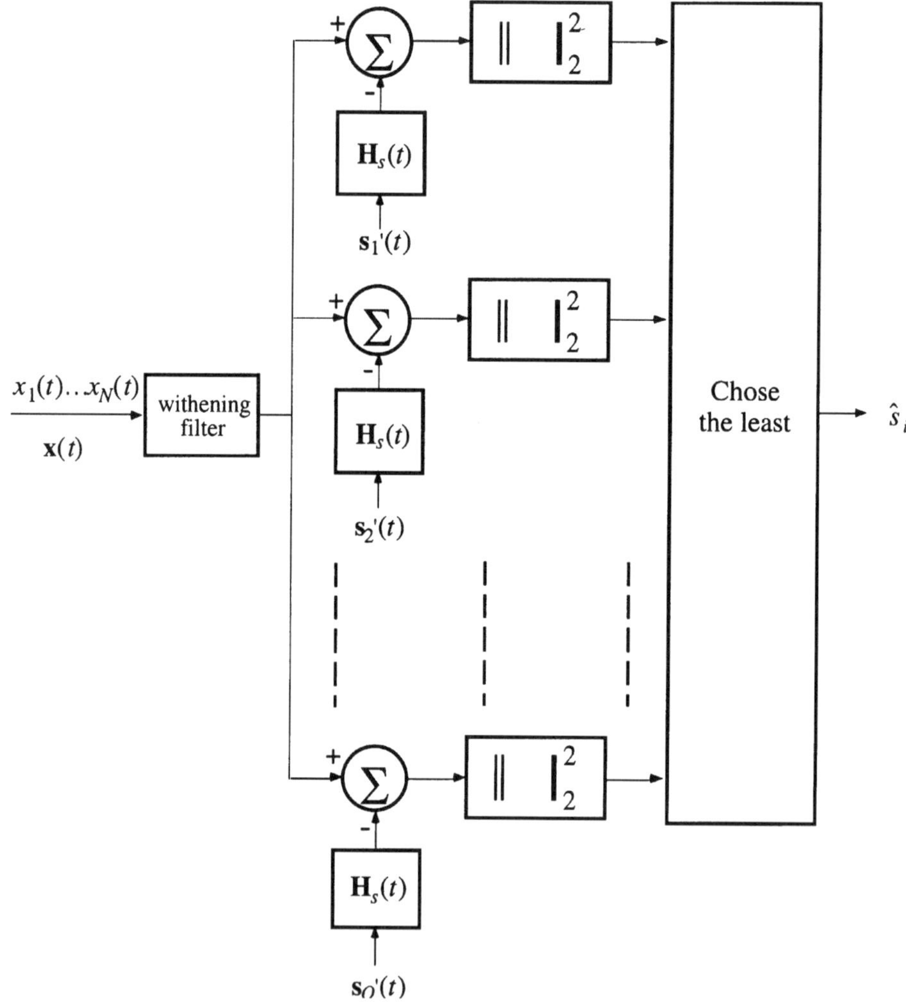

Figure 8.13   ST-MLSE receiver. Q represents the number of signal alternatives.

makes the decisions more reliable. The disadvantage is that for a large time dispersion, the signal must be observed over several symbol intervals and this leads to a higher degree of complexity at the receiver.

For the receiver which is described below, we return to the idea of inverse filtering of the channel, in this case both in space and time.

8.3.3 ST-MMSE

In cellular based systems the range and capacity are limited by co-channel interference. In many kinds of systems only one or a few sources of interference dominate. In such cases there are not enough signals making the total added interference to have a Gaussian distribution. An alternative solution, which in some cases can be favourable, is a spatio-temporal equaliser consisting of a combination of the adaptive antenna discussed earlier in this chapter and the adaptive equaliser discussed in Chapter 7 (cf. Figure 8.14). The output signal from each antenna element is passed through an FIR filter. The filter coefficients or weights, **W**, are optimised according to the minimum mean square criterion, MMSE, so that the output signal of the filter resembles the signal sent from the transmitter as much as possible. The decisions are then made on a symbol by symbol basis, unlike the case of the MLSE receiver.

The weights, **W**, in a spatio-temporal equaliser can be written (cf. eq. (8.13))

$$\mathbf{W}(k) = \begin{bmatrix} w_{11}(k) & \cdots & w_{1\Lambda}(k) \\ \vdots & \ddots & \vdots \\ w_{N1}(k) & \cdots & w_{N\Lambda}(k) \end{bmatrix} \tag{8.62}$$

where $\Lambda - 1$ is the number of delays after each antenna unit. In the ST-MMSE

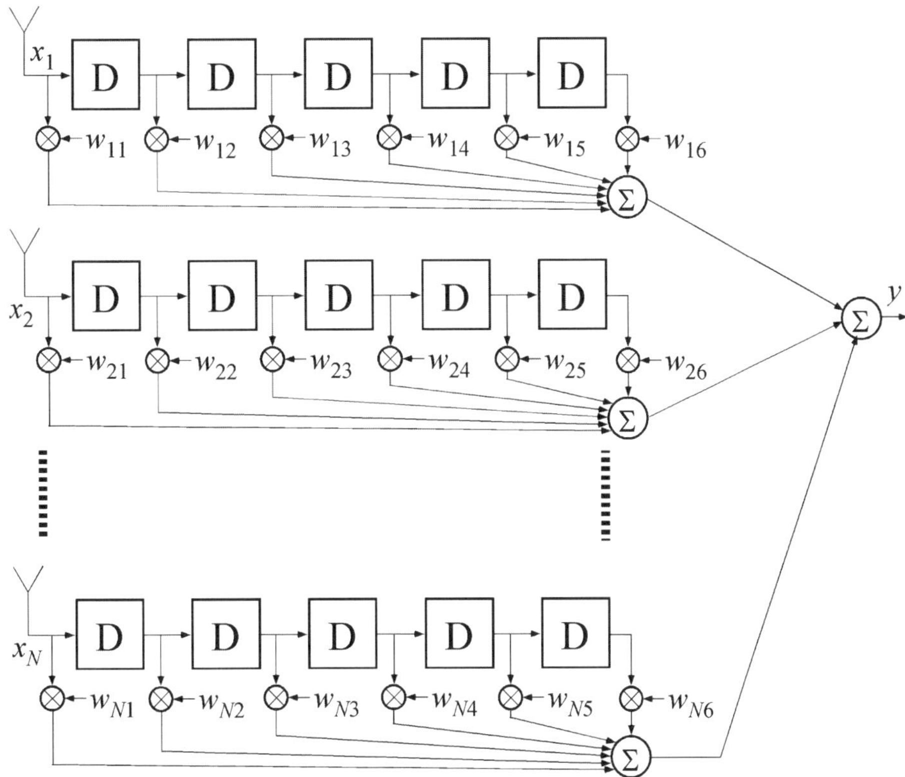

Figure 8.14 Array antenna with a FIR filter after each antenna element. $\Lambda = 6$.

filter, the spatio-temporal filter weights giving the MMSE solution are selected by

$$\underset{W}{\arg\min}\, E\left\{||\mathrm{tr}\{\mathbf{W}^H\mathbf{X}(k)\} - u_s(k - \zeta)||_2^2\right\} \tag{8.63}$$

where $\mathrm{tr}\{\mathbf{A}\}$ represents the trace of $\mathbf{A}$, i.e. the sum of the diagonal elements. The parameter ζ centres the reference signal of the impulse response and is very important for the quality of the solution, i.e. in the end the symbol error rate. In the same way as before, eq. (7.27), the least square solution becomes

$$\mathbf{W} = \mathbf{R}_X^{-1}\mathbf{r}_{Xs} = \mathbf{R}_X^{-1}E\{\mathbf{X}\,u_s(k - \zeta)\} \tag{8.64}$$

Suppose now that the information symbols are mutually independent, and that the noise is white and independent in relation to the wanted signal. Remember that the input signal can be written as $\mathbf{X} = \mathbf{HU} + \mathbf{N}$ (cf. eq. (8.59)), where $\mathbf{H}$ represents the multi-path propagation. Assign a vector to each column of the matrix $\mathbf{H}$, which gives $\mathbf{H} = [\mathbf{h}_1\ \mathbf{h}_2\ \cdots\ \mathbf{h}_N]$. Form a new matrix by introducing null vectors and reverse the order of the vectors $\mathbf{h}_x$:

$$\left[\begin{array}{ccccccccc} \mathbf{0} & \cdots & \underbrace{\mathbf{0}\ \ \mathbf{h}_N\ \ \cdots\ \ \mathbf{h}_1}_{\tilde{\mathbf{H}}_{(N-\zeta)}} & \mathbf{0} & \cdots & \mathbf{0} \end{array}\right] \tag{8.65}$$

$\tilde{\mathbf{H}}_{(N-\zeta)}$ is an $N \times N$ matrix, with the column $\mathbf{h}_{(N-\zeta)}$ to the right and the other columns taken consecutively towards the left from eq. (8.65). When $\zeta \geqslant N$, the vector $\mathbf{h}_{(2N-\zeta-1)}$ is instead the leftmost column and the others are taken consecutively towards the right. Which part of the matrix (8.65) that is selected is in other words determined by the value of ζ. The steering vector therefore becomes

$$\mathbf{r}_{Xd} = \tilde{\mathbf{H}}_{(N-\zeta)}E\{u_s(k - \zeta)u_s^*(k - \zeta)\}$$

The optimal solution for the weights then becomes

$$\mathbf{W} = \mathbf{R}_X^{-1}\tilde{\mathbf{H}}_{(N-\zeta)}E\{u_s(k - \zeta)u_s^*(k - \zeta)\} \tag{8.66}$$

The output signal from the spatio-temporal equaliser above is

$$y(k) = \mathrm{tr}\left\{\mathbf{W}^H\mathbf{X}\right\} \tag{8.67}$$

8.3.4 ESTIMATION OF CHANNEL PARAMETERS

In order to be able to use the ST-MLSE method, the channel matrix, $\mathbf{H}$, has to be estimated and to use ST-MMSE, $\mathbf{W}$ has to be determined. The matrix $\mathbf{W}$ can, however, be determined from $\mathbf{H}$, as can be seen from eq. (8.66). In other words, for the above-mentioned methods to work well it is very important that a good channel estimate is available. There are two main ways of estimating $\mathbf{H}$. The first includes the reference signal methods (cf. Sections 7.1.2 and 8.2.1) and the other consists of blind methods, named after the fact that they work without a training sequence. Instead blind methods optimise to some other property of the signal. Most communication systems have some kind of reference sequence, known by the receiver, which is transferred for synchronisation, channel identification, user identification, etc. Only the reference signal method is therefore described below.

The sequence used to identify the channel, i.e. to determine the parameters $\mathbf{H}$, is called the training sequence or reference signal. If the reference signal is called $\mathbf{T}$, the received signal becomes $\mathbf{X} = \mathbf{HT} + \mathbf{N}$. The problem is now formulated to give an equation system suitable for solving by the least square method. This gives the estimation

$$\hat{\mathbf{H}} = \mathbf{X}\mathbf{T}^{\dagger} \tag{8.68}$$

where $\mathbf{T}^{\dagger} = \mathbf{T}^{H}\left(\mathbf{TT}^{H}\right)^{-1}$, i.e. the so-called pseudo-inverse. Note that the length of the training sequence influences the number of channel coefficients that can be estimated and by what accuracy this can be done.

A way of creating a reference signal of as high a quality as possible without having to continuously use training sequences, is to use the same technique as for the decision directed equaliser (Section 7.1.2), and first perform a rough estimation of a known sequence and then feed back the decisions of the detector. In order to function, this requires that the feedback decisions are correct and that the delay can be kept at a low level in relation to the channel variations. For block-wise signal processing, it is also possible to make a successively improved channel estimation by detecting the complete block and then using this as a reference for a new channel estimation. A detection over the same block is then performed once more. This process can be repeated a few times, until a limit is reached where the symbol errors no longer decrease.

ST equaliser with decision feedback

As mentioned above, a version of the ST-MMSE equaliser can be used with feedback; see an example in Figure 8.15.

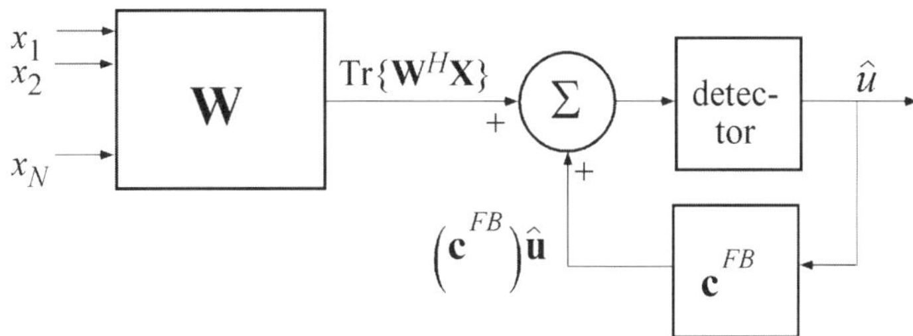

Figure 8.15 ST equaliser with decision feedback.

In the same way as in the temporal case, it can be shown that the ST-DFE with feedback performs better than the corresponding version without feedback.

Simplified ST-MMSE

If the number of antenna elements is large, the demands on the calculations can be high for the optimal equalisers described above. An example of a simplified ST-MMSE is shown in Figure 8.16. This can be seen as several separate array antennas, which individually optimise for different delays in the input signal.

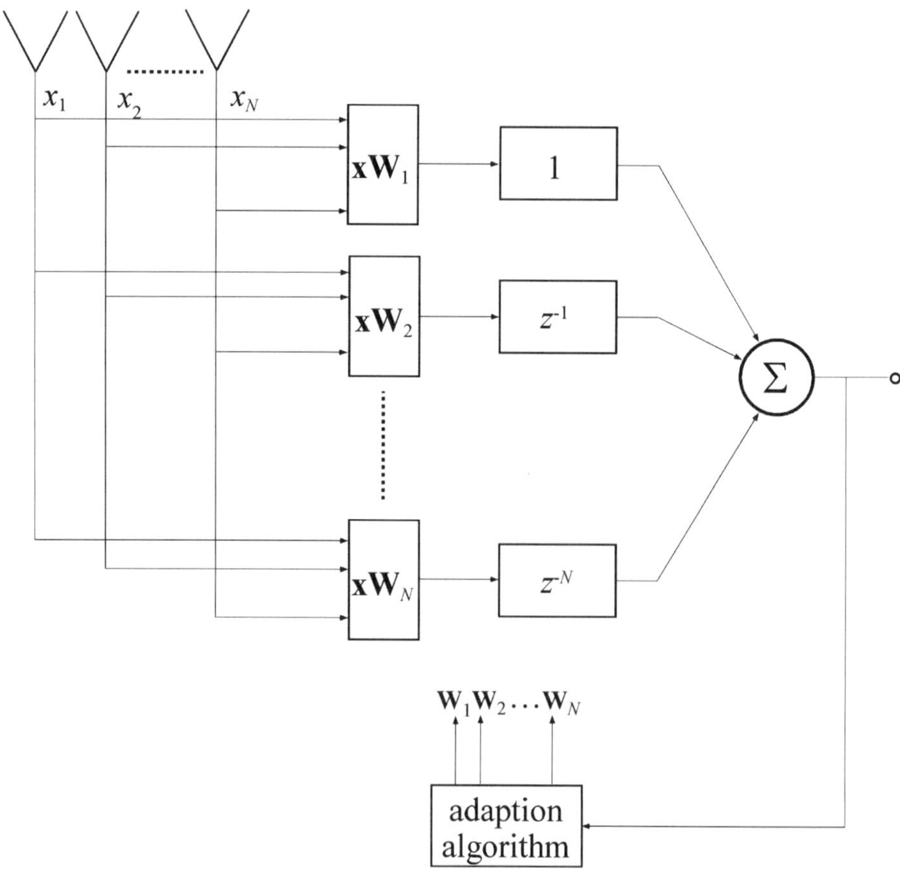

Figure 8.16 Simplified ST-equaliser. z^{-n} represents an n-step delay of the signal.

9 CDMA: CODES AND DETECTORS

The multiple access technique called code division multiple access (CDMA), requires a more detailed description than FDMA and TDMA. From a signal processing point of view, there are mainly two aspects which are different from the other systems. These are related to spreading codes and detectors, which are discussed in this chapter. Before starting reading, the description of CDMA in Chapter 1 should be studied.

The greatest advantage with CDMA is its considerable flexibility. When an FDMA system has active users on every frequency channel and a TDMA system has users in every time slot the systems are full and no further user can be accepted. In CDMA there is a gradual increase of the system load. Each additional user only means that the connections of all other users become slightly deteriorated. The new user can be admitted if the current ones allow it. In a CDMA system it is also fairly easy to arrange an exchange of error probability versus transmission rate.

Another advantage with CDMA is the possibility of so-called soft handover. In a radio network based on CDMA technology, all base stations can operate on the same frequency, which means that the signal from a cellular phone can be picked up by two or more base stations simultaneously. The base station temporarily receiving the strongest signal handles the transmission. This is called soft handover.

As mentioned in Chapter 1, a CDMA system works with a signal whose bandwidth is much larger than what is motivated by the data rate. The ratio between the bandwidth of the spread spectrum signal and the original bandwidth is called the spreading factor. The processing gain is the increase in signal to noise ratio obtained because of the spreading and has the same value as the spreading factor. Because of the noise-like properties of the spreading sequence, the sent signal appears as noise to all but the intended receiver. To the intended receiver the signal shape is, apart from the information content, deterministic.

For a system to be able to host several simultaneous users, several orthogonal spreading codes must be available. Methods for obtaining codes with low cross correlation and good autocorrelation properties are therefore vital. How to generate some fundamental spreading sequences for different applications in the CDMA system is described in the next section.

In an environment with multipath, the signal propagates along several paths. It is scattered and reflected in different ways so it reaches the receiver as a bundle of signals coming from different directions and with different time delays. Depending on the wavelength of the signal and the variation in distance for the different paths, some of the received signals will be exposed to constructive interference, others to destructive interference. This can be seen as a kind of filtering. For a broadband signal certain parts of the spectrum will be suppressed while others remain. This means the signal is not completely attenuated because of the multipath propagation, instead some power still remains at the receiver. A broadband signal is therefore more resistant to multipath, which is an incentive to use signals with a broad spectrum, for example CDMA signals. (Another way to obtain the same advantage is to let the carrier jump in frequency. This is called a frequency hopping system and is used for example in the GSM system.) A condition for the different users in a CDMA system not to interfere with each other, is that the spreading codes are orthogonal. The filtering effect of the channel increases the cross correlation between the spreading sequences, which in turn increases the total interference level of the system. The cross correlation can even become larger than the autocorrelation with the desired transmitter. It is therefore of great importance to limit the negative influence of the cross correlation. A few methods of doing this are described in the section on multi-user detectors.

9.1 Spreading codes

The spreading code is very important for the properties of the CDMA system. To obtain the desired properties in the communication system, the spreading sequence consists of a combination of several different sequences. These are applied to the signal in several steps. A few sequences and examples of how they can be applied to a message signal are described below.

An example of how codes can be used to spread the spectrum of a signal is shown in Figure 9.1. The long code is mainly used to make the signal look as noise-like, i.e. random, as possible and is also excellent for synchronisation purposes. The Walsh code is mainly used to make the different users individually orthogonal, so they do not interfere with each other.

As has been pointed out previously, the filtering effect of the transmission channel makes the signals mutually correlated. This in turn makes it difficult to achieve enough separation between the different users to handle the large dynamic range required in a radio system when one user is close to the receiver and another far away. The system therefore ensures that each user transmits with such power that all signals are received at the same level. This is achieved by the

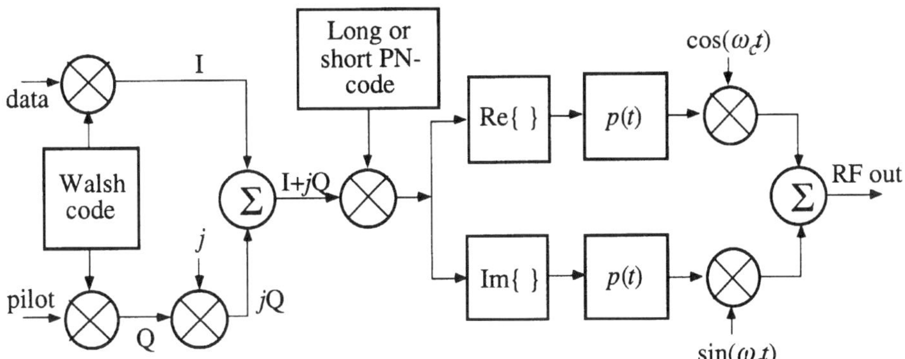

Figure 9.1   Example of a DS-CDMA transmitter for the down link, i.e. the signal to the mobile. The block $p(t)$ represents pulse shaping. Pilot is a channel for the control messages and a training sequence.

receiver measuring the received signal power and then asking the transmitter to increase or decrease its power. The power control is a key element for the function of a CDMA system.

9.1.1 WALSH CODES

Walsh codes consist of a set of orthonormal functions (cf. Section 6.3.2). This can be used for series expansion of a signal or a function. In CDMA systems, however, Walsh functions are used to achieve orthogonality between different users. The receiver can then identify the desired user and because of the orthogonality the other messages will be zero and will not interfere.

The Walsh functions are denoted $WAL(n,t)$ and have rectangular shaped wave forms, which only assume two values, $+1$ and -1 (see Figure 9.2). The functions are defined over a limited time interval T, called the time base and which is given a desired value. A description of the Walsh functions can be obtained from the matrix $\mathbf{H}$ where the rows consist of the consecutive values of the function. There are different ways of designing the $\mathbf{H}$ matrix. The Sylvester design constitutes a simple way of achieving Walsh functions. It produces functions of even lengths from a Hadamard matrix:

$$\mathbf{H}_2 = \begin{bmatrix} 1 & 1 \\ 1 & -1 \end{bmatrix} \tag{9.1}$$

Note that both rows and columns of the matrix are orthogonal. For the rows, for instance, we have $[1 \quad 1][1 \quad -1]^T = 0$. Matrices of a higher order are obtained through the recursion

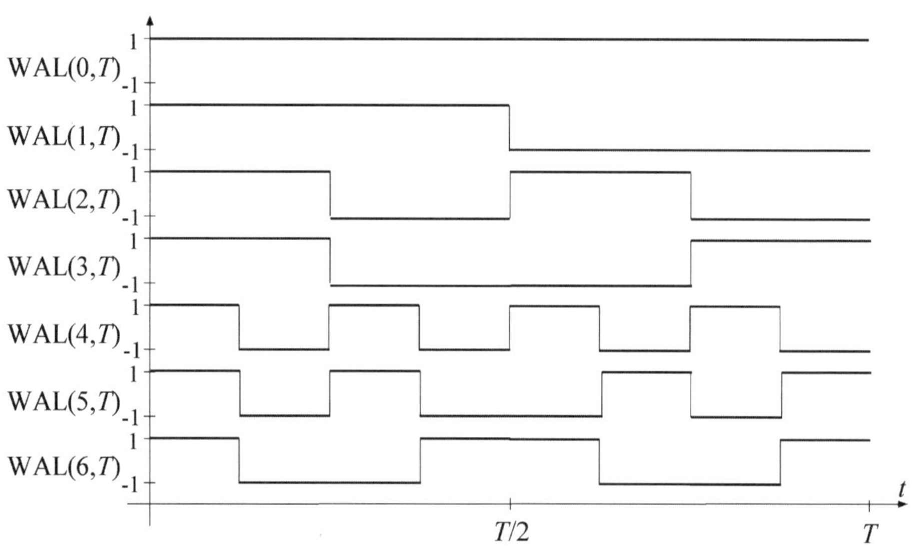

Figure 9.2   The first seven Walsh functions.

$$\mathbf{H}_q = \mathbf{H}_{q/2} \otimes \mathbf{H}_2 \tag{9.2}$$

where $\otimes$ in this chapter means the Kronecker product. This means that each element of the matrix $\mathbf{H}_{q/2}$ is replaced by the matrix $\mathbf{H}_2$, multiplied by the respective element in $\mathbf{H}_{q/2}$. Index q, the length of the sequence, should be a power of 2, i.e. $q \in \{4, 8, ..., N\}$. Number the matrix rows from zero to $q - 1$. The Walsh function $WAL(n,t)$ consists of row m in the matrix, where n is the 'bit reversed' value of m. Bit reversion means that if row number m is expressed in binary form, for instance $m = (11)_{10} = (1011)_2$, then $n = (1101)_2 = (13)_{10}$. This means that $WAL(13,t)$ is found in the 11th row. Two ways of ordering the Walsh functions are used, natural and sequential order. Here, the natural order is described.

Example 9.1

Design a set of 8 Walsh functions.

Solution   Start with the Hadamard matrix specified in eq. (9.1), which gives

$$\mathbf{H}_4 = \mathbf{H}_2 \otimes \mathbf{H}_2$$

After multiplication and substitution we obtain from (9.1)

$$\mathbf{H}_4 = \begin{bmatrix} 1 & 1 & 1 & 1 \\ 1 & -1 & 1 & -1 \\ 1 & 1 & -1 & -1 \\ 1 & -1 & -1 & 1 \end{bmatrix} \tag{9.3}$$

The next step of the recursion gives

$$\mathbf{H}_8 = \mathbf{H}_4 \otimes \mathbf{H}_2 = \begin{bmatrix} 1 & 1 & 1 & 1 & 1 & 1 & 1 & 1 \\ 1 & -1 & 1 & -1 & 1 & -1 & 1 & -1 \\ 1 & 1 & -1 & -1 & 1 & 1 & -1 & -1 \\ 1 & -1 & -1 & 1 & 1 & -1 & -1 & 1 \\ 1 & 1 & 1 & 1 & -1 & -1 & -1 & -1 \\ 1 & -1 & 1 & -1 & -1 & 1 & -1 & 1 \\ 1 & 1 & -1 & -1 & -1 & -1 & 1 & 1 \\ 1 & -1 & -1 & 1 & -1 & 1 & 1 & -1 \end{bmatrix} \tag{9.4}$$

Take, for instance, WAL(6,t). Number 6 in binary form is 110, which in bit reversed form becomes 011. Conversion to decimal form gives us that WAL(6,t) is in the 3rd row of the matrix and is therefore given the pulse sequence $\{1,1,-1,-1,1,1,-1,-1\}$.

For the series expansion of the function $f(t)$ a Walsh series can be written

$$f(t) \cong a_0 \mathrm{WAL}(0,t) + \sum_{n=1}^{N-1} a_n \mathrm{WAL}(n,t), \quad a_n = \frac{1}{T} \int_0^T f(t)\mathrm{WAL}(n,t)\, dt \tag{9.5}$$

where N is the number of Walsh functions. For CDMA systems, however, it is the orthogonality which is utilised and not the possibility of expressing a signal as a series expansion.

9.1.2 M-SEQUENCES

In order to achieve noise-like but yet deterministic signals, in communication, navigation, etc., binary feedback maximum length shift registers are used. To an outside observer the signal appears to be a random sequence of ones and zeros. These sequences are therefore generally called pseudo noise or PN sequences (pseudo = false). Another name is m-sequences because the shift registers are connected to give the sequence a maximum length. The sequences are utilised to

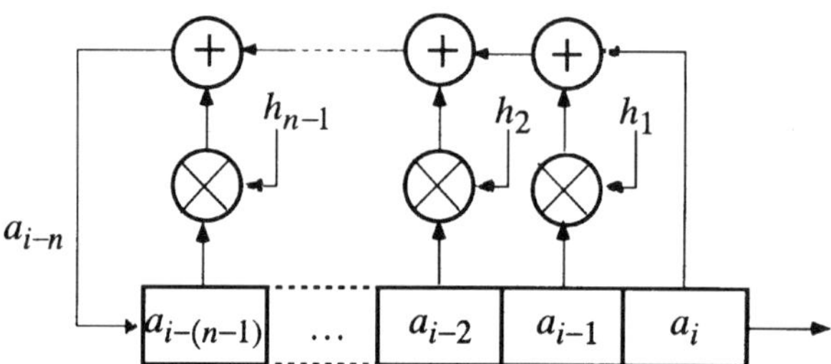

Figure 9.3   Feedback shift register.

obtain a signal with a broad spectrum. Their noise-like properties give a narrow autocorrelation peak which is beneficial for the synchronisation. The connection of the shift registers is determined by the coefficients h_i. The generator polynomial of the sequence, $h(p) = h_n p^n + h_{n-1} p^{n-1} + \cdots + h_1, +h_0$, is a binary polynomial of degree n, where the coefficients $h_0 = h_n = 1$ and the rest $h_i \in \{0, 1\}$.

The design of the sequence generator is shown in Figure 9.3. The signal a_i can be written

$$a_{i-n} = h_0 a_i + h_1 a_{i-1} + h_2 a_{i-2} + \cdots + h_{n-1} a_{i-n+1} = \sum_{k=0}^{n-1} h_k a_{i-k} \tag{9.6}$$

In this section the calculations are performed modulo 2. In the same way as for cyclic codes in Section 4.3.3, the delay operator p and a polynomial description can be used to describe the output signal of the shift register. Suppose the output signal consists of an infinite sequence of binary symbols $\{a_i\}_{i=0}^{\infty}$. A polynomial description of this sequence can be realised through the sum

$$A(p) = a_0 + a_1 p + a_2 p^2 + \cdots = \sum_{i=0}^{\infty} a_i p^i \tag{9.7}$$

This can be compared to the z-transform. By using eqs. (9.6) and (9.7), $A(p)$ can be written as a recursive expression. Remember that the shift a_{i-n} corresponds to $p^n A(p)$:

$$p^n A(p) = \sum_{i=0}^{\infty} a_{i-n} p^i = \sum_{i=0}^{\infty} \sum_{k=0}^{n-1} h_k a_{i-k} p^i = \sum_{k=0}^{n-1} h_k p^k \left[\sum_{i=0}^{\infty} a_{i-k} p^{i-k} \right]$$

$$= \sum_{k=0}^{n-1} h_k p^k \left[a_{-k} p^{-k} + a_{-k+1} p^{-k+1} + \cdots + a_{-1} p^{-1} + A(p) \right] \tag{9.8}$$

The coefficients a_{-n} to a_{-1} represent the initial values, i.e. the values loaded into the shift register cells before start up. If $A(p)$ is taken out of eq. (9.8), the polynomial description of the output sequence can be expressed as the ratio between two polynomials

$$A(p) = \frac{\sum_{k=0}^{n-1} h_k p^k \left[a_{-k} p^{-k} + \cdots + a_{-1} p^{-1} \right]}{\sum_{k=0}^{n} h_k p^k} = \frac{g_0(p)}{h(p)} \tag{9.9}$$

According to the description of binary arithmetic in Section 4, $+$ and $-$ are the same in modulo 2 operations. Furthermore we have $h_n = 1$, which makes the denominator a sum from 0 to n. The denominator $h(p)$ is also called the characteristic polynomial of the feedback shift register. As can be seen, this polynomial only depends on the feedback of the shift register. The polynomial $g_0(p)$ determines the phase of the sequence, or if $h(p)$ does not give an m-sequence, it also determines which of the possible sequences will be generated.

A necessary but not sufficient condition for an m-sequence to be generated is that $h(p)$ cannot be factored. A non-factorable polynomial which generates an m-sequence is called a primitive polynomial. These are difficult to find but are listed in the appendix of this chapter.

Since the number of combinations of zeros and ones in the cells of the shift register is a finite number, the sequence a_i will be periodic. For a shift register with n steps the length of the m-sequence becomes $N = 2^n - 1$, before the sequence is repeated. The shift register will therefore go through all possible symbol combinations in the shift register cells except from the combination where all cells contain the value zero. This is a locked state that the shift register never can get out of and must therefore never happen.

For use in CDMA systems, the output bits a_i are transformed by $u_i = 1 - 2a_i$ so that a one from the shift register becomes -1 and a zero becomes $+1$.

The transformed autocorrelation function, $\Phi_{uu}(k)$, of the sequence has one distinct peak, and in other words is very similar to that of white noise (see Figure 9.4). The advantage with a distinct peak is that it is easy for the receiver to synchronise its sequence to the received sequence in order to retrieve the original sequence. Since the noise sequency is repeated periodically, the autocorrelation is also periodic with period N, i.e. $\Phi_{uu}(k) = \Phi_{uu}(k + N)$. The pattern is therefore repeated for those values of k which lie outside the region shown in Figure 9.4. It is also possible to calculate the correlation using only one period of the sequence and a cyclic shift for each new value of k.

Suppose that u and v are two different m-sequences. The cross correlation, $\Phi_{uv}(k)$, between the sequences is not identically zero. A plot of the cross correlation would give a rough curve with a number of smaller and larger peaks. We know that for at least some value of the mutal time delay k, there is a cross correlation peak $|\Phi_{uv}(k)| > -1 + 2^{(n+1)/2}$. Furthermore, for the pair of m-

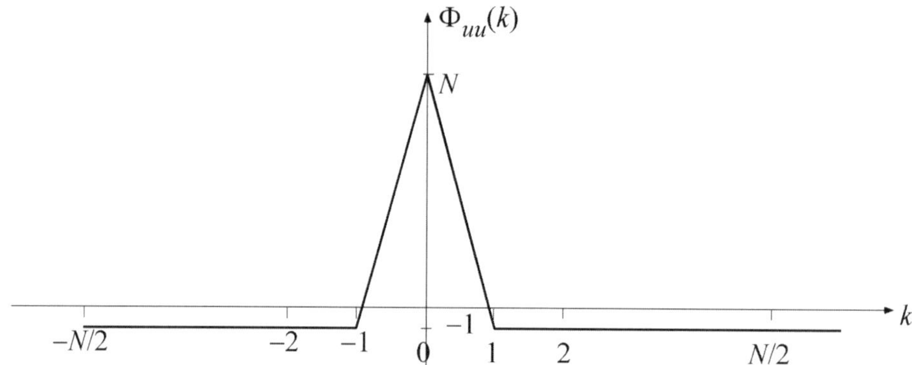

Figure 9.4   Autocorrelation function for m-sequence of length N.

sequences giving the highest cross correlation, the maximum peak is considerably larger than this value, especially if n is large. If these codes are used as spreading sequences, the signals are not orthogonal and will interfere with each other, something that obviously limits the performance of the system. It is therefore not possible to use only m-sequences to distinguish between different users.

m-Sequences with good cross correlation properties

As has been shown, the cross correlation between different pairs of sequences can vary considerably. In order to find good sequences with low cross correlation, the best ones can be found by using computer simulations. There are also some rules to guide the search for the good sequences.

A new sequence, v, can be generated by taking every qth bit out of a sequence u. The operation is written $v = u[q]$. If the greatest common divisor of the sequence length N and the decimation q is equal to one, i.e. $GCD(N, q) = 1$, the new sequence is also a maximum-length-sequence. A decimation by a prime number therefore always gives a new m-sequence. Generally, the length of the new sequence can be written $N/GCD(N, q)$.

A way of finding good pairs of sequences is to start from an m-sequence realised in a shift register using an odd number of steps, i.e. n is an odd number. A decimation q is then chosen such that $q = 2^l + 1$ or $q = 2^{2l} - 2^l + 1$ and that $GCD(n, l) = 1$. The variable l is a positive integer. For instance the sequence pairs u and $u[3]$, u and $u[5]$, as well as u and $u[13]$ will then have good cross correlation properties. For sequences generated in this way the cross correlation becomes $|\Phi_{uv}| \leq 1 + 2^{(n+1)/2}$, i.e. one can guarantee that the cross correlation is below this value.

Another way, leading to good code pairs when n is a multiple of 4, is to choose the decimation, $u[-1 + 2^{(n+2)/2}]$.

Thus, there are rules for finding the m-sequences that have the smallest cross correlation. These sequences can then be used for the generation of the Gold and Kasami sequences presented below.

9.1.3 GOLD SEQUENCES

The set of good sequences among the m-sequences is limited. This means that only a few users could use the CDMA systems unless there were ways of generating more good sequences. By generating Gold sequences a larger set of sequences is obtained, which have mutually good cross correlation properties. Assume a sequence for which the generator polynomial is the product of two generator polynomials, $h(p) = h_1(p) \cdot h_2(p)$. The polynomials $h_1(p)$ and $h_2(p)$ shall have no common factors and generate m-sequences of the same length. It can be shown that for the sequence generated by $h(p)$, each new relative shift of the sequences $h_1(p)$ and $h_2(p)$ gives a new sequence. An implementation of such a sequence generator is shown in Figure 9.5.

For each individual register the following holds: depending on which values are stored in the shift register cells at the beginning, it will run through the same single sequence but with different delays. Suppose that one shift register always starts with zeros and a single one in its cells. The other shift register can start with an arbitrary combination of zeros and ones in its cells. Each combination corresponds to a new delay of the sequence. This can be seen as if the sequence from shift register two has a variable delay of q steps. The combination of the sequences can then be written

$$y_q = u_1 \oplus u_2^{(q)} \tag{9.10}$$

After being combined with the sequence from generator one, each delay q leads to a new sequence. Since the sequences have a length of N symbols, the total number of sequences then becomes $N + 2$, where the two extra ones come from sequence u_1 or u_2 individually.

The set of sequences $\{y_q\}_{q=1}^{N+2}$ which have been generated according to eq. (9.10), where the included m-sequences have been formed by decimation in the

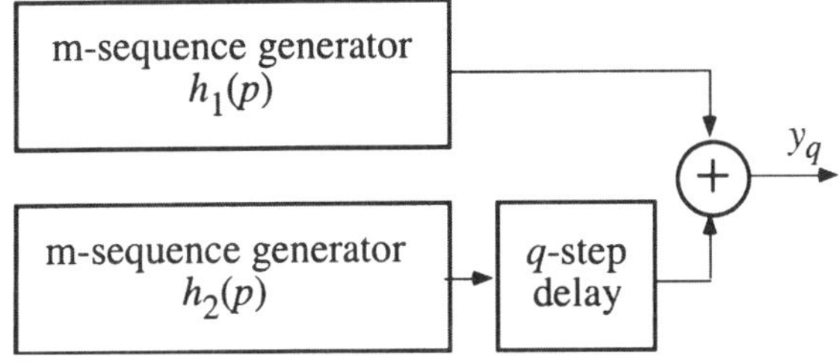

Figure 9.5 Gold sequence generator consisting of two feedback shift registers.

way described above, are called Gold sequences. It can be shown that the maximum cross correlation peaks of the complete set of Gold sequences become smaller than the cross correlation between u_1 and u_2,

$$\max_k \left\{ \Phi_{y^{(p)}y^{(q)}}(k) \right\} < \max_k \left\{ \Phi_{u_1 u_2}(k) \right\} \tag{9.11}$$

for all $p \neq q$. With this technique there are now far more spreading sequences with low cross correlation to choose from, which gives the possibility of more users in the CDMA system.

9.1.4 KASAMI SEQUENCES

Another set of sequences with good cross correlation properties are the Kasami sequences. These can be divided into two categories: 'the large set' and 'the small set'. How the small set of Kasami sequences are generated is described below. Assume that n is even and let u be an m-sequence generated by $h(p)$. Consider the sequence that has been generated through the decimation $u[2^{n/2} + 1]$. A factorisation of the expression for the length of the original sequence can be made, which gives $N = (2^n - 1) = (2^{n/2} - 1)(2^{n/2} + 1)$. Since $q = (2^{n/2} + 1)$, the length of the new sequence is $N/\text{GCD}(N, q) = (2^{n/2} - 1)$, which is shorter than the original. The new sequence can be seen as being generated by the polynomial $h'(p)$ of degree $n/2$. Now generate a sequence by using the polynomial $h(p) \cdot h'(p)$, which becomes a polynomial of degree $3n/2$. The sequences can be joined together in the same way as described for the Gold sequences (see Figure 9.5). This gives rise to a sequence of length $(2^n - 1)$. The number of different code sequences is: one for each phase state plus the code u separately, i.e. $(2^{n/2} - 1) + 1 = 2^{n/2}$. The maximum cross correlation between two pairs of sequences becomes $|\Phi_{uv}| = 1 + 2^{n/2}$. The number of Kasami sequences in the small set is less than the number of Gold sequences, but the maximum cross correlation, on the other hand, is only about half, which reduces the risk of different users interfering with each other.

9.2 Detectors for CDMA

When the CDMA signals reach the receiver, they have traversed through the transmission channel and are no longer orthogonal. This makes the coefficients $\rho_{jk} \neq 0$ also for $j \neq k$ in eq. (1.1). The different messages will therefore blend into each other and cause multiple access interference (MAI). Special detectors have been designed to reduce the influence of MAI. There are several ways of doing this and only a few can be described here. Two kinds of detection strategies are presented: decorrelation and interference cancelling. The time dispersion of the channel is assumed to be relatively small so that significant inter-symbol interference only appears in the received spreading code, which has a short symbol time. The information symbols, which have longer duration, will remain unaffected. Thus, this means that there is no intersymbol interference in the information sent. For simplicity it is also assumed that all transmitters are synchronised (the base stations are synchronised by means of the global positioning system), so that the symbols from the different users arrive simultaneously at the detector. In reality there are both synchronous and asynchronous systems. Asynchronous systems can be modelled in the same way as synchronous systems, although more users than the real number are assumed. The formulas below are therefore valid also for asynchronous systems but with larger matrices and vectors.

In order to simplify the understanding of the equations, vector notation is used. The presentation starts from Figure 1.11 and eq. (1.1) except that the received signals have different energy. To describe this a factor E_m is introduced. The output signal after each individual correlator is given by y_m and the output signal from all correlators can be summarised by the vector $\mathbf{y}$:

$$\mathbf{y} = \begin{pmatrix} y_1 \\ \vdots \\ y_M \end{pmatrix} = \mathbf{F}\sqrt{\mathbf{E}}\mathbf{x} + \mathbf{n} \tag{9.12}$$

where $\mathbf{E}$ is a diagonal matrix, $\mathbf{E} = \mathrm{diag}[E_1 E_2 ... E_M]$, which gives the symbol energy of the respective signal. The notation $\sqrt{\mathbf{E}}$ represents the square root of the matrix elements and $\sqrt{\mathbf{E}}(\sqrt{\mathbf{E}})^T = \mathbf{E}$. The elements of the vector $\mathbf{x} = [x_1 x_2 ... x_M]^T$ are the message symbols, and the vector $\mathbf{n} = [n_1 n_2 ... n_M]^T$ represents the noise. Note that $\mathbf{n}$ is not necessarily white noise, but is affected by the correlation with the spreading sequence. For a large enough spreading factor, all kinds of interference, including other users, can be regarded as Gaussian noise. The information symbols x_m and the noise n_m can be real or complex variables. Below we limit the model to $x_m \in \{-1, 1\}$ or $\sqrt{2} \times x_m \in \{-1-j, -1+j, 1-j, 1+j\}$ and to n_m, being a Gaussian distributed real or complex variable, respectively. The matrix $\mathbf{F}$ gives the cross correlation between the signals of the different users, see eq. (1.1), and is defined as

$$\mathbf{F} = \begin{pmatrix} 1 & \rho_{12} & \cdots & \rho_{1M} \\ \rho_{21} & 1 & & \rho_{2M} \\ \vdots & & \ddots & \vdots \\ \rho_{M1} & \rho_{M2} & \cdots & 1 \end{pmatrix} \tag{9.13}$$

Since the correlation coefficients $\rho_{im} = \rho_{mi}^*$, $\mathbf{F}$ is Hermitian. When $\rho_{im} = 0$

there is no cross talk between the messages i and m. Equation (1.1) can be seen as a row from eq. (9.12).

The decision of the detector is made independently for each y_m, i.e. $\hat{x}_m = \mathrm{sgn}(y_m)$ for all m, which also can be written for all channels together as $\hat{x} = \mathrm{sgn}(\mathbf{y})$. Figure 1.11 would in this case be completed by a limiter after each correlator. With such a detector the error rate would increase with the correlation of the spreading sequences. The detector should therefore be resistant to increased correlation between the signals. Two such detectors are described below. They are mainly used at the base station, since that is the only place where there is information about all spreading sequences and the need to detect all signals exists. From the viewpoint of total transmission capacity of the radio system, it is also more favourable to invest in a good detector at the base station than at the mobile.

9.2.1 LINEAR DETECTORS

Decorrelating detector

A natural and useful receiver strategy is to decorrelate the signals by multiplying the received signal by the inverse correlation matrix $\mathbf{F}^{-1}$. The decorrelating detector makes its decision based on

$$\hat{x} = \mathrm{sgn}\left(\mathbf{F}^{-1}\mathbf{y}\right) = \mathrm{sgn}\left(\underbrace{\mathbf{F}^{-1}\mathbf{F}}_{\mathbf{I}}\ \sqrt{\mathbf{E}}\mathbf{x} + \mathbf{F}^{-1}\mathbf{n}\right) \tag{9.14}$$

Since $\mathbf{F}^{-1}\mathbf{F} = \mathbf{I}$, the signal vector $\mathbf{x}$ is now only preceded by diagonal matrices. The mutual influence between the different users is thus eliminated. The detector decision is made based on each message stream separately. The decorrelating detector corresponds to the maximum likelihood detector when the signal energies of the different users are unknown by the receiver. The detector is also near-far resistant, which means that the properties of the detector are independent of the differences in signal levels, caused by the non-desired signal source being close to the receiver and the desired one being far away. The decorrelation is in principle the same operation as is performed in the zero-forcing equaliser described in Section 7.1.1. The detector has thus inherited the equaliser disadvantage of noise gain. In the same way as in Chapter 7, this can be avoided by using an MMSE method as described in the next section.

If the receiver succeeds in completely decorrelating the signals, it is possible to calculate the error probability of the detector by using the Q-function. It can be seen from eq. (9.14) that the energy of signal m is E_m. Since we have antipodal signalling, the Euclidean distance between the signal alternatives becomes $d = 2\sqrt{E_m}$. We can also obtain the noise from eq. (9.14). The error probability of a two-user-system becomes

$$P_e = Q\left(\sqrt{\frac{E_m}{\bar{\rho}_{mm}N_0}}\right) \tag{9.15}$$

where $\bar{\rho}_{mm}$ represents element mm in the inverted cross correlation matrix $\mathbf{F}^{-1}$. It can be seen that the error probability increases when the cross correlation between the signals increases, although this increase is obviously smaller than if no decorrelation were performed. Since $\bar{\rho}_{mm} \geq 1$, the decorrelating detector will give a larger error probability than if only a single user were active.

Detector based on the least mean square error

Just as in the previous case, the decision of the least mean square error detector is based on a linear transformation of the signals, $\mathbf{y}$, from the outputs of the correlators. Suppose that the transformation is determined by the matrix $\mathbf{C}$:

$$\hat{\mathbf{x}} = \mathrm{sgn}(\mathbf{Cy}) \tag{9.16}$$

In Figure 9.6, one filter branch in the receiver is shown. A complete receiver consists of a total of M identical branches. The transformation means that a suitable portion of the interfering signals are subtracted from each receiver branch before the decision is made. The portion is determined by the filter coefficients $\mathbf{c}_m = [c_{1m} \quad c_{2m} \quad c_{3m} \cdots c_{Mm}]^T$, where index m represents the receiver branch in question. The values of the coefficients c_{jm} have to be determined in some way. Form the matrix $\mathbf{C}$ by means of the filter coefficients for all receivers: $\mathbf{C} = [\mathbf{c}_1 \quad \mathbf{c}_2 \quad \mathbf{c}_3 \cdots \mathbf{c}_M]$. By forming a signal consisting of the difference between

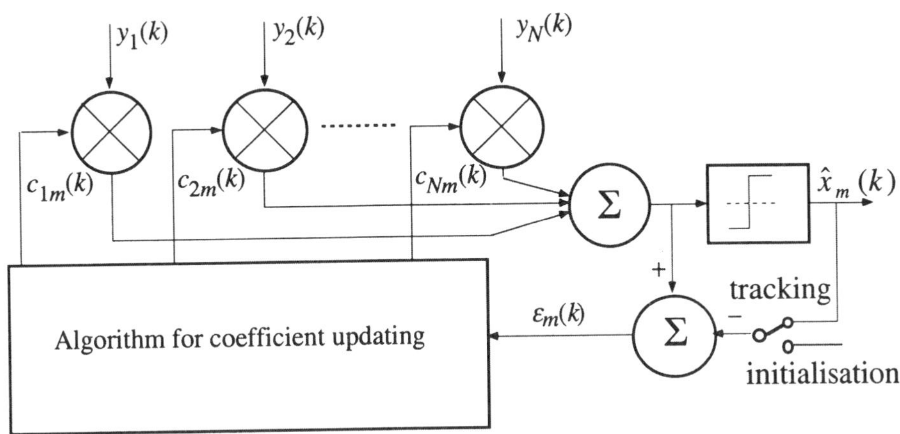

Figure 9.6 Branch m in an adaptive CDMA receiver, which transforms the input signal based on a MMSE criterion.

the received signal and the feedback decision, the mean square error (MSE) is obtained (cf. eq. (7.13)):

$$J = E\left\{|\varepsilon|^2\right\} = E\left\{(\mathbf{Cy} - \hat{\mathbf{x}})^H(\mathbf{Cy} - \hat{\mathbf{x}})\right\} \tag{9.17}$$

At the initialisation a reference signal is used instead of the feedback decision. The optimisation variable J_m of filter m becomes

$$J_m = E\left\{(\mathbf{c}_m\mathbf{y} - \hat{x}_m)^H(\mathbf{c}_m\mathbf{y} - \hat{x}_m)\right\} \tag{9.18}$$

As shown in Chapter 7, the optimum solution of a least square problem becomes (see eq. (7.27))

$$\mathbf{C} = \mathbf{R}_y^{-1}\mathbf{R}_{y\hat{x}} \tag{9.19}$$

where $\mathbf{R}_{y\hat{x}}$ is the cross covariance matrix between the received signal and the feedback decisions,

$$\mathbf{R}_{y\hat{x}} = \begin{bmatrix} E\{y_1\hat{x}_1\} & \cdots & E\{y_1\hat{x}_N\} \\ \vdots & \ddots & \vdots \\ E\{y_N\hat{x}_1\} & \cdots & E\{y_N\hat{x}_N\} \end{bmatrix} \tag{9.20}$$

Provided that the different messages are uncorrelated, the decisions of the detector are correct and the mean value of the noise is zero, $\mathbf{R}_{y\hat{x}}$ then becomes

$$\begin{bmatrix} E\{y_1\hat{x}_1\} & \cdots & E\{y_1\hat{x}_N\} \\ \vdots & \ddots & \vdots \\ E\{y_N\hat{x}_1\} & \cdots & E\{y_N\hat{x}_N\} \end{bmatrix} = \begin{bmatrix} \sqrt{E_1} & \cdots & \rho_{1N}\sqrt{E_N} \\ \vdots & \ddots & \vdots \\ \rho_{N1}\sqrt{E_1} & \cdots & \sqrt{E_N} \end{bmatrix} = \mathbf{F}\sqrt{\mathbf{E}} \tag{9.21}$$

If the corresponding solution of a single receiver branch is studied, the filter coefficients become $\mathbf{c}_m = \mathbf{R}_y^{-1}\mathbf{r}_{y\hat{x}_m}$ where $\mathbf{r}_{y\hat{x}_m}$ is the mth column in $\mathbf{R}_{y\hat{x}}$.

The covariance matrix $\mathbf{R}_y$ of the output signals from the code correlators is given by

$$\mathbf{R}_y = E\left\{\left(\mathbf{F}\sqrt{\mathbf{E}}\mathbf{x} + \mathbf{n}\right)\left(\mathbf{F}\sqrt{\mathbf{E}}\mathbf{x} + \mathbf{n}\right)^H\right\}$$

$$= \mathbf{F}\sqrt{\mathbf{E}}E\left\{\mathbf{xx}^H\right\}\sqrt{\mathbf{E}}^H\mathbf{F}^H + E\left\{\mathbf{nn}^H\right\}$$

$$= \mathbf{R}_c + \sigma^2\mathbf{I}_M \tag{9.22}$$

The different messages are uncorrelated, which gives the symbol cross correlation matrix $E\{\mathbf{xx}^H\} = \mathbf{I}_M$. Since $x_m = \pm 1$, the value of $E\{x_m x_m{}^*\} = 1$. Note that the noise is assumed to be white in eq. (9.22), in spite of what was earlier said in connection with eq. (9.13). As the noise is basically white, the diagonal elements in the covariance matrix of the noise will dominate, so it is a reasonable simplification. We can now define the covariance matrix $\mathbf{R}_c = \mathbf{F}\sqrt{\mathbf{E}}\sqrt{\mathbf{E}}^H\mathbf{F}^H$. The result from eq. (9.22) can then be summarised by writing the covariance of the output signals from the correlators as

$$\mathbf{R}_y = \mathbf{R}_c + \sigma^2\mathbf{I}_N \tag{9.23}$$

where σ^2 is the variance of the noise. The receiver then makes its decisions based on

$$\hat{\mathbf{x}} = \mathrm{sgn}(\mathbf{Cy}) = \mathrm{sgn}\left(\left(\mathbf{R}_c + \sigma^2\mathbf{I}_M\right)^{-1}\mathbf{F}\sqrt{\mathbf{E}}\,\mathbf{y}\right) \tag{9.24}$$

By using the matrix inversion lemma from eqs. (7.54) and (7.55) together with the fact that $(\mathbf{AB})^{-1} = \mathbf{B}^{-1}\mathbf{A}^{-1}$ when both $\mathbf{A}$ and $\mathbf{B}$ can be inverted, the transformation matrix $\mathbf{C}$, can be rewritten as

$$\begin{aligned}
\mathbf{C} &= \left(\mathbf{R}_c + \sigma^2\mathbf{I}_N\right)^{-1}\mathbf{F}\sqrt{\mathbf{E}} \\[6pt]
&= \left[\mathbf{R}_c^{-1} - \sigma^2\mathbf{R}_c^{-1}\left(\mathbf{R}_c + \sigma^2\mathbf{I}_N\right)^{-1}\mathbf{R}_c^{-1}\right]\mathbf{F}\sqrt{\mathbf{E}} \\[6pt]
&= \left(\mathbf{F}\sqrt{\mathbf{E}}\right)^{-1} - \sigma^2\mathbf{R}_c^{-1}\left(\mathbf{R}_c + \sigma^2\mathbf{I}_N\right)^{-1}\left(\mathbf{F}\sqrt{\mathbf{E}}\right)^{-1}
\end{aligned} \tag{9.25}$$

From the last equality we can see that if eq. (9.25) is compared to the decorrelating receiver, eq. (9.14), a term which is a function of the noise has been added. For a noiseless system ($\sigma^2 = 0$), the solutions would be identical apart from a normalisation. The channel is, in other words, not ideally inverted by the least square method, MMSE, but on the other hand there is no risk of having noise gain, as in the decorrelating receiver.

Since the correlation between the signals cannot always be assumed known to the receiver, the correlation has to be estimated by using an adaptive receiver. The MMSE solution has the advantage of being easy to make adaptive. By using the LMS algorithm from eq. (7.31), for instance, and making some rearrangements we obtain

$$\mathbf{C}(k + 1) = \mathbf{C}(k) + 2\mu\,\mathbf{y}(\hat{\mathbf{x}} - \mathbf{C}(k)\mathbf{y})^H \tag{9.26}$$

We can now investigate the structure of the algorithm for one of the filter coefficients by observing one matrix element from eq. (9.26). For a receiver having M correlators, the filter coefficient for detector m and filter tap n is determined by

$$c_{nm}(k + 1) = c_{nm}(k) + 2\mu\,y_n\left(\hat{x}_m - \sum_{l=1}^{M}c_{lm}(k)y_l\right) \tag{9.27}$$

It can be seen that the coefficients c_{nm} will be adjusted so that $E\{y_n(\hat{x}_m - \Sigma_l c_{lm}(k)y_l)\} = 0$. For instance, if there is no cross-talk between the messages, i.e. $\rho_{mn} = 0$ for all $m \neq n$, then c_{mm} is the only non-zero coefficient. For other values of the correlation coefficients, a linear combination of output signals, which also depends on the signal noise $\mathbf{y}$, is obtained.

An equivalent MMSE structure can be formulated by processing the incoming signal directly without a preceding correlator. The sampling period must then be equal to or shorter than the chip time. The number of coefficients in the receiver is in this case related to the spreading factor rather than to the number of active users.

9.2.2 INTERFERENCE CANCELLATION

Successive interference canceller

In the detector known as the successive interference canceller, SIC, the signals are arranged after signal strength. The strongest is given number 1 and the weaker ones successively higher numbers. In step 1 the strongest signal, x_1, is detected through the operation $\hat{x}_1 = \mathrm{sgn}(y_1)$, where y_1 is the output from correlator 1 (Figure 9.7). Since the rest of the signals are assumed to have less power, the error is small. The interfering signals are then subtracted in several steps, first the strongest one and then successively the weaker ones. In step two, for instance, the next strongest signal is determined through the operation

$$\hat{x}_2 = \mathrm{sgn}\left(y_2 - \rho_{12}\sqrt{E_1}\hat{x}_1\right) \tag{9.28}$$

The process continues until the message symbols for all users have been detected. The coefficients ρ_{im} and E_m must be known, however. The coefficient E_2 is obtained from the signal power of y_2, and again it is assumed that the rest of the signals are weak so the error of the estimation is small. The coefficients ρ_{im} can be determined by using an adaptive decorrelating detector or, for large enough SNRs, by using the MMSE described above. Provided that all symbol decisions are correct, the SIC detector will give the same error probability as if only one signal exists in the transmission channel.

A disadvantage with this receiver is that the decisions must be correct, otherwise the following decisions may be incorrect. For a SIC to work, the signal to noise ratio must therefore be large enough. Since the signal levels are often changed because of the movements of the mobiles, the ranking of the signals must be continuously updated.

Parallel interference cancellation

The interference can also be subtracted in parallel by using a parallel interference

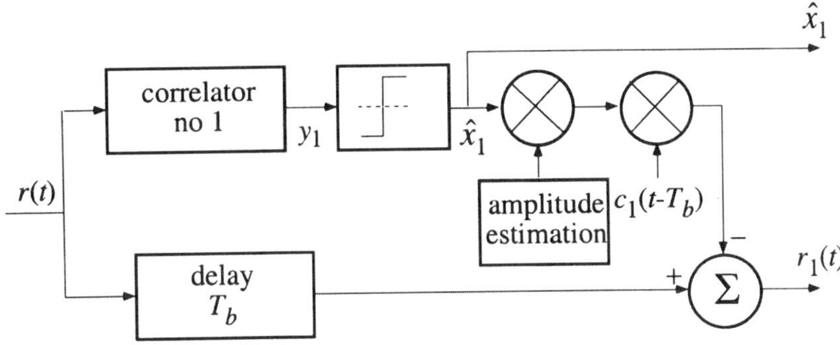

Figure 9.7 The first step of a SIC. Each new step is identical apart from the index of the signal notations.

canceller (PIC). The detector consists of M identical branches (see Figure 9.8). A first detection, iteration $k = 0$, is made for each of the received signals separately, possibly by using a decorrelating detector. This results in a number of detected symbols $\{\hat{x}_m(0)\}_{m=1}^{M}$. Now look at branch m, for instance. After the detection, all detected versions of the interfering signals are subtracted, i.e. $\hat{x}_p(0)$ where $p \neq m$, weighted by their particular signal strengths and correlation coefficients (see Figure 9.8). If all decisions are correct, only signal m now remains in branch m, and a renewed detector decision can be made as for a single signal. The same procedure is gone through in all branches.

Of course all decisions are not correct on the first occasion. That would make interference cancellation unnecessary. Instead, successively new decisions are being made, $k = 0, 1, 2...$ etc., according to

$$\hat{\mathbf{x}}(k + 1) = \text{sgn}\left(\mathbf{y} - \left(\mathbf{F}\sqrt{\mathbf{E}} - \mathbf{I}\right)\hat{\mathbf{x}}(k)\right) \tag{9.29}$$

until the detected symbols no longer vary between iterations. Hopefully, the error density has then levelled out at a lower level than after the first detection. How rapidly this is achieved depends, among other things, on the relative energies of

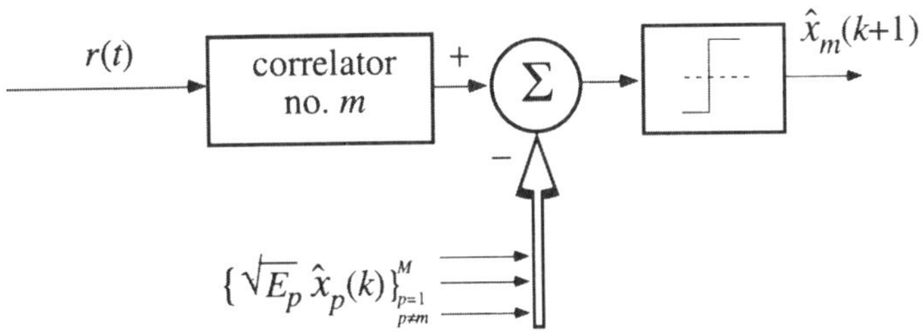

Figure 9.8 A branch of PIC. A complete PIC consists of M identical branches.

the signals. As for the SIC detector, the correlation matrix and the signal energies must be measured.

There are several versions of the PIC detector. It is, for instance, possible to make a soft decision, i.e. the sgn operator is replaced by some kind of metric which can assume many values. It is also possible to partly eliminate the interference in each step, but with a successively increasing fraction for each detection step.

9.3 Appendix: Generator polynomials for m-sequences

The coefficients in the table are arranged so that the most significant number corresponds to h_n in the polynomial $h(p) = h_n p^n + h_{n-1}p^{n-1} + \cdots + h_1 p + h_0$. This means, for instance, that for $(45)_8 = 100101$ $h_5 = 1, h_4 = 0, h_3 = 0, h_2 = 1, h_3 = 0, h_0 = 1$. It is always true that $h_n = h_0 = 1$.

Length, L	Generator polynomial, $h(p)$ (octal)
15	23, 31
31	45, 75, 67, 51, 57, 73
63	103, 147, 155, 141, 163, 133
127	211, 217, 235, 367, 277, 325, 203, 313, 345, 221, 361, 271, 357, 375, 253, 301, 323, 247
255	435, 551, 747, 453, 545, 543, 537, 703, 561, 455, 717, 651, 515, 615, 765, 607
511	1021, 1131, 1461, 1423, 1055, 1167, 1541, 1333, 1605, 1751, 1743, 1617, 1553, 1157, 1715, 1563, 1713, 1175, 1725, 1225, 1275, 1773, 1425, 1267, 1041, 1151, 1063, 1443, 1321, 1671, 1033, 1555, 1207, 1137, 1427, 1707, 1533, 1731, 1317, 1473, 1517, 1371, 1257, 1245, 1365, 1577, 1243, 1665
1023	2011, 2415, 3771, 2157, 3515, 2773, 2033, 2443, 2461, 3023, 3543, 2745, 2431, 3177, 3525, 2617, 3471, 3323, 3507, 3623, 2707, 2327, 3265, 2055, 3575, 3171, 2047, 3025, 3337, 3211, 2201, 2605, 2377, 3661, 2627, 3375, 3301, 3045, 2145, 3103, 3067, 2475, 2305, 3763, 2527, 3615, 2347, 3133, 3427, 3117, 3435, 3531, 2553, 2641, 2767, 2363, 3441, 2503, 3733, 2213
2047	4005, 4445, 4215, 4055, 6015, 7413, 4143, 4563, 4053, 5023, 5623, 4577, 6233, 6673, 7237, 7335, 4505, 5337, 5263, 5361, 5171, 6637, 7173, 5711, 5221, 6307, 6211, 5747, 4533, 4341, 6711, 7715, 6343, 6227, 6263, 5235, 7431, 6455, 5247, 5265, 4767, 5607, 4603, 6561, 7107, 7041, 4251, 5675, 4173, 4707, 5463, 5755, 6675, 7655, 5531, 7243, 7621, 7161, 4731, 4451
4095	10123, 15647, 16533, 16047, 11015, 14127, 17673, 13565, 15341, 15053, 15621, 15321, 11417, 13505, 13275, 11471, 16237, 12515, 12255, 11271, 17121, 14227, 12117, 14135, 14711, 13131, 16521, 15437, 12067, 12147, 14717, 14675, 10663, 16115, 12247, 17675, 10151, 14613, 11441, 10321, 11067, 14433, 12753, 13431, 11313, 13425, 16021, 17025, 15723, 11477, 14221, 12705, 14357, 16407, 11561, 17711, 13701, 11075, 16363, 12727

10 SYNCHRONISATION

10.1 Introduction

When the modulated signal is received it has propagated a distance which is unknown to the receiver. This means that the phase of the carrier, the phase of the symbol clock and the start time of the message are all unknown. Furthermore, the receiver does not usually have a time reference in common with the transmitter, so that the time parameters may vary during communication due to the variations in the electrical components. For mobile communication the Doppler effect also has to be considered, which may lead to a shifted carrier frequency when transmitter and receiver are in relative motion. Altogether, these factors mean that a number of time parameters need to be estimated by the receiver in order to make detection with the minimum possible error rate.

The problem can be solved by transferring a special signal together with the message, such as a pilot tone carrying the time parameter information. Another way is to use a type of modulation whose spectrum has frequency components at the carrier and symbol clock frequencies. These components can be filtered out by the receiver using devices for carrier and clock recovery, respectively. From a transmission efficiency point of view it is obviously beneficial if no signal power is taken by a carrier component that does not contain any part of the message. A way of avoiding arrangements for finding the carrier phase is to use non-coherent detection or differentially coded modulation. As is clear from Chapter 6, the offered simplification has its price, however, in the form of increased bit error probability in relation to coherent detection.

The objective of this chapter is partly to point out problems and partly to provide some examples of solutions. Some general solutions are presented here. Certain fundamental parts appear in most systems. In the following we mainly focus on such frequently appearing components as, for example, phase locked loops, Barker codes, etc., so the reader can easily continue to more specific systems and get a good idea of the synchronisation methods.

The sent signal is shown in eq. (5.86). The modulation is included in $a(t)$, $\phi(t)$ or both. The received signal is a delayed variant of the sent signal,

$$s(t) = a(t-\tau_0)\cdot\cos[\omega_c t + \phi_0 + \phi(t-\tau_0)] \tag{10.1}$$

where τ_0 is the time lag and ϕ_0 is an unknown phase angle where the phase shift due to the delay, $-\omega_c\tau_0$, is also included. For the moment ignore the fact that the signal amplitude has been changed and that noise has been added in the transmission channel. Depending on the type of modulation, $a(t)$ can be equal to x_k, or $\phi(t) = 2\pi(x_k-1)/M$ for instance. Apart from estimating the message, the receiver must, in other words, be synchronised to the received signal. Synchronisation contains:

1. carrier synchronisation, i.e. find the phase shift ϕ_0
2. clock synchronisation, i.e. find the delay τ_0
3. frame synchronisation, i.e. find a special sequence of x_k.

Synchronisation is a complex and extensive field and here only an orientation is given. The first part of the chapter looks closer at the phase locked loop which is an important part of many synchronising units. The following sections discuss carrier, clock and frame synchronisation.

10.2 Fundamentals of phase locked loops

A phase-locked loop consists of a phase detector, a loop filter and a voltage controlled oscillator whose frequency is controlled by the voltage level at the input (VCO). The phase-locked loop is used to lock the output signal of the VCO to the phase of another signal, whose phase can be more or less stable, for example the received carrier. In some cases the phase of the VCO should track the phase of the input signal and in other cases it should only track the long term changes. If the output signal of the VCO is to be used as the carrier reference for demodulation, a signal with a stable phase, which does not track the noise variations of the input signal, is desired. The carrier reference must, on the other hand, have the ability to respond to carrier phase shifts caused by, for instance, a change in the distance or other conditions of the wave propagation between transmitter and receiver. It is therefore important to be able to analyse the loop response to the phase variations of the input signal.

10.2.1 ANALYSIS OF AN ANALOGUE PHASE-LOCKED LOOP

The phase of an input signal is used as the input in the analysis, rather than the signal itself. The phase detector becomes a component obtaining the difference between the phase of the input signal and the output signal of the VCO (see Figure 10.1).

The phase of the input signal, $\theta_i(t)$, is time variant. Its Laplace transform is

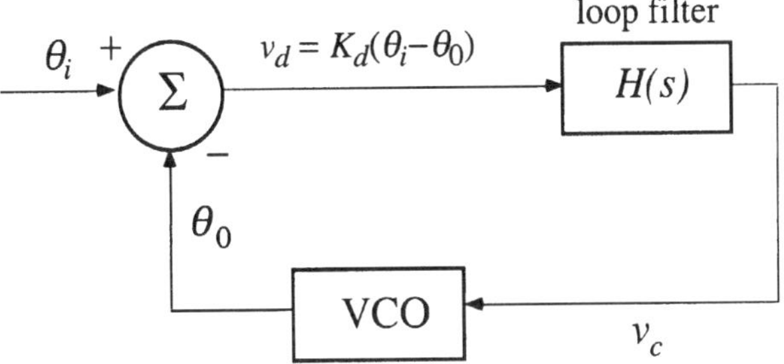

Figure 10.1 Phase-locked loop.

$$L\{\theta_i(t)\} = \theta_i(s) \tag{10.2}$$

The output signal of the loop filter is denoted v_c and is

$$v_c(s) = H(s) \cdot v_d(s) = H(s) \cdot K_d(\theta_i(s) - \theta_0(s)) \tag{10.3}$$

The parameter K_d is a gain constant of the phase detector with the dimension (volt/radian). The VCO is, in other words, an oscillator for which the frequency of the output signal is proportional to the input signal, $v_c(s)$

$$\omega_0 = K_0 v_c(s) \tag{10.4}$$

K_0 is a sensitivity parameter, [rad/(second volt)]. Since $d\theta/dt = \omega$ the Laplace transform of the frequency becomes

$$L\{\omega_0(t)\} = L\left\{\frac{d\theta_0(t)}{dt}\right\} = s\theta_0(s) = K_0 v_c \tag{10.5}$$

where $\theta_0(s)$ is the transformed phase of the output signal. Equations (10.3) and (10.5) then give

$$\theta_0(s) = \frac{K_0 K_d}{s} H(s)(\theta_i(s) - \theta_0(s)) \tag{10.6}$$

A transfer function, $K(s)$, as a relation between the phase of the output signal and the phase of the input signal, can then be formed

$$K(s) = \frac{\theta_0(s)}{\theta_i(s)} = \frac{K_0 K_d H(s)}{s + K_0 K_d H(s)} \tag{10.7}$$

When the loop filter is of the first order, for instance of the kind shown in Figure 10.2, the transfer function of the loop filter is

$$H(s) = -\frac{sCR_2 + 1}{sCR_1} = -\frac{s\tau_2 + 1}{s\tau_1} \tag{10.8}$$

Eq. (10.8) inserted in eq. (10.7) gives the transfer function of the loop, which in

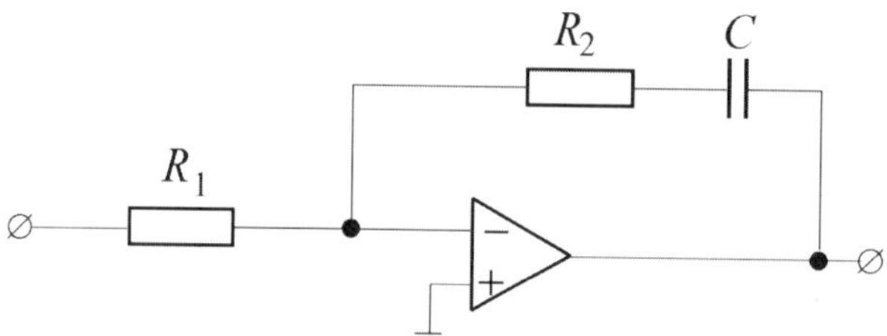

Figure 10.2 Active first order loop filter.

this case gives a second order system,

$$K(s) = \frac{2\zeta\omega_n s + \omega_n^2}{s^2 + 2\zeta\omega_n s + \omega_n^2}$$

(10.9)

where

$$\omega_n = \sqrt{\frac{K_0 K_d}{\tau_1}} \quad \text{and} \quad \zeta = \frac{\tau_2 \omega_n}{2}$$

(10.10)

The phase of the VCO signal therefore responds to the phase of the input signal as a second order system having a cut off frequency and a damping factor given by eq. (10.10). In spite of a first order loop filter, the response of the loop corresponds to a second order system, due to the fact that the VCO increases the order by one.

Implementation of the phase detector

The phase detector (cf. Figure 10.1), is usually implemented as a multiplier.

When the input signals of the multiplier are those given in Figure 10.3, its output signal becomes

$$K_d\sin(\theta_i - \theta_0) + K_d\sin(2\omega_0 t + \theta_i + \theta_0)$$

(10.11)

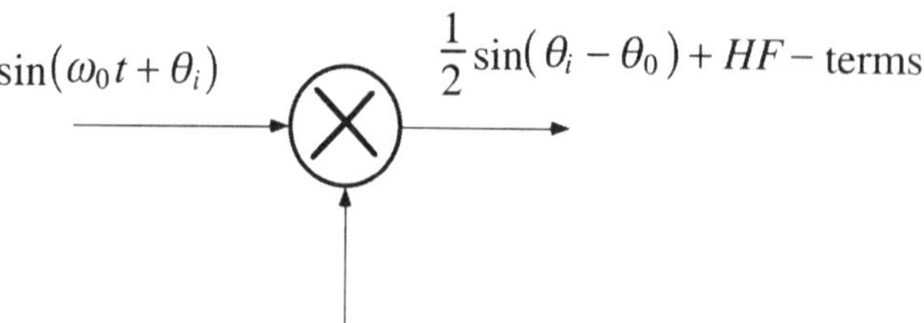

Figure 10.3 Phase detector implemented as a multiplier.

Let the factor 1/2 from the multiplication be included in the detector sensitivity K_d. The high frequency signal at $2\omega_0$ is cancelled by the loop filter whose upper cut off frequency is considerably lower. We are then left with

$$v_d = K_d\sin(\theta_i - \theta_0) \approx K_d(\theta_i - \theta_0) \tag{10.12}$$

where the approximation holds for small values of $\theta_i - \theta_0$, a requirement fulfilled when the loop is locked or close to being locked. In such cases, the multiplier will operate as the phase detector needed for the system to work in the way shown in Figure 10.1. Note, however, that with this phase detector the input and output signals will be 90° out of phase when the loop is locked. A double-balanced mixer, implemented with diodes or FET amplifiers, is an example of a circuit which may well work as a multiplier within a limited amplitude range.

Phase-locked loop with noisy input signal

Since the received signal is usually contaminated by noise, it is of interest to analyse the properties of the loop when the input signal is noisy. Assume sinusoidal input signals and a phase detector made up of a multiplier as above. The noise is a narrowband process and is described by its quadrature components. The input signal is then

$$v_{in}(t) = V_s\sin(\omega_0 t + \theta_i) + n_c(t)\cos(\omega_0 t) + n_s(t)\sin(\omega_0 t) \tag{10.13}$$

After multiplication and lowpass filtering we obtain, in the same way as before,

$$v_d(t) = K_d\sin(\theta_i - \theta_0) + \frac{n_c(t)\,K_d}{V_s}\cos(\theta_0) + \frac{n_s(t)\,K_d}{V_s}\sin(\theta_0) \tag{10.14}$$

Define

$$n'(t) = \frac{n_c(t)\,K_d}{V_s}\cos(\theta_0) + \frac{n_s(t)\,K_d}{V_s}\sin(\theta_0) \tag{10.15}$$

The variance of n' becomes

$$\sigma_{n'}^2 = \frac{1}{V_s^2}\left(E\{n_c^2\cos^2\theta_0\} + E\{n_s^2\sin^2\theta_0\} + E\{n_c n_s\cos\theta_0\sin\theta_0\}\right) \tag{10.16}$$

The sine and cosine terms are deterministic and are separately averaged. Since $\sigma_n^2 = E\{n_c^2\} = E\{n_s^2\} = E\{n^2\}$ and the noise terms are orthogonal, i.e. $E\{n_c n_s\} = 0$, the variance of the noise after the multiplier becomes

$$\sigma_{n'}^2 = \frac{\sigma_n^2}{V_s^2} \tag{10.17}$$

The input signal can then be regarded as noise free. The noise can be modelled as

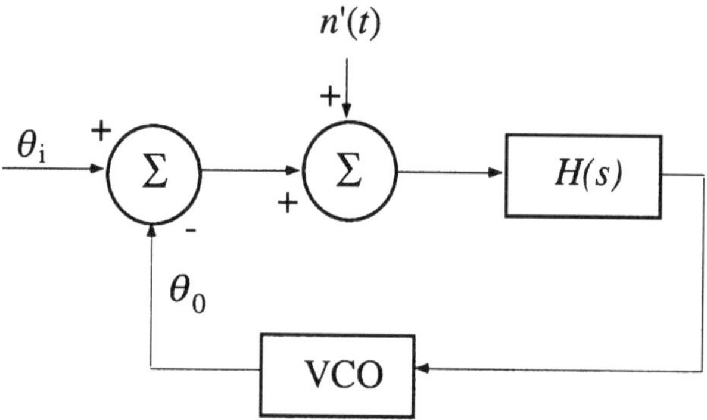

Figure 10.4   Model of phase-locked loop with noise.

an additive noise source located after the multiplier, having the variance $\sigma_{n'}^2$ and with the same amplitude distribution as at the input (cf. Figure 10.4).

Since no filtering has been done and the phase detector, apart from the frequency translation, can be seen as a linear component, the power density spectrum of the noise is the same after the phase detector as before:

$$P_{n'}(f) = \frac{1}{2V_s^2}\left(P_n(f_0-f)+P_n(f_0+f)\right) \tag{10.18}$$

After the loop filter the power density spectrum of the signal becomes

$$P_{v_c}(f) = P_{n'}(f)|H(f)|^2 \tag{10.19}$$

Given the fact that the multiplier can be seen as a linear phase detector, the same transfer function holds for both the input signal and the noise n', provided the noise level does not become too high. (An exact analysis would lead to solving a Fokker–Planck equation, which would go beyond the scope of this book.) Under the assumption of linearity we obtain the power density spectrum of the output signal in the same way as before:

$$P_{\theta_0}(f) = P_{n'}(f)|K(f)|^2 \tag{10.20}$$

where $K(f)$ is defined in eq. (10.9). The variance of the phase for the VCO output signal is given by

$$E\{\theta_0^2\} = \int_{-\infty}^{\infty} P_{n'}(f)|K(f)|^2 \, df \tag{10.21}$$

For white noise with the double sided noise density $N_0/2$ at the input we get

$$\sigma_{\theta_0}^2 = E\{\theta_0^2\} = \frac{N_0}{2V_s^2}\int_{-\infty}^{\infty}|K(f)|^2 \, df \tag{10.22}$$

which gives a mathematical expression for the variance of the reference signal phase as a function of the loop parameters and the noise level at the input.

Example 10.1

(a) Determine the response of a phase-locked loop with an active first order loop filter when the input signal makes a phase step by $\Delta\theta$ (rad).
(b) Determine the stationary phase lock error when the input signal makes a phase step by $\Delta\theta$, when the input signal makes a frequency step by $\Delta\omega$ (rad/s) and when the frequency of the input signal increases linearly as $d(\Delta\omega)/dt$ (rad/s^2). Neglect possible non-linear effects, such as saturation of the amplifier or exceeding the frequency range of the VCO.

Solution   (a) It is either the phase of the VCO output signal or the analogue voltage constituting the input signal of the VCO, which represents the response of the phase-locked loop. A linear relation holds between these parameters, since $s\theta_0(s) = K_0 v_c(s)$. Let the response be represented by $\theta_0(s)$ from eq. (10.6). The Laplace transform of the phase step is $\Delta\theta/s$, which gives

$$\theta_0(s) = L\{\theta_i(t)\}K(s) = \frac{\Delta\theta}{s} \frac{2\zeta\omega_n s + \omega_n^2}{s^2 + 2\zeta\omega_n s + \omega_n^2} \tag{10.23}$$

By using the inverse Laplace transform, the time function of the response becomes

$$\theta_0(t) = L^{-1}\{\theta_0(s)\}$$

$$= \Delta\theta\left[1 + \frac{\zeta\exp[-(\zeta\omega_n t)]\sin\left(\omega_n t\sqrt{1-\zeta^2}\right)}{\sqrt{1-\zeta^2}} - \exp[-(\zeta\omega_n t)]\cos\left(\omega_n t\sqrt{1-\zeta^2}\right)\right] \tag{10.24}$$

Equation (10.24) is plotted in Figure 10.5.
(b) The error can be written as

$$\varepsilon(s) = \theta_i(s) - \theta_0(s) = \theta_i(s)(1 - K(s)) \tag{10.25}$$

The error at a phase step then becomes

$$\varepsilon(s) = \theta_i(s)\left(1 - \frac{2\zeta\omega_n s + \omega_n^2}{s^2 + 2\zeta\omega_n s + \omega_n^2}\right) = \theta_i(s)\frac{s^2}{s^2 + 2\zeta\omega_n s + \omega_n^2} \tag{10.26}$$

The steady state error can be obtained by using the final value theorem

$$\lim_{t\to\infty} f(t) = \lim_{s\to 0} sF(s)$$

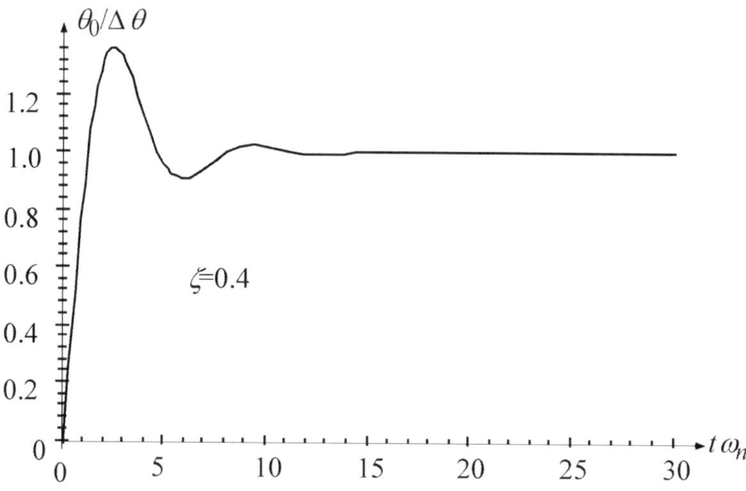

Figure 10.5 The response of the loop to a step in the phase of the input signal.

where $f(t)$ and $F(s)$ form a Laplace transform pair. This gives

$$\varepsilon_{ss} = \lim_{s\to0} s\varepsilon(s) = \lim_{s\to0} s\frac{\Delta\theta}{s}\frac{s^2}{s^2+2\zeta\omega_n s+\omega_n^2} = 0 \tag{10.27}$$

Since $\theta = \int \omega\,dt$, the frequency step corresponds to a phase ramp. By using the Laplace transform of a ramp, we obtain

$$\varepsilon_{ss} = \lim_{s\to0} s\varepsilon(s) = \lim_{s\to0} s\frac{\Delta\omega}{s^2}\frac{s^2}{s^2+2\zeta\omega_n s+\omega_n^2} = 0 \tag{10.28}$$

This means that in tracking a frequency change of the input signal, the phase-locked loop adapts its frequency to the new one, without any errors. A linearly increasing frequency gives the error

$$\varepsilon_{ss} = \lim_{s\to0} s\varepsilon(s) = \lim_{s\to0} s\frac{d(\Delta\omega)/dt}{s^3}\frac{s^2}{s^2+2\zeta\omega_n s+\omega_n^2} = \frac{d(\Delta\omega)/dt}{\omega_n^2} \tag{10.29}$$

Thus, a frequency which changes linearly with time will cause a constant phase error of the loop given by eq. (10.29).

10.2.2 DIGITAL PHASE LOCKED LOOP

The different parts in a digital phase locked loop (DPLL), is the same as in its analogue counterpart (cf. Figure 10.3). Instead of analogue voltages, digital signals flow in the connections between the blocks. The analysis is identical to

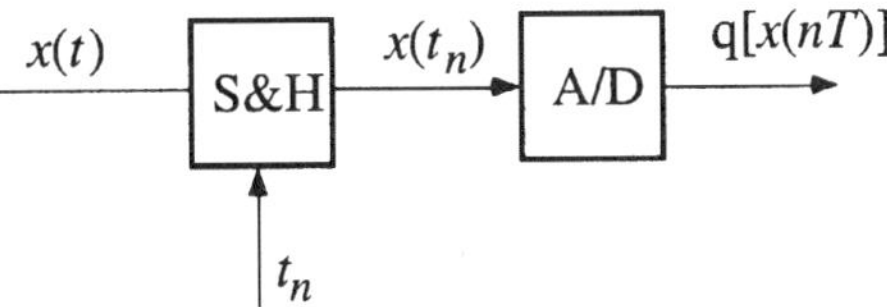

Figure 10.6 Phase detector joined with the A/D conversion.

the analogue loop except that the z-transform is used instead of the Laplace transform.

The voltage controlled oscillator (VCO), is substituted by a digitally controlled oscillator (DCO). Its function is the same as for the analogue counterpart except that the input and output signals are represented by digital signals. The periodicity of the output signal is proportional to the value of the input signal in the way that the next sampling instant occurs at the time $t_{n+1} = t_n + T - K_0 v_c(nT)$, which gives the z-transform of the phase (cf. Figure 10.6),

$$Z\{\theta_0(nT)\} = \frac{K_0}{z-1} Z\{v_c(nT)\} \tag{10.30}$$

The phase detector can be accomplished by a multiplier. It is also very common that the phase detector is combined with the analogue-to-digital conversion. The sampling instants are then controlled by the phase locked loop internal oscillator, the DCO. The sampled signal $x(t_n)$ is dependent on the phase angle between the analogue input signal $x(t)$ and the sampling signal. One sample is taken every time the DCO signal crosses zero in the positive direction. Therefore this loop is often called the zero crossing DPLL.

With the same reasoning as before the transfer function becomes

$$K(z) = \frac{K_0 K_d H(z)}{z - 1 + K_0 K_d H(z)} \tag{10.31}$$

where K_d denotes the sensitivity of the phase detector to the phase difference between the signals. It is desirable that the response from the digital loop conforms to the analogue loop (cf. the analogue loops transfer function in eq. (10.9)). Assume that the loop filter is implemented by a first order digital filter as shown in Figure 10.7.

The transfer function of the loop filter is

$$H(z) = \frac{c_0 z + c_1}{z - 1} \tag{10.32}$$

Inserted in eq. (10.31) this gives the total transfer function of the loop, which is of second order:

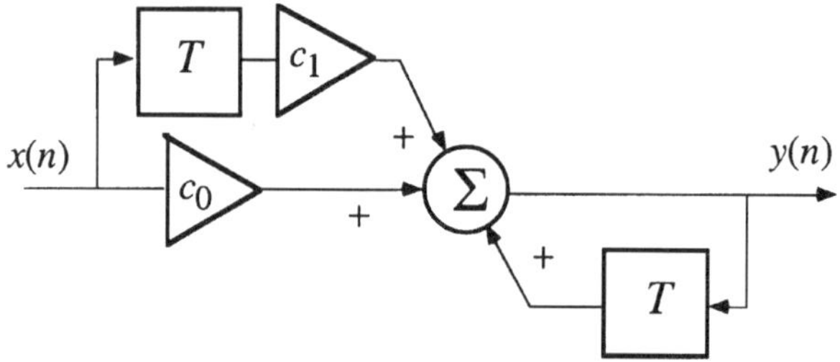

Figure 10.7   Loop filter for a second order DPLL.

$$K(z) = \frac{K_0 K_d (c_0 z + c_1)}{z^2 + z(K_0 K_d c_0 - 2) + K_0 K_d c_1 + 1} \tag{10.33}$$

If the filter coefficients are chosen as

$$c_0 = \frac{2}{K_0 K_d}\left(1 - \exp(-\zeta\omega_n T)\cos\left[\omega_n T\sqrt{1-\zeta^2}\right]\right), \quad c_1 = \frac{\exp(-2\zeta\omega_n T) - 1}{K_0 K_d}$$

a large similarity with the analogue loop is obtained if the sampling frequency T is chosen in such a way that $\omega_n T < 0.1$.

The noise, or the variance, of the output signal phase is

$$\sigma_{\theta_0}^2 = \frac{1}{2\pi j} \oint_{|z|=1} P_{n'}(z) K(z) K(z^{-1}) z^{-1} dz \tag{10.34}$$

where $P_{n'}(z)$ is defined in the same way as in eq. (10.18) though expressed by the z-transform. The noise level is increased by decreased sampling frequency $1/T$, which is one of the reasons for not making the sampling period too long.

10.3 Carrier synchronisation

To be able to carry out coherent demodulation, the carrier phase needs to be known. The phase may vary due to several factors, which have been previously mentioned. In certain modulation schemes the carrier is included in the modulated signal. The carrier can then be easily recovered by using a narrowband filter or a phase-locked loop. In other modulation schemes there is no component of the carrier as a part of the modulated signal. For such systems a carrier component must somehow be recreated before the carrier can be recovered in the receiver. How this is achieved depends on the type of modulation. A non-linear method is described below.

10.3.1 SQUARING LOOP

A so-called squaring loop is an example of carrier recovery for BPSK. The modulation is made up by a variable coefficient, $x(t) = \pm 1$, multiplying the carrier. Taking the square of the received signal will eliminate the modulation, i.e.

$$V_x(t) = x^2(t)\cos^2(\omega_c t + \phi_0) = \frac{1}{2}x^2(t)[1 + \cos(2\omega_c t + 2\phi_0)] \tag{10.35}$$

It can be seen from eq. (10.35) that the signal $V_x(t)$ is independent of the modulation since $x^2(t) = 1$ for both symbols, $+1$ or -1. The phase and frequency have been doubled and a DC component has been added. The latter is easily eliminated by a highpass filter. The doubled phase argument is eliminated by letting the signal go through a frequency divider, whose output signal then can provide a carrier reference to the receiver. As illustrated in Figure 10.8, however, a phase-locked loop, or a phase locked oscillator, is usually used to generate the recovered carrier signal. One of the advantages with this arrangement is that the phase locked loop can tolerate a possible shift in the carrier frequency and still suppress disturbances, so a stable reference can be obtained. Note that $\omega_i = \omega_c + d\phi_0/dt$.

Since the argument of a periodic function has been doubled, an ambiguity of $180°$ in the phase is obtained with this method. This ambiguity makes it impossible to determine whether a received signal is $+1$ or -1 without making special arrangements, such as using an initial reference sequence.

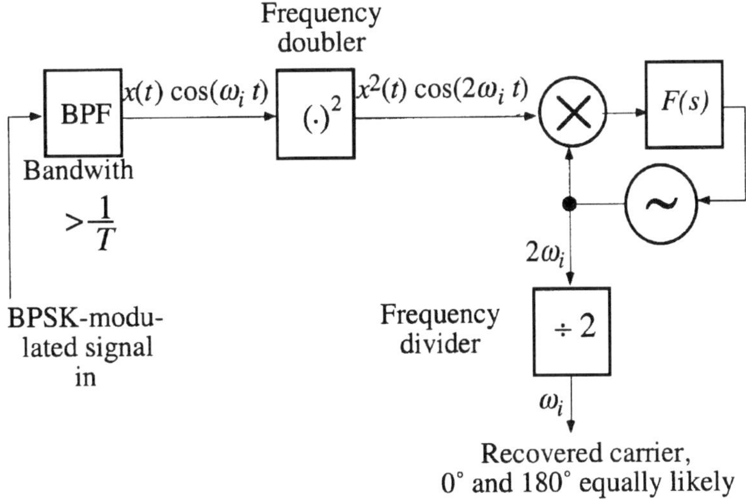

Figure 10.8 Squaring loop for carrier recovery.

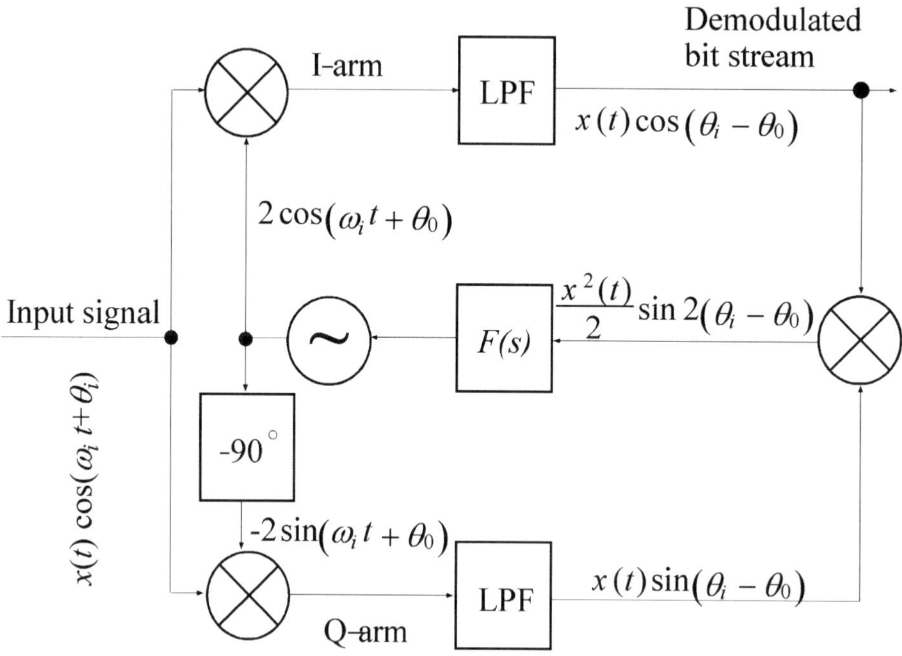

Figure 10.9   Costas loop. θ_0 and θ_i are phase angles which can be either constant or time dependent.

10.3.2 COSTAS LOOP

The technique for removing the modulation in order to recover the carrier can be further developed. The demodulated symbol sequence can be fed back and multiplied by the received signal, which has been delayed by the time corresponding to the time lag in the demodulator. This technique is commonly named remodulation. Another method of recovering the carrier is by using a so-called Costas loop (see Figure 10.9).

The Costas loop can be described in the following way. In the absence of modulation the Q-arm acts as an ordinary phase locked loop and the I-arm as a demodulator. When the signal is modulated the output signal from the Q and I arms will arbitrarily shift by 180°, which could interfere with the functioning of the loop. The modulation is, however, eliminated by the multiplier on the right-hand side in the figure.

The methods discussed above are intended for BPSK. For QPSK, for instance, the non-linearity in the squaring loop can be changed into a fourth order non-linearity, which then eliminates the modulation. The frequency divider then has to divide by four if the correct frequency of the recovered carrier reference is to be obtained. For the rest of the described methods there are also alternative circuit solutions available.

10.4 Clock synchronisation

The methods of clock synchronisation vary considerably depending on the type of modulation. In this section we give an example of clock recovering in BPSK. Since the clock rate is relatively slow, a coarse adjustment can be made by means of some special clock synchronisation sequence, followed by tracking during the information transmission. The early-late gate is shown in Figure 10.10. This consists of two branches each with an integrator, sample and hold, and a circuit which takes the magnitude of the input signal.

As mentioned earlier, the input signal of a linearly modulated signal is

$$s(t) = \sum_k x_k\, p(t-kT) \tag{10.36}$$

Assume that the estimated symbol clock is τ time units ahead of the real clock. Since the pulse shape is rectangular, the signal after the integrator in the upper arm (early) becomes

$$\alpha_{kE} = \frac{1}{T}\left[\left(\frac{1}{2}T-\tau\right)x_{k-1}+x_k\tau\right] \tag{10.37}$$

Given that $0 \le \tau \le T/2$. For the lower branch (late) we obtain

$$\alpha_{kL} = \frac{1}{T}\frac{x_k T}{2} \tag{10.38}$$

At the sampling times the error signal $\varepsilon(t)$ becomes

$$\varepsilon_k = |\alpha_{kE}|-|\alpha_{kL}| = \frac{1}{T}\left[\left|\left(\frac{1}{2}T-\tau\right)x_{k-1}+x_k\tau\right|-\frac{1}{2}T|x_k|\right] \tag{10.39}$$

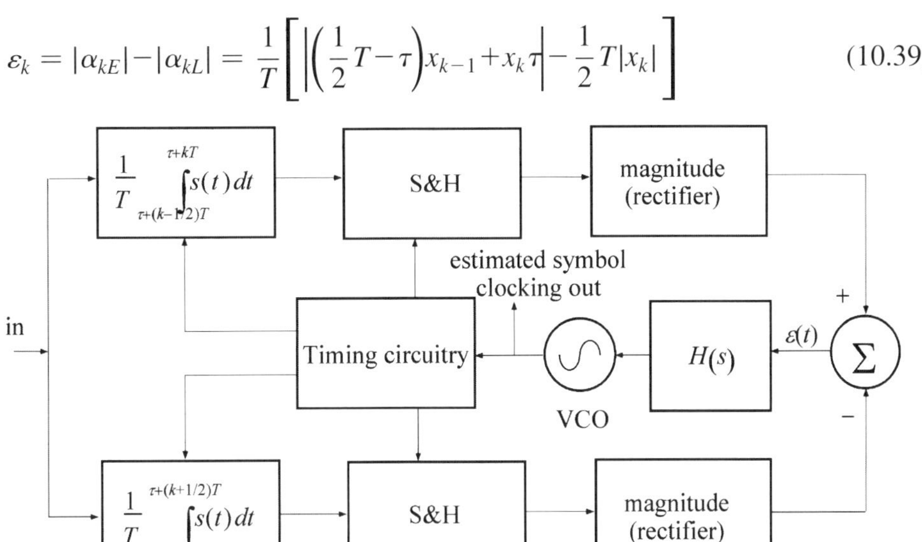

Figure 10.10 Early-late gate.

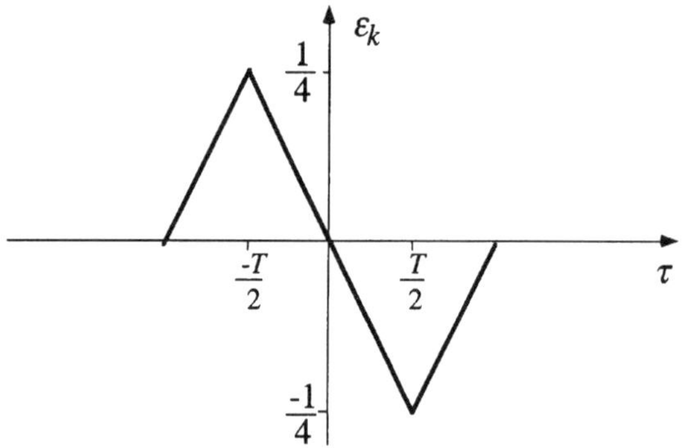

Figure 10.11   Detector characteristic for the early-late gate given the symbols $\pm$ 1.

For $\tau > 0$ the error signal is zero or negative. Taking the average by means of the filter $H(s)$ then gives a negative value which reduces the frequency of the VCO output signal. Equivalently $\tau < 0$ would give a positive error signal, which would push the receiver symbol clock in the opposite direction. Synchronisation to the transmitter symbol clock can then be achieved. The detector response is shown in Figure 10.11.

10.5 Frame synchronisation

Frame synchronisation is necessary for the separation of the received symbol stream into frames, which, for instance, represent the ASCII code of a character, the parameters of a speech encoder, etc. A good synchronisation method must be fast, have a reliable indication of the synchronised state and introduce a minimum of redundancy in the transmitted signal. Synchronisation is made possible by introducing a marker or synchronising sequence at the transmitter.

10.5.1 SYNCHRONISATION PROBABILITY

The usual way of achieving and maintaining frame synchronisation is to transmit a known symbol sequence at regular intervals. Assume that the symbol sequence is divided into blocks of D information symbols and a synchronisation sequence consisting of M marker symbols. This leads to a periodic sequence where the next synchronisation sequence is initiated after every $D+M$ number of symbols. The emitted synchronisation sequence is a vector, **m**, consisting of M known elements. A sequence of the same length as the synchronisation word is extracted out of the received symbol stream. The received sequence forms a vector

$\mathbf{r}_k = [r_k, \; r_{k+1} \cdots r_{k+M-1}]$, at time kT. Consider the Hamming distance between the recovered sequence, $\mathbf{r}_k$, and the synchronisation word, $\mathbf{m}$. The Hamming distance between $\mathbf{m}$ and $\mathbf{r}_k$ is denoted $d_H(\mathbf{m}, \mathbf{r}_k)$. When the Hamming distance is below a threshold h, synchronisation is said to be achieved. If synchronisation is not achieved, a 'one' is added to the value of k and a new sequence is extracted which is shifted one step relative to the previous one. The synchronisation algorithm for a binary symmetrical channel (BSC) can be summarised as follows:

1. set $k = 1$
2. if $d_H(\mathbf{m}, \mathbf{r}_k) \le h$, go to 4
3. set $k = k+1$ and go to 2
4. synchronisation achieved at time kT.

The threshold h determines the tolerance to symbol errors at the detection. The probability of acquire locking within a given synchronisation period is

$$P_A = \Pr\{\text{no locking for } k < M+D\}$$

$$\cdot \Pr\{\text{locking at } k = M+D \mid \text{no locking for } k < M+D\} \tag{10.40}$$

For binary symbols the probability of locking to the data section of the synchronisation period, i.e. the probability that in a random stream of ones and zeros a sequence is found which coincides with the synchronisation word in at least h positions, is

$$P_F = \frac{1}{2^M} \sum_{i=0}^{h} \binom{M}{i} \tag{10.41}$$

The sequence is, however, not completely random, since for each new symbol the considered window is only shifted one step. Therefore, each new symbol interval does not give a completely new sequence, only one new symbol is added and one disappears. Equation (10.41) can therefore be seen as an upper limit. Synchronisation is achieved if no more than h errors are introduced in the synchronisation sequence during the BSC transmission. Given the probability of bit error, P_e, the probability of correct locking at the marker sequence is

$$P_T = \sum_{l=0}^{h} \binom{M}{l} (1-P_e)^{M-l} P_e^{l} \tag{10.42}$$

This probability is certainly not conditioned on no locking for $k < M+D$, which is wanted in eq. (10.40). If the chosen synchronisation sequence is good enough, however, this probability is small, and eq. (10.42) can be seen as a good approximation. Inserted in eq. (10.40) the following limit for locking probability can be calculated.

$$P_A \geq [1-(D+M-1)P_F]P_T \tag{10.43}$$

The equations given above are derived under the condition that the chosen synchronisation sequence is good enough. This means that the probability of an arbitrary sequence of M information symbols looking like a synchronisation sequence, must be smaller than the probability of a synchronisation sequence looking like arbitrary data, when the channel has introduced errors in the sent synchronisation sequence. Such sequences can be found by a computer search.

10.5.2 SYNCHRONISATION CODES

Barker codes

A well known class of appropriate synchronisation sequences are Barker codes. Their primary property is that when the synchronisation word is clocked through a matched filter, the highest peaks in the output signal of the filter become ± 1, except for the peak which appears when the synchronisation word is in the middle of the filter. For this peak the output signal is M. This large difference in amplitude makes it easy to find the correct synchronisation point. The seven existing Barker codes are shown in Table 10.1.

The side lobe level is the maximum level relative to the peak level which is obtained when the synchronisation word is shifted through an analogue matched filter having an impulse response corresponding to the same synchronisation word. This is illustrated in Figure 10.12.

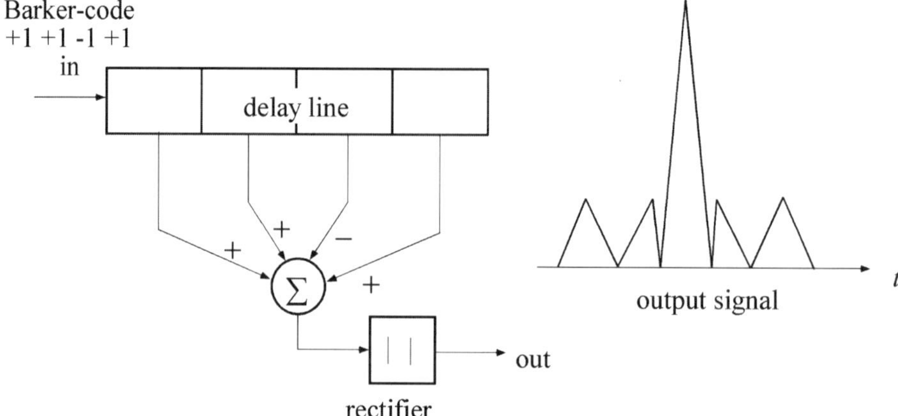

Figure 10.12   The response of a filter which has been matched to a certain synchronisation sequence in this case a Barker code.

Table 10.1   Barker codes

Length, M	Code, $\mathbf{m}$	Sidelobe level (dB)
2	$+1, -1;\ +1, +1$	-6.0
3	$+1, +1, -1$	-9.5
4	$+1, +1, -1, +1;\ +1, +1, +1, -1$	-12.0
5	$+1, +1, +1, -1, +1$	-14.0
7	$+1, +1, +1, -1, -1, +1, -1$	-16.9
11	$+1, +1, +1, -1, -1, -1, +1, -1, -1, +1, -1$	-20.8
13	$+1, +1, +1, +1, +1, -1, -1, +1, +1, -1, +1, -1, +1$	-22.3

Table 10.2

Length	Maury and Styles	Turyn	Length	Maury and Styles	Turyn
7	130	130	21	7351300	7721306
8	270		22	17155200	17743312
9	560		23	36563200	23153702
10	1560		24	76571440	77615233
11	2670	2670	25	174556100	163402511
12	6540		26	372323100	
13	16540	17465	27	765514600	664421074
14	34640	37145	28	1727454600	1551042170
15	73120	76326	29	3657146400	3322104360
16	165620	167026	30	7657146400	7754532171
17	363240	317522	31		16307725310
18	746500	764563	32		37562164456
19	1746240	170735S	33		77047134545
20	3557040	3734264	34		177032263125

Synchronisation code designed around ASCII characters

A synchronisation word can be composed by the idle character of the ASCII code which is labelled SYN and is represented by the bit sequence 0010110 or by 'end of transmission', which is represented by 1011100. Two sets of longer code words suitable as synchronisation words are the Maury–Styles and Turyn codes, respectively. These are shown in Table 10.2 in octal form and with the left bit in each code word always a one.

10.5.3 LOCKING STRATEGIES

The locking performance of a communication system depends not only on signal processing, but also on the routines in connection with the locking procedure. It is therefore a reason to say a few words about this.

The locking system can be in one of four possible states.

Locked states:

1. Correct locking has occurred and the system does not attempt to repeat the locking process.
2. Correct locking has occurred but the system interprets the locking as incorrect and seeks a new locking. This generates symbol errors in the message since a new locking procedure is initiated.

Unlocked states:

3. The system is not locked and attempts to achieve locking.
4. The system is not locked but the system assumes locking has occurred and does not attempt to achieve locking. This is a dangerous state where a large number of symbol errors can occur since the receiver can remain unlocked for a long period of time even though information transfer could take place.

The system should be designed so that it always aims to be in state 1. States 1 and 3 are functional while 2 and 4 are erroneous and counteract the function of the receiver. It is therefore important to design the locking procedure to minimise the probability of getting in the erroneous states.

The locking procedure can be divided in two separate processes: searching and confirmation. When the system during start-up or owing to high symbol error rate has ended up in state 3, it must be able to return to state 1 as quickly as possible. The locking process has been described above and entails searching for and finally finding a state which appears locked. The search stops when locking has been confirmed a number of times. The receiver is then in a locked state. Confirmation is important in order to avoid ending up in state 4, which can be reached when interference creates an incorrect symbol pattern which coincides with the synchronisation sequence. The search time must be short compared to the expected time between two false synchronisation sequences so that several attempts to lock can be carried out between each false synchronisation sequence.

When locking has occurred each new synchronisation word is checked. If the detection of a synchronisation sequence fails, the receiver waits and finds that the detection of several consecutive synchronisation sequences has failed. The search for a new synchronisation position then begins. The risk of ending up in state 2 occurs when the synchronisation of the receiver is correct, but the number of incorrect symbols due to interference is so large that the received synchronisation sequence is not interpreted as a synchronisation sequence. Symbol errors often appear in bursts. The time to declare the receiver as being unlocked must therefore be long compared to the length of the error bursts.

APPENDIX

As application examples of methods treated in this book, a selection of common radio communication systems are presented in this appendix.

A.1 Review of mobile telephone systems

In this section a review of some of the largest mobile telephone communication systems in the world is presented.

GSM stands for global system for mobile communications. The standard was developed in Europe and is used over large parts of the world. IS-136 TDMA is a digital mobile phone system that originates from the USA. The system is significantly more narrowband than GSM to be able to co-exist with older analogue systems. PDC is read as Pacific digital cellular and is a digital mobile phone system only used in Japan. IS-95 CDMA is the first mobile phone system built on the CDMA technology and is used in many places in the world. An evolution of GSM to better suit data transmission is EDGE, which means enhanced data rates for global evolution. The third generation of mobile phone systems has been designed from the start to handle data traffic in a better way than generation two, with for example a larger flexibility in the data rate. The European standard is called UMTS, universal mobile telephony system, and the corresponding American standard is called CDMA 2000.

Uplink means communication from the mobile station to the base station. Downlink is the communication from the base station to the mobile station. The mean output power from the transmitter in the mobile is 0.1–0.6 W and for the base station in the order of 1–100 W.

A.1.1 SECOND GENERATION SYSTEMS

	GSM	IS-136 TDMA	PDC	IS-9/5 CDMA
Multiple access method	TDMA 577 μs pulse	TDMA 6.67 ms pulse	TDMA 6.67 ms pulse	CDMA, maximum spreading factor 64
Frequency of uplink (MHz)	876–915; DCS 1800, 1710–1785; PCS 1900, 1850–1910	824–849, 1850–1910	940–956, 1429–1453	824–849, 1850–1910
Frequency of downlink (MHz)	921–960; DCS 1800, 1805–1880; PCS 1900, 1930–1990	869–894, 1930–1990	810–826, 1477–1501	869–894, 1930–1990
Channel separation	200 kHz	30 kHz	25 kHz	1.25 MHz
Talking users per carrier	8–16	3–6	3–6	10–20
Modulation	GMSK	$\pi/4$ DQPSK	$\pi/4$ DQPSK	QPSK
Pulse shaping	$B_bT = 0.3$ Gaussian	$\beta = 0.35$ raised cosine	$\beta = 0.50$ raised cosine	Windowed sinc function
Speech coding	RPE-LTP, VSELP[a] or ACELP	VSELP or ACELP (IS 641)	VSELP or ACELP	RCELP (IS 127)
Bit rate for speech	4.75–12.2 kbits/s	7.4 kbits/s	6.7–8.0 kbits/s	2, 4 or 8 kbits/s
Total bit rate	270.8 kbits/s	48.6 kbits/s	42 kbits/s	1.2288 Mc/s
Diversity method	Frequency hop			
All have interleaving and antenna diversity in uplink				
Required C/I	9dB	16dB	13dB	
Reuse factor per sector	≤ 9	≤ 21	≤ 12	1

[a] VSELP, vector sum excited linear predictor; Mc/s, megachips per second.

A.1.2 THIRD GENERATION SYSTEMS

	UMTS, WCDMA	CDMA 2000 1X	EDGE
Multiple access method	CDMA, maximum spreading factor 256	CDMA, maximum spreading factor 128	TDMA 577 µs pulse
Frequency of uplink (MHz)	1850–1910, 1920–1980	824–849, 1850–1910	824–849, 876–915, 1710–1785, 1850–1910
Frequency of downlink (MHz)	1930–1990, 2110–2170	869–894, 1930–1990	869–894, 921–960, 1805–1880, 1930–1990
Channel separation	4.2–5 MHz	1.25 MHz	200 kHz
Talking users per carrier	50–100	15–30	See GSM
Modulation	QPSK	QPSK	Offset 8PSK
Pulse shaping	$\beta = 0.22$ raised cosine	Windowed sinc function	Linearised GMSK
Speech coding	ACELP	RCELP (IS 127)	Only data
Bit rate for speech	4.75–12.2 kbits/s	2, 4 or 8 kbits/s	
Total bit rate	3.84 Mc/s[a]	1.2288 Mc/s	270.8 kb/s
Diversity method, all have interleaving and antenna diversity in uplink	Antenna diversity at the transmitter in down link	Antenna diversity at the transmitter in down link	Frequency hop
Reuse factor per sector	1	1	≤ 9

[a] Mc/s, megachips per second.

A.1.3 ROAMING

A key technology for mobile telephone communication systems is the so-called roaming technology. An overview of a typical mobile telephone system is depicted in Figure A.1.

The mobile subscriber has a number stored on his SIM card (subscriber iden-

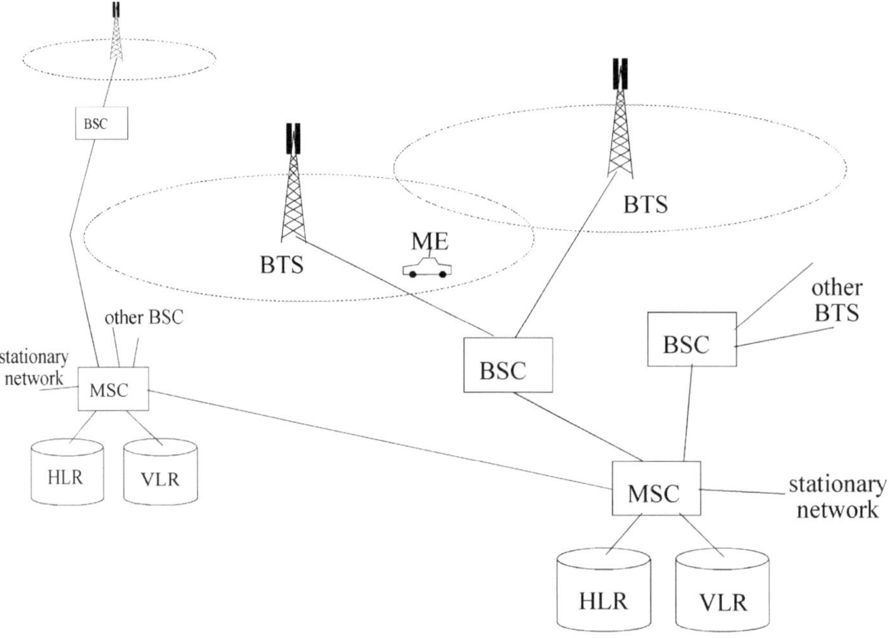

Figure A.1   Design of a mobile telephone communication system. ME, mobile equipment; BTS, base transceiver station; BSC, base station controller; MSC, mobile services switching centre; HLR, home location register; VLR, visitor location register.

tity module) which is inserted into a mobile terminal. This number comes from a number sequence belonging to the main exchange (MSC, mobile service switching) covering the geographical home district of the subscriber. The subscriber is registered in the home location register (HLR) of the exchange. When the mobile terminal is in operation, it occasionally transmits its identity which is received by those base stations that are located in the vicinity of the mobile terminal position. This is done during a special time slot called the RACH channel (random access channel). The base stations report the reception of the identity of a certain mobile to the main exchange. The position of the mobile station is then known to the main exchange. When a call is to be connected to the mobile terminal, a request, paging, is sent by those base stations which are closest to its current location. The paging of the subscriber is carried out on a special channel called PCH (paging channel).

Assume that a mobile terminal is making a journey which takes it outside the area of its home exchange, i.e. the MSC that has the mobile terminal stored in its HLR. The identity of the mobile terminal is then received and reported to another main exchange, MSC. The terminal is then registered in the VLR (visitor location register) of this exchange. A message is also sent to the home exchange, which in its HLR registers the information that the terminal is now to be accessed through the new host exchange. A call to the mobile terminal is first connected to the MSC given by its telephone number. In the HLR of the home switch there is then a notation saying that the call should be connected forward to the BSC, the host exchange, responsible for the area where the mobile terminal is presently located. The signalling between the exchanges is very fast and done by a signal system called SS7. The geographical area for which a main exchange is responsible depends on how many subscribers an exchange can handle (normally 50,000–60,000). In large cities these areas are therefore much smaller than in sparsely populated areas.

The connection problem is smaller when the mobile terminal is calling, since the terminal does not need to be located, but information about charges needs to be forwarded to the home exchange.

When a mobile terminal leaves the area covered by a base station, BST, during a connected call, it must be handed over to another base station. This process is called handover (handoff) and can be done both internally within the same BSC, to another BSC within the same MSC, or externally to another MSC. Handover can also be done because of heavy traffic load.

At each MSC there are two more registers, not given in the figure. One is the equipment identity register (EIR), which contains a list of all mobile terminals. Since a terminal is not equivalent to a subscriber in the GSM system, the EIR register is, for instance, required to make it possible to block stolen terminals. The other register is the authentication centre (AuC), which contains a copy of the encryption key stored on each SIM card.

A.2 Review of systems for cordless telephones

In contrast to a mobile telephone, which can be used anywhere within for instance a whole continent, is the so-called cordless telephones. They are intended for local use around a certain point, as an alternative to a traditional telephone connected via a wire.

System	DECT[a]	Personal handyphone
Used in	Europe	Japan
Frequency (MHz)	1897–1913	1895–1907
Channel separation	1.728 MHz	300 kHz
Multiple-access method	Multi-carrier TDMA-TDD	Multi-carrier TDMA-TDD
Modulation method	GMSK	$\pi/4$ QPSK
Symbol rate in radio-transmission	1.152 Mbits/s	384 kbits/s
Speech coding (kbits/s)	ADPCM 32	ADPCM 32 kbits/s

[a] DECT, digital enhanced cordless telephone.

A.3 Review of television systems

TV system	CCIR PAL	OIRT	French VHF	FCC (USA)
Band I (MHz)	47–68	48.5–100	50–70	54–88
Band II (MHz)	174–230	174–230	160–215	174–216
Band IV/V (MHz)	470–853			470–890
Separation, picture and sound carrier (MHz)	5.5	6.5	11.15	4.5
Sound modulation/swing	FM 50 kHz	FM 50 kHz	AM	FM 25 kHz
Number of lines	625	625	819	525
Number of pictures/s	50	50	50	60
Line frequency (Hz)	15.625	15.625	20.475	15.750
Video signal bandwidth (MHz)	5	6	10.6	4.2

The luminance in the different parts of the picture makes up a baseband signal in the total video signal and is called the luminance signal. It utilises the whole video signal bandwidth. The luminance signal gives only a black and white image. The colour information is modulated on a sub-carrier, which forms a narrowband signal with a bandwidth of 1.3 MHz. In the CCIR and NTSC systems the sub-carrier has an odd multiple of half the line frequency. The reason for this is that, because the neighbouring lines look approximately the same, the spectrum of the luminance signal will have its strongest components around even multiples of the line frequency. By choosing an odd multiple for the colour sub-carrier this will not interfere with the luminance signal. Thus, the frequency of the colour sub-carrier for the CCIR system ended up at a frequency of 4.4296875 MHz. The colour information is achieved by separating the spectrum of light into its three fundamental colours: red (R), green (G) and blue (B). The luminance signal (Y) is then a weighted sum of all three primary colours:

$$V_Y = 0.30R + 0.59G + 0.11B$$

The colour components of the luminance signal are weighted to correspond to the colour sensitivity of the human eye. Furthermore, two colour-difference signals are formed by the combinations

$$V_{R-Y} = 0.70R - 0.59G - 0.11B$$

$$V_{B-Y} = -0.30R - 0.59G + 0.89B$$

These are QAM modulated (AM-DSBSC) on the quadrature and in-phase components of the colour sub-carrier to form the chrominance signal. The three signals defined above form an equation system which is solved by the colour TV receiver in order to recover the colour signals R, G and B. The difference between the PAL system and the NTSC system is that in the PAL system one of the quadrature components of the chrominance signal is phase shifted 180° for every second line. This is done to compensate for phase errors appearing during the transmission chain. These phase errors will then alternatively shift the phase angle backwards and forwards for every other line. By adding the chrominance signal of two consecutive lines, which again are assumed to be nearly identical, the phase errors will cancel, or at least become very small and the reproduction of colours will improve.

Some systems:

PAL   phase alternation by line (adopted 1967)
NTSC   National Television System Committee (adopted 1953)
SECAM   Systeme Electronique Coleur Avec Memoire (adopted 1967)

A.4 Digital television

Even if the analogue TV standards are several decades old they still dominate the distribution of television programs. There are three standards for terrestrial TV distribution, ATSC, ISDB-T and DVB-T, which are much younger. The DVB system, digital video broadcasting, appears in three variants. They are DVB-C for video distribution via cable, DVB-S for distribution via satellite and DVB-T, a system for terrestrial transmitters, which is described below. The standard was accomplished in 1996 and specifies a single frequency network, which means that a group of adjacent transmitters are using the same frequency. In the standard there are many different parameters to tune, which is why a DVB signal does not need to look exactly the same for different TV distributors. For example two modes are specified. One is the 2k mode suitable for small networks, single transmitter and mobile reception due to smaller Doppler sensitivity. The other is the 8k mode suitable for nationwide networks of all sizes but also for single transmitters (Figure A.2).

A.4.1 SOURCE CODING

The source coding is done by the MPEG-2 standard (cf. the appendix in Chapter 3). There are four different categories of image quality:

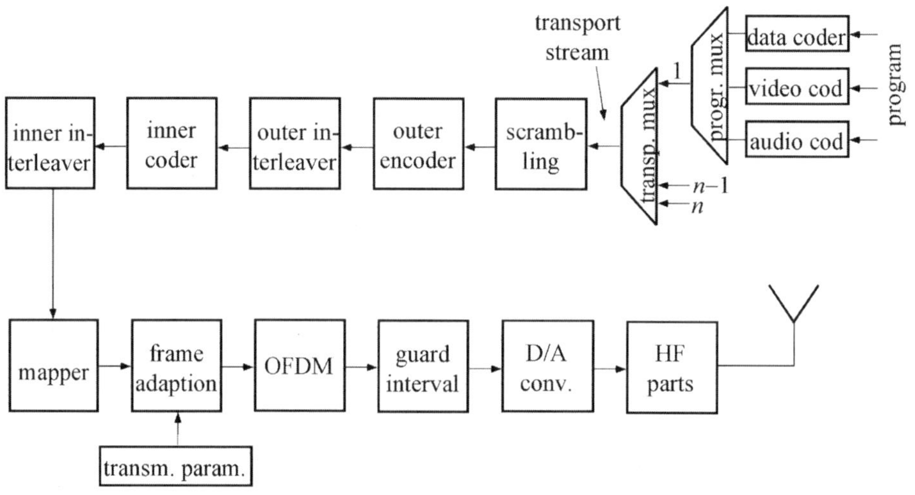

Figure A.2 Functional block diagram of a transmitter in the DVB system.

- standard-definition TV (SDTV) which is approximately the same quality as analogue systems;
- limited-definition TV (LDTV) that has approximately half the resolution of SDTV;
- enhanced-definition TV (EDTV) that approximately corresponds to analogue TV when it is best;
- high-definition TV (HDTV), giving the best image quality, it has double the resolution compared to SDTV. The relation between the image sides is 16:9.

Depending on the TV company requirements the program is distributed with suitable quality. A source encoded SDTV program generates a bit stream at approximately 5 Mbits/s. The double resolution of HDTV gives a bit stream at approximately $2 \times 2 \times 5 = 20$ Mbits/s. To satisfy the demands of different viewers, for example one SDTV and one HDTV variant of the programs can be distributed simultaneously (so-called simulcast).

A.4.2 MODULATION

In the DVB system OFDM is used (see Section 5.4.6). For the 8k mode 6817 carriers are used and for the 2k mode 1705 carriers are used. Each carrier is modulated by either QPSK, 16-QAM, 64-QAM, non-uniform 16-QAM or non-uniform 64-QAM. The mapping is done by Gray coding so adjacent signal points only differ one binary bit. In the non-uniform modulation the signal points in each quadrant have been somewhat displaced away from origin to reduce the risk of

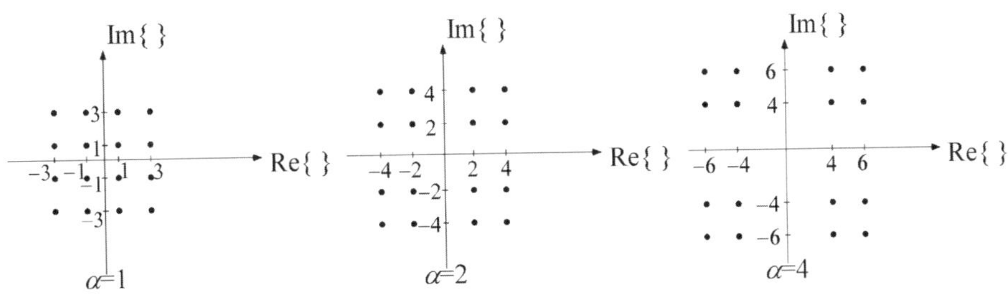

Figure A.3 Non-uniform 16-QAM modulation.

the detector choosing a quadrant other than the sent one, due to noise and other interference. At high signal to noise ratios, each individual signal point can be identified by the receiver and at lower signal to noise ratios only the quadrant at the signal point is located (Figure A.3).

The system supports hierarchic modulation, which means that data is ordered into two independent streams: one with high priority and one with low priority. The stream with high priority decides from which quadrant the signal points should be chosen and/or is given stronger protection by error correcting code, for example rate 1/2 convolutional code. This stream can for example contain the TV programs or applications that must operate under conditions with low SNR. The low priority stream decides which of the signal points within each quadrant shall be chosen and/or has a weaker error protection, for example rate 5/6. This stream can for example be used for data transmission or applications which only need to operate under high SNR conditions.

The bandwidth for the terrestrial transmission and the cable distribution is 8 MHz (can be changed to 7 or 6 MHz). The number of transmitted programs in this bandwidth is determined by the resolution, channel coding and modulation chosen. In, for example, 64-QAM and rate 2/3 coding the transport stream (cf. Figure A.2), can be up to 24 Mbits/s, which corresponds to 4 TV programs with SDTV quality. Moreover, as DVB is a single frequency system the spectrum efficiency becomes very large compared to analogue systems.

A.4.3 CHANNEL CODING

To prevent too long sequences of ones or zeroes the bit stream is multiplied by a PRBS (scrambling) generated in a feedback shift register with the generator polynomial $h(p) = 1 + p^{14} + p^{15}$.

The outer code is a shortened (cf. Figure 4.22) Reed–Solomon code RS(204, 188, $t = 8$) in GF(2^8) with the generator polynomial $G(p) = (p - \alpha)(p - \alpha^2)...(p - \alpha^{15})$, where $\alpha = 2$. The original code is a RS(255, 239, $t = 8$).

In the outer interleaving the bytes (1 byte = 8 bits) are permuted by a convolutional interleaver. The stream is divided into twelve different streams where the first is not delayed at all, the second is delayed 17 bytes, the third 34 bytes, etc. Thereafter the bytes are all put together into a single stream again.

The inner code is one of the convolutional codes with rates 1/2, 2/3, 3/4, 5/6 or 7/8. The rate 1/2 code, which is the basic code, has the generator polynomials $\mathbf{g}_1 = (171)_8$ and $\mathbf{g}_2 = (133)_8$.

The inner interleaving is done in two steps and is only briefly described here. The bit stream is divided into at most six different streams, depending on which of the modulation alternatives is chosen. Each new bit stream is partitioned in frames consisting of 125 symbols. Assume that for the bit stream i, the frame $\mathbf{B}_i = [b_{i,0}...b_{i,w}...b_{i,125}]$ is formed, which composes an input signal. The interleaved bits is the vector $\mathbf{A}_i = [a_{i,0}...a_{i,H(w)}...a_{i,125}]$. The mapping from $\mathbf{B}_i$ to $\mathbf{A}_i$ is done by a non-linear transform of index, $H_i(w)$, where the symbol number w is permutated and gets a new index: $H_0(w) = w; H_1(w) = (w + 63) \mod 126; H_2(w) = (w + 105) \mod 126; H_3(w) = (w + 42) \mod 126; H_4(w) = (w + 21) \mod 126$ and $H_5(w) = (w + 84) \mod 126$ for each stream, respectively. The next step is to map these bit streams on the various carriers. Depending on which mode is in use, the bits are distributed on 6048 or 1512 carriers for data, respectively, according to a given permutation scheme which is not described here to save space. The bit streams therefore hop between the different carriers, which means that no stream will be completely cancelled if one carrier happens to be in a fading dip. The other carriers are used for pilot cells and transmission parameter signalling (TPS). The pilot signals are reference signals consisting of one PRBS, known by the receiver. Some reference signals continuously remain on special carriers, for carrier synchronisation. Other reference signals are used for estimating the channel and are spread among the carriers. In the receiver the channel is equalised by multiplying each carrier with a complex weight. The TPS carriers are utilised to transmit parameters concerning the transmission, for example the error correction and modulation method presently in use.

The system has a sharp limit under which the bit error rate drastically increases. Given that one exceeds this limit the system is dimensioned such that the receiver after error correction has a bit error rate of $< 10^{-11}$, which is extremely low for a radio channel. For an AWGN channel, 64-QAM and rate 3/4 coding the limit is in practice around 20 dB. For mobile reception fading makes the limit increase to approximately 25 dB.

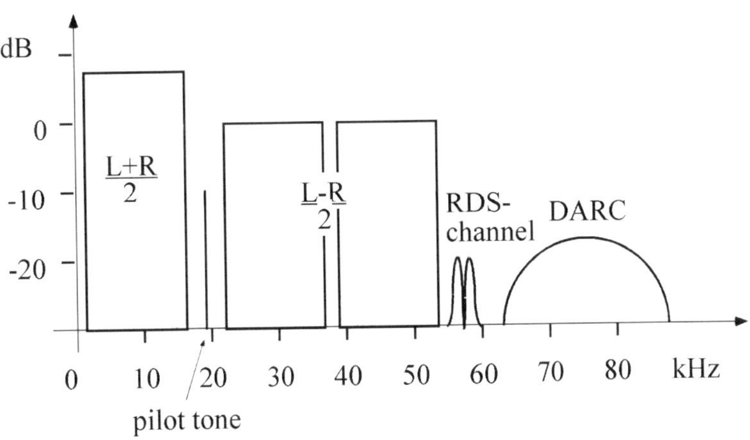

Figure A.4 Spectrum for a FM broadcasting signal.

A.5 FM broadcasting and RDS

Stereo broadcasting is transmitted on the VHF band 87.5–108 MHz except in Russia and Japan. The stereo signal consists of two separate channels, left (L) and right (R). These are added together, L + R, to form a mono channel taking up the frequency band 50–15,000 Hz (see Figure A.4). A difference signal, L − R, is AM modulated with a suppressed carrier on a 38 kHz sub-carrier. This signal covers the frequency range 23–53 kHz. The sum and difference signals form an equation system, which is solved by the receiver so that the L and R signals are recovered. Since the receiver has to recover the sub-carrier, a $38/2 = 19$ kHz pilot tone is included, which the receiver can extract and then double its frequency. This gives a sub-carrier which has the correct phase and thus can be used for the demodulation of the L − R signal. The RDS, radio data system, signal is modulated on a sub-carrier with frequency 57 kHz (3×19). The complete signal packet is then FM modulated with a maximum frequency swing of ±75 kHz for the main carrier. The amplitude of the frequency components over 3.2 kHz (corresponding to a time constant of 50 μs) is increased by 6 dB/octave up to 15 kHz, so-called pre-emphasis. The frequency response then levels out again. This can be achieved by, for instance, a first order RC network. A first order system has a derivative effect in the sloping region of the frequency response. In this region the frequency modulation then shifts to PM. The different sub-signals are allowed to modulate the transmitter with differing strengths. The sum signal is the strongest and is therefore least sensitive to noise. Within the normal coverage area of the transmitters, the demodulated audio signal gets an S/N of around 50 dB. The unmodulated RDS signal is allowed to modulate the main carrier to a swing of ±2.0 kHz.

The RDS channel is a special channel for programme identification, programme type, identification of traffic messages, alternative frequencies, radio text, various data transmissions, etc. The data rate of the RDS channel is 1187.5 bits/s (57/48). The sub-carrier BPSK is modulated by the data sequence, which is differentially coded in such a way that the previous bit is added modulo two to the current one. This makes the demodulated bit stream correctly decoded even if an error of 180° is made in the decoding. To prevent interference, for instance with the 38 kHz recovering, the BPSK signal spectrum is changed so that the components at and around 57 kHz are attenuated. This is achieved by using a pulse shape which is asymmetric around $t = 0$. This means that the spectrum of the pulse will have no DC component.

The RDS information is transmitted in groups of 4 blocks, each containing 26 bits. An alternative frequency, for instance, is transmitted in block 3 of a certain group of messages, where 11001100 means 107.9 MHz. Each block is a code word in a systematic error correction code. The first 16 bits are information bits and the following 10 are check bits. The generator polynomial is

$$G(p) = p^{10} + p^8 + p^7 + p^5 + p^4 + p^3 + 1$$

The code makes it possible to detect all single and double errors in a block or to detect 99.8% of all error bursts of 11 bits or longer. The binary code for the letters A, B, C (C') and D is added to the code words in each block. This is done for synchronisation purposes as the receiver subtracts these letters and detects if the result becomes a valid code word. If this is not the case, the message is shifted successively one bit each time. Synchronisation is achieved when accepted code words are detected.

SWIFT (system for wireless infotainment forwarding and teledistribution) is a service for transmission of different types of digital information. SWIFT is used for both civilian and military applications. SWIFT uses DARC (data radio channel) which is a transmission system for digital information over the usual FM broadcast radio transmitters. The data transmitted is MSK modulated on a sub-carrier whose frequency is 76 kHz. The sub-carrier amplitude is within certain limits proportional to the stereo signal level and is adjusted to between 4 and 10% of the maximal input signal, corresponding to a deviation of ± 3.0–7.5 kHz in the main carrier. The modulation scheme is therefore named level controlled minimum shift keying (LMSK). The data rate in the transmission is 16 kbits/s which gives a total bandwidth of approximately 35 kHz. The frames are 288 bits long and contain 16 bits for block identification, 176 bits for information, 14 for error detection with CRC and 82 check bits for the error correction coding. Two different types of error correction coding, which give different degrees of error protection, are used. One of the methods is based on a (272,190) × (272,190) product code (the information is ordered in a 190 × 190 matrix instead of a 1 × 190 matrix, the check bits are computed on both lines and columns in the matrix (cf. Figure 4.3); this is another way to improve the burst error properties). The other coding is based on a (272,190) block code. Depending on which method is used the net flow of information is between 6.8 and 9.8 kbits^{-1}. With a *C/N* of 23dB a bit error rate of less than 10^{-2} can be obtained.

A.6 Digital audio broadcasting

Digital audio broadcasting (DAB) is a system for digital broadcasting. The sound quality is equally good as for CDs, i.e. the bandwidth of the sound signal is better than 15 kHz and the *S/N* is better than 70 dB. The system is flexible and designed for many different transmission rates and carrier frequencies up to 3 GHz. In this section we only introduce the type of transmission which is intended to replace (or complement) FM broadcasting. This is the so-called Mode I, which is suitable for a nationwide ground-based network. The whole network broadcasts on one single frequency and the different programmes have their own individual time slot for the transmission. In the DAB system the frequency band 223–240 MHz is used.

A brief description of the methods for source coding, channel coding and modulation used by the system is given below.

A.6.1 SOURCE CODING

The original signal from the broadcasting studio is a stereo signal where each channel has been sampled at a rate of 48 kHz and with a resolution of 16 bits. This means 768 kbits/s. The speech and music signal will then be coded. This coding method is called MPEG-1, layer II. The coding is adapted to the human ear which has the property that strong frequency components mask neighbouring weaker frequency components. The coding is achieved by dividing the sound signal into 32 frequency bands. Each one is given the number of bits which is presently required regarding the current masking situation. The bit rate is thereby reduced to 128 kbits/s. Depending on the desired sound quality a higher or a lower bit rate can be chosen (Figure A.5).

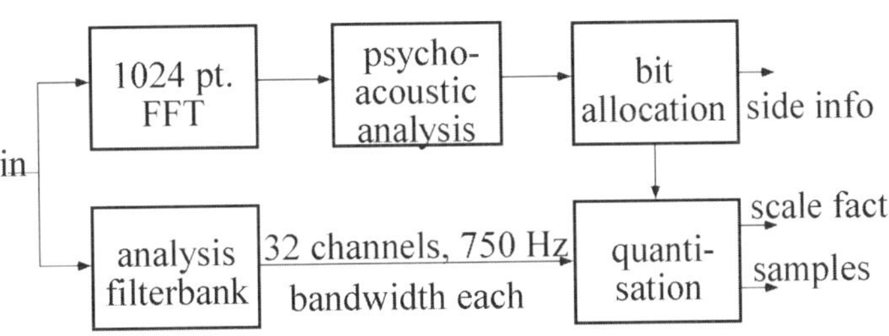

Figure A.5   Outline of a MPEG-1 encoder.

A.6.2 CHANNEL CODING

The transmitted bit stream is protected by using a convolutional code. Depending on the importance of the message or on its interference sensitivity, different code rates ranging from 1/3 to 3/4 are used. Which particular code rate is to be used for a specific message is decided by the unequal error protection (UEP) profile. A lower level of error protection is used for radio programmes distributed via cable than for programmes distributed via wireless broadcasting.

In order to spread out the error bursts, interleaving is used, which delays the signal by 384 ms.

The data flow is then divided into frames of 24 ms. Frames from all programme channels follow each other consecutively and constitute a bit stream with the rate 2.3 Mbits/s. The number of information bits depends on the chosen coding and varies between 0.6 and 1.7 Mbits/s.

A.6.3 MODULATION

The system utilises differential QPSK modulation with several orthogonal carriers, i.e. OFDM. The signal reaches the receiver along different paths partly because of multi-path propagation and partly through other transmitters in the same network. The data stream is divided into 1536 parallel sequences. The data rate in each sequence can then be made low enough to allow a symbol length of 1 ms and a guard interval of 0.25 ms between each symbol. This makes the differences in travelling time between the different paths negligible for the decoder. Due to the guard interval between each symbol, no inter-symbol interference will occur. The multi-path propagation will cause constructive interference for some carriers and destructive interference for others. This is mainly a problem for stationary receivers for which a certain bit stream has a risk of being continuously cancelled. The problem can be avoided by interleaving across the carriers, i.e. the data stream is shifted among the carriers. The total bandwidth of the DAB signal is $1536/10^{-3} \approx 1.5$ MHz, that is the number of carriers/symbol length.

ANSWERS TO EXERCISES

2.1 Without preamplifier $R_{max} = 2.06 \times 10^4$ m. With preamplifier $R_{max} = 3.9 \times 10^4$ m.

3.1 $H(X) = -\alpha\log_2\alpha - (1 - \alpha)\log_2(1 - \alpha)$.

3.2 0.811 bit.

3.3 $C = 2 \times 10^3\left(1 + \log_2(1 - p) + p\left(\log_2\dfrac{p}{1 - p}\right)\right)$

3.4 $H(X) = 1.9$; $H(ABABBA) = 9$; $H(FDDFDF) = 28.9$, six symbols 11.6.

3.5 Predictive coding with the coefficients $c_1 = 5/3$, $c_2 = -4/3$ and $c_3 = 1/2$.

3.6 Huffman coding, one of several solutions:

Symbol	Code word
A	1
B	011
C	010
D	001
E	0001
F	00001
G	00000

$\overline{N} = 2.44$; $H(X) = 2.37$, $\eta = 0.974$.

3.7 $H(X) = 1.65$ bit/symbol; $\eta_1 = 96.9\%$; $\eta_2 = 97.9\%$; $H_{markov}(X) = 1.26$; $P(*) = 0.38$; $P(!) = 0.14$; $P(\phi) = 0.09$; $P(\Delta) = 0.39$.

3.8 Set up a state diagram. $P_{good} = 10/11$.

3.9 Use variational calculus. One multiplier λ is needed and if $\Delta = x_2 - x_1$ is $\lambda = \log_2(\Delta) - 1$. The result becomes an even distribution $p_X(x) = 1/\Delta$ for $x_1 \le x \le x_2$.

4.1 Change the 1st and 3rd rows. The Hamming weights are 4 for all code words except for $\{0000000\}$, which gives $d_{\min} = 4$.

$$\mathbf{H} = \begin{bmatrix} 1 & 0 & 1 & 1 & 0 & 0 & 0 \\ 1 & 1 & 1 & 0 & 1 & 0 & 0 \\ 1 & 1 & 0 & 0 & 0 & 1 & 0 \\ 0 & 1 & 1 & 0 & 0 & 0 & 1 \end{bmatrix}$$

4.2 The generator matrix becomes

$$\mathbf{G} = \begin{bmatrix} 1 & 0 & 0 & 0 & 0 & 0 & 0 & 0 & 0 & 0 & 0 & 1 & 0 & 0 & 1 \\ 0 & 1 & 0 & 0 & 0 & 0 & 0 & 0 & 0 & 0 & 0 & 1 & 1 & 0 & 1 \\ 0 & 0 & 1 & 0 & 0 & 0 & 0 & 0 & 0 & 0 & 0 & 1 & 1 & 1 & 1 \\ 0 & 0 & 0 & 1 & 0 & 0 & 0 & 0 & 0 & 0 & 0 & 1 & 1 & 1 & 0 \\ 0 & 0 & 0 & 0 & 1 & 0 & 0 & 0 & 0 & 0 & 0 & 0 & 1 & 1 & 1 \\ 0 & 0 & 0 & 0 & 0 & 1 & 0 & 0 & 0 & 0 & 0 & 1 & 0 & 1 & 0 \\ 0 & 0 & 0 & 0 & 0 & 0 & 1 & 0 & 0 & 0 & 0 & 0 & 1 & 0 & 1 \\ 0 & 0 & 0 & 0 & 0 & 0 & 0 & 1 & 0 & 0 & 0 & 1 & 0 & 1 & 1 \\ 0 & 0 & 0 & 0 & 0 & 0 & 0 & 0 & 1 & 0 & 0 & 1 & 1 & 0 & 0 \\ 0 & 0 & 0 & 0 & 0 & 0 & 0 & 0 & 0 & 1 & 0 & 0 & 1 & 1 & 0 \\ 0 & 0 & 0 & 0 & 0 & 0 & 0 & 0 & 0 & 0 & 1 & 0 & 0 & 1 & 1 \end{bmatrix}$$

4.3 $d_{\text{free}} = 4$.

4.4 The state diagram becomes

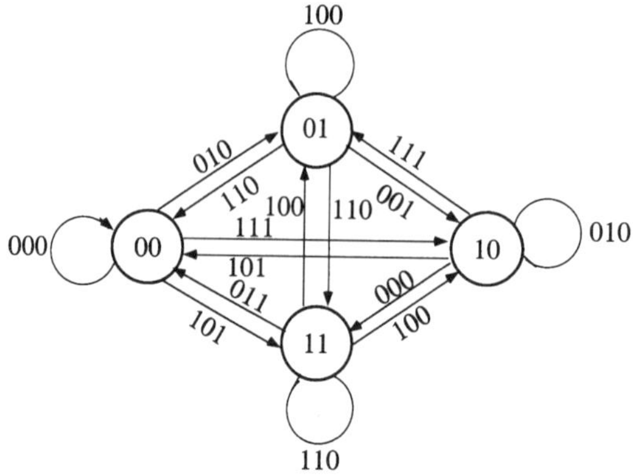

4.5 The constraint length of the code is 4. d_{free} is 6, which is clear already from the modified state diagram.

4.6 $S(p) = E(p) = p^2$.

4.7 $d_{free} = 21$.

4.8 The most probable code word sent is $\mathbf{x} = [1\ 0\ 0\ 1\ 1\ 1]$, since $P_b < 0.5$.

5.1 Use the complex description of the signal. For AM-DSBSC it is clear that the demodulated signal will be multiplied by the factor $\cos(\Delta\omega t + \phi)$. For AM-SSB you first have to formulate a complex baseband signal with only, e.g. positive frequencies. After that it is shown that the demodulated signal spectrum becomes $X(\omega - \Delta\omega)$.

5.2 Utilise eq. (5.111). Because of the precoding, the symbols are not independent $R(0) = 2$, $R(\pm 1) = 1$ and $R(n) = 0$ otherwise. Which gives the spectrum

$$U(f) = \frac{1}{4T}\left(\frac{\sin(\pi f T)}{\pi f T}\right)^2 \cos^2(\pi f T)$$

5.3 $S_{x,max} = 0.185$

5.4 A PM or FM signal. For FM $\beta = 200$ and the bandwidth is 8×10^5 rad/s.

5.5 $d_{\min} = \sqrt{8}\,d_0$

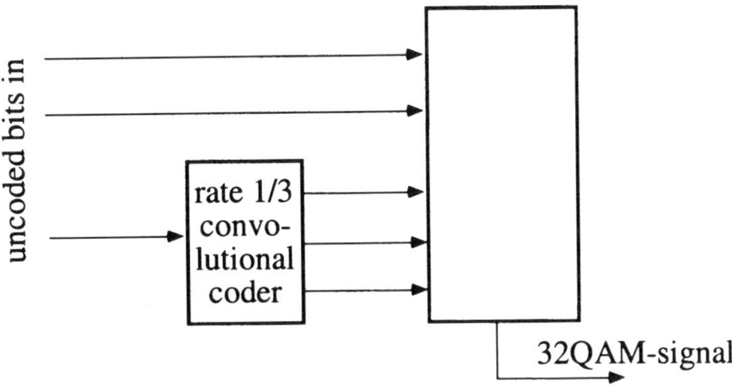

5.6
$$y(t) = \begin{cases} \dfrac{A^2}{2}\, t\, \cos(2\pi f_c t) & 0 \le t \le T \\[2mm] A^2\!\left(T - \dfrac{t}{2}\right)\! t\, \cos(2\pi f_c t) & T \le t \le 2T \\[2mm] 0 & \text{otherwise} \end{cases}$$

5.7 $\quad g(t) = \dfrac{1}{2}\exp\!\left(-\dfrac{t}{2}\right)\!\left(1 + \dfrac{j}{\sqrt{3}}\right)\exp\!\left(j2\pi\!\left(\dfrac{\sqrt{3}}{4\pi} - f_c\right)t\right)$

5.8 The equivalent noise bandwidth is: BPSK $= 1/T$, QPSK $= 1/2T$ and MPSK $= \pi^2/16T \approx 0.62/T$.

5.9 For BPSK $U(f) = (T/2)\mathrm{sinc}^2(fT)$. For QPSK the effective bit rate can be reduced to one half, as $T_{QPSK} = 2T$.

6.1 $\quad P(e|j) \le Q\!\left(\sqrt{\dfrac{2E}{N_0}}\right) + 2Q\!\left(\sqrt{\dfrac{E}{N_0}}\right)$

Only d_{min},

$$P(e|j) \le 3Q\!\left(\sqrt{\dfrac{E}{N_0}}\right)$$

6.2 The error probability is reduced from 3.87×10^{-6} to 2.02×10^{-7}, that is by a factor of 19.

6.3 SNR becomes 38.1 dB, which gives the bit error rate ≈ 0.

6.4 (a) QPSK or QAM. The decision regions are separated by lines which emanate from the origin by $45°$, $135°$, $225°$ and $315°$, see e.g. Figure 6.36. $E_b = A^2 D/(2\log_2(M)) \Rightarrow A = 8.76$.

6.5 Bandwidth 210 kHz. Modulation gain 26.5 dB.

6.6 8PSK ought to be used. $E_b/N_0 \approx 13$ dB $\Rightarrow S/N \ge 21$dB.

6.7 $M = 32$ orthogonal signalling, for example orthogonal FSK. $S/N \ge 17$dB.

6.8 $P_e = 0.023$.

6.9 Dimension 2. Basis functions:

$$\phi_1(t) = \sqrt{2/T}\cos(2\pi f_c t) \text{ and } \phi_2(t) = \sqrt{2/T}\sin(2\pi f_c t)$$

$$\mathbf{s}_1 = \left[\sqrt{E/2} \quad -\sqrt{E/2}\right], \ \mathbf{s}_2 = \left[0 \quad -\sqrt{E}\right]$$

$$\mathbf{s}_3 = \left[-\sqrt{E/2} \quad -\sqrt{E/2}\right], \ \mathbf{s}_4 = \left[-\sqrt{E} \quad 0\right]$$

6.10 64-FSK.

6.11 SNR in is 13 dB, after the demodulator 3.7 dB.

6.12 P_b without error correcting coding is 0.18×10^{-3}. When one error path is assumed the error probability with coded transmission is:

$$P_b \approx \frac{2^7}{2}\left(\exp\left(\frac{-15.85}{2 \times 2}\right)\right)^{7/2} = 0.061 \times 10^{-3}$$

INDEX